W0112396

GRAVITATIONAL *N*-BODY PROBLEM

ASTROPHYSICS AND SPACE SCIENCE LIBRARY

A SERIES OF BOOKS ON THE RECENT DEVELOPMENTS OF SPACE SCIENCE AND OF GENERAL GEOPHYSICS AND ASTROPHYSICS PUBLISHED IN CONNECTION WITH THE JOURNAL SPACE SCIENCE REVIEWS

VOLUME 31

GRAVITATIONAL N-BODY PROBLEM

PROCEEDINGS OF IAU COLLOQUIUM No. 10
HELD IN CAMBRIDGE, ENGLAND
AUGUST 12–15, 1970

Edited by

MYRON LECAR
Smithsonian Astrophysical Observatory,
Harvard College Observatory, Cambridge, Mass.

D. REIDEL PUBLISHING COMPANY

DORDRECHT-HOLLAND

Library of Congress Catalog Card Number 72-154740

ISBN-13:978-94-010-2872-1 e-ISBN-13:978-94-010-2870-7
DOI: 10.1007/978-94-010-2870-7

Softcover reprint of the hardcover 1st edition 1972

INTRODUCTION

This volume contains the proceedings of the third IAU conference on the Gravitational N-Body Problem. The first IAU conference [1], six years ago, was motivated by the renaissance in Celestial Mechanics following the launching of artificial earth satellites, and was an attempt to bring to bear on the problems of Stellar Dynamics the sophisticated analytical techniques of Celestial Mechanics. That meeting was an outgrowth of the 'Summer Institutes in Celestial Mechanics' initiated by Dirk Brouwer. By the second IAU conference [2], our interest had been captured by the attempts to simulate stellar systems on the computer. Computer simulation is now an essential part of stellar dynamics; journals of computational physics have started in the United Kingdom and in the United States and symposia on computer simulation of many-body problems have become a perennial event [3, 4, 5]. Although our early hopes that the computer would 'solve' our problem have been tempered by experience, some techniques of computer simulation have now matured through five years of testing and use. A working description of the six most popular methods is appended to this volume.

During the past three years, stellar dynamicists have followed closely the developments in the related field of Plasma Physics. The contexts of Plasma and Stellar Physics are deceptively similar; at first, results from Plasma Physics were bodily transferred to stellar systems by 'changing the sign of the coupling'. We are more sophisticated and more skeptical now. Still the size and vitality of the Plasma effort commands our (sometimes envious) attention. We are grateful to John Dawson for organizing the tutorial session on methods and results from Plasma Physics which are presented in Chapter III.

From the viewpoint of stellar dynamics, stellar systems fall naturally into two classes, depending on whether or not encounters (the analog to collisions in gas kinetic theory) contribute to their dynamical evolution. Stellar Associations, Galactic Clusters, Globular Clusters and Clusters of Galaxies belong to the first class and are referred to here as 'Collisional Systems'. These were the first systems to be computer simulated; the 'classic' work was reported by Sebastian Von Hoerner in 1960 [6]. This is still the most active field, and Chapter I (which contains almost half of the contributions to these proceedings) is devoted to this topic.

A more recent development is the simulation of Galaxies, whose dynamics are dominated by collective ('Collisionless') effects. This work excites us not only because Galaxies are the most dramatic objects in the astronomers sky, but also because the theory of collective interactions has made enormous progress in the past few years, and has produced a successful gravitational theory of the Spiral Arms. Recent developments on the Spiral Arm problem are contained in the Proceedings of IAU Colloquium

No. 38 [7], and in Chapter II where we deal with Collisionless Systems, we have not repeated that material.

The present colloquium was organized by George Contopoulos (who has been the guiding spirit of all three of the IAU Colloquia), Sverre Aarseth and myself, with help from Michel Hénon and Victor Szebehely. Dr Aarseth was also the host and local organizer at the Institute of Theoretical Astronomy, Cambridge University. I would like to thank Jane C. Ackland and Eliza Collins (and her staff of the Editorial and Publications Division of the Smithsonian Astrophysical Observatory) for editorial and secretarial assistance. Finally, at times when I questioned my ability to make sense out of one more abstruse contribution (my own not excepted), my courage was bolstered by remembering the patient but dogged tutelage I underwent at the hands of Professor Rupert Wildt at Yale University.

December, 1970 MYRON LECAR
Cambridge, Mass.

References

[1] Contopoulos, G. (ed.): 1966, *The Theory of Orbits in the Solar System and in Stellar Systems,* Proceedings of IAU Symposium No. 25, held at Thessaloniki, Greece in August, 1964, Academic Press, New York.

[2] 'Colloque sur le problème des *N* corps', Proceedings of an IAU Colloquium held at Paris, France in August, 1967,
Bull. Astron., Tome **3**, Fascicules 1 and 2.

[3] Nahon, F. and Henon, M. (eds.): 1967, 'Colloque sur les méthodes nouvelles de la Dynamique stellaire', Proceedings of a Symposium held at Besançon, France, in September, 1966,
Bull. Astron., Tome **2**, Fascicule 1.

[4] 'Symposium on Computer Simulation of Plasma and Many Body Problems', Proceedings of a Symposium held at Williamsburg, Virginia, in April, 1967, National Aeronautics and Space Agency Special Report SP-153.

[5] 'Computational Physics', Proceedings of a Symposium held at Culham Laboratory, U.K., in July 1969,
UKAEA Culham Laboratory and the Institute of Physics and The Physical Society, CLM-CP (1969).

[6] Von Hoerner, S.: 1960, *Z. Astrophys.* **50**, 184.

[7] Becker, W. and Contopoulos, G. (eds.): 1970, *The Spiral Structure of Our Galaxy,* Proceedings of IAU Symposium No. 38, held at Basel, Switzerland, in August, 1969,
D. Reidel Publishing Company, Dordrecht, Holland.

TABLE OF CONTENTS

PART II / COLLISIONLESS SYSTEMS

A. *Analytic Treatments*

B. *Numerical Experiments*

PART III / NUMERICAL EXPERIMENTS AND ANALYTICAL TREATMENTS IN PLASMA PHYSICS

PART IV / SUMMARY OF THE COLLOQUIUM

PART V / APPENDIX: METHODS OF COMPUTER SIMULATION OF THE GRAVITATIONAL *N*-BODY PROBLEM

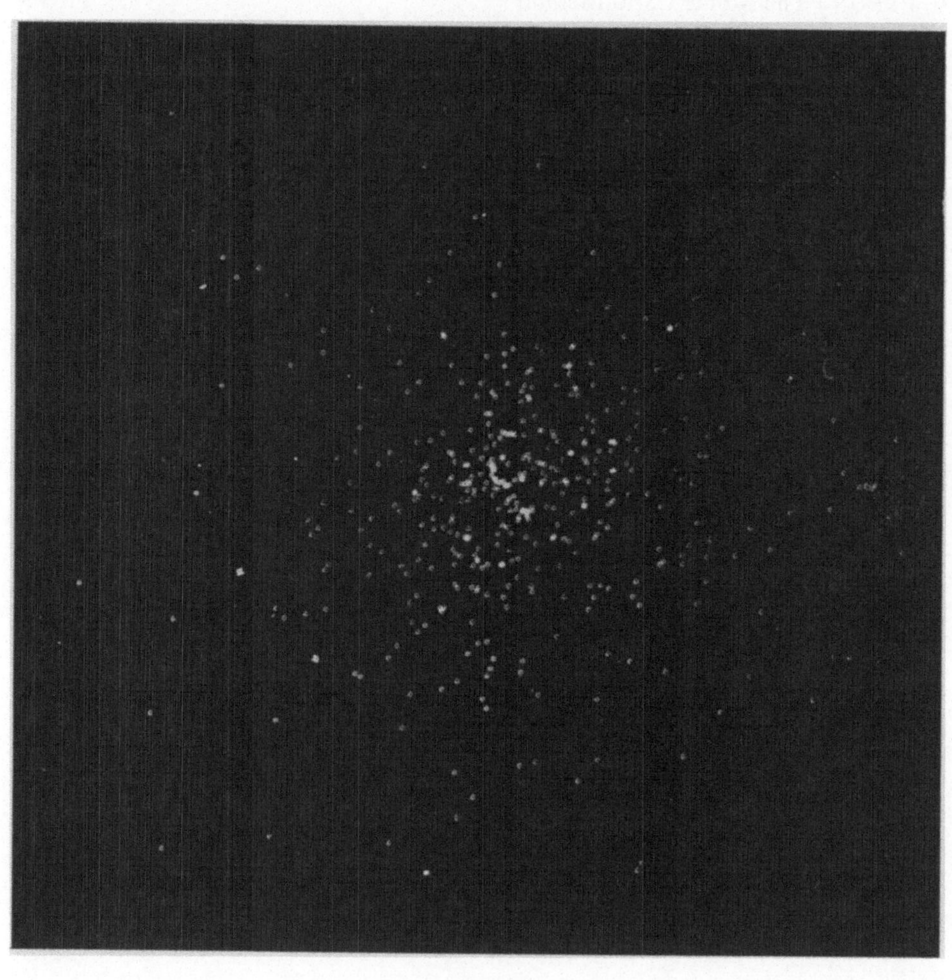

Computed star cluster with 500 initial members after 15 crossing times.
(By courtesy of S. J. Aarseth, Institute of Theoretical Astronomy, Cambridge.)

Real star cluster M 67 photographed in blue with a 17″ Schmidt telescope.

(By courtesy of A. N. Argue, The Observatories, Cambridge.)

PART I

COLLISIONAL SYSTEMS

A. ANALYTIC TREATMENTS

COLLISIONAL PROCESSES IN STELLAR SYSTEMS

IRA H. GILBERT

Dept. of Physics, Brandeis University, Waltham, Mass. 02154, U.S.A.

Abstract. The theory of collisional relaxation in stellar systems is discussed in terms of an expansion in powers of $1/N$, the inverse of the total number of stars. The results are expressed in terms of the concept of gravitational polarization.

1. Probability in Stellar Dynamics

At first glance stellar dynamics appears to be a remarkably simple subject. One deals with well defined physical systems, collections of idealized point masses, interacting according to the laws of classical mechanics and Newtonian gravitation. One need only prescribe the initial state of the system and integrate the equations of motion in order to determine the state at any future time. The catch, of course, is that the integration of these equations is totally impractical when the number of stars is large enough to represent a real system. Furthermore, the information contained in the precise dynamical state of the system is too detailed to be useful.

The way around these difficulties lies in replacing certainty by chance, i.e. by settling for a description in terms of probabilities instead of precise dynamical states. The general idea is that a statistical description should be simpler and involve fewer variables than an exact one. Furthermore, as the number of stars grows increasingly large the statistical description should improve in accuracy as statistical fluctuations diminish according to the law of large numbers.

Although there is universal agreement on how probabilities are to be handled mathematically there are differences as to their meaning. In my opinion probabilities are introduced in stellar dynamics to represent states of partial knowledge. From this point of view their time dependence, governed by the Liouville equation, reflects the evolution of one's information concerning the dynamical state of the system. If one is provided with some imprecise description of an initial state the Liouville equation allows one to make best possible guesses about the resultant state at some future time.

2. Alternative Statistical Descriptions

Let us represent the Liouville probability density for a system of N stars by $D(1, 2,..., N)$, where the argument i represents both the position and velocity of particle i. In this compact notation $D(1,..., N)\,\mathrm{d}(1)\,\mathrm{d}(2)...\mathrm{d}(N)$ is the probability of finding particle 1 within $\mathrm{d}\mathbf{r}_1$ of position $\mathbf{r}_1$ and within $\mathrm{d}\mathbf{v}_1$ of velocity $\mathbf{v}_1$ etc. For simplicity it is convenient to take D to be symmetric in the various stars. If the stars are equally massive this presents no difficulty. If they are not we may still construct a symmetric function by considering the mass of each star to be a random variable.

M. Lecar (ed.), Gravitational N-Body Problem, 5–12. All Rights Reserved

Of course each mass, once chosen, remains constant in time. With this extension, the argument (i) represents the position, velocity, and mass of the ith star.

The Liouville equation, a partial differential equation in $6N+1$ independent variables, is not any easier to solve than the original dynamical equations. The problem is that the Liouville probability density still contains too much information. To obtain a less detailed description one may integrate D over the phase space of all but s particles and so obtain $f^{(s)}$, the s-particle distribution function.

$$f^{(s)}(1, \ldots, s) = \int D(1, \ldots, N)\, \mathrm{d}(s+1) \ldots \mathrm{d}(N). \tag{1}$$

This quantity gives the probability distribution for the coordinates and velocities of a set of s stars, averaged over all possible states of the remaining stars. In particular, the single-particle function,

$$f(1) = \int D(1, \ldots, N)\, \mathrm{d}(2) \ldots \mathrm{d}(N), \tag{2}$$

which will be written for simplicity without a superscript, is the quantity of central interest in stellar dynamics. It is the least detailed of the distribution functions, depending on the coordinates and velocities of a single star. The higher order distribution functions provide increasingly detailed descriptions. Notice that if $s>s', f^{(s)}$ contains all the information in $f^{(s')}$ and more. Indeed, integrating Equation (1) over the phase space of stars $s'+1$ through s yields the identity

$$f^{(s')}(1, \ldots, s') = \int f^{(s)}(1, \ldots, s)\, \mathrm{d}(s'+1) \ldots \mathrm{d}s, \tag{3}$$

which connects these functions.

Now the original idea in introducing probabilities was to simplify the problem by ignoring the finer details while retaining the main macroscopic features. The lower order distribution functions do just this. If we can determine f as a function of the time we will have achieved our goal. The trouble, however, is that the equation which determines the time dependence of f turns out to involve $f^{(2)}$. Quite generally, the equation for $f^{(s)}$ involves $f^{(s+1)}$. The entire set of equations, called the BBGKY hierarchy for Bogolioubov, Born, Green, Kirkwood, and Yvon, are therefore coupled. In order to solve for f without solving for all the distribution functions at once (which would again be equivalent to integrating the original equations of motion) these equations must be decoupled. This is accomplished with the aid of certain combinations of the distribution functions called correlation functions. They are symbolized by $g^{(s)}(1, \ldots, s)$ $s=2, \ldots, N$, and defined by equations of the form

$$\begin{aligned} f^{(2)}(1,2) &= f(1)\, f(2) + g(1,2), \\ f^{(3)}(1,2,3) &= f(1)\, f(2)\, f(3) + f(1)\, g(2,3) + f(2)\, g(1,3) + \\ &\quad + f(3)\, g(1,2) + g^{(3)}(1,2,3), \end{aligned} \tag{4}$$

where for simplicity the superscript of the two-particle correlation function, which appears frequently in the following discussion, has been suppressed. In general $f^{(s)}$

is expressed as a sum of products, each corresponding to a different partition of the integers 1 through s.

Unlike the f's the g's are essentially independent. Equation (3), expressed in terms of the g's, becomes

$$\int g^{(s)}(1, \dots s)\,\mathrm{d}(s) = 0 \qquad s = 2, \dots, N. \tag{3'}$$

It no longer connects different functions but constrains each correlation function separately. The only reason the various g's are not completely independent is that the distribution functions, being probability densities, are non-negative. This means that the g's must satisfy a set of inequalities such as

$$f(1)\,f(2) + g(1, 2) \geqslant 0, \tag{5}$$

etc.

3. Correlations, Dynamics, and Initial Conditions

The correlation functions, as their name would indicate, represent multi-particle correlations. If one *independently* distributes stars according to some single-particle distribution function all of the g's will vanish. On the other hand, if one independently distributes *pairs* of stars $g^{(2)}$ will not necessarily vanish but all higher order correlation functions will.

Statistical correlations between stars are determined both by the initial probability distribution and the gravitational dynamics. It seems plausible that in most cases the disorganized motions of the stars will disrupt groups which were initially nearby and quickly erase the initial correlations. The correlations which result will be determined by the dynamics and the single-particle distribution function alone. In order to obtain a closed theory in which the single-particle function is the only variable we will assume this to be the case. More precisely, we will assume probability distributions which have evolved from initially uncorrelated states.

It is, however, possible to imagine situations which are quite different. For example, if one randomly distributes tightly bound binaries it will take a very long time for encounters to disrupt them and so eliminate the initial correlations. We will not consider this possibility.

4. Expansion in Powers of $1/N$

Up to this point we have not used the important piece of information that N is a large number. It is just this fact, however that results in the decoupling of the BBGKY hierarchy which was the reason for introducing the correlation functions in the first place.

The Liouville equation involves N both explicitly, through the summations that appear, and implicitly, through the interstellar force. This implicit dependence comes about because we are considering M, the total mass of the system, to be fixed and therefore the individual stellar masses to be proportional to $1/N$. For simplicity assume all the stars to have the same mass, M/N. The Liouville equation may then be written

$$\frac{\partial D}{\partial t} + \sum_{i=1}^{N} \mathbf{v}_i \cdot \frac{\partial}{\partial \mathbf{r}_i} D + \frac{1}{N} \sum_{i,j=1}^{N}{}' \mathbf{a}_{i,j} \cdot \frac{\partial}{\partial \mathbf{v}_j} D = 0, \tag{6}$$

where

$$\frac{1}{N} \mathbf{a}_{i,j} = -\frac{MG}{N} \frac{(\mathbf{r}_i - \mathbf{r}_j)}{|\mathbf{r}_i - \mathbf{r}_j|^3} \tag{7}$$

is the acceleration of star i due to star j. All dependence on N is now explicit.

A set of equations for the correlation functions may be obtained from the Liouville equation in a number of ways, the simplest being by means of a generating functional first introduced by Bogolioubov. The advantage of this technique is that the single functional equation that results is entirely equivalent to the set of N differential equations for the g's. This makes it easy to establish certain general results.

It turns out that there are solutions of these equations for which $g^{(s)}$ is of order $(1/N)^{s-1}$ for all s. Thus if one starts with initial correlations with this dependence on N the dependence is maintained by the dynamics. In particular, if one starts from an initially uncorrelated state, as previously discussed, then at later times $g^{(s)}$ will be of order $(1/N)^{s-1}$.

The mean total acceleration at $\mathbf{r}_1$ is obtained by averaging $\mathbf{a}_{1,2}/N$, the acceleration due to a single star at $\mathbf{r}_2$, over all possible values of $\mathbf{r}_2$ and then multiplying by N.

$$\mathbf{A}_1 = \int \mathbf{a}_{1,2} f(2)\, \mathrm{d}(2). \tag{8}$$

The first two equations for the correlation functions turn out to be:

$$\frac{\partial f(1)}{\partial t} + \mathbf{v}_1 \cdot \frac{\partial}{\partial \mathbf{r}_1} f(1) + \frac{N-1}{N} \mathbf{A}_1 \cdot \frac{\partial}{\partial \mathbf{v}_1} f(1) = -\frac{\partial}{\partial \mathbf{v}_1} \cdot \int \mathbf{a}_{1,2}\, g(1,2)\, \mathrm{d}(2), \tag{9}$$

and

$$\begin{aligned}\frac{\partial g(1,2)}{\partial t} &+ \left(\mathbf{v}_1 \cdot \frac{\partial}{\partial \mathbf{r}_1} + \mathbf{v}_2 \cdot \frac{\partial}{\partial \mathbf{r}_2}\right) g(1,2) + \frac{N-2}{N} \left(\mathbf{A}_1 \cdot \frac{\partial}{\partial \mathbf{v}_1} + \mathbf{A}_2 \cdot \frac{\partial}{\partial \mathbf{v}_2}\right) \times \\ &\times g(1,2) + \frac{N-2}{N} \left(\frac{\partial f(1)}{\partial \mathbf{v}_1} \cdot \int \mathbf{a}_{1,3}\, g(2,3)\, \mathrm{d}(3) + \frac{\partial f(2)}{\partial \mathbf{v}_2} \times \right. \\ &\left. \times \int \mathbf{a}_{2,3}\, g(1,3)\, \mathrm{d}(3)\right) + \frac{1}{N} \left[(\mathbf{a}_{1,2} - \mathbf{A}_1) \cdot \frac{\partial}{\partial \mathbf{v}_1} + \right. \\ &\left. + (\mathbf{a}_{2,1} - \mathbf{A}_2) \cdot \frac{\partial}{\partial \mathbf{v}_2}\right] f(1) f(2) \\ &+ \frac{1}{N} \left(\mathbf{a}_{1,2} \cdot \frac{\partial}{\partial \mathbf{v}_1} + \mathbf{a}_{2,1} \cdot \frac{\partial}{\partial \mathbf{v}_2}\right) g(1,2)\end{aligned} \tag{10}$$

$$+\frac{1}{N}\left(f(2)\frac{\partial}{\partial \mathbf{v}_1}\cdot\int \mathbf{a}_{1,3}g(1,3)\,\mathrm{d}(3)+f(1)\frac{\mu}{\partial \mathbf{v}_2}\times\right.$$
$$\left.\times\int \mathbf{a}_{2,3}g(2,3)\,\mathrm{d}(3)\right)+\frac{N-2}{N}\int\left(\mathbf{a}_{1,3}\times\right. \tag{10}$$
$$\left.\times\frac{\partial}{\partial \mathbf{v}_1}+\mathbf{a}_{2,3}\cdot\frac{\partial}{\partial \mathbf{v}_2}\right)g^{(3)}(1,2,3)\,\mathrm{d}(3)=0\,.$$

These equations are horribly complicated but exact. Equation (10) may be simplified by taking advantage of the smallness of $1/N$ and neglecting terms of higher order than the first. Recalling that $g^{(s)}$ is of order $(1/N)^{s-1}$ we see that the last three lines of the equation may be dropped and the factor $N\text{-}2/N$ replaced by unity. We find

$$\frac{\partial g(1,2)}{\partial t}+\left(\mathbf{v}_1\cdot\frac{\partial}{\partial \mathbf{r}_1}+\mathbf{v}_2\cdot\frac{\partial}{\partial \mathbf{r}_2}+\mathbf{A}_1\cdot\frac{\partial}{\partial \mathbf{v}_1}+\mathbf{A}_2\cdot\frac{\partial}{\partial \mathbf{v}_2}\right)g(1,2)$$
$$+\frac{\partial f(1)}{\partial \mathbf{v}_1}\cdot\int \mathbf{a}_{1,3}g(2,3)\,\mathrm{d}(3)+\frac{\partial f(2)}{\partial \mathbf{v}_2}\cdot\int \mathbf{a}_{2,3}g(1,3)\,d(3) \tag{11}$$
$$+\frac{1}{N}\left[(\mathbf{a}_{1,2}-\mathbf{A}_1)\cdot\frac{\partial}{\partial \mathbf{v}_1}+(\mathbf{a}_{2,1}-\mathbf{A}_2)\cdot\frac{\partial}{\partial \mathbf{v}_2}\right]f(1)\,f(2)=0\,.$$

Equations (9) and (11) serve to determine f and g and so achieve the desired closed description in terms of low order distribution functions.

We may note that these equations are similar but not identical to the corresponding equations of plasma physics. There, the factor $N\text{-}1/N$ does not appear in the equation for f and the inhomogeneous term in the equation for g does not involve A_i. This difference comes about because the plasma equations are based upon an expansion in powers of the inverse of the number of electrons in a sphere whose radius is the Debye length. This number is independent of the total number of electrons in the system, usually taken as infinite. On the other hand the analog of the Debye length for a stellar system such as a globular cluster is the radius of the system itself. Thus, in a stellar system the number of stars in a Debye sphere and the total number of stars are necessarily equal and N, as it appears here, has a dual significance.

5. The Kinetic Equation

Equations (9) and (11) may, in principle, be simultaneously solved for f and g as functions of the time. The simplest situation occurs when the system is in equilibrium with respect to purely collective motions and the only time dependence is through the slow, secular, effects of collisions. In that case the zeroth order terms in Equation (9) vanish and $\partial f/\partial t$ is seen to be of first order in $1/N$. It is then reasonable to assume that $\partial g/\partial t$ will be of second order allowing us to drop the first term in Equation (11). This equation then serves to determine g as a functional of f which, when inserted into the right hand side of Equation (9) yields a kinetic equation for f alone.

The actual elimination of g may be explicitly carried out for homogeneous plasmas because the spatial homogeneity allows the use of the spatial Fourier transform, resulting in an essentially algebraic relation between f and g. The result is the Balescu-Lenard equation. In stellar systems, which are necessarily inhomogeneous, this technique fails. Although an explicit elimination of g in terms of f has not been achieved it is still possible to construct a formal solution and to interpret it in terms of the underlying physical processes.

6. Gravitational Polarization

The physical content of the formal solution previously mentioned is most easily understood in terms of the auxiliarly concept of gravitational polarization. This quantity represents the response of the system to the gravitational field of a selected star moving in a specified orbit. In calculating this response one ignores collisional effects entirely and treats the field of the selected star as a small, externally applied perturbation. The polarization is the change in the single-particle distribution function that this perturbation induces. Symbolized by $\tilde{f}\,(1/2)$ it represents the change in the probability density for finding a star at $\mathbf{r}_1$ with velocity $\mathbf{v}_1$ given that another star is known to be located at $\mathbf{r}_2$ and moving with velocity $\mathbf{v}_2$.

There is a very interesting relation, first noted by Rostoker, between the polarization and the two-particle correlation function. It is

$$g(1,2) = \tilde{f}\,(2/1)\, f(1) + \tilde{f}\,(1/2)\, f(2) + N \int \tilde{f}\,(1/3)\, \tilde{f}\,(2/3) f(3)\, \mathrm{d}(3)\,. \tag{12}$$

The final result of this analysis is that collisional effects in stellar systems, i.e. dynamical effects of order $1/N$, may be divided into two distinct processes. The first is the gravitational force exerted on each star by the polarization it induces. This may be termed 'polarization drag'. The second is the effect upon each star of the random, fluctuating field resulting from the superposition of the fields of the other stars, each modified by its own polarization. These stars are to be considered to move in unperturbed orbits and not to respond to the influence of the field star under consideration. This last effect may be termed statistical acceleration.

The effects of these two kinds of processes are quite different. The statistical acceleration of a star will, on the average, increase its energy. Because of the identity of inertial and gravitational mass the statistical acceleration affects all stars in the same way. On the other hand, since the polarization induced by a given star is proportional to its mass we may expect heavy stars to be slowed more effectively by polarization drag than light ones. The competition between polarization and statistical effects should lead to a relative concentration of heavier stars in the central portions of a stellar system and lighter ones in the outer portions, i.e. an approach to equipartition.

7. Divergences

The theory sketched here completely takes into account the effects of collective interactions and spatial inhomogeneity which are absent from more elementary treatments. As a consequence, no long-range divergence appears as it does in the Fokker-Planck theory. It is interesting that although the equations of plasma physics and stellar dynamics are very similar, the mechanism for the elimination of this divergence is different in the two cases.

In plasmas the repulsive interparticle force results in Debye shielding which cuts the force off, eliminating the divergence. In stellar systems the attractive interstellar force results in 'anti-shielding' or amplification of the bare gravitational force of a star. This tends to make the divergence worse. It is only the limited spatial extent of the system that finally removes it.

There is, however, in the present theory a divergence at small distances. It comes about because the expansion in powers of $1/N$ is non-uniformly convergent. The physics behind this is quite simple.

In simplifying Equation (10) we assumed g to be small compared with f. But g actually measures the deviations of stars from the paths they would follow if subject only to the mean field. Thus, when two stars approach so closely that they are substantially deflected the assumption that g is small breaks down. It turns out that Equation (11) yields accurate values for g only for interstellar distances which are large compared with the impact parameter for 90° scattering. As this distance goes to zero the equation yields an unphysical infinity.

This divergence is precisely equivalent to the one occurring in elementary Fokker-Planck treatments in which the momentum transferred in a collision is approximated by integrating the force exerted by one particle upon another while assuming them to move in straight lines. At vanishingly small impact parameters one then erroneously calculates infinite momentum transfer. One can eliminate this error simply by replacing the straight line approximation by the correct two-body orbits.

In the present theory the rigorous treatment of two-body dynamics requires retaining all terms in g appearing in Equation (10). The term in $g^{(3)}$ may still be dropped because it is important only for collisions in which *three* stars are simultaneously deflected through large angles. The result is again a closed set of equations for f and g. This method of eliminating the divergence in g is, however, quite complicated. A simpler approach seems preferable. One merely replaces the true inverse-square force by a fictitious force carefully chosen so that the momentum transfer it yields under the straight line assumption is approximately equal to the momentum transfer calculated for the original force using rigorous two-body orbits. The fictitious force differs substantially from the true one only at distances less than the impact parameter for 90° scattering, distances much smaller than the mean separation between stars.

It should be mentioned that by dropping the three-particle correlations one necessarily ignores the formation and disruption of binaries in close, three-body encounters.

8. Conclusion

The theory outlined here has not yet been used to provide accurate numerical estimates of collisional relaxation in stellar systems. It has, however, provided considerable physical insight into the dynamical processes involved. It seems likely that with further development it will turn into a useful calculational tool.

POLARIZATION CLOUDS AND DYNAMICAL FRICTION

AGRIS J. KALNAJS

Harvard College Observatory, Cambridge, Mass., U.S.A.

Abstract. We argue that dynamical friction can be viewed as the drag exerted on a 'test' star by the wake it induces in the field stars. We compute the wakes for the uniform infinite medium and a flat rotating sheet. In the first case we obtain a result which differs by a factor of two from the classical result. In the second case the drag vanishes.

1. Introduction

The classical model for the study of dynamical friction is due to Chandrasekhar (1943). It consists of a uniformly moving test star which was introduced a long time ago in a homogeneous infinite distribution of field stars. By assuming that the system has settled into a steady state Chandrasekhar reduced the calculation of dynamical friction to a sum of momentum exchanges by binary encounters.

It is interesting to visualize a particular realization of this process at some instant in time. Since the interaction between the field and test star is gravitational, the force on the latter can be inferred from the spatial distribution of field stars. If we calculate the expectation value of this force by averaging over an ensemble, we conclude that the density of field stars cannot be uniform if there is dynamical friction. Thus we can say that dynamical friction is the drag exerted by the wake induced by the test star. Furthermore from the statistical point of view we would say that dynamical friction must be due to a nonvanishing correlation between stars.

The approach of Chandrasekhar has a distinct computational advantage since it involves the knowledge of complete as opposed to partial encounters, but it is restricted to the classical model.

Gilbert (1968) has developed a theory of collisional relaxation for an N-star system. By expanding the dynamics in powers of $1/N$ he shows that the leading collisional terms consist of two parts: a polarization drag and a fluctuating gravitational field. The polarization drag on a star is the force exerted by its wake and can therefore be identified with dynamical friction. Gilbert's calculation of the polarization drag differs from most previous work for it takes into account collective effects. It differs also in an important practical way: the calculation is equivalent to an understanding of the behaviour of a finite system to all perturbations, which is still an unsolved problem.

Only for homogeneous infinite systems is the perturbation problem completely soluble. Although such systems are unrealistic in many respects, they can be used for illustrative purposes. We shall consider two cases: the classical uniform infinite medium and a rotating flat sheet. In the first case we cannot take the collective effects properly into account since the model is unstable. The second case is a stable periodic system in which collective effects suppress dynamical friction.

M. Lecar (ed.), Gravitational N-Body Problem, 13–17.

We shall deduce the polarization drag by finding the wake behind the test star with the help of the linearized Vlasov and Poisson equations (Gilbert, 1968). On linearizing the Vlasov equation one introduces the assumption that the perturbation is small and therefore its effect can be calculated using the unperturbed orbits. This is equivalent to the straight-line assumption introduced by Hénon (1958).

2. Dynamical Friction in the Classical Model

Let the classical model consist of stars of equal mass m distributed uniformly through all space with a particle density n and having a gaussian distribution of velocities characterized by a dispersion σ. The test star moves with velocity $\mathbf{w}$ along the z-axis and is assembled adiabatically in a characteristic time $1/\varepsilon$. The density associated with the test star is a moving δ-function,

$$me^{\varepsilon t}\delta(\mathbf{x}-\mathbf{w}t) = m\iiint \frac{\mathrm{d}^3\mathbf{k}}{(2\pi)^3}\exp\left[i(\mathbf{k}\cdot\mathbf{x}-\mathbf{k}\cdot\mathbf{w}t-i\varepsilon t)\right]. \tag{1}$$

The identity (1) shows that we can represent the test star by a superposition of plane waves. Each plane wave component $m\exp[i(\mathbf{k}\cdot\mathbf{x}-\omega t)]$, with $\omega\equiv k_z|\mathbf{w}|+i\varepsilon$, induces a plane wave among the field stars. The amplitude of the induced plane wave is $mR(k,\omega)$, where $k\equiv|\mathbf{k}|$, and

$$R(k,\omega) = \frac{4\pi Gmn}{k^2\sigma^2}\left[1+\frac{\omega}{(2\pi\sigma^2)^{\frac{1}{2}}k}\int_{-\infty}^{\infty}\frac{\mathrm{d}u\,e^{-u^2/2}}{u-\omega/(k\sigma)}\right]. \tag{2}$$

The above expression for the response R can be obtained by a simple adaptation of the analogous longitudinal oscillation results for a hot plasma (Stix, 1962, Chapter 7).

The wake obtained from the superposition of the induced waves with amplitudes $mR(k,\omega)$ ignores the contribution due to collective effects. The latter arise from the fact that the actual perturbing field is the applied test star field plus the field from the wake. A bit of simple algebra shows that the amplitudes of the wake are $mR/(1-R)$. The potential associated with a plane density wave of unit amplitude is also a plane wave with amplitude $-4\pi\,G/k^2$. Hence, the potential associated with the wake is

$$\phi(\mathbf{x},t) = -\frac{Gm}{2\pi^2}\iiint\frac{\mathrm{d}^3\mathbf{k}}{k^2}\frac{R(\mathbf{k},\omega)}{1-R(\mathbf{k},\omega)}\exp\left[i(\mathbf{k}\cdot\mathbf{x}-\omega t)\right], \tag{3}$$

which exerts a force on the test star; and

$$-m\frac{\partial}{\partial\mathbf{x}}\phi(\mathbf{x},t)\Bigg|_{\mathbf{x}=\mathbf{w}t} = me^{\varepsilon t}\frac{Gm}{2\pi^2}\iiint\frac{\mathrm{d}^3\mathbf{k}}{k^2}\frac{i\mathbf{k}R(k,k_z\mathbf{w}+i\varepsilon)}{1-R(k,k_z\mathbf{w}+i\varepsilon)}. \tag{4}$$

By symmetry the x and y components of the force vanish.

If we try to evaluate the above integral we soon discover that the denominator $(1-R)$, which incorporates the collective effects, can vanish if $k^2\leq 4\pi\,Gnm/\sigma^2$. The source of this difficulty is well known – it reflects the fact that the classical model is unstable in Jeans's sense, and if we try to perturb it, the consequences are grave – it collapses. We can salvage part of the result (4) if we omit the contributions from the small wavenumbers on the physical grounds that the system should be stable and therefore finite. In this process we must forgo any estimate of the contribution from collective effects, for they are important precisely at the small wave numbers. Thus we will delete the $(1-R)$ term in the denominator.

In the remaining expression we see that since the force is real, only the imaginary part of R will contribute to the final result. We let $\varepsilon \to 0$, and obtain the imaginary part, which is

$$= \frac{4\pi Gmn}{k^2\sigma^2} \frac{k_z|\mathbf{w}|}{k\sigma\sqrt{2\pi}} \pi \exp\left[-\left(\frac{k_z|\mathbf{w}|}{k\sigma}\right)^2/2\right]. \tag{5}$$

If we introduce spherical polar coordinates in $\mathbf{k}$-space, calling $|\mathbf{k}|=k$, $k_z=k\cos\theta$, and $|\mathbf{w}|/\sigma=v$, the expression for the force becomes

$$-\frac{4\pi G^2m^3nv}{\sigma^2\sqrt{2\pi}} \int_{k_{\min}}^{\infty} \frac{dk}{k} \int_0^{\pi} d\theta \sin\theta \cos^2\theta \exp\left[-(v^2\cos^2\theta)/2\right]. \tag{6}$$

Because we have not excluded close encounters the k-integral diverges logarithmically. To secure convergence we can either modify the interparticle force law for distances shorter than Gm/σ^2, or simply ignore the contribution to the integral from wave-numbers greater than $k_{\max}=\sigma^2/Gm$.

By a change of variable, $y=v\cos\theta$, the above result can be put in a form which is very similar to Chandrasekhar's result (Chandrasekhar, 1943)

$$= -\frac{8\pi G^2m(mn)^2}{n\sigma^2v^2\sqrt{2\pi}} \log\left(\frac{k_{\max}}{k_{\min}}\right) \int_0^{v} y^2 e^{-y^2/2}\, dy. \tag{7}$$

The essential difference between two results is that Chandrasekhar's factor $(m_{\text{test}} + m_{\text{field}})$ is replaced by m_{test}. This arises from the fact that in the polarization calculation the field stars are treated as a fluid.

We note that the deceleration of test star decreases as $1/n$, if nm is kept fixed.

With minor modifications we can compute the dynamical friction for the analogous two-dimensional sheet. The equilibrium is now characterized by a surface density (nm), the wave-number space is two dimensional, and since the potential is calculated from a sinusoidal surface density we must replace $4\pi G/k^2$ by $2\pi G/|k|$ in Equations (2)–(6). Instead of spherical polar coordinates we now use polar coordinates in $\mathbf{k}$-space and obtain the corresponding expression for the polarization drag

$$= -\frac{\pi G^2m^3n}{\sigma^2\sqrt{2\pi}} \int_{k_{\min}}^{k_{\max}} dk \int_0^{2\pi} d\theta \cos^2\theta \exp\left[-(v\cos\theta)^2/2\right]. \tag{6a}$$

However now the deceleration of the test star is proportional to $k_{\max} \approx \sigma^2/Gm$ and if we keep the surface density (mn) fixed the deceleration is independent of n. This result was first pointed out by Rybicki (this volume). We note that in the numerical experiments $k_{\max}$ is usually determined by the approximate solution of Poisson's equation. In general it is much smaller than σ^2/Gm.

3. Dynamical Friction in a Rotating Sheet

For the equilibrium model we take an infinite sheet which is uniformly rotating around an axis normal to it. We choose a co-rotating reference frame, and suppose that in this frame the velocity distribution is gaussian with a dispersion σ. The surface density will be denoted by S_0. The orbits of stars will be circles, and they can be described by giving the center, the radius ϱ, and the angular position of the star on the circle. The angular revolution rate of a star is denoted by Ω.

A completely equivalent way of describing the velocity distribution is to give the distribution of the radii ϱ of stars around each point. This will be of the form $\exp[-(\varrho/\bar{\varrho})^2/2]\ \varrho\ \mathrm{d}\varrho/\bar{\varrho}^2$.

The model is an idealization which incorporates some of the local properties of our Galaxy.

The normal modes of this system are again waves, and the dynamics of small disturbances is summarized by the response R of the sheet to a plane wave. The calculation of R in this case is essentially the same as that for the propagation of waves perpendicular to a uniform magnetic field in a hot plasma (Stix, 1962, Chapter 8).

In our case

$$R(k, \omega) = \frac{2\pi G S_0}{\Omega^2 \bar{\varrho}} \frac{1}{k\bar{\varrho}} \sum_{l=-\infty}^{\infty} \frac{l\Omega \exp[-k^2\bar{\varrho}^2]\, I_l(k^2\bar{\varrho}^2)}{l\Omega + \omega}, \tag{8}$$

where $I_l(z)$ is the Bessel function of imaginary argument.

Toomre (1964) showed that if $(\cdot 535)\, 2\pi G\, S_0/(\Omega^2\bar{\varrho}) < 1$ the sheet is stable. We assume that this condition is satisfied.

Rather than calculate the force exerted on a test star by its wake, we shall show that the sum of the wake and the density representing the test star, or the 'dressed' particle, is time independent and symmetric around the center of the orbit.

We choose as our origin the center of the test star's orbit. The radius of the latter is ϱ. As before we assume that the test star is assembled slowly, or that its mass grows as $me^{\varepsilon t}$. For $t \leqslant 0$ we represent the circling and slowly growing mass point in terms of plane waves. In this case the amplitude $A(\mathbf{k}, \omega)$ of each plane wave component will be of the form

$$\sum_{m=-\infty}^{\infty} \frac{a_m(\mathbf{k})}{\omega - m\Omega + i\varepsilon} \tag{9}$$

and the corresponding amplitude of the 'dressed' particle is $A(\mathbf{k}, \omega)/(1-R(k, \omega))$. If we multiply this expression by $\exp(i\omega t)/(2\pi)$ and integrate over all ω we obtain the time dependence of the **k**-th spatial Fourier component. It will consist of a sum of terms, the typical one being

$$\frac{ia_m(\mathbf{k}) \exp[\varepsilon t + im\Omega t]}{1 - R(k, \omega)}, \quad \text{if} \quad t \leqslant 0. \tag{10}$$

From the expression (8) we see that R has a pole at $\omega = m\Omega$, if $m \neq 0$. Therefore if we let $\varepsilon \to 0$ the only term (10) that does not vanish is the time independent $m=0$ term. Adding up all the time independent spatial components we obtain a time independent function. We can argue further that it must also be symmetric around the origin since it does not depend on the ever changing angular position of the test star*.

The conclusion we draw from this example is: if we have a periodic system and are willing to wait long enough for weak effects from successive periods to accumulate we can reach a state in which dynamical friction vanishes.

References

Chandrasekhar, S.: 1943, *Astrophys. J.* **97**, 255.
Gilbert, I. H.: 1968, *Astrophys. J.* **152**, 1043.
Hénon, M.: 1958, *Ann. Astrophys.* **21**, 186.
Stix, T. H.: 1962, *The Theory of Plasma Waves*, McGraw-Hill, New York.
Toomre, A.: 1964, *Astrophys. J.* **139**, 1217.

* The actual expression for the 'dressed' particle can be calculated, and is

$$= \frac{m}{2\pi} \int_0^\infty k \mathrm{d}k J_0(k\varrho) J_0(kr) [1 - R(k, 0)]^{-1}. \tag{11}$$

The series expression (8) for $R(k, 0)$ can be summed, and written as

$$R(k, 0) = \frac{2\pi G S_0}{\Omega^2 \varrho^2 k} [1 - \exp[-k^2\varrho^2] I_0(k^2\varrho^2)]. \tag{12}$$

J_0 is the Bessel function of first kind.

A CERTAIN DISCONTINUOUS MARKOV PROCESS IN STELLAR DYNAMICS

WERNER TSCHARNUTER
*Inst. für Theoret. Astronomie, Wien, Austria**

1. Chandrasekhar's Theory

The basic assumption in Chandrasekhar's approach of statistical stellar dynamics (Chandrasekhar, 1942) is the postulate that a test star within a stellar system being stationary in the sense of collisionless continuum theory suffers random displacements in velocity space generated by the fluctuating part of the gravitational field in a manner that can be described in terms of a random walk. This is equivalent to the assertion that the increments of velocity are regarded as stochastically independent in disjoint time intervals. From this Chandrasekhar derived a diffusion process in velocity space. The equation of motion of the probability density $W(r, u, t)$ in the whole 6-dimensional phase space is then written in the form of a Fokker-Planck-type equation:

$$\frac{\partial W}{\partial t} + u \cdot \nabla_r W + \nabla_r \Phi \cdot \nabla_u W = \nabla_u \left(q \nabla_u W + \eta W u \right) \tag{1}$$

r, u=position, velocity vector; Φ=gravitational potential of the 'smoothed out' distribution of matter; q=diffusion coefficient; and η=coefficient of dynamical friction.

The coefficients q and η are related by the condition that a given Maxwellian distribution f of velocities remains invariant for all times, i.e. $(\partial f / \partial t) \equiv \emptyset$.

It is well known that the sample paths of every diffusion process are continuous with probability one. Thus the random velocity of the test star varies continuously in the course of time. Because of this fact the failing of the theory is obvious, since a close encounter of the test star with a field star is able to produce a large change of velocity within a small time interval, clearly contradicting the notion of continuity. Numerical studies on the gravitational n-body problem show that even a velocity higher than the escape velocity can be reached as a result of a single encounter, (Wielen, 1967).

2. Discontinuous Model

The aim of the following considerations is to show that Chandrasekhar's basic assumption is more general than the choice of a diffusion process. Thus the phrase 'stochastically independent increments in disjoint time intervals' is not equivalent to the property 'diffusion', since there are infinitely many Markov processes that share only the first property but not the second. If the stochastic variations in velocity space

* Now at Universitäts Sternwarte, Göttingen, West Germany.

M. Lecar (ed.), Gravitational N-Body Problem, 18–21. *All Rights Reserved*

are approximately describable as a Markov process at all, the question arises whether it is uniquely determined by the properties of the fluctuating part of the gravitational field and whether the jump phenomena mentioned above can be explained by an analysis of its sample paths alone without employing additional assumptions.

To begin with, let us reformulate more definitely the conditions of a random walk. The total increment of velocity within the time interval $(\emptyset, t)$, $t \gg$ characteristic time T during which an elementary fluctuation of the random gravitational field takes place, can be written as a sum of a great number of independent random variables representing the (at least on the average) small displacement in velocity space after the short amount of time T has passed. Now the crucial point is the determination of the distribution law of this sum. This problem, however, is solved exhaustively by the socalled extended central limit theorems of probability theory which were essentially established by Lévy and Khintchin in the 1930's. In general, a convergence to the normal distribution and hence a diffusion is expected, but the probability distribution of the random gravitational field is shown to be asymptotically the Holtsmark-distribution, the characteristic function (Fourier transform) of which is

$$h(\omega) = \exp(-\alpha|\omega|^{3/2}), \qquad \omega = (\omega_1, \omega_2, \omega_3). \tag{2}$$

The Holtsmark-distribution is a stable distribution and belongs therefore to its own domain of attraction (e.g. Feller, 1966). This leads necessarily to a distribution law for the total increment of velocity within $(\emptyset, t)$ with characteristic function

$$\varphi(\omega) = \exp(-\sigma t|\omega|^{3/2}), \qquad \sigma = \alpha\sqrt{T} \tag{3}$$

φ is the Fourier transform of a transition function belonging to a Markov process which is called the stable process with characteristic exponent $\frac{3}{2}$. So far, dynamical friction acting in a purely systematic manner has been ignored. Taking it into account an equation of motion analogous to (1) can be written down:

$$\frac{\partial W}{\partial t} + u \cdot \nabla_r W + \nabla_r \Phi \cdot \nabla_u W = -(-\Delta_u)^{3/4}(\sigma W) + \nabla_u(\eta W u) = \mathbf{A}^* W \text{ (say)} \tag{4}$$

σ appears to play a similar rôle as the diffusion coefficient q in (1). The $\frac{3}{4}$-power of the Laplace operator Δ_u is uniquely defined in the sense of its spectral representation. This operator acts on a given function f in the sense that the Fourier transform of $(-\Delta)^{3/2} f$ equals to the function $|\omega|^{3/2} \hat{f}(\omega)$, $\omega = (\omega_1, \omega_2, \omega_3)$ where $\hat{f}$ denotes the Fourier transform of f. The correct derivation of this term in (4) is not easy and requires sophisticated functional analytic techniques.

From general theorems on Markov processes follows the rightcontinuity of the sample paths. It can also be shown that the mean number of jumps increases to infinity as their heights converge to zero and conversely. This is a very important property because, if one identifies these jumps as the results of far and close encounters respectively, the importance of the far encounters is emphasized on the one hand but spontaneous large changes in velocity due to close encounters are also possible on the other hand. These characteristic features are in full agreement with results of numerical experiments on the gravitational n-body problem (e.g. Wielen, 1967).

3. Relaxation Time, Escape Probability

In the following we shall restrict ourselves to the spatially homogeneous case, i.e. $\nabla_r W \equiv \emptyset$, $\nabla_r \Phi \equiv \emptyset$, because it seems impossible to solve Equation (4) in the full 6-dimensional space analytically as well as numerically. Unfortunately the assumption of spatial homogeneity is a very serious simplification, because any segregation of mass at the center of a star cluster and the setting up of a halo which are both characteristic phenomena in the dynamical evolution of a stellar system and which will influence its further evolution for their part considerably are entirely ignored. In particular, the escape probability and the number of escaping stars will extensively depend on the space distribution of the field stars.

The following definitions for a relaxation time and an escape probability are even valid, if the stochastic process in velocity space is non-Markovian. Thereafter the relaxation time is defined to be the expected amount of time $\bar{\varrho}$ a test star with initial velocity equal to zero requires to reach the mean velocity of the field stars for the first time. The escape probability is defined to be the probability distribution of the random instant of time τ the test star gets a velocity greater than the critical escape velocity v_∞ for the first time. Clearly, the escape probability is a function of the initial velocity of the test star. By 'stopping' the Markov process at the random 'exit' times ϱ or τ which are both socalled Markov times for the process all desired quantities can actually be computed.

4. Numerical Methods and Results

To the abstract theory as well as to numerical computations the infinitesimal generator **A** defined by

$$\mathbf{A}f = -\sigma(-\Delta_u)^{3/4} f - \eta u \cdot \nabla_u f = \frac{\partial f}{\partial t} \tag{5}$$

is much more appropriate – by the way, **A** has no physical meaning in contrast to its dual **A***! Both, however, **A*** and **A** generate the same Markov process. Now it is easy to write down the corresponding equation for the stopped process:

$$\frac{\partial f}{\partial t}(u, t) = \begin{cases} \mathbf{A}f(u, t) \\ \emptyset \end{cases} = \mathfrak{A} \quad \begin{matrix} |u| < V \\ |u| \geqq V \end{matrix} \tag{6}$$

where V denotes either the mean velocity of the field stars or the critical escape velocity. Thus boundary conditions are introduced and the whole theory of Fourier series applies. Infinitesimal and transition operators are then approximated by matrices and functions by vectors.

The coefficients σ, i.e. α, T in (3), and η are given by Chandrasekhar (1942). It is easily shown that the Maxwell distribution of velocities is not an invariant distribution of the given Markov process. This fact causes trouble, since the relation between σ and η cannot directly be established. Numerically it can be defined with the aid of the dual eigenfunction which corresponds to the absolutely least eigenvalue of the operator $\mathfrak{A}$, the velocity V in (6) being chosen sufficiently high. This eigenfunction may be taken as an approximation of the invariant distribution, but the computational difficulties appeared to be enormous. Thereby a reduction of σ by a factor 0.85 is suggested having η fixed, the problem, however, is far from being solved.

Some relaxation times T_R of different star clusters were computed by using the same input parameters as Chandrasekhar (1942) has taken. The quotient $Q=T_R/T_E$ of corresponding relaxation times were found to depend only on the total number of stars N ($Q=2.5$ for $N=50$ monotonically decreasing to $Q=0.07$ for $N=10^8$) but almost not on the radii of the clusters.

References

Chandrasekhar, S.: 1942, *Principles of Stellar Dynamics*, New York.
Feller, W.: 1966, *An Introduction to Probability Theory and Its Applications*, Vol. II, New York.
Wielen, R.: 1967, *Veröff. Astron. Rechen-Inst. Heidelberg*, No. 19.

RELAXATION TIMES IN STRICTLY DISK SYSTEMS

GEORGE B. RYBICKI

Smithsonian Astrophysical Observatory, Cambridge, Mass. 02138, U.S.A.

Abstract. It is shown that the time of relaxation by particle encounters of self-gravitating systems in the plane interacting by $1/r^2$ forces is of the same order of magnitude as the mean orbit time. Therefore such a system does not have a Vlasov limit for large numbers of particles, unless appeal is made to some non-zero thickness of the disk. The relevance of this result to numerical experiments on galactic structure is discussed.

The use of the collisionless Boltzmann equation (Vlasov equation) for investigating galactic structure has been justified by simple estimates of the ratio of the relaxation time by particle encounters to the typical orbit time in the mean field (Chandrasekhar, 1939; Hénon, 1958; Ostriker and Davidsen, 1968). This ratio is of the order of $N/\log N$, where N is the number of particles, so that for large N the effects of particle encounters can be neglected on a mean field time scale.

A galaxy, even though it may be highly flattened, is still a three-dimensional system, and it is to three-dimensional systems that these estimates of relaxation time apply. However, galaxies are often approximated as strictly disk systems in which stars are still assumed to interact by $1/r^2$ forces, but are constrained to move in a plane. Therefore, it is of some interest to consider the problem of relaxation times for strictly disk systems.

A simple order of magnitude estimate of relaxation time may be made as follows. For a system of N particles, each of mass m, which is of typical size R, the virial theorem gives an estimate of a typical total particle velocity V from the relation

$$V^2 = GN\, m/R, \tag{1}$$

where G is the gravitational constant. It is useful to distinguish between this total velocity and the typical random velocity v of particles relative to a local frame of rest. This random velocity is some fraction of the total velocity:

$$v = \lambda V, \quad 0 < \lambda < 1. \tag{2}$$

In a collision between two particles with relative velocity v and with impact parameter b, the change of velocity is easily estimated to be

$$\Delta v \sim Gm/(bv). \tag{3}$$

This formula holds for $\Delta v \lesssim v$. It breaks down at the point where $\Delta v \gtrsim v$, which is the condition for a *close encounter*. This occurs when $b \lesssim b_0$, where

$$b_0 = Gm/v^2 \sim R/(\lambda^2 N), \tag{4}$$

using the virial theorem.

M. Lecar (ed.), Gravitational N-Body Problem, 22–26.

The relaxation time t_R may now be estimated as the time required for a typical particle to suffer a close encounter. Since this neglects the cumulative effect of long-range encounters, this will be an overestimate of the true relaxation time. In a time t_R the motion of the particle takes it a distance vt_R relative to neighboring particles. An estimate of the number of close encounters during this time is equal to the surface density of particles

$$\varrho \sim N/R^2 \tag{5}$$

times the area $(vt_R)\cdot(2b_0)$ within which another particle will cause a close encounter with the given particle. Setting this equal to unity leads to the result

$$t_R \sim R^2/(2Nb_0 v). \tag{6}$$

The orbit time for a typical particle in the mean field is defined by

$$t_M = R/V. \tag{7}$$

Combining these results yields

$$t_R/t_M \sim \lambda/2. \tag{8}$$

The relaxation time is seen to be at most the same order of magnitude as the mean orbit time, independent of the number of particles. *Therefore the collisionless Boltzmann equation can never be an adequate description of a strictly disk system, however large.*

A refined derivation will now be given that includes the cumulative effects of long-range encounters. In this case the relaxation time is defined as the time at which the root mean square velocity change due to encounters is of the same order of magnitude as the typical random velocity v. Thus, the condition is

$$v^2 \sim (\Delta v)^2_{\text{total}} \sim (G^2m^2/v^2)\,(vt_R\varrho)\int (2\,\mathrm{d}b)/b^2. \tag{9}$$

The quantity $vt_R\varrho\cdot 2\mathrm{d}b$ represents the number of particle encounters during the time t_R having impact parameters in the range b to $b+\mathrm{d}b$. The lower limit of the integral is taken to be b_0 where the formula for Δv breaks down and the divergence must be cut off. Since the integral converges rapidly for large b, the upper limit may be taken as ∞. Then

$$t_R \sim v^3 b_0/(2G^3 m\varrho)^2 \sim v^3R^3/(2G^2M^2N^2\lambda^2), \tag{10}$$

where the results (4) and (5) have been used. With the virial theorem (1) and Equations (2) and (7) this gives

$$t_R/t_M \sim \lambda/2. \tag{11}$$

The fact that this estimate is identical with the previous one (8) indicates that the relaxation is substantially due to close encounters, and that the cumulative effect of long-range encounters is of no more than the same order of magnitude.

The relaxation time is seen to be proportional to λ, implying that the rate of relaxation is greater when the system is cooler. However, as the system relaxes it heats up, decreasing the rate of relaxation; therefore a cool system still takes longer to reach any given stage of relaxation than a hotter one. If the parameter λ is included in a similar three-dimensional derivation the relaxation time is proportional to λ^3.

Both of the above derivations suffer from the deficiency of assuming that the number of encounters in a certain impact parameter range may be computed simply on the basis of mean particle densities and typical relative velocities. The second derivation further assumes independence of encounters. The use of 'typical' velocities could be avoided by introducing velocity distribution functions, but we do not expect the order of magnitude estimates obtained here to be seriously altered. The assumption of independence of encounters does not present any difficulties in this two-dimensional case, since the relaxation is principally due to close encounters, and these encounters would be expected to occur independently. In this regard it is interesting to note that a long-range divergence occurs in the three-dimensional case when simple derivations of this sort are used. However this divergence does not occur here, and there is no need to introduce a long-range cutoff.

The use of mean particle densities implies lack of certain correlation effects, and this assumption would be difficult to validate rigorously. However, we offer the following rough argument: Because the relaxation is due to close encounters, in order for correlation effects to be serious there would have to be a severe alteration in the distribution of colliding particles having small impact parameters. Since the unperturbed orbits in the mean field are quite smooth this would require rather special synchronizations of the phases to avoid close encounters at orbit crossings. This seems unlikely in view of the great number of orbit crossings (order N^2) that must occur in a strictly plane geometry. Furthermore, it is difficult to imagine correlation effects producing the synchronizations necessary to avoid a close encounter when this is to occur at some great distance along the orbit.

It is of some interest that the difference between the two- and three-dimensional results may be traced to the different statistical weighting of impact parameters. In the three-dimensional case this is $2\pi b\,\mathrm{d}b$ while in the two-dimensional case it is $2\mathrm{d}b$. The extra factor of b in the three-dimensional case tends to suppress the effect of close encounters and to weight the longer range encounters more. An interesting comparison between two- and three-dimensional systems is the average number of close encounters occurring in one mean field period. For two-dimensional systems it is of order N, while for three-dimensional system it is of order unity. This is the difference between *each* particle and *one* particle having a close encounter each mean field period.

These arguments concerning strictly disk systems of course do not apply to actual galaxies, which are three-dimensional; the validity of the Vlasov equation is well established in this case. However, strictly disk system approximations are commonly made in analytical and numerical treatments of galaxies, and it is necessary to judge these approximations in the light of the preceding results. For the analytical treatments there is no such problem of the relaxation time at all; the use of Vlasov theory is first established in view of the non-zero thickness of the disk, and then it is simply a question whether a two-dimensional form of the Vlasov equation is a good approximation to the three-dimensional form. Although this question is not trivial, at least it can be answered within the framework of Vlasov theory.

The situation regarding the numerical calculations (Miller and Prendergast, 1968; Miller *et al.*, 1969; Hohl and Hockney, 1969), on the other hand, is not so straightforward. Here there is a strict zero thickness of the disk, and the relaxation results presented here apply. Therefore such calculations, *if done sufficiently precisely*, are not faithful simulations of the Vlasov equation and thus do not apply to actual galaxies.

Fortunately, numerical calculations are themselves subject to further approximations that tend to reduce the severity of this difficulty. The grid size on which the potential (or force) is calculated effectively limits the close encounters that cause rapid relaxation. This *discretization* effect may be estimated as follows: Let h be the distance over which the gravitational force is cut off by grid size effects or by some other effect, such as purposely altering the force law at short distances (Miller and Prendergast, 1968). Then a derivation of relaxation time may be given as before (see also Hohl, 1970), except that the lower limit on the integral in Equation (9) is taken to be h rather than b_0.

This yields

$$t_R/t_M = \lambda^3 Nh/(2R).$$

For a typical numerical experiment $N \sim 10^5$ and $h/R \sim 4 \times 10^{-2}$, so that $t_R/t_M \sim \sim 2 \times 10^3 \lambda^3$. For large values of random velocity ($\lambda \sim 1$) the system might be followed for at least hundreds of mean periods, but for relatively 'cool' systems, say $\lambda \sim 0.1$ only several mean periods might be followed. For 'cold' systems λ is zero, and relaxation would proceed rapidly, although the random velocities produced would then tend to stabilize the system against further rapid relaxation. By comparison a simple $N/\log N$ estimate would lead to an expectation of many mean periods of validity (10^3–10^4), at least for relatively hot systems.

The numerical experiments of Hohl and Hockney (1969) were done with cold and relatively cool disks for up to five mean periods. The manner of breakup into condensations was interpreted by them in terms of the stability theory of Toomre (1964). It would seem, however, that the relaxation time effect described here might be operating as well, at least for the cold disks. An examination of the velocity distributions in these cases might be helpful in deciding whether the random velocities that developed were due to particle relaxation or mean field relaxation. Repeating the calculations with a mass spectrum might also be helpful. The experiments done with relatively cool disks, on the other hand, probably are not affected. This seems to be also in accord with recent calculations (Hohl, 1970).

The numerical experiments of Miller *et al.* (1969) are more difficult to assess, since their computations included a 'gas' component as well as a 'star' component. The gas component was artificially 'cooled' after each time step. One reason given for this cooling was that real gas clouds would be subject to the cooling mechanism of inelastic collisions. Aside from this physical reason they also state that such cooling is necessary to prevent rapid development of random velocities, which make it impossible to model a cool galaxy. However, in light of the discussion presented here one must question whether this latter presumed necessity for artificial cooling may simply be a result of the artificially rapid relaxation in strictly disk systems. Also the hot star populations which evolved in their calculations might be spurious for the same reason. Again, experimental checks can be done to answer these questions.

On the positive side, the estimates of relaxation time given here suggest some ideas for making numerical experiments more realistic and reliable. For example, a force law cutoff, such as used by Miller and Prendergast (1968) and Hohl (1970), would seem desirable in all such experiments. How much of a cutoff can be tolerated could be decided by suitable experimentation. In any case it is definitely clear that there is nothing to be gained by attempting to treat close encounters more exactly than has already been done.

Another way of minimizing the difficulty with relaxation time is to include, whenever possible, a fixed central force field, such as the galactic nucleus might produce (Hohl, 1970). This decreases the mean field time, leaving the relaxation time roughly unchanged; therefore more mean periods may be followed.

One experimental method of testing a computational procedure to see whether it is sensitive to the undesirable particle relaxation effects described here is to repeat a calculation using a mass spectrum. If the system is being correctly modeled as a Vlasov system there should be no effect; if not then there will be a tendency towards energy equipartition between the various masses.

References

Chandrasekhar, S.: 1939, *Principles of Stellar Dynamics*, Dover Publ., New York.
Hénon, M.: 1958, *Ann. Astrophys.* **21**, 186.
Hohl, F. and Hockney, R. W.: 1969, *J. Comp. Phys.* **4**, 306.
Hohl, F.: 1970, NASA TR R-343.
Miller, R. H. and Prendergast, K. H.: 1968, *Astrophys. J.* **151**, 699.
Miller, R. H., Prendergast, K. H., and Quirk, W. J.: 1969, Univ. of Chicago Rept. COO-614-72.
Ostriker, J. P. and Davidsen, A. F.: 1968, *Astrophys. J.* **151**, 679.
Toomre, A.: 1964, *Astrophys. J.* **139**, 1217.

B. NUMERICAL EXPERIMENTS

NUMERICAL EXPERIMENTS ON THE N-BODY PROBLEM

S. J. AARSETH

Institute of Theoretical Astronomy, Cambridge, England

Abstract. This review first discusses the different types of numerical methods available for integrating the equations of motion of N-body systems. It is desirable to supplement ordinary integration schemes with special treatments of close encounters using a two-body perturbation description or introducing regularizing transformations of the co-ordinates and time. Direct methods are at present limited to the study of a few hundred particles but larger systems may be investigated using Monte Carlo techniques or the Boltzmann moment equations.

N-body computations have been performed for a whole range of initial conditions and the general results are summarized. Numerical investigations have already clarified a number of important aspects of cluster evolution and the qualitative behaviour of small stellar systems is now quite well understood. Recent theoretical modifications have reduced the disagreement with experiments but further improvements are still needed.

1. Introduction

The study of N-body systems by numerical methods celebrates its tenth anniversary this year (von Hoerner, 1960). This relatively new development in stellar dynamics owes its existence entirely to modern technology and progress is therefore closely linked to the availability of bigger and faster computers as well as improved methods of solution. The simplicity of Newton's law of gravitation lends itself naturally to a numerical attack on the cluster problem. Given the initial distribution of individual masses, co-ordinates and velocities, the task is very well defined; i.e., to calculate in detail the behaviour of the system as a function of the time. Complete solutions to this problem can only be obtained by numerical methods based on time series expansions.

Repeated integrations of the equations of motion can readily be made by exploring a whole range of parameters for different values of the particle number. In addition, more realistic star cluster models may be simulated by allowing for stellar evolution effects and the influence of an external gravitational field. It is the aim of the direct approach to provide a better understanding of the behaviour of self-gravitating systems and use the results for dynamical interpretation of actual star cluster evolution. A considerable theoretical effort has also been directed towards the solution of this problem and the different methods of attack make it necessary to distinguish between the branches of theoretical and experimental stellar dynamics.

Although numerical studies are free from simplifying assumptions, the direct approach is limited to the simulation of small stellar systems because of heavy demands on the computing time. It is nevertheless encouraging that already the range of particle numbers has been extended to 500 which is comparable to the membership in typical galactic clusters and clusters of galaxies. The theoretical treatment becomes more reliable as the number of particles increases and a semi-empirical attack may therefore

M. Lecar (ed.), Gravitational N-Body Problem, 29–43.

be made on the more formidable globular cluster problem. In addition, collisional effects are less important in rich systems which may be studied by fast numerical methods.

The problem of dynamical interpretation of the results presents considerable difficulties since it has been shown that the numerical solutions are strongly divergent (Miller, 1964; Standish, 1968). Numerical integrations of the same system at different levels of accuracy lead to an increasing separation in phase space as defined by the individual evolution. It is therefore impossible to obtain the unique solution for all time specified by a set of initial conditions. On the other hand, time reversal tests have demonstrated that reliable individual orbits can be calculated for a few crossing times, after which the numerical solutions begin to depart significantly from the original behaviour. This interval is sufficiently long to establish general features of the evolution which are consistent with results obtained over much longer times. Small systems may exhibit rather large fluctuations, however, and many cases must be studied in order to define meaningful properties. Numerical experiments can therefore be considered as repeatable under different conditions only in a statistical sense.

The evolution of richer clusters can be described with increasing confidence since each particle orbit is subject to a greater number of encounters. Thus the actual distribution of impact parameters resembles more closely the theoretical expectations which are represented by continuous expressions and more reliable comparisons can therefore be made in single cases. Cluster simulation by numerical means makes it possible to study individual interactions in considerable detail and suggest modifications of the dynamical assumptions which form the basis of theoretical treatments. The direct approach has already yielded extensive results which permit a qualitative description of the long term cluster evolution and this hopeful development should provide further stimulus for theoretical improvements.

2. Direct Integration Methods

A wide variety of direct methods have been used to integrate the equations of motion of the N-body problem

$$\ddot{\mathbf{r}}_i = -\sum_{\substack{j=1 \\ j \neq i}}^{N} \frac{m_j (\mathbf{r}_i - \mathbf{r}_j)}{|\mathbf{r}_i - \mathbf{r}_j|^3}, \tag{1}$$

where the scaled mass and co-ordinates of a particle is denoted by m_i and $\mathbf{r}_i$, respectively, and dots represent differentiation with respect to the time t. These second-order differential equations are usually written as $6N$ equivalent equations of first order which can be solved numerically by step-wise integration. It is evident that the acceleration calculation (1) becomes very time-consuming when the particle number N is large and for this reason high-order difference schemes are preferable since past information may then be used with very little additional effort.

Writing the force per unit mass as an extrapolating polynomial of degree n through $n+1$ fitting points in Newtonian form we have

$$\mathbf{F}_i(t) = \sum_{\lambda=0}^{n} \mathbf{A}_\lambda (t - t_0)^\lambda . \tag{2}$$

The coefficients A_λ are given by divided backwards differences weighted by constant coefficients (Wielen, 1967) and the expansion is valid over an interval $\Delta t = t - t_0$ which may be determined from the convergence property of Equation (2) itself. Integrating the polynomial twice we obtain the corresponding expression for the predicted position

$$\mathbf{r}_i(t) = \mathbf{r}_i(t_0) + \dot{\mathbf{r}}_i(t_0)(t - t_0) + \sum_{m=2}^{n+2} \frac{1}{m(m-1)} \mathbf{A}_{m-2}(t - t_0)^m , \tag{3}$$

with the two first terms representing the position and velocity at time $t = t_0$. The new velocity is obtained in a similar manner.

The predicted co-ordinates and velocities may be improved by taking the difference between the n'th-order extrapolating polynomial (2) and an interpolating polynomial of order $n+1$ which is obtained after the calculation of the new force. This procedure may be referred to as a semi-iteration, since the main part of the improvement is achieved without recalculating the force based on the predicted position. In this way almost one extra order of integration is included at very little additional effort. Practical experience indicates that the gain in efficiency beyond $n=3$ or $n=4$ does not justify the use of higher orders unless extreme accuracy is demanded.

High-order schemes also require special starting procedures which are sometimes more complicated than the integration method itself. In the present formulation the derivatives may be obtained directly by explicit differentiation of Equation (1), involving relative co-ordinates and velocities only. The coefficients A_λ can readily be expressed in terms of the corresponding Taylor series derivatives. This method is very convenient for starting purposes but the multiple summations required at each stage makes it rather inefficient for continuing the integration of large particle numbers (Gonzalez and Lecar, 1968).

The second essential requirement of an efficient numerical method consists of adopting a wide range of individual integration intervals. Thus the time-step appropriate to close encounters must be reduced sufficiently in order to preserve the accuracy while new positions on smooth particle orbits may be recalculated much less frequently. Suitable criteria for selecting individual steps Δt_i may readily be obtained by considering the convergence of the force polynomial (2). This procedure necessitates separate prediction of all co-ordinates because of the force calculation (1) but the order used may be lower than employed by the full integration. Hence only a small proportion of the total time is used for the additional predictions but there is a considerable gain in efficiency by avoiding unnecessary force calculations. The computing time requirement is still proportional to N^2 operations per crossing time, however, and more efficient procedures should be investigated.

The idea of individual time-steps may be carried one stage further by recalculating the contributions to Equation (1) at different times rather than simultaneously. Again the times for individual force computations may be determined from the rate of change of the corresponding contributions. For convenience all intervals are quantized into a relatively small set of categories where the steps Δt_i may change by a factor of two either way. This idea has only been tried for the case $n=1$ (Hayli, 1967) and its full advantage has therefore not been realized yet.

Alternatively, distant force contributions may be calculated by the use of Legendre polynomials rather than individual summations (Aarseth, 1967). This simplification would be particularly useful for the integration of particles near the cluster centre where the net force produced by halo members is usually very small. Evidently a considerable gain in efficiency may result when the effect of distant particles is included by approximate methods, but any modifications which are introduced must be dynamically consistent with the collisional nature of the direct approach.

Numerical integrations of the equations of motion (1) cannot yield the exact solution for all time specified by a set of initial conditions. Although the error at each interval can be controlled by choosing suitable time-steps, the strong non-linearity of gravitational interactions leads to an exponential growth with a rather short time-scale (Miller, 1964; Standish, 1968).

The error amplification introduced by successive encounters may be reduced by more powerful methods to be discussed subsequently but cannot be entirely removed. Results may still be used for statistical purposes, however, provided that the evolution rate depends mainly on the overall structure and systematic errors do not enter into the calculation of individual encounters.

Practical applications make use of the ten general constants of the motion as integration tests. The conservation of the total energy, angular momentum and centre of mass motion does not by itself guarantee the accuracy of numerical solutions since errors may cancel but this is not likely to occur in general. An additional and more detailed check is provided by the time reversibility of the equations of motion. It is therefore prudent to investigate the correlation between a meaningful integration and the corresponding accuracy of the conserved quantities by time reversal tests before deciding on the final integration parameters.

3. Special Treatments of Close Encounters

Direct integrations of the equations of motion are very time-consuming for systems which develop high-density cores. This difficulty reflects the shortening of the central relaxation time due to the strong interaction of close neighbours. As the force fluctuations increase in strength it also becomes more difficult to conserve the integrals of motion. The choice therefore lies between terminating the calculations or making use of more powerful methods.

Detailed examinations of critical stages of evolution usually reveal the presence of close binaries or sub-groups where the members are strongly bound. The numerical problem becomes more serious if the binaries have long life-times since the calculation of each revolution typically takes a few hundred time-steps even if the basic elements do not change significantly over many periods. Close binaries may also suffer further changes of the semi-major axis and the replacement of the two components by the centre of mass motion would not be dynamically consistent. Instead it is natural to introduce two-body perturbation methods, replacing the dominant term by the analytical solution which is modified by the effect of the external field.

A classical variation of parameters method (Pines, 1961) has been introduced for studying close binaries in simulated clusters (Aarseth, 1970). The equation of motion for the relative co-ordinates $\mathbf{R} = \mathbf{r}_k - \mathbf{r}_l$ can be derived by writing Equation (1) for the two mass points m_k, m_l. Subtraction gives

$$\ddot{\mathbf{R}} = -\frac{m_k + m_l}{R^3}\mathbf{R} + \mathbf{F}, \tag{4}$$

where $\mathbf{F}$ denotes the relative perturbation acceleration.

The unperturbed solution of Equation (4) may be written in terms of the Lagrangian representation coefficients f and g as

$$\mathbf{R} = f\,\mathbf{R}_0 + g\dot{\mathbf{R}}_0\,. \tag{5}$$

The complete motion of both components is obtained by calculating the corresponding centre of mass motion. Equation (5) and its derivative can be inverted to express the initial position and velocity vector $\mathbf{R}_0$, $\dot{\mathbf{R}}_0$ as functions of $\mathbf{R}$ and $\dot{\mathbf{R}}$. The relation (5) is then maintained in the presence of perturbations by integration of the variational equations

$$\begin{aligned} \mathbf{R}_0^{`} &= \dot{g}^{`}\mathbf{R} - g^{`}\dot{\mathbf{R}} - g\mathbf{F}\,, \\ \dot{\mathbf{R}}_0^{`} &= -\dot{f}^{`}\mathbf{R} + f^{`}\dot{\mathbf{R}} + f\mathbf{F}\,. \end{aligned} \tag{6}$$

This notation uses dots to represent the two-body variation which would remain in the absence of perturbations while the perturbative variation of a quantity h is denoted by $h^{`}$. The auxiliary variables are derived from basic elements and the perturbation effect enters through the scalar products $\mathbf{R}\cdot\mathbf{F}$ and $\dot{\mathbf{R}}\cdot\mathbf{F}$ which should remain small for the method to be efficient.

An invariant parameter

$$\gamma = \frac{|\mathbf{F}|R^2}{m_k + m_l} \tag{7}$$

may be introduced to indicate the relative importance of the external force field. Hence the perturbation treatment is limited to values of γ which produce slowly varying elements and the two-body description must be replaced by direct integrations of both components if γ exceeds a few per cent. On the other hand, it is permissible to make use of the unperturbed formulation (5) if the external effect is too small to change the relative binding energy by a significant amount. It is interesting to note that the unperturbed approximation may describe the actual motion more accurately than the corresponding orbit calculated by direct means because of numerical errors introduced during critical encounters. The appropriate two-body reflection in the centre of mass frame may therefore be used with advantage for the closest encounters, provided that the relative perturbation is sufficiently small.

More recently a general regularization method has been developed for the perturbed two-body problem where close encounters are treated very accurately (Kustaanheimo and Stiefel, 1965). The basic idea of regularization is to transform the equations of motion (1) into a form which removes the singularity of two-body collisions. This is achieved by first introducing generalized coordinates $\mathbf{u}$ for the relative motion $\mathbf{R}=(X_1, X_2, X_3)$ by the transformation

$$\begin{aligned} X_1 &= u_1^2 - u_2^2 - u_3^2 + u_4^2, \\ X_2 &= 2(u_1u_2 - u_3u_4), \\ X_3 &= 2(u_1u_3 + u_2u_4), \end{aligned} \tag{8}$$

which satisfies the relation

$$R = u_1^2 + u_2^2 + u_3^2 + u_4^2. \tag{9}$$

The singularity is removed by the regularizing substitution

$$\mathrm{d}t = R\,\mathrm{d}\tau, \tag{10}$$

where τ is the new fictitious time. The resulting equation of motion then becomes (Stiefel, 1967)

$$u_j'' = \tfrac{1}{2}(h_0 + W)\,u_j + \tfrac{1}{4}Rq_j \qquad (j = 1, 2, 3, 4), \tag{11}$$

in which h_0 is the energy constant of the two-body motion and W is the work done by the external force field. The generalized perturbations q_j are calculated from the actual components of $\mathbf{F}$ by the relations

$$q_j = \sum_{i=1}^{3} \frac{\partial X_i}{\partial u_j} F_i. \tag{12}$$

Hence the second term becomes arbitrarily small as $R \to 0$ and the solution is regular for collision orbits.

It has been demonstrated that the regularized solution is more powerful than the direct method at relatively large separations for three interacting particles (Peters, 1968). This early attempt also showed a significant improvement when regularizing the closest pair at any time for one case $N=25$ starting from rest. Only minor modifications are necessary for an efficient treatment of systems with large particle numbers.

The combined term $h_0 + W$ in Equation (11) represents the actual binding energy per unit mass of the relative motion which reduces to

$$h = \left[2\sum_{j=1}^{4} u_j'^2 - (m_k + m_l)\right]\frac{1}{R}. \tag{13}$$

Although not applicable to collision orbits this expression has the advantage of being very simple, whereas the original application involves an N^2 term which becomes prohibitive for large particle numbers.

The modified regularization treatment may be used to study the general cluster problem without experiencing the numerical difficulties due to close two-body encounters. This approach is more efficient than the perturbation method since much greater external effects can be included. Again the complete motion of both components is obtained by introducing the corresponding centre of mass. In addition, the relation (10) must be integrated in order to provide a connection between the fictitious time and the global reference time. Finally, the singularity of Equation (13) may be avoided by introducing an equation of motion for the binding energy itself, making 16 equations in all. The simultaneous use of two different methods necessitates additional programming for decision-making and reorganization but the resulting saving of time-steps and gain of accuracy is substantial (Aarseth, 1972).

The two-body regularization method is less efficient when dealing with multiple encounters since only the dominant interaction can be regularized. Although such events are less common it is nevertheless desirable to seek an improved treatment. One promising alternative consists of using the time transformation

$$\mathrm{d}t = \frac{1}{V}\,\mathrm{d}\tau \tag{14}$$

for an arbitrary number of closely interacting particles giving rise to the potential V. Although less powerful than the full two-body regularization, the resulting equations of motion are again non-singular since Equation (14) reduces asymptotically to the form (10) in the event of a binary collision. It is also natural to include all strongly interacting particles in the transforming function V, leaving more distant members to be treated by the standard method (Heggie, 1972).

The recent introduction of special integration procedures is very promising for further numerical explorations of the collisional cluster problem. It is perhaps fitting that the rapid increase of available computing power should be matched by improvements in technique. Even so the task of the direct attack is formidable and alternative ideas must be considered for an extension to larger particle numbers.

4. Simplified Methods

In view of the severe computing time requirement for all direct methods it is desirable to explore faster alternatives which reproduce the general behaviour of N-body systems at particle numbers already studied. The increased speed of calculation may then be exploited to simulate the evolution of richer clusters where the adopted approximations have greater validity. Clearly the collisional approach must not sacrifice too much of the essential dynamics; i.e., encounter effects between neighbouring stars must be included. The introduction of spherical symmetry already implies a considerable simplification without loss of dynamical consistency. In addition, equal masses may be assumed if the more general case cannot be treated. It may be noted that neither of these assumptions would make much difference to the direct method as long as the particle description is maintained.

A fast Monte Carlo method has recently been introduced to study the evolution of spherical star clusters (Hénon, 1966). Each particle is represented by two quantities only; the binding energy and angular momentum per unit mass. Subsequent changes of the fundamental parameters are calculated by selecting a position along the orbit at random and letting the particle encounter another body chosen at random. The effect of the two-body interaction is then multiplied by an appropriate factor to give the new velocity at the end of the encounter and the next particle is treated similarly. This approach replaces the summation of N-1 terms in Equation (1) by the interaction of one typical member only and therefore leads to considerable saving of time. Furthermore, the time intervals Δt_i which at present are equal can be a substantial fraction of the relaxation time and may easily exceed the crossing time for rich systems.

The assumption of spherical symmetry also permits a fast calculation of the potential, hence the new binding energy is readily obtained. A wide variety of steady-state initial conditions may be studied; recent results are discussed elsewhere in this volume (Hénon, 1972). It may be emphasized here that effects of multiple encounters are neglected in this procedure which can be shown to be mathematically equivalent to a solution of the corresponding Fokker-Planck equation. Direct integrations indicate that such effects are important for all cases studied so far, but the dependence on particle number is not yet known.

A second new method for computing cluster evolution is based on the numerical solution of moment equations derived from the Boltzmann equation (Larson, 1970). Again spherical symmetry is assumed and the treatment is most suitable for equal-mass cases. Four moments are used to characterize the velocity distribution which is expanded in Legendre polynomials about a Maxwellian. Expressions for the corresponding collision terms in the Fokker-Planck equation are derived on the assumption that the velocity distribution is nearly Maxwellian. The relaxation effects in the outer parts are therefore not correctly described but encounters are also less important in regions of low density. The numerical integration of the moment equations proceeds by well tried hydrodynamical methods which give physically acceptable solutions (Larson, 1972).

Both the Monte Carlo technique and the Boltzmann moment approach are very promising for attacking the cluster problem. The results obtained so far are encouraging and there is no doubt that such methods must be used to describe the evolution of richer systems when encounter effects are included. In the first instance, however, it is desirable to explore the region of overlap with direct methods ($N \lesssim 1000$) in order to gain more confidence in the results.

5. Definitions and Initial Conditions

N-body integrations are most conveniently performed with scaled quantities and the results may be discussed in terms of a well-defined mean crossing time. Natural units are introduced by taking the gravitational constant equal to unity; in addition, the total mass and binding energy of a bound system are scaled by the relations

$$\sum_{i=1}^{N} m_i = N, \qquad E = -\tfrac{1}{4}N^2. \tag{15}$$

This energy scaling has the advantage that results for different starting values of the virial theorem parameter may be compared at the same time once the initial evolution is over. Using the rms velocity the units (15) then define the mean crossing time

$$t_{\mathrm{cr}} = \left(\frac{8}{N}\right)^{1/2}. \tag{16}$$

The relevant physical time in years is related to the scaled time by

$$T \simeq 1.5 \times 10^7 \left(\frac{\bar{R}^3}{\bar{M}}\right)^{1/2} t, \tag{17}$$

where $\bar{R}$ corresponds to $r=1$ as given in pc and $\bar{M}$ is the mean particle mass in solar units.

The dynamical state of systems with small particle numbers is best described in terms of the individual binding energy and angular momentum rather than the complete distribution function $f(m, \mathbf{r}, \dot{\mathbf{r}})$ used by theoretical considerations. In practice these distributions are often discussed separately; in addition, integral properties such as space density and mean velocity are studied as functions of central distance. The binding energy per unit mass is defined with respect to the inertial frame by

$$E_i = -\sum_{\substack{j=1 \\ j\neq i}}^{N} \frac{m_j}{|\mathbf{r}_i - \mathbf{r}_j|} + \tfrac{1}{2}\dot{\mathbf{r}}_i^2, \tag{18}$$

using the convention $E_i<0$ for bound orbits. Although the binding energy is of fundamental interest in stellar dynamics, additional knowledge is often needed in order to give a better description of the overall evolution. For instance, the presence of binaries cannot be deduced directly from the energy distribution. Instead it is useful to introduce an invariant parameter λ giving the fraction of total energy absorbed by the relative motion of m_k and m_l as

$$\frac{m_k m_l}{2a} = \tfrac{1}{4}\lambda N^2, \tag{19}$$

where a is the semi-major axis.

A wide variety of initial conditions have been studied for particle numbers in the range $N=3$ to $N=500$. Simple starting models can be obtained from random distributions of co-ordinates and velocities for a given mass spectrum. The velocities are then scaled to the desired value of the virial theorem parameter. This procedure does not introduce any special constraints on the initial configuration and simulated clusters are free to set up a dynamically consistent structure.

The restriction of starting with constant density systems is of little consequence for initial non-equilibrium configurations because of the short time-scale for significant redistribution of mass. Alternatively, more specific cases may be considered with starting conditions based on theoretical or astrophysical models. In particular it is of interest to study centrally concentrated stationary systems based on theoretical solutions of the Liouville equation (Wielen, 1967). Of the many possible mass distributions, the function $f(m) \propto m^{-2}$ is representative of the mass spectrum in young galactic clusters and is also sufficiently steep to be of general dynamical interest.

Individual particles suffer changes in the orbital parameters primarily by close encounters and the combined effect is obtained by dynamical theories for assumed distributions. Corresponding expressions may be introduced for relaxation times derived from numerical studies. Empirical definitions usually depend on the size of the sampling interval; furthermore, the contributions from two close particles may dominate unless many averages are taken. It can readily be shown that the instantaneous rate of change of the binding energy of one particle is given by

$$\dot{E}_i = - \sum_{\substack{j=1 \\ j \neq i}}^{N} \frac{m_j \dot{\mathbf{r}}_j \cdot (\mathbf{r}_i - \mathbf{r}_j)}{|\mathbf{r}_i - \mathbf{r}_j|^3}. \tag{20}$$

The expression (20) may be used to calculate mean relaxation times which only depend on the co-ordinates and velocities. In addition to the weighted mean defined by

$$\bar{t}_E = \frac{\sum_{i=1}^{N} m_i \dot{\mathbf{r}}_i^2}{\sum_{i=1}^{N} m_i |\dot{E}_i|}, \tag{21}$$

it is instructive to study the r-dependence of the relaxation time since the former may be characteristic of the central region only. Repeated averages reduce spurious effects of close hyperbolic encounters but large contributions from permanent binaries would still be included for small perturbations. Instead it would be more meaningful to include the contribution from the centre of mass motion only.

Given the mean relaxation time of the whole system or a group of similar particles, the corresponding escape rate Q or $Q(m)$ may be derived from the relation

$$\Delta N = \frac{Q N \Delta t}{\bar{t}_E}, \tag{22}$$

where ΔN denotes the number of dynamical escapers during the time interval Δt. Events leading to escape are usually well defined for isolated systems. The uncertainty in the derivation of Q is therefore mainly due to statistical fluctuations if $\Delta N \ll N$. On the other hand, it is difficult to preserve uniform relaxation time definitions over the long intervals which are necessary in order to achieve significant escape rates.

6. General Discussion of Results

Cluster simulations are of limited value unless it can be established that different systems exhibit a characteristic behaviour when studied over significant times. On theoretical grounds it is expected that the evolution of all cases considered by direct methods are dominated by close encounter effects, except during the early stages of non-equilibrium configurations when mass motions are important. Stellar encounters therefore provide a regulating mechanism for structural readjustments which tends to decrease initial condition differences. During this process of energy exchange some particles achieve escape velocity as the result of close encounters and are lost from isolated systems. The continuous production of escapers prevents clusters from reaching a steady state, but the evolution may nevertheless be described by a sequence of quasi-equilibrium states provided that the escape rate is small.

The overall evolution of simulated clusters proceeds in the direction of more pronounced core-halo type mass distributions. Only a small proportion of the halo members have sufficient kinetic energy to escape the system altogether, but more distant parts of the halo are gradually populated by highly eccentric orbits. The latter are mainly ejected from the central regions and the relatively long periods ensure that there is a net outward mass flux. The additional loss of fast particles from the inner region decreases the retardation on bound halo orbits and prolongs the phase of outward motion. Conditions of strict equilibrium implies an equal number of positive and negative radial velocities at all energies. Such configurations cannot be reached in practice but it is useful to study the gradual approach to equilibrium from simple initial states.

Halo orbits are approximately collisionless on the time-scale of most numerical investigations. Significant modifications of the binding energy and angular momentum therefore takes place on subsequent passages through the central region, thereby maintaining the predominance of radial motion for distant particles. The velocity anisotropy increases outwards from an approximate Maxwellian central distribution and is closely linked to the overall cluster structure. Although the transition is gradual the isotropic region can be considered completely relaxed by encounters. Consequently the halo velocity distribution may be used as a dynamical age indicator provided that distant particles were not present originally. Alternatively, it may be possible to ascribe an approximate dynamical age to the central nucleus if violent initial conditions can be ruled out.

The simulation of systems with a general mass distribution introduces many interesting features and is also more realistic. Relative evolution rates may be estimated by comparing similar equal-mass cases. Numerical half-lives of 5–50 mean crossing times have been obtained for repeated experiments with $N \leqslant 24$ (van Albada, 1968). The disruption rate is considerably increased for small clusters with moderate mass dispersions, whereas larger mass ratios do not decrease the life-times correspondingly. Encounters between unequal masses tend to promote equipartition of kinetic energy and light particles therefore have a higher probability of achieving escape velocity.

Mass segregation takes place at the same time, however, and the longer intervals between significant encounters for light bodies counteracts the greater probability of escape in a given event.

Particles which lose kinetic energy become more strongly bound to the centre and may eventually form part of a dense nucleus dominated by heavy bodies. The strong central force field generates high velocities in order to maintain approximate equilibrium, whereas the mean velocity decreases at greater distances. This paradoxical development does not necessarily contradict the tendency towards equipartition during single encounters since the velocity of light halo bodies exceed the mean central value during passages through the nucleus. Invoking the equipartition effect, it is also possible to explain qualitatively the higher escape probability of light particles from the centre since a moderate mass ratio would be sufficient to raise the equipartition value above the local escape velocity. Further increases of the mass dispersion do not alter the character of this process, since the greater relaxation efficiency is partly compensated by a reduced number of central bodies.

The mass segregation of bound members may be estimated by comparing the mean central distance or binding energy for a given mass interval. Although differences in the mean quantities increase with time on the average, the dispersion within one group is often considerable. Thus light halo particles often become bound to the nucleus by temporary capture. Conversely, heavy bodies may be found at large central distances; this is particularly common in highly evolved systems which have expelled a significant proportion of members originally in the nucleus. The presence of heavy bodies in the outer region leads to enhanced relaxation of halo orbits but the time-scale for randomization of velocity components is usually too long to be reached by actual calculations.

Escapers carry away excess kinetic energy and therefore leave the remaining system more strongly bound. In addition, the expansion of the outer region must be compensated by a corresponding contraction of the core which itself is losing particles. Since no general configurations with more than two bodies are known to be stable it may be conjectured that the final state of evolution would tend towards one close binary with the remaining members at infinity. Evidently the time-scale for complete disruption of an isolated system may be arbitrarily large since it depends on the actual distribution of binding energies. For practical applications it is therefore more useful to estimate the half-life which is subject to less uncertainty and also within reach of machine calculations.

The final binary state conjectured above is of more than theoretical interest since the binary phenomenon appears to be fundamental to most phases of cluster evolution. At first, short lived pairs are formed at small central distances where the probability of favourable multiple encounters is relatively high. Once formed, a binary may either increase its binding or suffer disruption by further encounters. The former process dominates for heavy pairs interacting with field particles, and leads to an energy sink behaviour. A more detailed discussion is given elsewhere in this volume.

Binary activity is intimately connected with the halo expansion and escaper formation. It has been established that most energetic escapers are associated with binary interactions (van Albada, 1968; Aarseth, 1968; Hayli, 1970). The efficiency of this mechanism favours the formation of one close central binary where the evolution measure λ as defined by Equation (19) may eventually exceed unity (Allen, 1968; van Albada, 1968; Aarseth, 1968). Such extreme configurations have been reached after about 20–30 initial crossing times for systems containing up to 250 particles.

The addition of a galactic tidal field also speeds up the escape rate and should be included in the simulation of real star clusters. In this case many halo members which would remain bound in isolated systems escape when passing close to the Lagrangian equilibrium points (Hayli, 1970). The escape rate is increased further by the disruptive effect of passing interstellar clouds (Bouvier and Janin, 1972). Mass loss effects in heavy stars during advanced stages of physical evolution have also been considered (Wielen, 1968); more recent results of realistic cluster simulations are given elsewhere in this volume (Wielen, 1972).

7. Comparison with Theory

Many attempts have been made to construct theoretical cluster models which are dynamically consistent. Although the numerical experiments are restricted to rather small particle numbers, there is nevertheless some hope that a meaningful comparison can be made with theoretical predictions. As yet no theory can be said to give a completely satisfactory description of cluster evolution as understood from the numerical results. In particular it is desirable that improved treatments should include mass segregation effects as well as velocity anisotropy in the outer region. The presence of one or more close binaries also leads to increased relaxation and the production of energetic escapers not predicted theoretically.

Numerical calculations are particularly well suited for testing theoretical assumptions which are made in order to obtain complete solutions to the time dependent problem. The usual approach is to proceed from equal-mass systems to the general case by assuming the velocity distribution of different mass groups. The expected equipartition of kinetic energy does not occur for reasons discussed above. In addition, the total velocity distribution is not Maxwellian even in the simpler equal-mass case, but this approximation may still be used in the inner regions where encounters are more effective.

It is encouraging that the mean relaxation time for centrally concentrated systems is in substantial agreement with the corresponding numerical values for cases with $N=100$ (Aarseth, 1966; Wielen, 1967). A systematic dynamical comparison with the predicted dependence on particle number has not yet been attempted because of the uncertainty in defining a satisfactory numerical procedure. Instead the equivalent relaxation time due to deflections of one particle moving in a stationary field has been calculated for many constant density systems (Standish and Aksnes, 1969). The agreement with the classical Chandrasekhar theory is extremely good in the range $N=25$ to $N=2500$. This result therefore supports the assumption of relaxation by two-body encounters in homogeneous systems, but dynamical calculations have shown that multiple encounters are important under more realistic conditions.

Calculated relaxation times may be used to derive the mean escape rate of different mass groups as defined by Equation (22). In the absence of a standard definition of the relaxation time, however, it is more meaningful to compare the mass dependent relative escape rate per crossing time

$$L_m = \frac{\Delta N_m t_{cr}}{N_m \Delta t}. \tag{23}$$

Although statistical fluctuations still enter through the uncertainty in the number of escapers, this effect can be minimized by repeated calculations or, alternatively, one system can be studied over longer time intervals. The two procedures are not equivalent, however, because the structure may change significantly in the latter case.

Experimental determinations of the relative escape rate have been made for several cases with $N=100$ giving $L_m \simeq (0.7 \pm 0.1) \times 10^{-2}$ for three mass groups (Wielen, 1967). The conclusion of nearly constant relative escape over a wide mass range is supported by one case with $N=250$ integrated over 28 mean crossing times (Aarseth, 1968); viz. $L_m \simeq (0.9 \pm 0.2) \times 10^{-2}$ for three groups containing 90% of the particles. The increased escape rate is mainly due to the steeper mass spectrum and the presence of a close heavy binary which dominates the evolution. Corresponding escape rates for equal-mass systems are significantly smaller even if temporary binaries are present. Thus an early calculation with two cases $N=25$ gives $L_1 \simeq 2 \times 10^{-3}$ based on one escaper only (von Hoerner, 1963) but this value is consistent with more recent studies of small particle numbers (van Albada, 1968).

The numerical escape rates may be compared with theoretical predictions for similar systems. Analytical expressions have been derived in the case of Plummer's model for a general mass distribution with spherical symmetry and velocity isotropy (Hénon, 1969). Although mass segregation effects are not included, it is more important that the interaction between different masses has been calculated analytically.

The resulting agreement between theoretical and experimental escape rates is now better than a factor of two for the cases $N=100$ and $N=250$ discussed above. Some allowance should also be made for the effect of binaries and the higher central density in many cases, both of which would tend to decrease the discrepancy further and might in fact lead to a theoretical over-estimate. On the other hand, improved treatments of the mass segregation effect should reduce the number of light escapers in order to be consistent with the numerical escape rates. Finally, we note that the predicted escape rate remains small for equal-mass systems; i.e., $L_1 \simeq 2.6 \times 10^{-2} N^{-1}$ in good agreement with numerical results for small particle numbers.

References

Aarseth, S. J.: 1966, *Monthly Notices Roy. Astron. Soc.* **132**, 35.
Aarseth, S. J.: 1967, *Bull. Astron.* **2**, 47.
Aarseth, S. J.: 1968, *Bull. Astron.* **3**, 105.

Aarseth, S. J.: 1970, *Astron. Astrophys.* **9**, 64.
Aarseth, S. J.: 1972, this volume, p. 373.
Albada, T. S. van: 1968, *Bull Astron. Inst. Neth.* **19**, 479.
Allen, C.: 1968, Ph.D. Thesis, Mexico University.
Bouvier, P. and Janin, G.: 1972, this volume, p. 71.
Gonzalez, C. C. and Lecar, M.: 1968, *Bull. Astron.* **3**, 209.
Hayli, A.: 1967, *Bull. Astron.* **2**, 67.
Hayli, A.: 1970, *Astron. Astrophys.* **7**, 17.
Heggie, D. C.: 1972, this volume, p. 148.
Hénon, M.: 1966, *Compt. Rend. Acad. Sci. Paris* **262**, 666.
Hénon, M.: 1969, *Astron. Astrophys.* **2**, 151.
Hénon, M.: 1972, this volume, p. 44.
Hoerner, S. von: 1960, *Z. Astrophys.* **50**, 184.
Hoerner, S. von: 1963, *Z. Astrophys.* **57**, 47.
Kustaanheimo, P. and Stiefel, E.: 1965, *Math.* **218**, 204.
Larson, R. B.: 1970, *Monthly Notices Roy. Astron. Soc.* **147**, 323.
Larson, R. B.: 1972, this volume, p. 60.
Miller, R. H.: 1964, *Astrophys. J.* **140**, 250.
Peters, C. F.: 1968, *Bull. Astron.* **3**, 167.
Pines, S.: 1961, *Astron. J.* **66**, 5.
Standish, E. M.: 1968, Ph.D. Thesis, Yale University.
Standish, E. M. and Aksnes, K.: 1969, *Astrophys. J.* **158**, 519.
Stiefel, E.: 1967, *NASA Report* CR–769.
Wielen, R.: 1967, *Veröff. Astron. Rechen-Inst. Heidelberg*, No. 19.
Wielen, R.: 1968, *Bull. Astron.* **3**, 127.
Wielen, R.: 1972, this volume, p. 62.

MONTE CARLO MODELS OF STAR CLUSTERS

M. HÉNON
Observatoire de Nice, France

Abstract. The dynamical evolution of spherical star clusters under the effect of internal encounters is followed numerically using a Monte Carlo procedure. Successive states of the system are computed, separated by a time step which is a fraction of the relaxation time. In any given state, each star is characterized by its total energy and its angular momentum with respect to the centre. Changes in these two quantities from one state to the next are computed by randomly selecting the position of the star on its orbit, randomly choosing a field star, letting the two stars interact, and multiplying the effect by an appropriate factor. This procedure can be shown to reproduce correctly the behaviour of the system as given by the Fokker-Planck equation. The computation is much faster than the exact N-body integration. Multiple-encounter effects are neglected, but this is probably not of serious consequence when N is large.

Some provisional results are presented. Once more it is found that N-body systems develop a very high central density peak. The velocity distribution becomes isotropic in the central parts, radially elongated in the halo. Models started with widely different initial conditions tend to become similar after a few relaxation times. The presence of a tidal field, or a distribution of masses, accelerate the evolution of the system.

A companion paper gives a detailed technical description of the method.

1. Introduction

In a previous note (Hénon, 1966), a brief description was given of a new computational scheme for the dynamical evolution of a spherical stellar system, under the effect of encounters. The present paper describes more fully the ideas underlying the scheme, and presents some results. A companion paper (Hénon, 1972, hereafter called II) contains the details of the computation.

A similar scheme has already been presented for the case of an infinite homogeneous system (Hénon, 1967). Here we shall be interested in the more realistic case of a finite, inhomogeneous system; therefore, although the basic principle is the same, the practical scheme will be quite different, and more complicated.

We shall make three basic assumptions:

(1) The system has spherical symmetry.

(2) The number N of its stars is large: $N \gtrsim 1000$.

(3) We are interested in the evolution of the system over a period of time comparable to or larger than the relaxation time.

This description fits well most globular clusters, and it is primarily for these objects that the scheme has been developed. An application to open clusters and to clusters of galaxies is also possible, if they are sufficiently rich. On the other hand, an application of the present scheme to galaxies is excluded, since both assumptions (1) and (3) are violated.

Assumption (3) means that we must take into account the effect of encounters between stars. On the other hand, assumptions (2) and (3) taken together imply that the age of the system is much larger than its crossing time t_c, since the two basic time

M. Lecar (ed.), Gravitational N-Body Problem, 44–59.

scales t_r (relaxation time) and t_c are related by

$$t_r/t_c = C_1 N/ln\, N\,, \tag{1}$$

where C_1 is a numerical constant (Chandrasekhar, 1942, Equation (5.227)). We can therefore assume that collective motions have long since died out (Hénon, 1964). At any given time, the system is practically in a steady state. It does not change significantly during one crossing time t_c. It does, however, slowly evolve under the effect of encounters, with the time scale t_r. It is this evolution which will be studied here.

2. Outline of the Method

Two main approaches have been used for the study of the dynamics of stellar systems. The first one is the straightforward numerical integration of the equations of motion of the N bodies, starting from given initial conditions. Unfortunately, this requires a tremendous amount of computer time. At each step, the computation of the mutual forces takes a time proportional to N^2; as N increases, the time step must be made smaller for a given accuracy, because close approaches are more frequent, and as a result the number of time steps per crossing time is roughly proportional to N; finally, the ratio of relaxation time to crossing time is approximately proportional to N according to (1). Thus the total computing time is proportional to N^4 (von Hoerner, 1960). This situation is somewhat improved by the use of individual or doubly-individual time steps (Aarseth, 1963; Hayli, 1967); then for the majority of the stars the time step will not depend on N any more, but it must still be a fraction of the crossing time, so that the total computing time is still proportional to N^3 at least. The largest value reached to date is $N=500$ (Aarseth, 1971). (The much larger values of N which have been achieved in the numerical simulation of galaxies are possible only because relaxation effects can be neglected in that case).

The second approach is a statistical description of the system by a *distribution function* $f(\mathbf{r}, \mathbf{v}, m, t)$, where $\mathbf{r}$ is the position, $\mathbf{v}$ the velocity and m the mass of a star. The evolution of f with time is given by a Fokker-Planck equation (Rosenbluth *et al.*, 1957), which is much too complicated for a numerical solution in the general case; in order to reduce it to a manageable form, one is forced to make a number of more or less arbitrary simplifications.

We shall describe here a third approach, which is in a sense intermediate between the previous ones, and which tries to combine their advantages. We start with a critical examination of the exact N-body integration, asking: why does it require so much computing time?

We remark first that, in the N-body integrations, no use is made of the theoretical knowledge on the problem: the computation would proceed in just the same way if no theory existed at all. It is true that the theory is still in a somewhat rudimentary state; some results, however, are reasonably well established. Could not we use them to provide short cuts in the computations?

A second point is that the N-body integrations make no use of the peculiarities of a system, such as spherical symmetry. It takes as much time to compute the evolution of a spherical system as it would in the most general case of a system without any symmetry. In the theoretical approach, by contrast, the assumption of spherical symmetry brings about tremendous simplifications. Could we not also use in some way this symmetry to shorten the numerical computations?

Our third remark is that the numerical experiment gives us too much information: after all, we are not interested in the detailed motion of individual stars, but rather in the evolution of the system as a whole. Moreover, the excess information which we get is in fact useless, as the comparison studies have shown (Lecar, 1968): because of the exponential build-up of errors, the computed orbits soon become completely different from the exact ones. The only meaningful information is contained in the statistical properties of the system; all that is required, therefore, is that we set up our computation in such a way that the computed system behaves statistically as the real system.

We note also that, since we are interested in the slow evolution of the system under the effect of encounters, the natural time scale is t_r, not t_c; thus, the fact that the exact integration has to operate on the time scale t_c, or less, can be taken as an indication that this method is not quite adapted to the physics of the problem.

Finally, since N is of the order of 10^5 in globular clusters, a proper scheme should be designed so as to be valid for large values of N, or mathematically speaking, for the limit $N \to \infty$. This means that one should be able to eliminate N entirely from the equations, with a proper choice of units. If this could be achieved, an added advantage would be that the results of one particular computation could be applied to a real system with any number of stars.

These considerations lead naturally to the basic ideas of the Monte Carlo scheme. First, because N is large, we can divide the gravitational field in two parts: a main smoothed-out field, or mean field, and a small irregular, fluctuating field. We consider the motion of a star during an interval of time Δt, such that

$$t_c \ll \Delta t \ll t_r. \tag{2}$$

During that interval, the effect of the fluctuating field can be neglected in a first approximation. The motion of the star is then governed by the mean field. This field is spherically symmetric, and changes only with the time scale t_r, so that it can be taken as time-independent over Δt. The star has then a plane rosette motion, described by simple analytical formulas. The detailed numerical integration along the orbit becomes thus unnecessary: the formulas give at once the motion over the whole interval Δt, which includes a number of revolutions according to (2).

However, the fluctuating field, although small, is not entirely negligible. Its effect is to change slowly and randomly the parameters of the orbit. This effect is small over Δt, but it builds up and becomes significant over a time of the order of t_r; we must therefore take it into account.

It would seem at first view that we must consider the effect of all other stars (which we shall call *field stars*), at all points of the orbit during the interval Δt. We would then be practically back at the exact N-body integration. To avoid this, we apply here the basic Monte Carlo tactics. First, instead of integrating the perturbations along the orbit, we shall select randomly just one point of the orbit, and we shall compute the perturbation only at this point. Second, instead of considering the effect of all field stars, we shall select randomly just one of them, and compute only the perturbation from that star. Finally, we shall multiply this perturbation by an appropriate fixed factor in order to account for all the time points and all the field stars which have not been considered.

Of course, this procedure will not give the exact perturbation in the motion of the star. But, since this perturbation is a random quantity, it is not its exact value which matters, but its statistical properties; more specifically, its moments of the first and second order. If the procedure is correctly set up, these moments will be correctly reproduced. The evolution of the whole artificial system of stars will be statistically the same as the evolution of the real system, and this is all that is required.

The computation reduces to a fixed number of operations for each star during each time step Δt, so that the total computing time is proportional to N only, instead of N^3 in the exact computation. In effect we save a first factor N by computing the perturbations at only one point during Δt, or in other words by using a time step Δt which is a fixed fraction of t_r rather than a fixed fraction of t_c, and we save a second factor N by selecting one field star out of N. Actually the computation does not work with stars, but with *superstars* (see II), and the computing time is proportional to the number n of superstars, which is usually even less than N.

We have presented the Monte Carlo scheme as an adaptation of the exact N-body integration. But it can also be seen from a different angle, as deriving from the theoretical approach: it can be simply considered as a convenient algorithm for the numerical solution of the Fokker-Planck equation. The samplings effected along the orbit and among the field stars are then merely an application of the classical Monte Carlo trick for the quick evaluation of multiple integrals.

This interpretation makes clear one important point: the assumptions which lie behind the Monte Carlo scheme are essentially the same as the assumptions involved when one writes the Fokker-Planck equation; namely, that a stellar system can be adequately represented by a one-particle distribution function, and that the evolution is only due to binary encounters. Thus, multiple encounters are ignored; in particular the formation and disruption of binaries is not considered. This omission is probably not serious when N is large. It is true that numerical experiments have shown that a tight binary usually forms at the centre of the system. However, this binary appears to be a consequence rather than a cause of the evolution of the system; its main function seems to be to provide an energy sink in which the negative energy flowing towards the centre can be accumulated. In the Monte Carlo scheme, a similar sink exists: when the radius of a superstar tends to zero, its potential energy tends to $-\infty$, so that a single superstar suffices to store any amount of negative energy. Experiments show that accumulation of energy actually takes place in this way.

3. Results

The results presented here should be considered as provisional, inasmuch as the program is still in a state of development and should be improved on some points (see II).

A. COMPARISON WITH EXACT COMPUTATIONS

Figure 1 is a comparison with an exact N-body computation, as a check on the Monte Carlo method. The abscissa is time, reduced to a non-dimensional quantity by the use of the unit (see II)

$$\frac{N_0}{\ln N_0} G M_0^{5/2} (-4\mathscr{E}_0)^{-3/2}, \tag{3}$$

where N_0 is the number of stars, M_0 the mass and $\mathscr{E}_0$ the energy of the system for

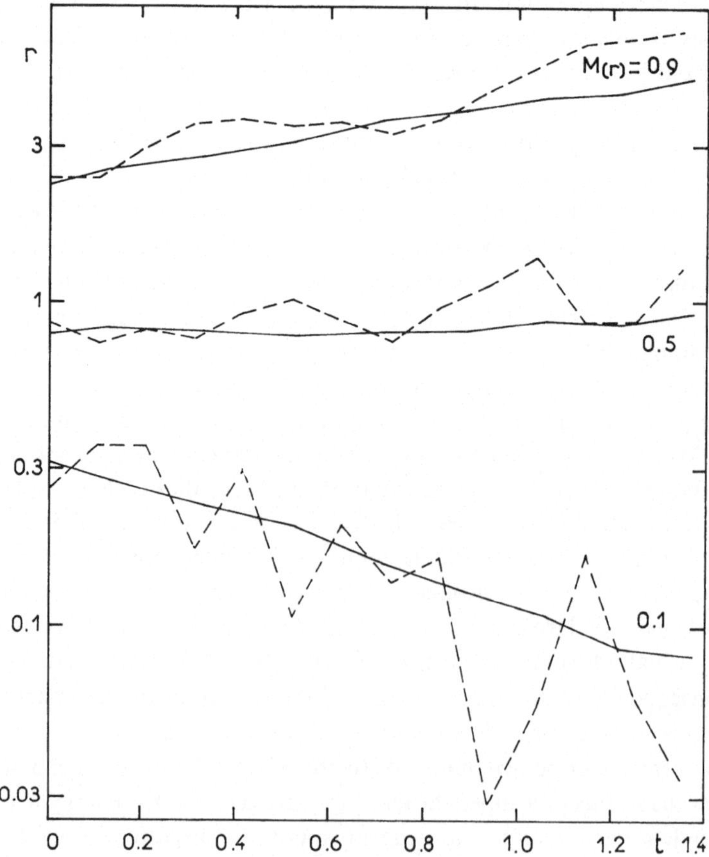

Fig. 1. Comparison between the exact N-body computation (dashed lines) and the Monte Carlo computation (full lines).

$t=0$; this unit is roughly equal to 4 times the initial mean relaxation time. For a typical globular cluster, with $N_0=10^5$ stars, it is of the order of 5×10^9 yr. The curves represent the radii of the spheres which contain a definite fraction of the total mass: 0.1, 0.5 or 0.9; the unit of length is

$$GM_0^2\,(-4\mathscr{E}_0)^{-1}. \tag{4}$$

The dashed curves represent the evolution of Wielen's model P (1967); this is an exact computation for a system of 100 stars, with different masses. The initial state is a polytrope of index 5, with an isotropic distribution of velocities and no mass

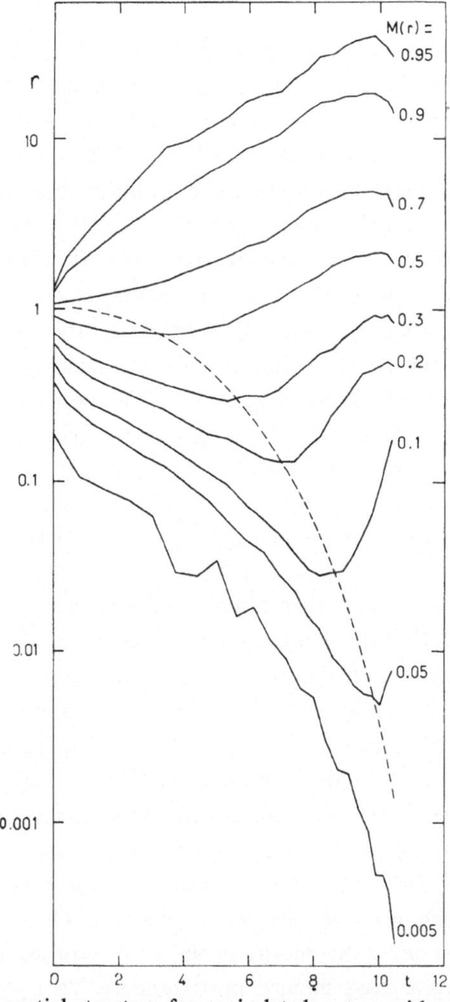

Fig. 2. Evolution of the spatial structure for an isolated system with stars of equal mass. Each curve represents the radius of the sphere containing a given fraction of the total mass. Dashed curve: approximate boundary between core and halo.

segregation. The full curves represent the evolution of the same system computed by the present Monte Carlo scheme with 1000 superstars. The agreement is satisfying; it should be noted that there is no adjustable parameter in this comparison. The Monte Carlo computation took about 6 min on a CDC 6400 computer.

B. EVOLUTION OF THE SPATIAL STRUCTURE

Figure 2 represents the evolution of a system of stars of equal mass. The initial state is a polytrope of index 1; this was chosen because it is not very different from a homogeneous sphere, and thus affords a good opportunity to observe the formation of a central concentration. The initial distribution of velocities is isotropic. Each curve again represents, as a function of time, the radius of the sphere which encloses a given fraction of the total mass of the system. The number of superstars is again $n=1000$.

As time goes on, the curves diverge from each other. The radius enclosing $\frac{5}{1000}$ of the mass decreases by three orders of magnitude, which means that the central density has increased by a factor 10^9 at least. The phenomenon is accelerating; the central density increases at a faster than exponential rate. There is a strong suggestion that the central density will become infinite at some finite time, in agreement with the results obtained by a number of other methods (Aarseth, 1966; Hénon, 1961, 1965; King, 1966; Larson, 1970a, b; Lynden-Bell and Wood, 1968; von Hoerner, 1963, 1968). Unfortunately, the Monte Carlo method in its present state cannot quite reach that critical time; it breaks down when the central density becomes too high (see II).

Simultaneously, the radius enclosing $\frac{95}{100}$ of the mass goes up by a factor 30, indicating the formation of an extended halo. The upper curves tend to become parallel; this is a confirmation of the hypothesis that the halo tends to evolve in a homologous manner (Hénon, 1961, 1965). The slight drop of the curves after $t=10$ is probably due to numerical errors caused by the strong central condensation (see II).

Let us consider one of the curves of Figure 2 with $M(r)\leqslant 0.5$: first it goes down, accompanying the collapse of the core; then it reaches a minimum, at a time which varies from curve to curve, and thereafter it follows the expanding motion of the halo. This simply means that the mass of the core decreases. If we draw a line through the minima (the dashed curve on Figure 2), we plot the radius of the core as a function of time, and from the intersections of this line with the solid curves we can also read the mass of the core as a function of time. The energy of the core can also be estimated from the numerical results (see also Figure 3 below); it is negative, and increases in absolute value with time. Thus the evolution can be summarized as follows: the core contains less and less of the total mass, and more and more of the total energy; and conversely for the halo. The system tends towards a final state in which all the mass is in the halo, and all the negative energy is in the core. Of course this extreme state cannot be quite reached since the energy in the core requires a material support; at least two stars, forming a close binary, must remain. This picture is confirmed by the results of exact computations, where as a rule a central binary forms and absorbs an increasing fraction of the total energy.

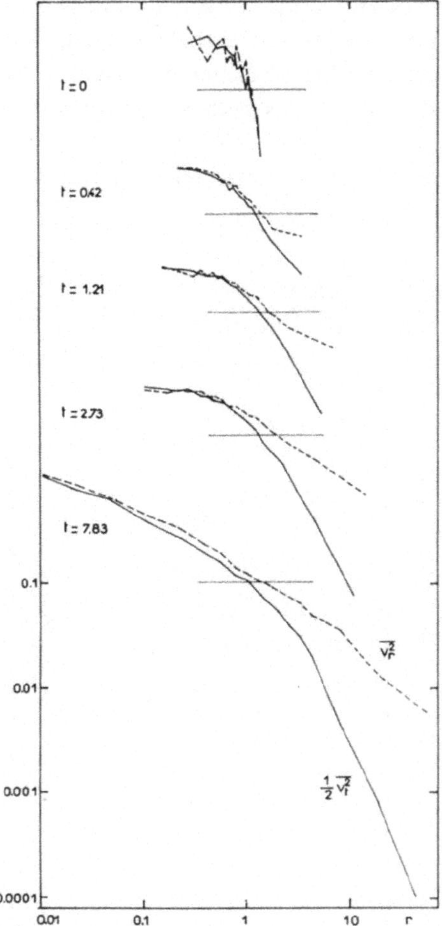

Fig. 3. Dispersion of radial velocities (dashed lines) and transverse velocities (full lines) against radius, at different times.

C. EVOLUTION OF THE VELOCITIES

Figure 3 represents the dispersion of the velocities in the radial direction, $\overline{v_r^2}$ (dashed curves) and in a transverse direction, $\overline{v_t^2}/2$(full curves), as a function of radius, at different times. The curves for $t=0$ to $t=2.73$ have been displaced upwards for clarity; the horizontal line indicates the level $\overline{v_r^2}=0.1$.

For $t=0$ the two curves are identical, apart from fluctuations, since the initial velocity distribution is isotropic. As the system evolves, two regions can be clearly distinguished: in the inner region, for $r<1$ approximately, the velocity distribution remains very nearly isotropic; in the outer region, $r>1$, the velocity distribution is more and more elongated as r increases. This is easily explained (Woolley and Robertson, 1956): in the inner region the density is high, therefore relaxation is strong and enforces isotropy. The outer region was initially void and is populated only by stars ejected from the inner region; this creates an elongated velocity distribution, which does not significantly alter afterwards because the density is low and relaxation is negligible.

In the outer region, $\overline{v_t^2}$ is approximately proportional to r^{-2}, so that the mean angular momentum is independent of r; this confirms the above explanation. $\overline{v_r^2}$ varies roughly as $r^{-0.6}$.

An explanation of the formation of a core-halo structure has been given by Lynden-Bell and Wood (1968) on the basis of thermodynamical arguments: the system evolves away from equilibrium, with the core becoming 'hotter' and the halo 'colder'. This

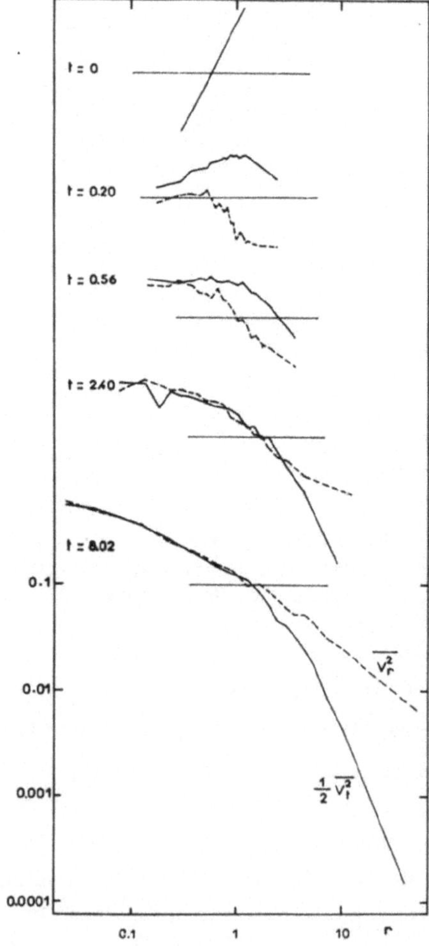

Fig. 4. Velocity dispersions against radius, for a model with purely circular orbits at $t = 0$.

explanation is fully confirmed by our results. Figure 3 shows that the 'temperature', measured by $\overline{v_r^2}+\overline{v_t^2}$, increases with time in the centre and decreases at the outside.

D. DISAPPEARANCE OF PECULIAR INITIAL CONDITIONS

Other experiments have been made to see how a system progressively forgets its initial state. Figure 4 represents the same quantities as Figure 3, for a different initial state: at $t=0$ all orbits are purely circular. The density is constant inside a sphere. Thus initially $\overline{v_r^2}=0$, and $\overline{v_t^2}$ is proportional to r^2. The system quickly relaxes and radial velocities build up; the process is faster in the inner regions. At $t=2.40$ isotropy has been established in most of the system, and the radial velocities have even become larger than the transverse velocities in the halo. Simultaneously there is a reversal in the slope of the curve for $\overline{v_t^2}$. At $t=8.02$, the curves closely resemble those of Figure 3: almost no trace is left of the peculiar initial conditions.

Another case was started with purely radial orbits: $\overline{v_t^2}=0$ initially. After a comparable time, the distribution of velocities is again very similar to the previous cases (Figure 5).

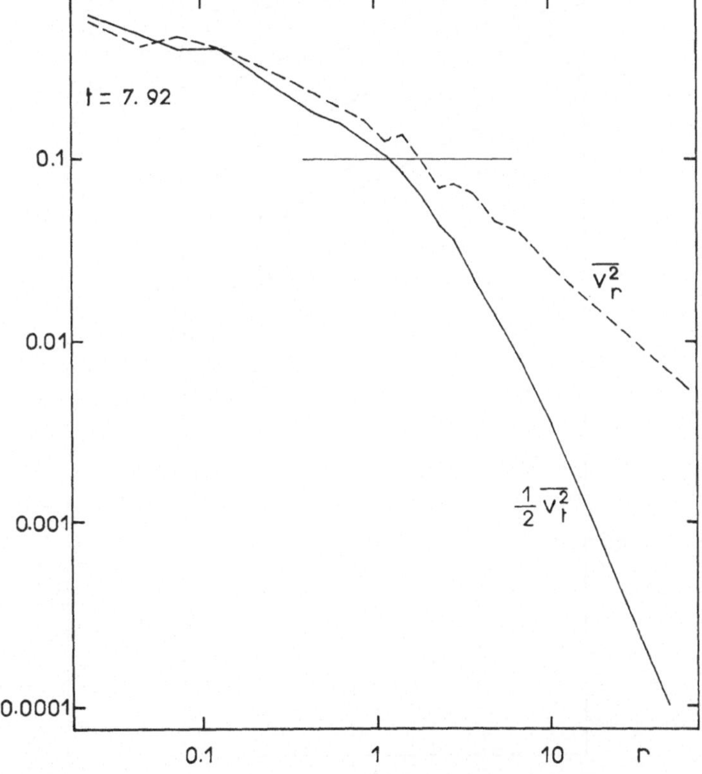

Fig. 5. Velocity dispersions against radius, at an advanced stage of evolution, for a model with purely radial orbits at $t=0$.

E. EFFECT OF AN EXTERNAL TIDAL FIELD

Figure 6 shows the spatial evolution of a non-isolated system. The tidal radius r_e is represented (see II, Equation (37)); its initial value is $r_{e0}=5$. The curves represent here a given fraction of the remaining mass. The evolution of the core is similar to the isolated case (Figure 2), but faster. The halo is suppressed, since stars going beyond r_e escape. Figure 7 shows the decrease of the mass of the system with time; this is also accelerating, and there is a suggestion that the system will disappear entirely shortly after the infinite central density has formed.

Figure 8 represents the dispersions of radial and transverse velocities as a function of radius, at a late stage of evolution. The velocity distribution is nearly isotropic throughout the system, in contrast to the isolated case (Figures 3 to 5); this is due to

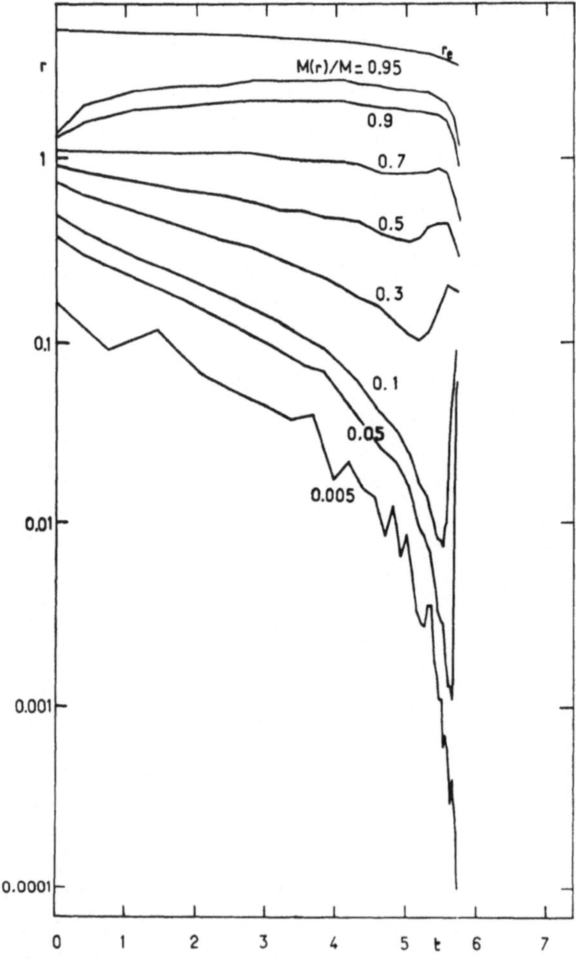

Fig. 6. Evolution of the spatial structure for a non-isolated system. r_e: tidal radius.

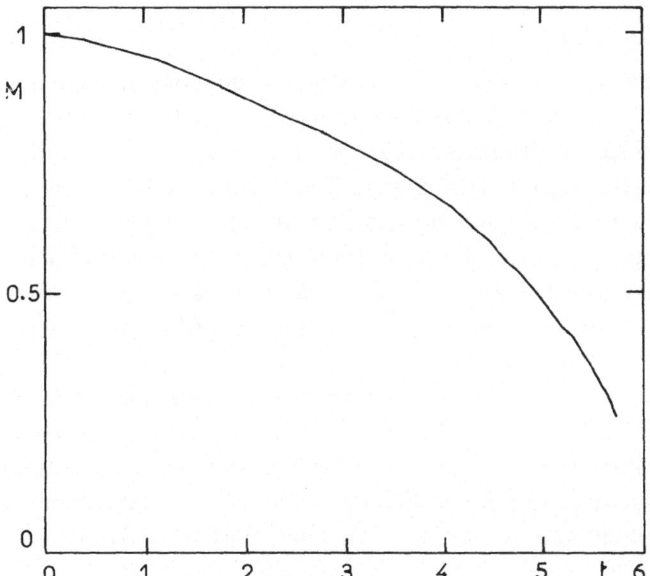

Fig. 7. Evolution of total mass with time for a non-isolated system.

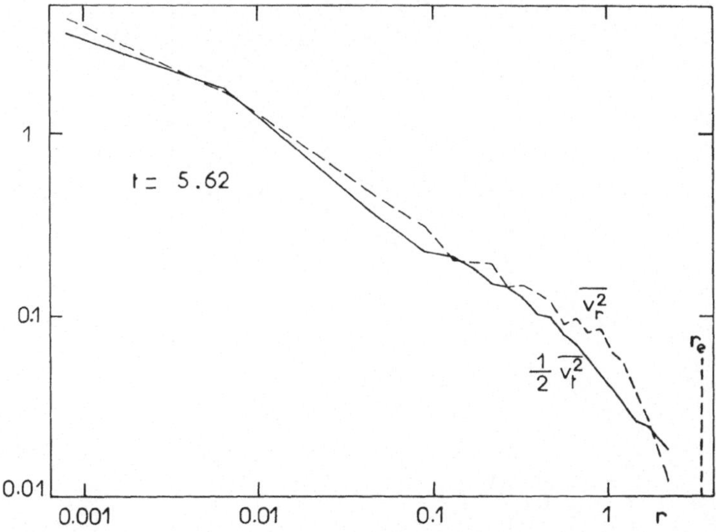

Fig. 8. Velocity dispersions for a non-isolated system, at an advanced stage of evolution. r_e: tidal radius.

the absence of a halo. Thus the assumption of isotropy, often made in theoretical models, seems to be justified in the case of non-isolated systems.

In the outer parts, the radial velocities drop faster than the transverse velocities. This is due to the presence of the tidal boundary: when r approaches r_e, the radial velocities must approach zero, while the transverse velocities can remain finite.

F. EFFECT OF UNEQUAL MASSES

Figure 9 represents the evolution of an isolated system of stars with different masses. One of Aarseth's (1968) mass distributions has been used: among 1000 stars there are 32 with mass 6.25, 64 with mass 3.125, 128 with mass 1.5625, 256 with mass 0.78125, 512 with mass 0.390625, 8 with mass 0. The initial state is a polytrope of index 1, with an isotropic distribution of velocities and no mass segregation. The curves correspond here to the radii of the spheres which contain a given fraction of the total number of stars. The evolution is similar to that of Figure 2, but faster by a factor 4 approximately. At $t = 1.46$ the computation was stopped because the spurious escape became too important (see II).

Figure 10 shows the progress of mass segregation with time. Each curve represents the mean distance from the centre for stars with a given mass. Figure 11 represents the mean quadratic velocities. The evolution is away from equipartition: the heaviest stars acquire the largest velocities. Figures 10 and 11 are in agreement with the results from exact N-body integrations (Aarseth, 1966; Wielen, 1967).

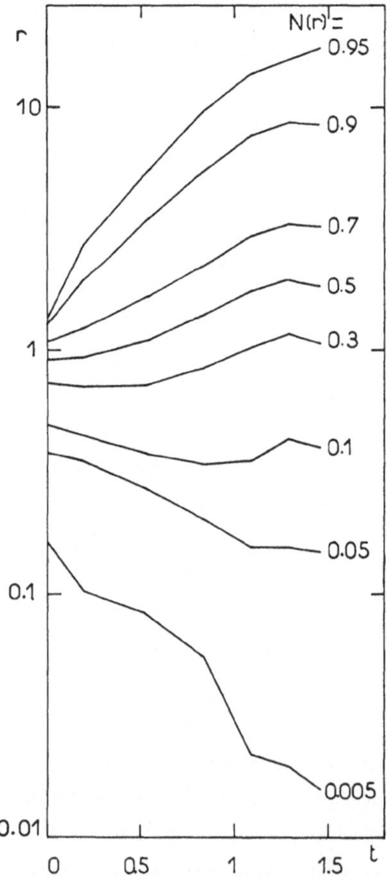

Fig. 9. Evolution of the spatial structure for an isolated system with unequal masses.

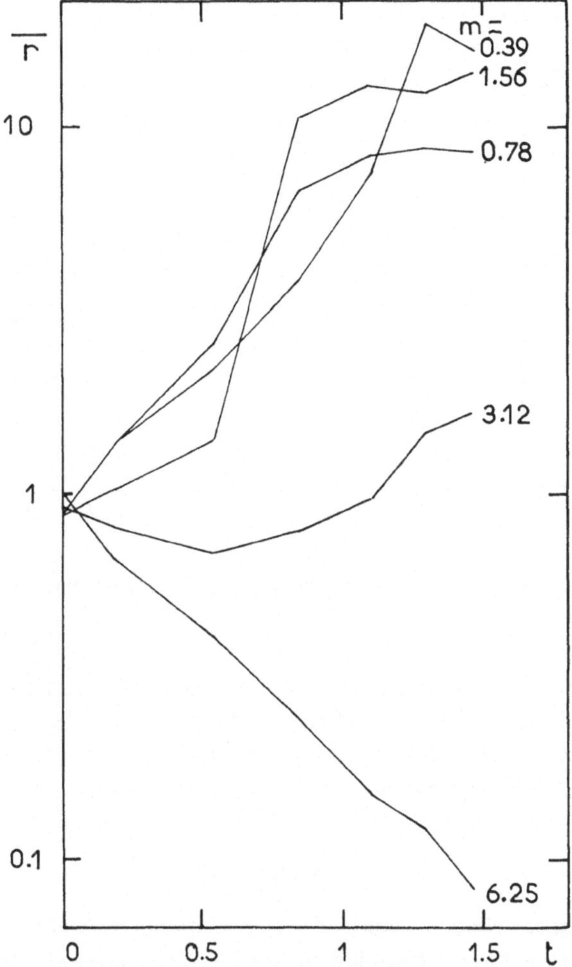

Fig. 10. Mean radius against time for different mass groups.

4. Possible Extensions

It would be interesting to try to extend the Monte Carlo method to other cases than the spherical symmetry, and in particular to the case of axial symmetry. However, before one can apply the method as presented here, the characteristics of the motion of a star in the mean field must be known; more specifically, in order to be able to select at random a new position of the star, one must know precisely which part of space is visited by the orbit, and what fraction of time the star spends in each volume element. Mathematically, this means that one must know explicitly all the isolating integrals of the motion. In the case of axial symmetry, therefore, serious difficulties would arise from the ill-defined nature of the third integral (Hénon and Heiles, 1964).

This problem can be eliminated by sacrificing the first half of the Monte Carlo method: that is, one integrates numerically the motion of the stars in the mean field,

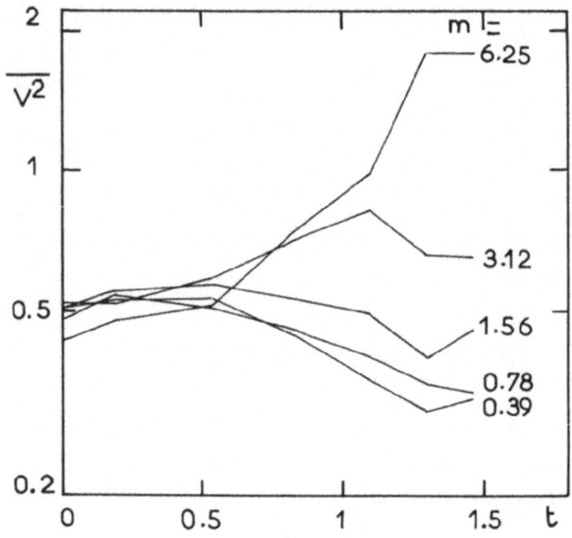

Fig. 11. Mean square velocity against time for different mass groups.

with a time step small compared to t_c. The second half of the method could be preserved: one would still compute the effect of encounters by selecting at random one field star. The computing time would be proportional to nN. The value of N would have to be specified, since both time scales t_c and t_r would appear in the computation.

Acknowledgements

This work was made while the author was a National Research Council Visiting Research Associate at the Smithsonian Institution Astrophysical Observatory, Cambridge, Massachusetts, from January to June 1968. I wish to thank Dr. Lecar for making this visit possible, as well as for several helpful discussions.

References

Aarseth, S. J.: 1963, *Monthly Notices Roy. Astron. Soc.* **126**, 223.
Aarseth, S. J.: 1966, *Monthly Notices Roy. Astron. Soc.* **132**, 35.
Aarseth, S. J.: 1968, private communication.
Aarseth, S. J.: 1972, this volume, p. 88.
Chandrasekhar, S.: 1942, *Principles of Stellar Dynamics*, Dover Publ.
Hayli, A.: 1967, *Bull. Astron. (Paris)* **2**, 67.
Hénon, M.: 1961, *Ann. Astrophys.* **24**, 369.
Hénon, M.: 1964, *Ann. Astrophys.* **27**, 83.
Hénon, M.: 1965, *Ann. Astrophys.* **28**, 62.
Hénon, M.: 1966, *Compt. Rend. Acad. Sci. (Paris)* **262**, 666.
Hénon, M.: 1967, *Bull. Astron. (Paris)* **2**, 91.
Hénon, M.: 1972, this volume, p. 406.
Hénon, M. and Heiles, C.: 1964, *Astron. J.* **69**, 73.

King, I. R.: 1966, *Astron. J.* **71**, 64.
Larson, R. B.: 1970a, *Monthly Notices Roy. Astron. Soc.* **147**, 323.
Larson, R. B.: 1970b, *Monthly Notices Roy. Astron. Soc.* **150**, 93.
Lecar, M.: 1968, *Bull. Astron. (Paris)* **3**, 91.
Lynden-Bell, D. and Wood, R.: 1968, *Monthly Notices Roy. Astron. Soc.* **138**, 495.
Rosenbluth, M. N., Mac Donald, W. M., and Judd, D. L.: 1957, *Phys. Rev.* **107**, 1.
von Hoerner, S.: 1960, *Z. Astrophys.* **50**, 184.
von Hoerner, S.: 1963, *Z. Astrophys.* **57**, 47.
von Hoerner, S.: 1968, *Bull. Astron. (Paris)* **3**, 147.
Wielen, R.: 1967, *Veröff. Astron. Rechen-Inst. Heidelberg*, No. 19.
Woolley, R. v. d. R. and Robertson, D. A.: 1956, *Monthly Notices Roy. Astron. Soc.* **116**, 288.

A FLUID-DYNAMICAL METHOD FOR COMPUTING THE EVOLUTION OF STAR CLUSTERS

RICHARD B. LARSON
Yale University Observatory, New Haven, Conn., U.S.A.

One way that can be used to study the development of a stellar system is the fluid-dynamical approach, whereby the stars are considered as the constituent particles of a continuous fluid whose behavior is described by moment equations derived from the Boltzmann equation. This method is most useful for systems with very many stars, where it complements the *n*-body technique which is feasible only for small systems. The fluid-dynamical approach begins by defining suitable moments of the velocity distribution at each point in space. In studying the evolution of a star cluster it is necessary to consider moments of at least fourth order in the velocities in order to represent all the essential physical effects, including an outward 'heat flow' caused by the escape of the most energetic stars, and an excess or deficiency of high velocity stars relative to a Maxwellian distribution. Allowing for unequal radial and transverse velocity dispersions, we find that six moments in all are required, for which six fluid-dynamical equations may be derived by taking the corresponding moments of the Boltzmann equation. The fluid-dynamical equations contain relaxation terms which may be evaluated from the Fokker-Planck equation, assuming that deviations from a Maxwellian velocity distribution are small. In the absence of other effects, the relaxation terms have the effect of making the various deviations from a Maxwellian velocity distribution decay exponentially with decay times which are closely related to the classical relaxation time. The resulting fluid-dynamical equations can then be solved numerically to yield the values of all quantities as functions of position and time.

This method has been used to compute in detail the evolution of several types of stellar systems, including (1) a globular cluster of mass $2\times10^5\ M_\odot$, (2) a galactic cluster of mass 100 $M_\odot$, and (3) a dense galactic nucleus of mass $10^8\ M_\odot$. The effect of a tidal boundary has been simulated in each case by assuming that the system is bounded in space by a perfectly absorbing wall. In each case the results show the qualitative type of behavior predicted by classical relaxation theory – a steadily increasing central concentration, a decreasing total mass, and an increasing anisotropy of the velocity distribution in the outer part of the system. Also, the time scales agree in a rough mean sense with those predicted classically, but are found to be quite sensitive to changes in the structure of the system. As the central part of the system relaxes toward a more nearly isothermal structure, the evolution time slows down relative to the relaxation time, although it continues to speed up in absolute terms. Differences in the way that

M. Lecar (ed.), Gravitational N-Body Problem, 60–61. *All Rights Reserved*

different systems evolve indicate that the ratio of relaxation time to dynamical time is an important parameter for the evolution of stellar systems. Finally, the results show clearly that the rate of mass loss, particularly for small systems, is strongly affected by the existence of a tidal boundary.

A detailed account of this work appears in *Monthly Notices Roy. Astron. Soc.* (1970), **147**, 323, and **150**, 93.

ON THE LIFETIMES OF GALACTIC CLUSTERS

R. WIELEN

Astronomisches Rechen-Institut, Heidelberg, Germany

Abstract. From the observed age distribution of galactic clusters within 1 kpc we deduce that the typical total lifetime of a galactic cluster is about 2×10^8 yr. The individual lifetimes vary between 10^8 and 10^{10} yr. The observed lifetimes are compared with the evaporation times which are found from numerical experiments with star cluster models. These models contain up to 250 stars with a realistic mass spectrum. The effect of the galactic tidal field is taken into account and enhances the rate of escape significantly. Escapers are identified by using the Jacobian integral. We give the evaporation time in years as a function of the median radius for different values of the total mass of a cluster. The agreement between the resulting theoretical lifetimes and the observed values is sufficiently good. We estimate that the tidal field of passing interstellar clouds should be in most cases less efficient in dissolving a galactic cluster than the internal evaporation process combined with the effect of the general galactic field.

1. Introduction

In numerical experiments on the gravitational N-body problem we can handle at present systems which contain a few hundred stars. This number N is representative of a small galactic cluster and is not too different from the number of stars in a typical galactic cluster where N is of the order of 1000 stars. Hence we can almost directly compare the results of numerical experiments with observational features of galactic clusters provided that our star cluster models take into account the relevant physical effects, e.g. the galactic tidal field.

2. The Observed Age Distribution of Galactic Clusters

The most direct information on the time scale of the dynamical evolution of galactic clusters is provided by the observed distribution of the cluster ages. In Figure 1 we give the observed frequency of the ages of 60 clusters nearer than 1000 pc. We have used the catalogues of well observed galactic clusters compiled by Becker (1969) and Lindoff (1968). There is no indication that the age distribution of the clusters within 1 kpc is seriously affected by selection effects. Under the assumption that the rate of formation of clusters has not changed with time, the age distribution reflects the finite lifetimes of galactic clusters (Oort, 1958). We deduce from the data that 50% of newly formed clusters disintegrate within 2×10^8 yr, 10% become older than 4×10^8 yr, 1% older than 1×10^9 yr, and 0.2% older than 2×10^9 yr. While the typical total lifetime of a galactic cluster is short, a few clusters survive for a much longer time. Both results should be explained by any theory of cluster dissolution.

3. Star Cluster Models

Table I gives a summary of the star cluster models for which we have studied the

M. Lecar (ed.), Gravitational N-Body Problem, 62–70.

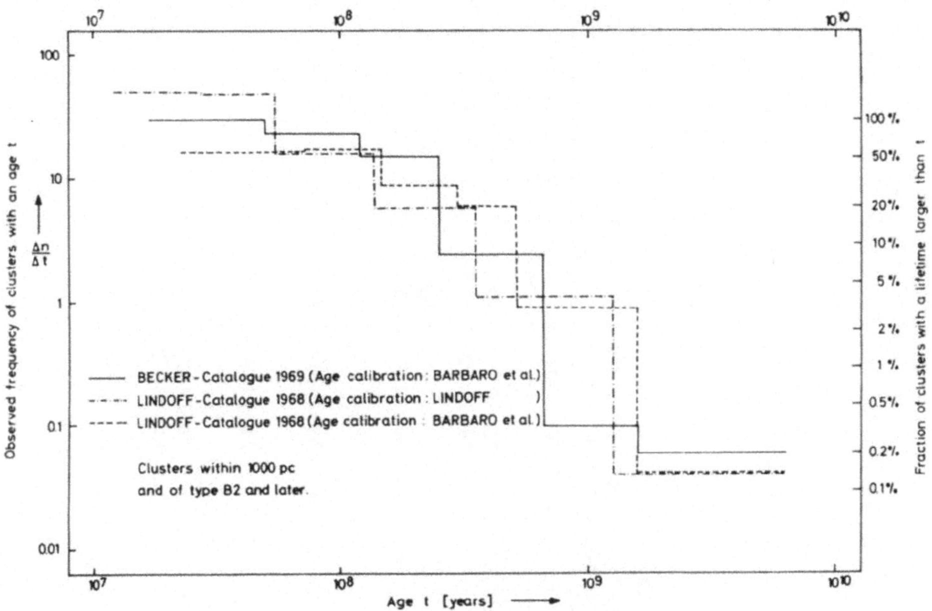

Fig. 1. Observed age distribution of galactic clusters.

dynamical evolution by numerical integrations of the N-body problem. For details of the method of integration, initial conditions, masses of the stars etc. we refer to earlier papers (Wielen, 1967a, b; 1968; 1969). The effect of passing interstellar clouds has been neglected in these models. The crossing time T_{cr} is defined as

$$T_{cr} = \frac{G\mathfrak{M}^{5/2}}{(-2E)^{3/2}}, \tag{1}$$

where $\mathfrak{M}$ is the total mass and E is the total energy of the cluster (e.g. van Albada, 1968). For a typical galactic cluster, T_{cr} is of the order of 10^7 yr; hence the integration times correspond typically to about 10^8 yr.

4. Identification and Number of Escapers

In most cases we have not followed the dynamical evolution of a star cluster model up to its total dissolution. The isolated models have lost less than 10% of their members during the time of integration, model G2 has lost 34% and model G nearly all its stars. We define the evaporation time of a cluster by

$$T_{ev} = -\frac{N}{\dot{N}} \sim -\frac{\mathfrak{M}}{\dot{\mathfrak{M}}}. \tag{2}$$

If the absolute rate of escape $\dot{N}$ is constant in time, then the instantaneous evaporation time is equal to the remaining lifetime of a cluster.

TABLE I

Properties of star cluster models

Initial state	Spectrum of stellar masses	External field		Designation and time of integration		
				$N=50$ stars	$N=100$ stars	$N=250$ stars
Plummer's model	Realistic	None		Model HP 16 T_{cr}	Model P 10 T_{cr}	Model DP 7 T_{cr}
					Model P2 6 T_{cr}	Model DP2 11 T_{cr}
		Galactic field	$R=0.08\,\xi_L$		Model G2 10 T_{cr}	
			$R=0.3\,\xi_L$		Model G 10 T_{cr}	
	Equal masses	None			Model E 18 T_{cr}	
Plummer's model with differential rotation	Realistic	None			Model R 13 T_{cr}	
Plummer's model with initial mass segregation	Realistic	None			Model M 16 T_{cr}	
Evolved state of models P, G or G2 (10 T_{cr})	Realistic but with mass loss of evolving stars	None			Model L 19 T_{cr}	
					Model L2 19 T_{cr}	
		Galactic field	$R=0.08\,\xi_L$		Model G2L 13 T_{cr}	
			$R=0.3\,\xi_L$		Model GL 13 T_{cr}	

In isolated clusters, escapers can be easily identified by their positive total energy, $E_i > 0$. When the galactic tidal field is added we have to consider the Jacobian integral C_i (per unit mass) instead of E_i/m_i. C_i is an integral of motion for a star under the combined influence of the internal and external fields if we neglect encounters and if some requirements on the internal potential and on the galactic orbit of the cluster are (approximately) fulfilled. The zero-velocity surfaces (or equipotential surfaces) of a cluster which is embedded in the galactic field are closed for $C_i \leqq C_L$. The critical surface which is the last closed one corresponds to the Jacobian constant

$$C_L = -\tfrac{3}{2}\left[G^2\mathfrak{M}^2 4A(A-B)\right]^{1/3} \tag{3}$$

(A, B are Oort's constants of galactic rotation). The Lagrangian points L_1 and L_2 lie on this critical surface at a distance (King, 1962) of

$$\xi_L = \left[\frac{G\mathfrak{M}}{4A(A-B)}\right]^{1/3} \tag{4}$$

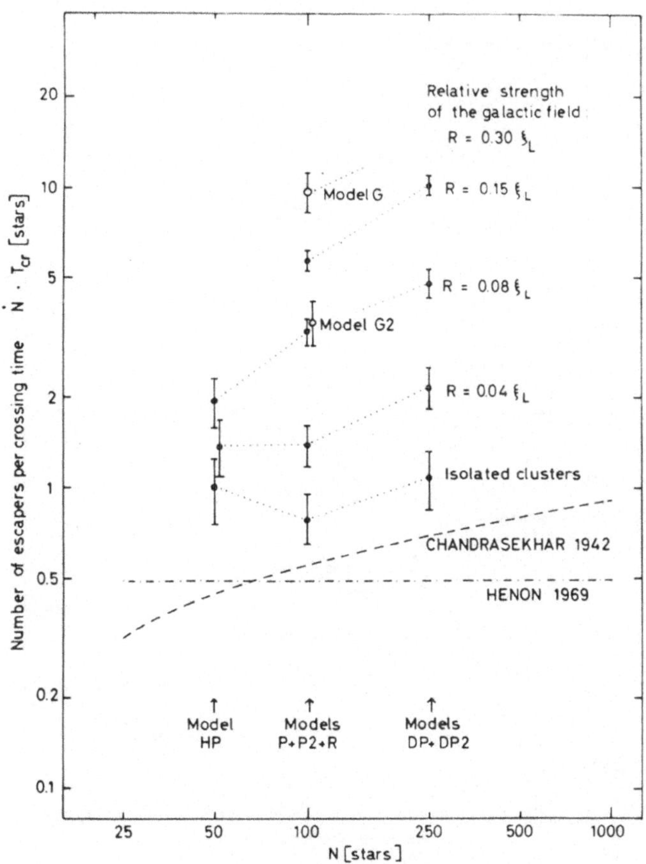

Fig. 2. Number of escapers in star cluster models.

from the cluster center. Stars with $C_i > C_L$ can escape because the corresponding surfaces are open. We find experimentally that those stars really escape in most cases. For isolated star cluster models we apply an approximative method in order to simulate the effect of the galactic field on the lifetime. We use an energy cut-off and count the stars with $E_i/m_i > C_L$ as escapers. This method produces reasonable results as long as the galactic field does not strongly affect the main body of the cluster.

Figure 2 summarizes the number of escapers per crossing time which we observe in our star cluster models. The ratio of the median radius R to the Lagrangian distance ξ_L measures the relative strength of the galactic field compared with the internal field of the cluster. The galactic field enhances the rate of escape significantly. The number of escapers per crossing time seems to increase with increasing N. This would conflict with Hénon's theory (1969) which, however, agrees well with the numerical experiments in many other respects.

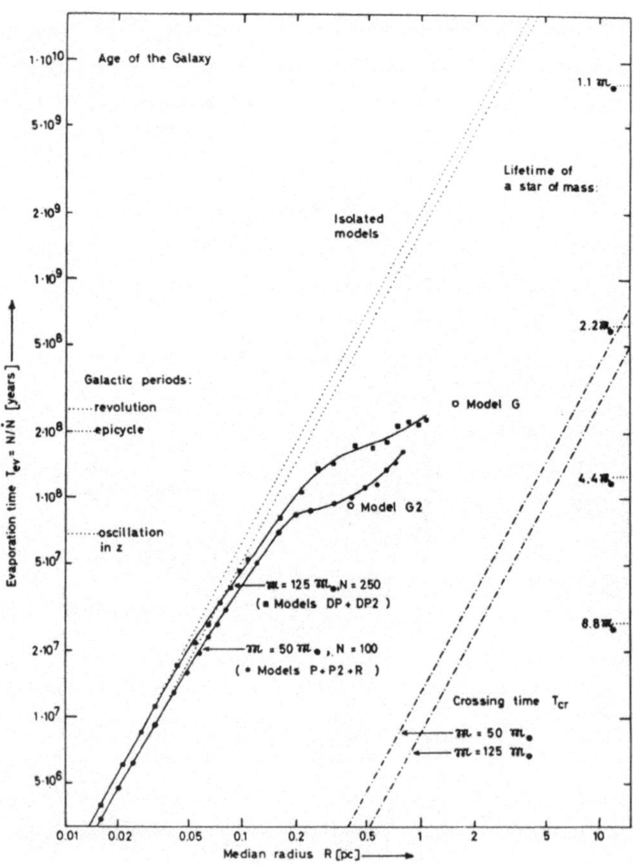

Fig. 3. Experimental evaporation times based on star cluster models.

5. Experimental Evaporation Times

The evaporation time T_{ev} in years depends mainly on three parameters of a cluster: (a) its median radius R in pc, (b) its total mass $\mathfrak{M}$ in $\mathfrak{M}_\odot$, and (c) its total number of stars N. For actual galactic clusters, $\mathfrak{M}$ and N are strongly correlated; we use $\mathfrak{M}=N\cdot 0.5\,\mathfrak{M}_\odot$ (van Altena, 1966). In Figure 3 we have plotted the evaporation time T_{ev} for our models with $N=100$ stars and $N=250$ stars as a function of the median radius R. For the evaporation time of a cluster with $N=250$ stars, $\mathfrak{M}=125\,\mathfrak{M}_\odot$ and $R=1$ pc we obtain 2×10^8 yr. Within the covered range of radii the evaporation time always increases with increasing R in spite of the galactic field.

Many galactic clusters contain more than 250 stars. Hence we shall try to extrapolate our experimental results to somewhat higher N. A typical galactic cluster has a total mass of some 500 $\mathfrak{M}_\odot$ and contains about 1000 stars. We divide our extrapolation process into two steps. First we increase the total mass $\mathfrak{M}$ while N is not

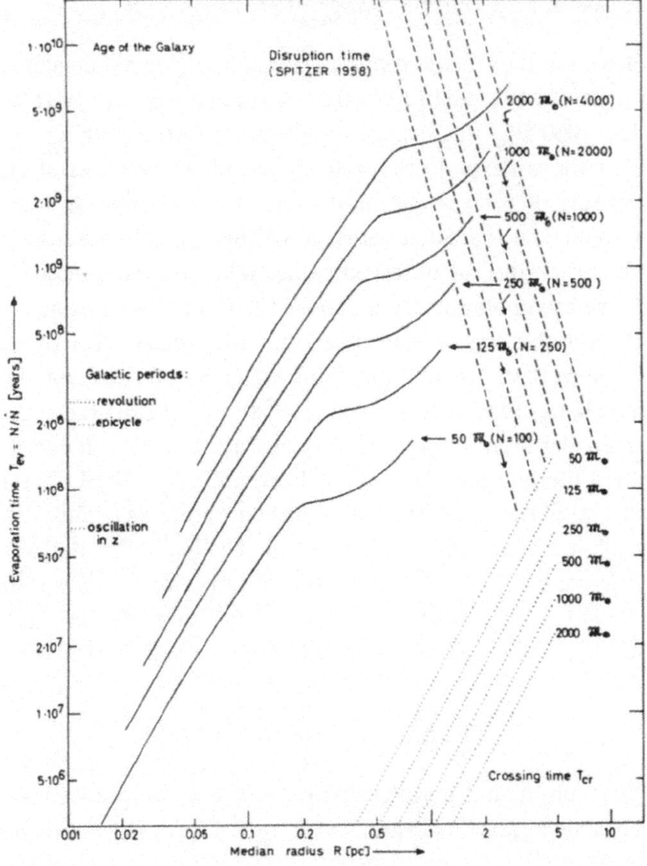

Fig. 4. Extrapolated evaporation times of galactic clusters and disruption times due to interstellar clouds.

changed. This is done by scaling which produces a horizontal shift of our experimental curve to higher radii in such a way that $\mathfrak{M}/R^3$ remains constant. The second step is the extrapolation to higher N while now $\mathfrak{M}$ is fixed. This means a vertical raise of the shifted curve. If we wish to extrapolate from 250 to 1000 stars, the following increase in the evaporation time is predicted for fixed $\mathfrak{M}$: (a) a factor of 4 by Hénon's theory (1969) which, however, is applicable in a strict sense only to isolated clusters, (b) a factor of 3.1 from Chandrasekhar's formula (1942) for the time of relaxation, (c) a factor of 2.5 by extrapolating the slope between our experimental data for 100 and 250 stars. Hence the evaporation time of a typical galactic cluster ($\mathfrak{M}=500\,\mathfrak{M}_\odot$, $N=1000$ stars, $R=1$ pc) should lie between 5 and 8×10^8 yr. In Figure 4 we show a family of extrapolated curves for a variety of total masses. We have used the data for $N=100$ stars as the starting point in this case; the extrapolation is based on Hénon's theory. The resulting figures may overestimate the evaporation times for $N>100$.

6. Comparison of Experimental Evaporation Times with Observed Lifetimes

From Figure 4 we see that in the range where actual galactic clusters are observed, namely in the intervals of median radii from 0.5 pc to 3 pc and of total masses from $100\,\mathfrak{M}_\odot$ to a few $1000\,\mathfrak{M}_\odot$, the evaporation times vary from 10^8 yr up to the age of our Galaxy. Thus we can explain the wide spread of the lifetimes of galactic clusters as due to the variety of total masses and radii of the clusters. A slight discrepancy (by a factor of about 3) is indicated between the theoretically predicted lifetime of an average galactic cluster and the observed value. What are the possible sources of this difference? (a) The extrapolation from 250 to 1000 stars may not be correct. (b) The dynamical evolution of a cluster may speed up with time so strongly that the evaporation time is longer than the lifetime. This effect is predicted on the basis of the classical theories (King, 1957; von Hoerner, 1958), but at least our models G and G2 show that this acceleration cannot be very drastic. (c) The deduced value of the 'observed' median lifetime may be too small because of undetected selection effects in the observational material or because of an inaccurate age calibration.

Considering all the uncertainties involved in the theoretical and observed values of the typical lifetime of a galactic cluster, I would say that we have found reasonable agreement between theory and observation. It seems quite certain that the observed age distribution of galactic clusters can be explained on a dynamical basis as presented here.

7. The Effect of Interstellar Clouds

Spitzer (1958) has pointed out that the tidal field of passing interstellar clouds may be efficient in disrupting a galactic cluster. Up to now I have neglected this effect in the numerical experiments. In order to estimate the importance of this effect we have plotted in Figure 4 Spitzer's results for the disruption time of a cluster due to clouds. An inspection of Figure 4 reveals that the disruption by clouds is effective only when the cluster is already at or beyond the verge of stability according to the galactic tidal field. If the disruption time as given by Spitzer does not underestimate the cloud effect, then we must conclude that the disintegration of most galactic clusters is caused by the evaporation process combined with the general galactic tidal field rather than by clouds. The interstellar clouds will help, however, to disperse the halo of escapers and nearly escaping stars around a stable cluster, and the clouds will enhance the dissolution of an already unstable cluster.

8. Final Remarks

Figure 5 gives a schematic representation of the lifetime of a galactic cluster of about 500 $\mathfrak{M}_{\odot}$ and 1000 stars as a function of its median radius R. For $R<0.1$ pc, the dissolution of the cluster is caused exclusively by individual encounters. For larger radii, the energy cut-off due to the galactic field enhances the rate of escape. If R is so large, that the cluster becomes unstable against the galactic field, the effect of

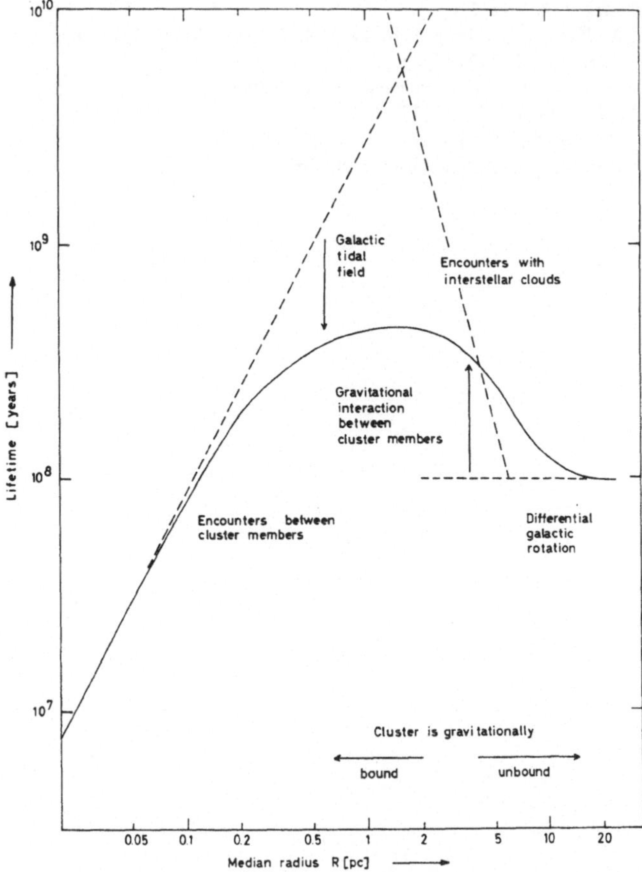

Fig. 5. Schematic representation of the lifetime of a galactic cluster as a function of its radius.

passing interstellar clouds may accelerate the disruption of the cluster. For $R > 10$ pc, the shearing effect of the differential galactic rotation will smear out the cluster. The longest lifetime occur at $R \sim 2$–3 pc. This favorable value of R depends only slightly on $\mathfrak{M}$ or N. Hence we may expect that the oldest galactic clusters have radii in this range. This is nicely verified by the observed median radii (van den Bergh and Sher, 1960) of M67 ($R = 2.2$ pc, 5×10^9 yr old) and of NGC 188 ($R = 2.4$ pc, 10^{10} yr old).

References

Albada, T. S. van: 1968, *Bull. Astron. Inst. Neth.* **19**, 479.
Altena, W. F. van: 1966, *Astron. J.* **71**, 482.
Barbaro, G., Dallaporta, N., and Fabris, G.: 1969, *Astrophys. Space Sci.* **3**, 123.
Becker, W.: 1969, private communication.
Bergh, S. van den and Sher, D.: 1960, *Publ. David Dunlap Obs.* **2**, 203.
Chandrasekhar, S.: 1942, *Principles of Stellar Dynamics*, Univ. Chicago Press, Chicago.
Hénon, M.: 1969, *Astron. Astrophys.* **2**, 151.
Hoerner, S. von: 1958, *Z. Astrophys.* **44**, 221.
King, I.: 1957, *Astron. J.* **62**, 144.
King, I.: 1962, *Astron. J.* **67**, 471.
Lindoff, U.: 1968, *Arkiv. Astron.* **5**, 1.
Oort, J. H.: 1958, *Stellar Populations* (ed. by D. J. K. O'Connell), Vatican Obs., p. 63.
Spitzer, L.: 1958, *Astrophys. J.* **127**, 17.
Wielen, R.: 1967a, *Veröffentl. Astron. Rechen-Inst. Heidelberg*, No. 19.
Wielen, R.: 1967b, *Bull. Astron.* (3) **2**, 117.
Wielen, R.: 1968, *Bull. Astron.* (3) **3**, 127.
Wielen, R.: 1969, Habilitationsschrift, Univ. Heidelberg.

DISRUPTION OF STAR CLUSTERS THROUGH PASSING INTERSTELLAR CLOUDS INVESTIGATED BY NUMERICAL EXPERIMENTS

P. BOUVIER and G. JANIN

Interstellar matter, often abundant along the Milky Way, is not uniformly distributed over the galactic disk but shows irregularities suggesting the concept of discrete clouds of finite mass and dimensions endowed with velocities of order of presumably 5 to 10 km/s.

Repeated passages of such clouds near a stellar cluster will tend, in general, to increase the total energy of the latter, causing it to expand and eventually to disintegrate. Spitzer had tackled this problem in 1958 and, after having made several simplifying assumptions, had calculated a disruption time t_d as a function of the cluster's mean mass density.

However, the latest data on interstellar clouds (Stone 1970, and others) lead for t_d to a value which in most cases appears larger than the lifetimes of clusters against the other possible causes of disintegration (Schmidt, 1962). Moreover, Spitzer had neglected very close encounters of the clouds with the cluster.

We therefore examined this problem with the help of numerical experiments performed over spherical clouds passing in the vicinity of a small stellar cluster. The initial state of the cluster was defined by distributing the positions of its 25 stars uniformly within a sphere of radius 2 pc and the stellar velocities uniformly up to a maximum value fixed by the virial theorem. Further a mass spectrum in m^{-2} was ascribed to the cluster stars. The clouds, divided into three classes (light, intermediate, heavy) were made to pass by the cluster in such a way that their number density remains unchanged, in order to simulate the disrupting effect undergone by the cluster.

The integration method used is that of Nordsieck (1962).

In such conditions, the disruption of the cluster occurred over a time scale of 450 million years, with a cloud concentration of 40 clouds per million cubic parsec. The disruption time t_d should have been, according to Spitzer's analysis, four times larger than the above value. The process seems therefore to be dominated by the very close encounters, occurring when the impact parameter is even less than the cloud radius; the situation is somewhat similar to the one met in the evaporation process where the accumulated effect of distant stellar encounters is unable to produce the escape of a star from the cluster while only close encounters may transfer sufficient energy to a star to allow it to evade.

In order to get some results which could be compared to observations, several different sets of initial states for the cluster should be considered and the continuous tidal action of the galactic field must also be included in the problem.

M. Lecar (ed.), Gravitational N-Body Problem, 71–72.

Note added in proof. The detailed account of the present paper is to be found in *Astron. Astrophys.* **9**, (1970), 461.

The problem has been rediscussed and extended lately by Bouvier (1971) in *Astron. Astrophys.* **14**, 341.

References

Nordsieck, A.: 1962, *Math. Comp.* **16**, 22.
Schmidt, R. H.: 1962, *Astron. Nachr.* **287**, 41.
Spitzer, L.: 1958, *Astrophys. J.* **127**, 17.
Stone, M. E.: 1970, *Astrophys. J.* **159**, 293.

NUMERICAL EXPERIMENTS ON THE ESCAPE FROM NON-ISOLATED CLUSTERS AND THE FORMATION OF MULTIPLE STARS

AVRAM HAYLI

Institut d'Astrophysique et Collège de France, Paris, France

Abstract. Isolated and non isolated clusters with a mass distribution have been studied by numerical techniques. The rates of escape of stars and of kinetic energy are compared with Hénon's theoretical expressions. Multiple encounters play a very important role in the escape phenomenon, at least for clusters with a small number of stars. This leads to a theoretical underestimate of the rates of escape when the stars have equal masses and to an overestimate when masses are unequal.

For non isolated clusters, the tidal field of the Galaxy is responsible for one half of the rate of escape of the stars. The energy of a star escaping because of the tidal effect grows slowly while that of a star escaping after an encounter increases very rapidly. The stars escaping because of the tidal effect leave the cluster in the vicinity of the equilibrium points.

Encounters and the tidal field are not efficient enough to explain why very old open clusters are not observed. Other escape mechanisms have to be considered.

Very stable subsystems are formed which are not destroyed under the influence of the galactic tide. Separation between stars can be as low as 1000 UA.

1. Introduction

It is well known that a phenomenon of escape takes place during the evolution of clusters. The different mechanisms of escape have been studied, using numerical techniques, by van Albada (1968) and their relative importance evaluated in the case of small isolated clusters. Numerous theoretical estimates of the escape rate have also been obtained. A study was made recently by Hénon (1969) for the general case where the stars of an isolated cluster have unequal masses.

In this paper we intend to give an account of numerical studies of the escape phenomenon, for isolated clusters and for clusters moving in the galactic field in a circular orbit, at the distance of the Sun. 67 examples have been studied, 57 of which with 15 stars. The stars have either equal or unequal masses. These examples have been followed over physical times which are on average 200×10^6 yr. Some clusters have been followed for longer, sometimes for as much as 500×10^6 yr.

We have compared the numerical results obtained with Hénon's theoretical formulae. We have also examined the mechanisms and the rate of escape from the clusters moving in the galactic field.

The integration method used, the galactic field chosen, the equations of the motion and the corresponding first integrals have already been described (Hayli, 1967). The units adopted are the parsec, 10^6 yr and the solar mass. In each case, the initial conditions are the following: the stars are distributed with a uniform density in a sphere of radius 1 parsec. The velocity distribution is isotropic and uniform. The quantity $2T/\Omega$ has been taken as equal to unity, where T is the total kinetic energy of the stars

M. Lecar (ed.), Gravitational N-Body Problem, 73–87.

in the reference system of the center of mass of the cluster and Ω the total potential energy of the stars in the field of the cluster.

A cluster contains N stars. When they all have the same mass, this mass is taken equal to that of the Sun. When they have unequal masses, these are chosen in such a way that if N_j is the number of stars of mass m_j, the product $N_j m_j$ does not depend on j. We have:

$$\sum_j N_j = N \quad \text{and} \quad \sum_j N_j m_j = M, \tag{1}$$

where M is the total mass of the cluster. Moreover,

$$m_j = 2^{j-1} M_\odot . \tag{2}$$

Thus the clusters of 15 stars of different masses contain 1 star of 8 $M_\odot$, 2 stars of 4 $M_\odot$, 4 stars of 2 $M_\odot$ and 8 stars of 1 $M_\odot$.

Table I gives a short description of the calculated examples.

TABLE I

Description of the calculated examples

Number of stars	Masses	Galactic field	Number of examples
15	equal	no	10
15	equal	yes	14
15	unequal	no	13
15	unequal	yes	20
25	equal	no	2
25	equal	yes	1
31	unequal	no	1
31	unequal	yes	1
32	equal	no	1
32	equal	yes	1
48	equal	no	1
48	equal	yes	2

The comparisons will be made with the theoretical formulae given by Hénon (1969) for self gravitating clusters. It should be recalled that Hénon's calculations were made only taking into account two-body interactions. The following formulae give the escape rate $\mathrm{d}N_i/\mathrm{d}t$ and the energy flux $\mathrm{d}E_i/\mathrm{d}t$ carried by the stars of mass m_i leaving the cluster in the case of Plummer's model:

$$\frac{\mathrm{d}N_i}{\mathrm{d}t} = G^{-1} M^{-9/2} |E|^{3/2} N_i \sum_j N_j \mathscr{F}\left(\frac{m_i}{m_j}\right) m_j^2 , \tag{3}$$

$$\frac{\mathrm{d}E_i}{\mathrm{d}t} = G^{-1} M^{-11/2} |E|^{5/2} m_i N_i \sum_j N_j \mathscr{G}\left(\frac{m_i}{m_j}\right) m_j^2 , \tag{4}$$

$\mathscr{F}$ and $\mathscr{G}$ being two functions tabulated by Hénon. G is the gravitational constant and E the total energy of the cluster.

One must note however, that the numerical studies show that during the evolution, the clusters develop central densities which, after a time, become greater than those forseen in Plummer's model. Nevertheless this model can be taken as a basis for comparison, at least during the first phases of the evolution.

Another consequence of the evolution of clusters is the formation of very stable subsystems. This question has already been studied in the case of isolated clusters. We shall see in the following the influence of the galactic field on the formation of double and multiple stars and what are the minimum separation distances it seems possible to reach for dynamically formed binaries.

2. Results

The comparisons with the theory have been made over physical times not exceeding 200×10^6 yr and often less than this. It is necessary that the escape should not modify too noticeably the composition of the clusters and that the central densities should not become too high compared to those of a Plummer's model.

2.1. Isolated Clusters

2.1.1. *Equal Masses*

The mean escape rate observed is about three times that predicted by theory. Let us explain how this result was obtained as the same method of comparison was used in other cases in order to see to what extent there was an agreement between theoretical predictions and experimental values.

For $N=15$, equal masses and without galactic field, the theoretical escape rate, deduced from Hénon's formula (3) is given by

$$\mathrm{d}N/\mathrm{d}t = 1.9 \times 10^{-2} |E|^{3/2} .$$

Table II describes the results obtained with a sample of eight different clusters. E_0 is the total energy of the cluster, $t_{\max}$ is the physical time, in units of 10^6 yr, during which the cluster was followed. v_p is the predicted number of escapes while v_e is the effective number of escapers. v_p and v_e are given for $t=t_{\max}$ and also for at $t_{100} = 100 \times 10^6$ yr,

As one can see, the mean values of v_p and v_e at $t=100 \times 10^6$ yr are

$$\bar{v}_p = 0.3 \qquad \bar{v}_e \geqslant 0.87 ,$$

so that the observed mean escape rate is about three times that predicted by theory. This difference is not surprising since when N is small (although one cannot give a precise limit for 'small') the multiple encounters play a very important role in the escape process. Multiple interactions are not taken into account in the theoretical calculation. The study of a great number of examples with different values of N shows that this disagreement decreases a little when N increases. It is difficult how-

TABLE II

Comparison between the theoretical and the experimental rate of escape from clusters for $N=15$, equal masses, no galactic field

Example No.	E_0	$dN/dt \times 10^2$	t_{max}	t_{max}		$t=100$	
				ν_p	ν_e	ν_p	ν_e
1	−0.20589	0.18	200	0.36	0	0.18	0
2	−0.28998	0.30	70	0.21	1	0.30	$\geqslant 1$
3	−0.29438	0.30	200	0.60	0	0.30	0
4	−0.40703	0.49	110	0.54	2	0.49	1
5	−0.31206	0.33	160	0.53	3	0.33	1
6	−0.23844	0.22	260	0.57	1	0.22	0
7	−0.26779	0.26	80	0.21	3	0.26	$\geqslant 3$
8	−0.27835	0.27	95	0.26	1	0.27	$\geqslant 1$

ever to be categorical about this. If we distinguish the escapes due to binary interactions from the other escapers, the agreement with the theory becomes somewhat better. But this time the rate observed is lower than the theoretical one. This can be explained by noting that a multiple encounter can also be considered as a binary one.

2.1.2. *Unequal Masses*

It is difficult to draw any conclusions on the escape rate of the more massive stars, as there are not enough of them in the examples studied. We have therefore limited ourselves to the study of escapers of 1 $M_\odot$ and 2 $M_\odot$.

In both cases, the escape rate observed is two times lower than the rate calculated by Hénon's formula. This result can be explained, at least qualitatively, by noting that a fraction of the lightest stars is captured at the beginning of the evolution, to form subsystems with the heaviest stars, or else to form a halo around them. These stars are partly removed from the encounter process and everything happens as if the cluster contained fewer light stars. This phenomenon is important, since it can completely hide the contribution of multiple encounters to the escape rate.

We have also compared the theoretical energy flux leaving the cluster, given by Hénon, with the flux observed. Generally, the theoretical flux is too small by a factor of the order of 10. But one notes that the escapes which are produced by events concerning more than two stars, remove more energy than the escapes produced by binary encounters. The agreement with the theory is quite good when one considers the energy removed only by the escapes following binary encounters, in clusters where all the stars have the same mass. When the clusters have stars of unequal masses, the agreement is rather poor.

2.2. Clusters in the Galactic Field

We have retained the same criterion for escape as for an isolated cluster where a star will escape if its energy becomes positive. We have kept this criterion for non-isolated clusters, as we do not have anything better.

We observe here a phenomenon that is very rare in the case of isolated clusters. It can happen that the total energy of a star which has already become positive (this is the sum of the kinetic energy in the system of reference of the center of mass of the cluster and the potential energy due to the other stars) can become negative again. This would mean, in terms of an isolated cluster, that the star has been recaptured by the cluster. So we shall consider as escapers those stars whose total energy becomes positive and remains positive to the end of the calculation. The positions of these stars showed, usually without ambiguity, that they had left the potential well due to the remaining of the cluster and that their motion was governed, in practice, by the galactic field alone.

To estimate the effect of the galactic field on the escape rate, we have compared the numerical results both with the results obtained in the isolated case, and with Hénon's formula, which does not take this field into account.

2.2.1. *Equal Masses*

The ratio of the escape rate observed to the theoretical rate as given by Hénon's formula is near 5. It was only 3 in the absence of the galactic field. Furthermore, if this ratio is considered during successive time intervals, it is seen to increase with time.

We have made a detailed study of the escapers taken one by one, to distinguish

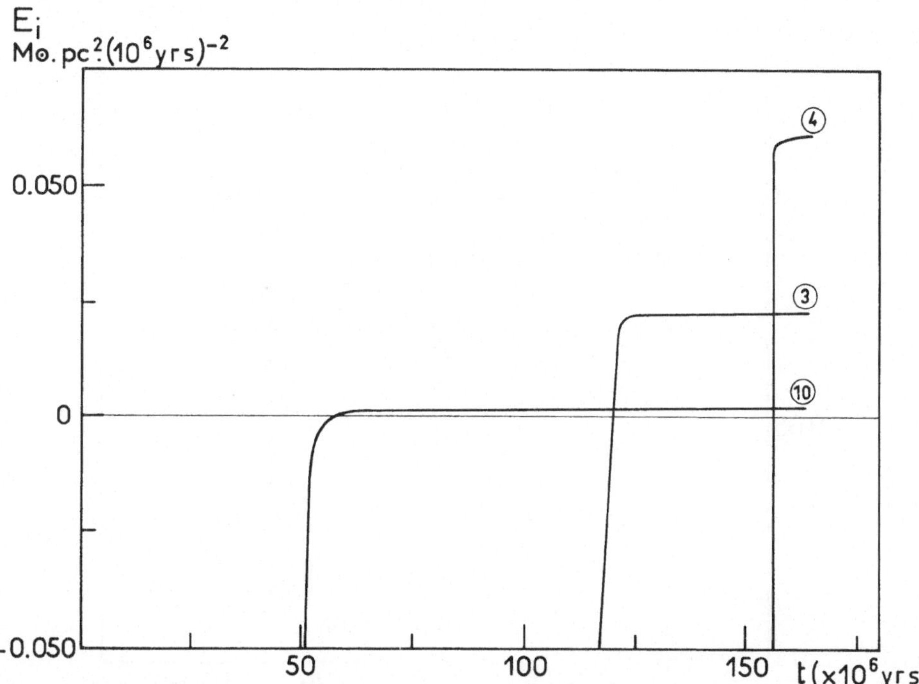

Fig. 1. Variation of the energies E_i of the stars escaping from an isolated cluster ($N = 15$, equal masses). The numbers written inside the circles correspond to the number given to each star. The variation of E_i before escape is not shown. Note that each E_i tends towards a limit.

escapes due to the tidal effect from those due to other causes. This study confirms that the tidal effect is the most efficient of all escape mechanisms, that is to say the one that causes the greatest number of escapers. It also shows that the contribution of encounters to the escape rate is not noticeably modified by the galactic field.

The stars escaping by tidal effect are characterised by the fact that their total energy increases slowly during the interval of time when it becomes positive, whereas the stars escaping through an interaction with one or more stars, undergo a very rapid increase of energy at the time of escape. Figure 1 shows, for an isolated cluster ($N=15$, equal masses) the variations of the energy E_i of the stars escaping between $t=0$ and $t=165\times10^6$ yr. Figure 2 shows the variations of E_i inside an interval of time containing the moment of escape, for the stars escaping from a cluster ($N=15$, equal masses) evolving in the galactic field. The stars are numbered from 1 to 15; in Figure 2 the stars 1, 9 and 7 escape because of interactions between two or more stars, whereas the stars 2, 15, 8, 11, 13, 5 and 4 escape under the influence of the galactic tidal forces. Figures 3a and 3b show, for this same example, the variations of the potential energy and of the total energy E_i of each escaping star, from the moment of

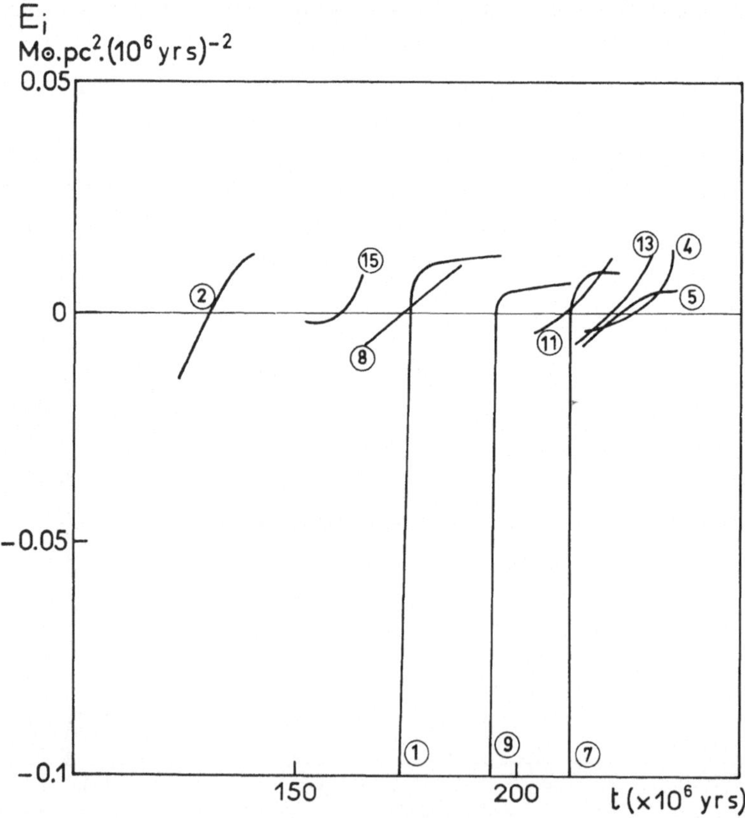

Fig. 2. Variation of the energies E_i of the stars escaping from a non isolated cluster ($N=15$, equal masses) evolving in the galactic field. Stars 1, 9, and 7 escape as a result of encounters, the others as a result of the galactic tide.

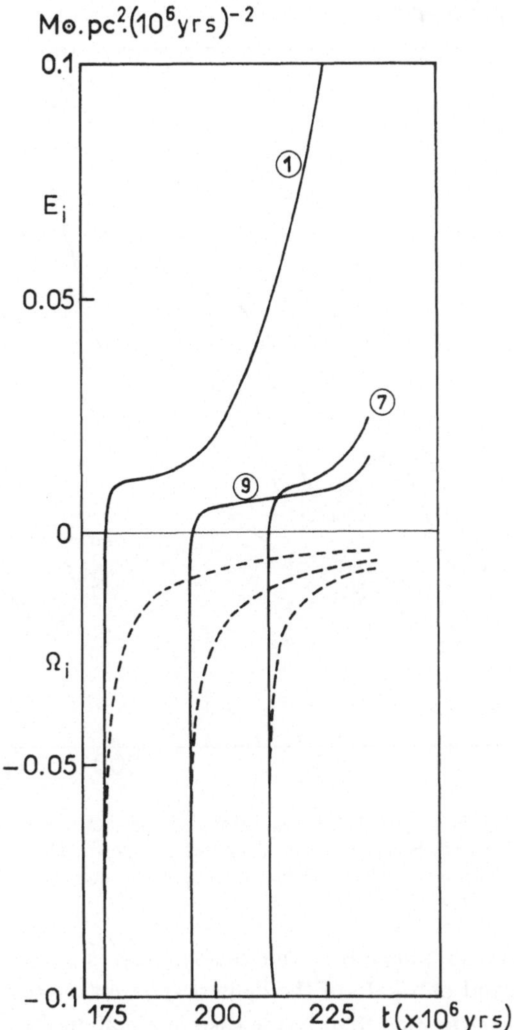

Fig. 3a. Same cluster as for Figure 2. The variations of the total energy E_i (full lines) and of the potential energy Ω_i (dotted lines) of stars 1, 9 and 7 escaping because of encounters are shown. The increase of E_i after the plateau corresponds to the acceleration by the galactic field.

escape until the end of the calculation, at $t = 235 \times 10^6$ yr. In contrast to what one observes for an isolated cluster (Figure 1), the total energy of each star does not tend towards a limit; the stars which have escaped from a non isolated cluster are then accelerated by the galactic field.

2.2.2. *Unequal Masses*

We have made the same comparisons as above for the escaping stars of 1 $M_\odot$ and 2 $M_\odot$. Here again, we have found that the tidal effect plays a very important role in the escape phenomenon, and this role increases with time, but at a slower rate than

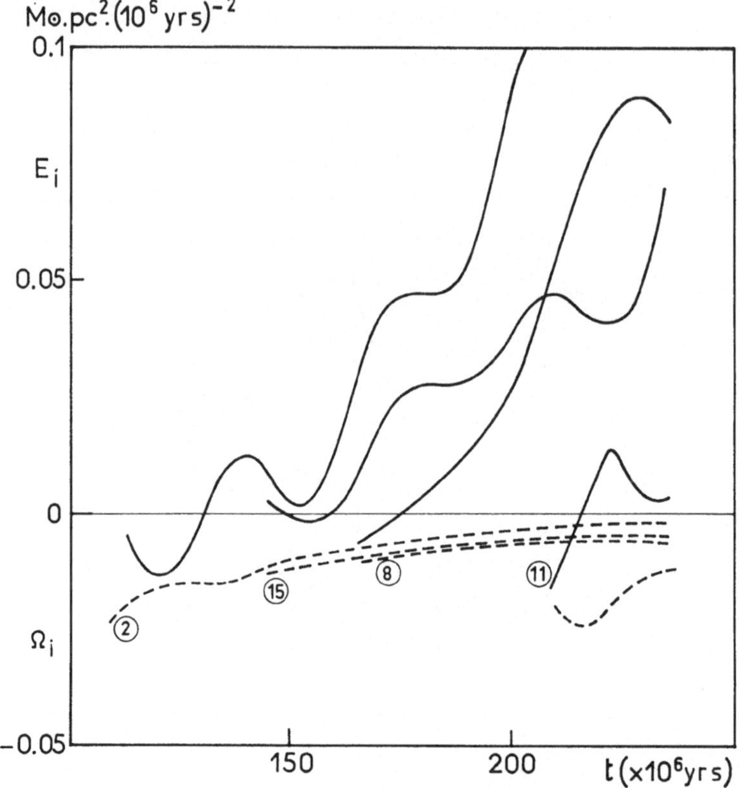

Fig. 3b. Same cluster as for Figure 2. The variations of the total energies E_i and the potential energies Ω_i of the stars escaping because of the tidal effect are shown. The variation of E_i for $E_i > 0$ corresponds to the acceleration through the galactic field.

for the clusters where stars have the same masses; this is because the formation, in time, of subsystems and of a halo of the lightest stars make these stars less sensitive to the effect of the galactic tide than in the case of equal masses.

As far as concerns the energy of the stars escaping by one or the other mechanism, the results are similar to those obtained in the case of equal masses.

3. Directions of Escape

The initial conditions adopted give no privileged directions of escape from isolated clusters. The cluster must remain spherically symmetrical, and this is what is observed within the limits of statistical fluctuations.

In the galactic field, the cluster becomes flat while the escapes take place in directions very close to the galactic plane (Hayli, 1968). But the stars that escape because of encounters are very different in their orbits from the stars escaping through tidal effect. In the first case a star receives enough energy during the interaction that provokes its escape, to allow it to leave the cluster in any direction. This is not the case for stars escaping as a result of tidal effect; these leave the potential well of the cluster in the close vicinity of the equilibrium points. These results have always been remarkably well confirmed by all the numerical experiments. They are illustrated in Figures 4 and 5.

Figures 4 and 5 show the family of curves formed by the intersection of the equipotential surfaces of the total field and the galactic plane, the total field being the sum of the galactic field and the field due to the cluster ($N=15$, $M=15\,M_{\odot}$ for Figure 4 and $N=31$, $M=80\,M_{\odot}$ for Figure 5). For each of these family of curves we have superposed the orbits of the escaping stars.

To find the family of equipotential curves, we have adopted the galactic field already given (Hayli, 1967). Moreover, to simplify things, we have supposed that all the invariant mass of the cluster was concentrated at the center of mass taken as the origin of the coordinates. x and y are the coordinates of a point in the galactic plane.

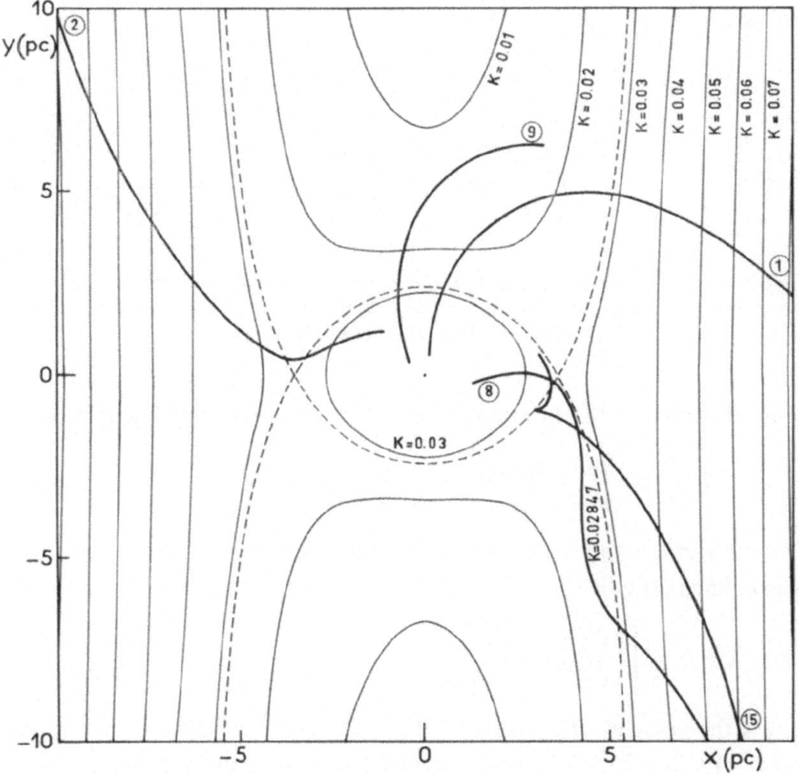

Fig. 4. Equipotential curves in the galactic plane and projection of the orbits of the escaped stars into the galactic plane for a cluster $N=15$, equal masses, $M=15\,M_{\odot}$. Stars 2, 8 and 15 have escaped as a result of tidal effect. They leave the cluster in the immediate neighbourhood of the equilibrium points.

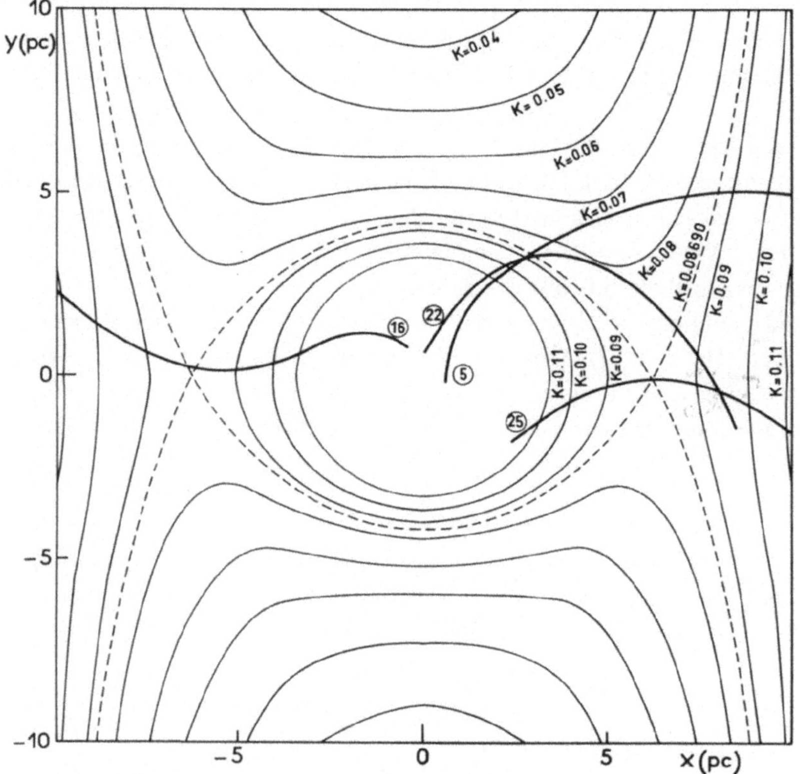

Fig. 5. Equipotential curves and projection of the orbits of the escaped stars into the galactic plane for a cluster $N=31$, unequal masses, $M=80\ M_\odot$. Stars 16 and 25 escape because of tidal effect, stars 5 and 22 by other mechanisms.

It is easily shown that the family of equipotential curves is defined by the relation

$$-\tfrac{1}{2}\lambda x^2 + \frac{GM}{r} - K = 0\,, \tag{5}$$

where $r=(x^2+y^2)^{1/2}$ and where K is a constant. A curve corresponds to each value of K. λ is defined (Hayli, 1967) as

$$\lambda = 2\frac{V_0}{R_0}\left[\left(\frac{\partial V}{\partial R}\right)_0 - \frac{V_0}{R_0}\right], \tag{6}$$

where V_0 is the velocity of rotation of the Galaxy at a distance R_0 from the centre. We have $\omega = V_0/R_0 = 25\times 10^{-2}$ km/s·pc, in the vicinity of the Sun, and $\lambda = -15\times 10^{-4}$ km^2/s^2·pc^2.

Writing $x=r\cos\theta$ and $y=r\sin\theta$ Equation (5) becomes

$$\cos^2\theta = 2\frac{(GM/r) - K}{\lambda r^2}, \tag{7}$$

which allows us to find, point by point, each curve of the family. The coordinates of the points of equilibrium are given by

$$x = \left(-\frac{GM}{\lambda}\right)^{1/3}; \qquad y = 0. \tag{8}$$

The equipotential curves intersect the x axis in points symmetrical with respect to the origin. These points satisfy the relation $\cos^2\theta = 1$. The points of intersection with a positive abscissa ($\xi > 0$) satisfy

$$\lambda\xi^3 + 2K\xi - 2GM = 0. \tag{9}$$

Such an equation is also found in the treatment of Hill's problem in celestial mechanics, and it is shown that for each value of K corresponds 0 or 2 values of ξ (see Szebehely, 1967), that is 0 or 4 intersections with the x axis.

The singular equipotential on which the equilibrium points occur corresponds to $K = K_0$ where $K_0 = \frac{1}{2}(G^2M^2\lambda)^{1/3}$. If $K < K_0$ Equation (9) has no roots. If $K > K_0$ it has two roots.

In Figure 4, the projection on the galactic plane of the orbits of the five stars which escape before $t = 200 \times 10^6$ yr are shown (see Figure 2). Until then, the cluster remains fairly compact about the origin and the number of escapers has not affected the positions of the equilibrium points too much. The part of each orbit shown starts at a point roughly corresponding to the moment when the energy of the star considered becomes positive, or even earlier when it is a star escaping through the tidal effect, and ends at the position occupied by the star at the end of the calculation, when its position is within the limits of the figure.

We see that stars 2, 15 and 8 escaping by the tidal effect pass extremely close to the equilibrium points. (We have checked that the corresponding z coordinate of each of these three stars was very close to zero). In fact the equilibrium points are not precisely defined, as much because of the motion of the stars in the cluster as the escapers.

Stars 1 and 9 escaping because of encounters have no reason to pass the equilibrium points and this is what is observed. A star could not escape by tidal effect, following an orbit like that of stars 1 or 9; the increase of its energy by the galactic field does not allow it to leave the potential well of the cluster except by 'spouts' in which the equilibrium points occur.

The cluster corresponding to Figure 5 has been followed for only 85×10^6 yr. Of the seven stars which had escaped at the end of the calculation, we considered only those that had left, at that time, the central regions of the cluster so as to show clearly the orientation of the orbits with respect to the family of equipotential curves. We see that stars 16 and 25, both of solar mass, have passed, in leaving the cluster, by the equilibrium points or in their immediate vicinity. A study of each individual case has shown that both escapes were due to the galactic tide. Stars 5 and 22 of 4 $M_\odot$ and 1 $M_\odot$ respectively left the cluster without passing near the equilibrium points;

here again, the study of the energy and of the previous history of these stars has shown that their escape was due to encounters, this term being taken as generally referring to the results of binary or multiple interactions. This second example is even more significant than the previous one: on one hand the approximation of the total mass at the center of mass is better than in the preceding example because the mutual distance between the equilibrium points is large compared to the size of the cluster; on the other hand the fraction of the total mass that has escaped is small – less than 10% – and therefore does not affect much the position of the equilibrium points with respect to the cluster, during the evolution.

Finally let us note that the centre of density (von Hoerner, 1960) has remained very close to the centre of mass throughout the cluster was followed.

More details can be found in Hayli (1969).

4. The Formation of Subsystems. Double and Multiple Stars

It was reminded in the introduction to this paper that one of the consequences of the redistribution of the energy between the stars is the formation of extremely stable subsystems inside the cluster. In fact, double or multiple stars are formed and disappear during the life of a cluster, their lifetimes being very different. A recent study by van Albada (1968b) concerned the formation of double stars in small systems ($N=10$), while Aarseth (1968) concluded that double stars were formed in a cluster with many more members ($N=250$), at various stages of its evolution.

The examples we have studied are closer to those considered by van Albada and our conclusions are very similar. We shall therefore summarize our results and give details only for one case: that of a sample of clusters evolving in the galactic field and whose stars do not have all the same mass.

Let us recall that two stars of masses m_i and m_j form a binary system when their binding energy is positive, that is when

$$\frac{m_i + m_j}{r_{ij}} - \tfrac{1}{2}v_{\text{rel}}^2 > 0, \tag{10}$$

where r_{ij} is the distance between the stars and v_{rel} the velocity of one star relative to the other. The expression is not valid when the binary system is not isolated. It is however a good approximation when the perturbing stars are sufficiently distant. Our classification of stars by categories permits us to recognize double and multiple stars, and to take into consideration the effects of perturbing stars.

4.1. Equal masses

When all stars have the same mass, many binaries are formed, and disappear. Some have lifetimes of more than 20×10^6 yr, while triple systems have much shorter lifetimes, of the order of 5×10^6 yr. For most of the computed examples, the lifetime of the binaries was shorter when the number of escapers was less, which is explicable because the binaries are in general destroyed during a close encounter with a third star. When a large fraction of the stars of a cluster has escaped, a very close binary is always formed for which the very long computation leads in the end to an unacceptable precision for the energy integral. The binaries at the end of the computation thus formed have a mean separation of 1000 AU their orbits having eccentricities in the range 0.6 to 0.9. In one case we observed a separation of 100 AU. Another interesting phenomenon was noticed which occurred only once: the formation of two double stars which separated from each other with a velocity of order 1.5 km/s (the mean velocity of the stars being 0.3 km/s) while all the other stars of the system had acquired a positive energy.

The effect of the galactic field is to advance the time of formation of finally stable binaries, while the separation of the components seems to be larger than in the no field case. These two effects can be due partly to the higher escape rate which reduces the encounter and hence the destruction probability of the binary, and partly due to the tidal forces of the galactic field which tend to separate the two components of the binary.

4.2. Unequal Masses

One notes that the galactic field has no noticeable effect on the time of formation of the subsystems. The presence of unequal masses results to the contrary, in a much more rapid formation of binaries, and very often of subsystems containing three or four stars. When a binary is formed it contains almost always the most massive star of the cluster. The binary systems destroyed, not considering those that are transformed into a system of more than two components, never contain the most massive star. All the clusters followed by us ended by becoming a binary containing in general the most massive stars, or a system with three or four components containing the most massive star, which did not escape in a single computed case.

TABLE III

Characteristics of the subsystems

Example no.	t_{max} (10^6 yr)	m_b, m'_b	t_f (10^6 yr)	n_r	d_{min} (AU)	r_f (pc)
1	255	8, 4	105	4(8, 4, 2, 2)	2800	0.30
2	120	8, 2	78	3(8, 2, 2)	1000	0.10
3	150	8, 4	38	2(8, 4)	1500	
4	455	8, 4	195	2(8, 4)	5000	
5	140	8, 4	65	2(8, 4)	1500	
6	125	8, 2	48	4(8, 4, 2, 1)	1000	0.25
7	275	8, 2	163	3(8, 4, 2)	2000	0.22
8	135	8, 4	111	2(8, 4)	1000	
9	175	8, 4	122	3(8, 4, 4)	2000	0.35
10	355	8, 4	119	2(8, 4)	6000	
11	55	8, 4	16	3(8, 4, 1)	2000	0.35

Table III gives some details on the formation of subsystems in the case of unequal masses, with $N=15$, a galactic field being present. t_{max} is the physical time for which the cluster was followed, m_b and m'_b the masses of the two stars closest to each other between time t_f and the end of the computation, d_{min} the minimum distance between these two stars in AU, n_r the number of stars in the final subsystem, and r the radius in parsecs of the sphere inside which the subsystem occurs.

This table shows that the binary finally formed is created around $t=100$, and that the subsystem contains 40% to 55% of the initial total mass of the cluster. Similar results were obtained for examples with $N=31$, while the final subsystem was formed much earlier.

Finally let us point that binaries and subsystems do accumulate most of the binding energy between stars. This fact was already known for the isolated case (Aarseth, 1968).

5. Conclusions

The tidal effect plays a very active role in the escape process of stars from a cluster evolving in the galactic field. It even seems that this role increases with time. The stars escaping as a result of the tidal effect have a dynamic behaviour very different from that of stars escaping because of encounters. These stars gain enough energy in the centre of the cluster to travel to regions far from the centre. In the isolated case such stars would not be able to gather enough supplementary energy to leave the cluster permanently; but in the non isolated case, the limits of the cluster are at a finite distance and a star arriving in the vicinity of an equilibrium point can easily be given energy by the galactic field and leave the cluster. This is the reason why the total energy of such a star can vary slowly during the escape process.

If we admit that the role of multiple encounters in the escape process decreases when N increases, but that of the galactic field remains the same, it is reasonable to give roughly the same importance to the encounters as to the galactic field as a cause of the escape of stars from a cluster with a mass distribution.

Under these conditions, a simple calculation shows that this leads us to give to open clusters a life much greater than the real ages. We conclude that it is necessary to find one or more supplementary escape mechanisms. Spitzer (1958) showed that the clouds passing near a cluster can give some energy to the stars, allowing them, in this way, to escape more easily. King (1958) indicated in what circumstances this mechanism was dominant. The first numerical experiments made by Janin (1969) on models seem to indicate that it will be possible in this way to obtain escape rates in better agreement with the ages of open clusters.

We can also consider that the formation of very stable subsystems inside a cluster is a very general phenomenon little influenced by the presence of the galactic field, at least when the stars have unequal masses as is the case in real clusters. One can thus form binary stars whose components are separated by distances varying between 500 and 15000 AU. These subsystems appear to contain a large fraction of the order of 50% of the total mass of the cluster.

References

Aarseth, S. J.: 1968, *Bull. Astron.* (3) **3**, 105.
Albada, T. S. van: 1968a, *Bull. Astron. Inst. Neth.* **19**, 479.
Albada, T. S. van: 1968b, *Bull. Astron. Inst. Neth.* **20**, 57.
Hayli, A.: 1967, *Bull. Astron.* (3) **2**, 67.
Hayli, A.: 1968, *Bull. Astron.* (3) **3**, 189.
Hayli, A.: 1969, Thèse de Doctorat d'Etat, Université de Paris.
Hénon, M.: 1960, *Ann. Astrophys.* **23**, 668.
Hénon, M.: 1969, *Astron. Astrophys.* **2**, 151.
Hoerner, S. von: 1960, *Z. Astrophys.* **50**, 184.
Janin, G.: 1969, private communication.
Spitzer, L.: 1958, *Astrophys. J.* **127** (1), 17.
Szebehely, V.: 1967, *Theory of Orbits*, Academic Press, London and New York.

BINARY EVOLUTION IN STELLAR SYSTEMS

S. J. AARSETH
Institute of Theoretical Astronomy, Cambridge, England

Abstract. Three new star cluster models containing 250 members and one case with 500 particles have been studied by numerical methods of direct integration. The evolution is dominated by one central binary in all systems with a realistic mass spectrum and more than 50% of the total energy is absorbed by one heavy pair after only 6–18 mean crossing times. General conditions for binary formation and disruption are discussed and a qualitative explanation is given for the energy sink behaviour. Strong interactions with close binaries lead to increased relaxation and the ejection of energetic escapers. At the same time the corresponding recoil kinetic energy is transferred to the other central members by two-body encounters which prevent a secondary phase of central contraction.

1. Introduction

Recent N-body calculations have established that the binary phenomenon plays an important role in the dynamical evolution of simulated clusters. It is already well known from theoretical considerations that bound sub-systems may form as the result of suitable three-body encounters where the third body is required to remove the binding energy. This process still operates in the presence of other particles and additional members may also be involved. Furthermore, the normal cluster evolution leads to high central densities which provide favourable conditions for binary formation by multiple encounters.

Once formed a binary may remain stable over many revolutions or be disrupted by subsequent encounters. Passing field stars may also gain kinetic energy from binary interactions and eventually produce very close pairs. Thus the continual escape of particles tend to leave behind more strongly bound binaries. Since no general configurations with more than two bodies are known to be stable, it may be conjectured that the final state of cluster evolution should approach one close binary with the remaining members at infinity. The time-scale associated with such extreme evolution may be arbitrarily large since the time for complete disruption of an isolated system depends on the detailed distribution of binding energies.

Actual calculations with small particle numbers have reached stages of evolution which are energetically similar to the projected final state (van Albada, 1968). Thus the time-scale for one binary to absorb a significant fraction of the available energy is found to be considerably less than the calculated half-life. It may be noted that this process is particularly efficient when particles of different masses are included since field stars tend to gain kinetic energy from heavy pairs.

Binaries may also be important in larger systems as suggested by one case of $N=250$ particles studied over 28 mean crossing times (Aarseth, 1968). Since then more powerful methods have been developed for dealing with binary configurations which are otherwise very time-consuming. Further results have been obtained for three new cases with $N=250$ and one example of $N=500$. The former have been studied using

M. Lecar (ed.), Gravitational N-Body Problem, 88–98.

a two-body perturbation method for close binaries (Aarseth, 1970) while critical encounters in the larger system are treated by the Kustaanheimo-Stiefel regularization procedure discussed elsewhere in this volume (Aarseth, 1972).

2. Initial Conditions

In the present series of investigations we adopt simple initial conditions for non-equilibrium configurations which quickly lead to the formation of core and halo type systems. Different cases are then generated by varying the mass spectrum as well as the particle number. We introduce natural units for the individual masses m_i and co-ordinates $\mathbf{r}_i$ by the scaling conventions

$$\sum_{i=1}^{N} m_i = N, \qquad E_0 = \tfrac{1}{4}N^2, \tag{1}$$

where E_0 is the initial total energy. It is convenient to discuss the results in terms of the mean crossing time at equilibrium defined by

$$t_{cr} = [8/N]^{1/2}. \tag{2}$$

Initial co-ordinates are obtained from a random spherical distribution while the corresponding velocities are selected from a truncated Maxwellian. In all cases the total kinetic energy is scaled to half the value required for approximate equilibrium. The adopted energy scaling then implies mean central distances close to unity during the early stages of evolution.

Four cases of $N=250$ have been studied and will subsequently be denoted by the numerals I–IV. Some results were discussed previously for case I in which the particles are divided into six groups with scaled masses 25.0, 5.0, 2.5, 1.25, 0.625, 0.3125 and a corresponding membership of 2, 8, 16, 32, 64 and 128, respectively. Apart from the two heavy members the adopted distribution is derived from the simple relation $n(m) \propto m^{-2}$ characteristic of the mass spectra in young galactic clusters.

The members of case II are chosen as above, except that the two heavy bodies are replaced by massless particles; the scaling (1) then implies a correction factor of 1.25. An equal-mass system was studied next in order to emphasize the dynamical importance of different masses. Individual bodies in case IV are selected from the continuous spectrum by the relation

$$\frac{1}{m_i} = \frac{1}{m_1} - \left[\frac{1}{m_1} - \frac{1}{m_N}\right]\left[\frac{i-1}{N-1}\right] \qquad (i = 1, 2, \ldots, N), \tag{3}$$

where the maximum and minimum mass is denoted by m_1 and m_N, respectively. We adopt the mass ratio $m_1 : m_N = 32:1$ consistent with the discrete distribution; this makes $m_1 \simeq 8.8$ scaled units for $N=250$. Finally, the effect of increasing the particle number is investigated in case *V* by choosing $N=500$ with the distribution (3) and the same mass ratio as in the previous case.

3. Binary Formation and Disruption

Before discussing binary mechanisms it may be useful to summarize some of the general features of cluster evolution. The early stages are dominated by mass motions unless special initial conditions have been constructed. Approximate equilibrium configurations, however, are reached after the first few crossing times. Subsequent structural changes take place on a much longer time-scale and are mainly due to the interactions of neighbouring members. During this process of energy exchange some particles exceed the velocity of escape as the result of close encounters and are lost from the system. A larger fraction of the high velocity members move into the halo in elongated orbits and subsequently return to the inner regions but most of the time is spent at large central distances where encounter effects are relatively unimportant. In the meantime the central core suffers further loss of particles and shrinks in size; the smaller cross-section for interaction with halo orbits is only partly compensated by high eccentricities. The long-term evolution is also dominated by mass segregation effects which at the same time promote favourable conditions for binary activity.

Three separate processes of binary formation may be distinguished. The mechanism involving multiple encounters is most common although the absolute formation rate is rather small at ordinary densities. Conditions for the formation of close binaries are therefore not favourable until a high density core has developed whereas wider pairs may form earlier or at greater distances. The continual loss of particles from the strongly bound nucleus eventually leads to the appearance of one close pair at the centre. This situation corresponds closely to the evolutionary run-away of isothermal systems (Lynden-Bell and Wood, 1968) and occurs typically on a time-scale of 10–40 crossing times, depending on the steepness of the mass spectrum. Two heavy bodies may also lose velocity rapidly through the equipartition effect and spiral inwards to the centre before the nucleus has developed. In that case the dominant components would form a wide binary, provided that the relative velocity is sufficiently small.

It is no longer adequate to define binaries by the usual binding energy condition when other particles are present. Instead the stability of the relative orbit depends on the size of the semi-major axis compared to the local inter-particle separation. It is useful to express the relative strength of the external field by the invariant perturbation

$$\gamma = \frac{|\mathbf{F}_k - \mathbf{F}_l| R^2}{m_k + m_l}, \tag{4}$$

where $\mathbf{F}_k - \mathbf{F}_l$ is the tidal acceleration of the two-body motion and R is the separation between the components m_k and m_l. We note that $\gamma \ll 1$ at maximum separation ensures a stable configuration over the subsequent revolution whereas $\gamma \simeq 1$ may lead to rapid disruption or exchange of companions.

Large perturbations may also change the binding energy in the direction of greater stability. According to the introductory considerations this process should dominate for at least one close pair. Thus the proportion of the total energy absorbed by the relative motion may be used as an evolution measure. An invariant energy parameter λ is introduced by the relation

$$\frac{m_k m_l}{2a} = \lambda E, \tag{5}$$

where a is the semi-major axis and E denotes the current total energy of the bound cluster.

Suitable conditions for binary disruption by an incoming third particle may be obtained from the reversal of all velocities at the time of formation. The two processes are not entirely symmetrical, however, since the velocities of field particles are normally less than the local escape velocity, whereas the latter may easily be exceeded after a strong interaction. This imbalance therefore provides a mechanism for increasing the evolution measure of close binaries once the threshold energy for dissociation has been reached. The latter condition may be estimated from

$$a \simeq \frac{m_k m_l}{2N m_\gamma}, \tag{6}$$

where the velocity of the incoming particle m_γ has been taken as twice the rms value and it is assumed that the binary absorbs the whole kinetic energy. Although this relation should be considered as giving a lower limit only, it is evident that relatively close pairs may be disrupted by fast field particles even in the equal-mass case. Conversely, the relations (4) and (6) demonstrate the greater stability of heavy pairs.

Bound triple systems may form if the binding energy exceeds the threshold condition (6). Such configurations are particularly unstable when the masses are different, and often result in the lightest member being ejected with high velocity. The equipartition effect is also important in close encounters involving light particles and heavy binary components. Interactions with lighter masses therefore provide an additional energy sink mechanism. In the subsequent discussion of close binaries we consider binding energies greater than the mean kinetic energy of field particles, this criterion corresponds to $\lambda > 1/N$.

The identity of binary components satisfying condition (6) may be altered by exchange of companions. This process is of particular interest for general mass distributions since an evolution in the direction of greater combined mass would at the same time increase the energy parameter λ. The frequency of suitable close encounters for exchange decreases with time as the nucleus is de-populated. Only actual calculations can therefore determine whether the energetically most favourable mass configuration can be reached before the cluster evolution is essentially completed.

4. Numerical Results

The five cases described previously have been studied over significant intervals of time yielding extensive binary information. General trends become apparent after detailed examinations but only the most interesting developments can be discussed here.

As noted elsewhere (Aarseth, 1968) the evolution of case I is completely dominated by the two heavy bodies which form a binary. This is also the only example of the third process discussed above since formation occurs after only two crossing times from typical initial conditions. The semi-major axis decreases by a factor of 12 from $a \simeq 0.30$ over the next four crossing times, when nearly 80% of the total energy has been absorbed by the relative motion. This rapid evolution is strongly correlated with the escape rate. Thus the heavy binary is more efficient in producing escapers while the separation is relatively large. During the next 20 crossing times the semi-major axis shrinks by a factor of two until the evolution measure is about 92% at $t=5$. Some of the 68 escapers are very energetic and their removal increases the total energy of the bound system by over 70%. One further close pair is also formed after two crossing times with an evolution measure $\lambda \simeq 0.005$ which stays nearly constant over 400 revolutions until disruption. The calculations were continued until $t \simeq 9$ using the perturbation method referred to above but no other significant binaries appear and the heavy binary only shrinks from $a \simeq 0.013$ to $a \simeq 0.011$.

The early evolution of case II proceeds rather differently since eight heavy bodies have the same mass. At first a stongly bound nucleus develops in which the particles are predominantly heavy. Only temporary binaries occur during the first 11 crossing times, although some configurations absorb as much as 10% of the total energy. The increasing central activity is rapidly resolved by the formation of one very close pair with $\lambda \simeq 0.41$, while the other heavy members become less bound to the nucleus. One subsequent energetic interaction increases the evolution measure to 56% with a corresponding eccentricity $e \simeq 0.70$ which later exceeds 0.98. At time $t=4.5$ one member from the second mass group is captured into a bound orbit with respect to the binary centre of mass, while the excess energy is transferred to a body from the third group which then escapes. Although the outer semi-major axis is 0.09 as compared to 0.002 for the heavy pair, the orbital eccentricity is sufficiently high to produce a significant three-body encounter. An exchange of companions takes place, reducing the combined mass from 12.5 to 9.37 scaled units, while the energy is increased from $\lambda \simeq 0.55$ to $\lambda \simeq 0.67$. The resulting recoil effect is quite strong and the new binary becomes much less bound to the cluster centre.

The second case also contains one wide pair initially with components belonging to the first and second mass group. At first the motion is confined to the halo but the semi-major axis is halved to $a \simeq 0.04$ after returning to the central region. This relatively wide binary is preserved over 17 crossing times until exchange by a heavy component takes place. The resulting energy parameter of 3% is subsequently increased to $\lambda \simeq 0.08$. One further exchange involves a bound triple configuration of heavy members. Again the share of binding energy increases, this time from $\lambda \simeq 0.10$ to $\lambda \simeq 0.18$, while the third particle escapes. Later a critical interaction takes place between the two binaries; viz. $\lambda \simeq 0.24$ for the second pair, while the smaller semi-major axis remains unchanged. The associated recoil effect is extremely strong and the lighter pair nearly escapes to infinity. No further significant changes are noted in the heavy binary which returns to the central region for some time before the calculations are halted.

Binaries are much less important for the evolution of case III at corresponding times. The nucleus itself contains on the average at least 20 strongly bound particles inside a small radius $r \simeq 0.2$ during the interval $t \simeq 1$ to $t \simeq 3$ but no significant binaries are formed near the centre. A few wide pairs with $a \simeq 0.02$ formed further out are quickly disrupted again. Close binaries with $a \lesssim 0.005$ begin to appear after about 20 crossing times but the life-times are usually short. As many as six wide binaries have been noted simultaneously but occasionally there are none.

The mechanism of exchange appears to be more important in the equal-mass case; this is partly due to the higher particle density in the nucleus where the kinetic energy is somewhat less than assumed in the derivation of Equation (6). One dominant pair finally forms after about 41 crossing times when the semi-major axis decreases rapidly to $a \simeq 5 \times 10^{-4}$. This is the smallest semi-major axis noted so far but even so the corresponding evolution measure has only reached 6%. The calculations are terminated soon afterwards, however, because the perturbation method is not suitable when a third body is captured to the binary.

The results of case IV are best compared to the similar second case. This time the first significant binary is formed after only five crossing times. Already the first and second heavy body are combined with an evolution measure $\lambda \simeq 0.01$ but some 16 revolutions later an exchange takes place, reducing slightly the mass and energy parameter. The latter is very quickly increased to $\lambda \simeq 0.32$ by the ejection of one strongly bound particle and the resulting eccentricity $e \simeq 0.98$ is characteristic of critical binary interactions. A further exchange after about 300 revolutions restores the original configuration but the third body remains bound to the massive binary. The binding of the triple system is first increased by the ejection of several escapers, whereupon the old binary component also escapes. Thus after only seven mean crossing times the heavy pair has absorbed 62% of the total energy with an eccentricity of 0.95. Subsequent interactions increase the evolution measure to $\lambda \simeq 0.80$ at $t \simeq 5.5$ but the main binary evolution is essentially completed at an early stage.

One additional close pair formed near the centre after 30 crossing times is ejected into the halo. The period is subsequently shortened by favourable encounters in the central region but the corresponding energy ratio is still below 1% because of the small mass of one component. In this case the calculations were extended to cover 40 crossing times without producing any further significant binaries. The binding energy evolution of the dominant pair is displayed in Figure 1 together with the total energy of the bound system.

The larger system also gives rise to some interesting binary developments. About ten wide pairs are present in the initial distribution; i.e., $a \simeq 0.05$ typically, but all are disrupted during the phase of collapse and subsequent re-expansion. The first close binary appears after only four crossing times with an evolution measure of 1.6% and combined mass 11.3 scaled units. This pair very soon suffers an exchange, increasing

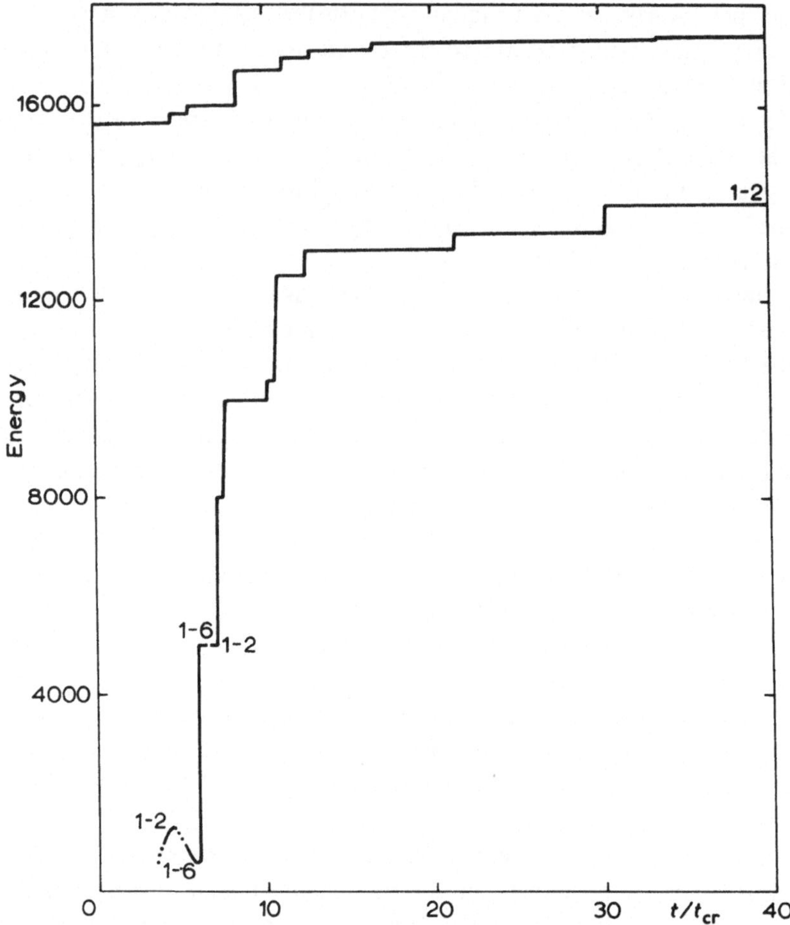

Fig. 1. Energy sink behaviour of close binaries; case IV. The lower curve illustrates the evolution of the dominant binding energy (5) together with exchange of companions. The upper curve gives the total energy of the bound system after distant escapers have been removed.

the energy to $\lambda \simeq 0.06$ with total mass 14.3. Another heavy body is captured to the binary, resulting in a strongly bound three-body configuration where successive exchanges take place during the next two crossing times.

The process of rapid binary formation and disruption by exchange within the nucleus is not resolved until time $t \simeq 1.12$ when two dominant pairs emerge as the result of close encounters which produce several energetic escapers. At this stage the respective binding energies correspond to $\lambda \simeq 0.17$ and $\lambda \simeq 0.15$ with combined masses of 14.6 and 16.3. No significant change occurs until $t \simeq 1.26$ when the former pair returns to the nucleus and interacts strongly with the second binary where now $\lambda \simeq 0.17$. The heavier binary increases its energy share to $\lambda \simeq 0.40$ with an eccentricity of only 0.10 while the incoming components are disrupted, but remain strongly bound to the centre. The additional part of the binding energy increase is accounted for by the loss of a heavy triple companion from the central pair. In spite of the critical nature of this event the total energy of the system is preserved to a relative accuracy of 4×10^{-5} after repeating the calculation.

The subsequent formation at $t \simeq 1.30$ of an extremely close triple system is rather troublesome for the two-body regularization method because of the large number of binary exchanges. This configuration consists of the three heaviest bodies, but the third member is only 11% lighter than the first and the bound sub-group behaves more like an equal-mass system where evolution is much slower than in the general case. Several of the exchanges originate from highly elongated orbits in which the kinetic energy of the returning body is sufficiently great to compensate for the smaller mass. The bound triple persists until time $t \simeq 1.47$ when the first and third member form a stable binary, ejecting the second body from the central region. At this stage $\lambda \simeq 0.50$, $e \simeq 0.70$ and no significant change is noted over the subsequent 5×10^3 revolutions until a transition at $t \simeq 1.77$ to $\lambda \simeq 0.55$. Finally, the evolution measure is increased to $\lambda \simeq 0.60$ at $t \simeq 2.52$ after about 2×10^4 nearly unperturbed revolutions in the central region. A further five crossing times have been studied during which the semi-major axis remains essentially constant. The main binary development is illustrated in Figure 2.

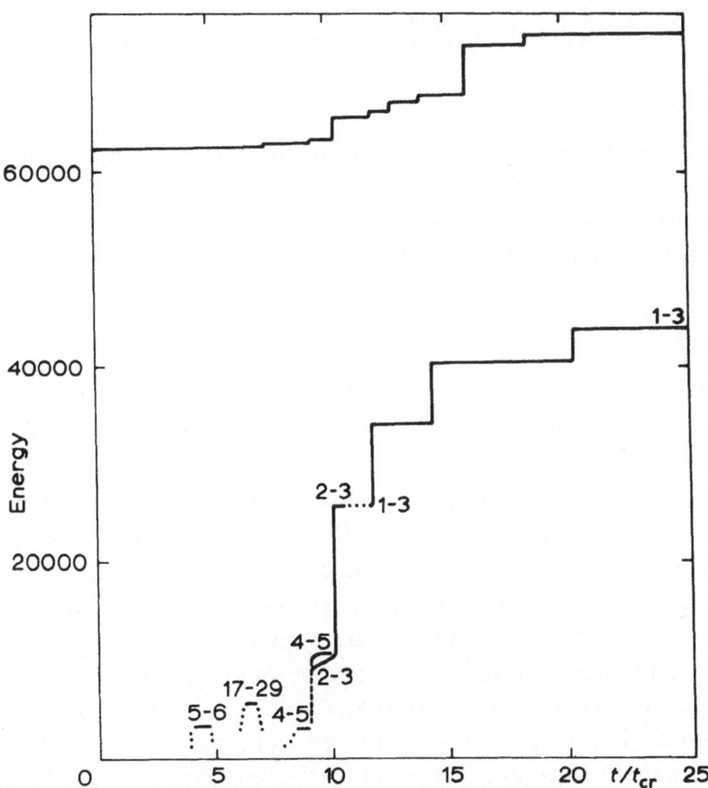

Fig. 2. Energy sink behaviour of close binaries; case V.

Secondary binaries show very little evidence of becoming strongly bound once the dominant pair has developed. Eventually another close pair of combined mass 10.7 forms near the centre at $t \simeq 1.8$. The evolution measure increases to $\lambda \simeq 0.01$ by the ejection of a fast escaper but the semi-major axis is approximately constant over the next 200 revolutions until slow disruption takes place at $t \simeq 2.7$ near the centre. In this example the nearly radial motion is confined to the inner cluster region; viz. $r \lesssim 0.6$, and the low rate of significant interactions over more than seven crossing times merely reflects the loss of particles at small central distances. Three additional wide binaries are noted at this stage with life-times of a few crossing times but there is no tendency for any semi-major axis to shrink.

TABLE I

Final binary parameters

Case	t_f/t_{cr}	$m_k + m_l$	a	e	λ	$(E_f - E_0)/E_0$	$nt_{cr}/2\pi$
I	50	50.0	0.011	0.96	0.92	0.80	170
II	59	9.4	0.0008	0.63	0.67	0.13	3700
II	–	12.5	0.0042	0.90	0.24	–	370
III	42	2.0	0.0005	0.92	0.06	0.01	3100
IV	40	16.6	0.0025	0.94	0.80	0.11	940
IV	–	4.2	0.0062	0.98	0.006	–	120
V	25	16.8	0.0008	0.77	0.60	0.18	3600

Parameters of all final close binaries are summarized in Table I together with the corresponding number of computed crossing times. The relative increase of total energy due to escaper removal has been added in column seven while the last column contains the number of unperturbed revolutions per crossing time, n being the final mean motion.

5. Discussion

The results displayed in Table I and Figures 1 and 2 indicate that the evolution of all the systems with different masses is dominated by one close binary. As for the equal-mass system, the equivalent development of advanced evolution is reached much later, firstly because the mean relaxation time is longer and secondly because of the smaller semi-major axis required. It may be noted that the final binaries already contain 50% of the total energy after only 6, 18, 7 and 12 crossing times, respectively, for the cases I, II, IV and V. At this stage the proportion of escapers is still small and the corresponding kinetic energy flux from the centre cannot account for the main part of the absorbed binding energy. Instead the time-scale for close binary formation is related to the rapid evolution of the nucleus. The shortening central relaxation time ensures that this process is completed in a relatively small number of mean crossing times, whereupon additional energy is absorbed by the central pair. A similar energy sink behaviour has also been predicted by theoretical models for cluster evolution without invoking the binary mechanism specifically (Hénon, 1961; Lynden-Bell and Wood, 1968).

The spatial segregation of different masses leads to preferential formation of heavy pairs in regions of high density. In addition, the exchange mechanism favours configurations of large mass which are also more stable. Special conditions, however, may bring about an exchange of companions where the combined mass actually decreases as noted in some examples above. It is nevertheless significant that the general evolutionary trends of the two continuous mass distributions are quite similar, leading to the maximum mass in case IV while the final configuration in case V consists of the first and third body. Strong interactions of short duration often produce very high eccentricities. The latter may also increase systematically at nearly constant semi-major axis due to the small perturbations of a captured third body; one extreme example of $e > 0.999$ has been noted. Corresponding pericentre distances of elongated close binaries are very small and may present difficulties for ordinary methods of integration.

The presence of heavy binaries leads to increased relaxation of field particles, firstly because the two-body relaxation efficiency is greater when two bodies are combined. Encounters with impact parameters of the order of the semi-major axis may also result in additional gain of kinetic energy where the binary acts as an energy sink. Thus the principle of energy conservation connects two important dynamical mechanisms. Except for the equal-mass case, the mean escape rate is typically between one and two particles per crossing time during the early stages but decreases with time. A considerable proportion of all escapers can be attributed to close encounters with binaries and some events are particularly energetic. The resulting velocities evaluated at infinity may readily exceed twice the rms value, while examples of four and five times the mean have been noted. In consequence the excess kinetic energy carried away by a few fast escapers contributes significantly to the total energy increase of the bound systems as shown by Figures 1 and 2.

Large values of the evolution measure λ imply a significant loss of particles from the central region. This process is continued by subsequent close encounters with the heavy binary but favourable interactions become less frequent because of the smaller mean density as well as the decreasing cross-section. Thus it would take an extremely long time to approach the theoretical limit $\lambda = 1$ arbitrarily close. Conversely, only the formation of hierarchical configurations can maintain the escape rate but no suitable triples have been noted so far*. The advanced stages of close binary evolution are dominated by two-body encounters where the internal binding energy remains approximately constant. The loss of sufficient kinetic energy by the centre of mass

* Subsequent integrations of case V to 30 crossing times reveal a significant hierarchical triple system involving the fifth heaviest body. The outer component has a binding energy corresponding to $\lambda = 0.06$ with respect to the inner pair and the configuration remains stable during the last half crossing time. A similar example with $\lambda = 0.03$ occurs in Case I during four final crossing times omitted from Table I because of difficulties experienced by the perturbation method of integration. In both examples the eccentricity of the inner binary changes significantly whereas the semi-major axis remains nearly constant.

motion would promote the formation of a secondary nucleus and hierarchical structure. On the other hand, the recoil energy associated with critical binary encounters acts in the opposite direction. Thus the additional kinetic energy derived from three-body interactions replaces the energy lost during two-body encounters. Energetic events, although rare, may therefore play an important role in the long-term cluster evolution. The results of the present investigations are consistent in showing no significant tendency for a secondary shrinkage of the central region.

The general problem of strongly interacting three-body systems cannot be treated in detail by analytical methods since it involves too many unknown parameters. Instead some progress may be achieved by investigating special configurations, combining numerical techniques with theoretical considerations. In this way it may be possible to obtain a better understanding of the general conditions leading to third body capture or the exchange of companions. Numerical explorations can also yield some information about the more difficult process of binary formation by three-body encounters (Agekian *et al.*, 1969). Qualitative considerations have been put forward here in an attempt to distinguish between different formation mechanisms and explain the systematic energy sink behaviour of heavy pairs, but as yet the binary phenomenon has received very little theoretical attention. At the same time the dynamical role of binaries in larger systems must be clarified before faster methods can be used with confidence.

References

Aarseth, S. J.: 1968, *Bull. Astron.* **3**, 47.
Aarseth, S. J.: 1970, *Astron. Astrophys.* **9**, 64.
Aarseth, S. J.: 1972, this volume, p. 373.
Agekian, T. A., Anosova, J. P., and Bezgubova, B. N.: 1969, Astrofiz. **5**, 637.
Albada, T. S. van: 1968, *Bull. Astron. Inst. Neth.* **19**, 479.
Hénon, M.: 1961, *Ann. Astrophys.* **24**, 369.
Lynden-Bell, D. and Wood, R.: 1968, *Monthly Notices Roy. Astron. Soc.* **138**, 495.

ON THE DISSOLUTION TIME OF A CLASS OF BINARY SYSTEMS

CARLOS CRUZ-GONZÁLEZ
Instituto de Astronomía, Universidad Nacional Autónoma de México

and

ARCADIO POVEDA
Instituto de Astronomía, Universidad Nacional Autónoma de México, and Observatorio Astrofísico Nacional, S.E.P.

Abstract. To test the various theories of the dissolution time of binary systems, we have performed a series of numerical experiments. The present simulations represent a number of binaries subject to perturbations by passing field stars. Various masses and velocities of the field stars were used to exhibit their effects on the dissolution time. Because of simplicity, the pairs considered consisted of a primary of one solar mass plus a secondary of negligible mass.

The computations gave, among other things, the evolution in time of the energies and eccentricities of the secondary components. We found from these results the need to redefine the concept of time of dissolution to represent more realistically the rate of loss of secondary components. The dissolution times found from the present computations do not agree with any of the existing theories, neither in the general behavior nor in actual numerical values. The times of dissolution found in the present calculations are between a factor of two and a factor of fifteen longer than predicted by existing theories.

1. Introduction

The motivation for the present study comes from the inconsistencies one finds in the literature for the dissolution times of binary systems subject to random perturbations produced by passing stars.

A class of binary systems of particular interest is the cloud of bound comets which surrounds the Sun, and which appears to be a steady source of 'new comets'. The 'new comets' are those members of the cloud that, as the result of stellar perturbations, acquire orbital elements allowing them to come to the vicinity of the Sun, where they become observable.

The list of comets with the most reliably determined orbital elements (Strömgren, 1914) shows that new comets come from distances between 30000 and 150000 AU. The cloud of bound comets around the Sun contains therefore a large number of members with semimajor axes of up to 150000 AU which have survived, as members of the cloud, throughout 5×10^9 yr of stellar perturbations. The preceding considerations, which are very hard to escape, appear to be in conflict with the times of dissolution for binary systems computed with the theories of Ambartsumian (1937) and Chandrasekhar (1944). To illustrate this point we list in Table I the theoretical times of dissolution T_D according to Ambartsumian (1937), Chandrasekhar (1944) and Oort (1950).

The times listed in Table I were obtained from Equations (1), (2) and (3) which follow from the theories after appropriate modifications, when needed. The latter

M. Lecar (ed.), Gravitational N-Body Problem, 99–113. *All Rights Reserved*

TABLE I

Author	T_D (yr)[a]
Ambartsumian (1937)	6×10^8
Chandrasekhar (1944)	6×10^7
Oort (1950)	2×10^9

[a] $a = 10^5$ AU.

take into account that one has to deal with 3 different masses, i.e. M_1: the mass of the primary, $M_2 \approx 0$, the mass of the secondary, and M_*, the average mass of the field stars. The equations for the time of dissolution are shown below:

$$T_{\text{Ambartsumian}} = \frac{(M_1 + M_2)\, v_*}{4\pi G M_*^2 \nu a \left\{ \ln\left[1 + \left(\dfrac{a v_*^2}{G(M_* + M_1)}\right)^2\right] + \ln\left[1 + \left(\dfrac{a v_*^2}{G(M_* + M_2)}\right)^2\right] \right\}} \tag{1}$$

$$T_{\text{Chandrasekhar}} = \frac{(M_1 + M_2)^{1/2}}{4\pi G^{1/2} M_* \nu a^{3/2}}, \tag{2}$$

$$T_{\text{Oort}} = \frac{M_1 v_*}{2\sqrt{6\pi}\, G M_*^2 \nu a \ln\left(\dfrac{a v_*^2}{2 G M_*}\right)^2}, \tag{3}$$

where a is the semimajor axis of the secondary, ν is the number density of field stars and v_* is the rms velocity of field stars.

The times in Table I are based on the following parameters for the solar vicinity: $\varrho = \nu M_* = 0.057\ M_\odot/\text{pc}^3$; $v_* = 20$ km/s; $M_* = 1/2\ M_\odot$; where ϱ is the mass density due to stars, and M_* is the mean mass of the stars in the solar vicinity.

An inspection of Table I immediately shows that if the dissolution time of 6×10^7 yr given by Chandrasekhar's theory is correct, then the present cloud of comets cannot be a relic from the days of formation of the planetary system. On the contrary, Oort's dissolution time is quite consistent with a cloud that has survived without substantial losses since the beginning of the planetary system.

In addition to the inconsistencies shown in Table I, the qualitative behavior of the expression for the dissolution time changes from theory to theory; it is particularly remarkable that Chandrasekhar's dissolution time does not depend on the velocities of the field stars nor on their individual masses, provided that the mass density νM_* is kept constant.

Before proceeding to discuss whether there should be any comets left in the cloud, clearly one should first investigate which of the theories is the correct one. The purpose of the present paper is, therefore, to test the validity of the theories for the dissolution time by means of numerical simulations of the astronomical problem, and to find out how the orbital elements of the comets in the cloud evolve with time.

2. Simulation of the Astronomical Problem

To simulate the astronomical problem in question, we begin by defining a spherical region $\mathcal{R}$ of 5.6 pc radius, containing the test binary and a number of field stars. The test binary consists of a primary of 1 $M_\odot$ and a secondary of zero mass. The primary is set initially in a circular orbit of 10^5 AU radius. The field stars are placed in random positions, uniformly throughout $\mathcal{R}$. Their numbers and masses are selected so that the mass density in $\mathcal{R}$ agrees with the mass density of stars in the solar vicinity. The field stars move in $\mathcal{R}$ always along rectilinear paths with uniform random velocities, which are initially obtained as follows: for each star, three random numbers between -1 and $+1$ are generated; these are then multiplied by the velocity modulus v_*, giving the three velocity components for the star. The equations of motion of the test binary, subject to perturbations caused by field stars, are now integrated numerically. The field stars are supposed not to interact with each other, and therefore their trajectories are not integrated numerically. When a field star, by virtue of its space velocity, leaves $\mathcal{R}$, a new star is generated to replace it. The new star is placed in a random position at the boundary of $\mathcal{R}$; its velocity is generated in the same way as the initial velocities; if, however, the new velocity vector happens to be directed away from the center of $\mathcal{R}$, the star is discarded and another one generated.

As a result of perturbations by field stars the primary gains kinetic energy and begins to drift away from the center of $\mathcal{R}$. Whenever the primary has drifted more than 0.56 pc this effect is compensated by shifting the coordinate system – and with it the center of $\mathcal{R}$ – by the amount necessary to put the primary back at the center.

In order to study the dependence of the time of dissolution on the masses and velocities of field stars, we computed series of cases with different combinations of these parameters. Field stars having equal masses of either 5 $M_\odot$ or 0.5 $M_\odot$ and velocities of 1 km/s or 3 km/s were taken. The number of field stars in $\mathcal{R}$ was set equal to 6 or 60, depending on their individual masses being 5 $M_\odot$ or 0.5 $M_\odot$, respectively. Table II summarizes the number of cases computed for each combination of mass and velocity. All cases of a given series were integrated to the same final time T_f. Table II contains the values of T_f (in non-dimensional units) for the various series.

TABLE II

Series	$M_*/M_\odot$	v_* [km/s]	T_f	Number of cases per series
$A(5, 1)$	5	1	500	84
$B(5, 1)$	5	1	1000	48
$A(5, 3)$	5	3	500	35
$B(5, 3)$	5	3	1000	40
$A(1/2, 1)$	1/2	1	1000	38
$B(1/2, 1)$	1/2	1	1800	9

The velocities and some of the masses chosen for the numerical experiments clearly do not correspond to the actual values for the field stars in the solar vicinity. To simulate the actual conditions in the solar vicinity a very large amount of computer time would have been necessary. However, for the purpose of testing the various theories of the dissolution time, it is just as convenient to choose values for the masses and velocities which reduce the computing time, provided they are consistent with the hypotheses underlying the theories. The selection of masses and velocities listed in Table II satisfies this condition.

3. The Numerical Solution

From the description of the astronomical problem given in the preceding section it is clear that the equations of motion of the primary and secondary stars are, respectively,

$$\frac{\mathrm{d}^2\mathbf{r}_1}{\mathrm{d}t^2} = \sum_{i=3}^{n} Gm_i \frac{\mathbf{r}_{1i}}{r_{1i}^3},$$

and

$$\frac{\mathrm{d}^2\mathbf{r}_2}{\mathrm{d}t^2} = Gm_1 \frac{\mathbf{r}_{2i}}{r_{2i}^3} + \sum_{i=3}^{n} Gm_i \frac{\mathbf{r}_{2i}}{r_{2i}^3},$$

whereas the equations of motion of the field stars are

$$\frac{\mathrm{d}^2\mathbf{r}_i}{\mathrm{d}t^2} = 0 \qquad (i \geqslant 3).$$

In the above equations, m_i is the mass of star i, $\mathbf{r}_i$ is the position vector of star i, $\mathbf{r}_{ij}=\mathbf{r}_j-\mathbf{r}_i$ and $r_{ij}=\mathbf{r}_{ij}$. For the primary star $i=1$; for the secondary star $i=2$ and for the field stars $i\geqslant 3$. The equations of motion of the test binary were integrated numerically by the method developed by Cruz and Lecar (1968). This method uses the explicit evaluation of the coefficients of the Taylor expansion of the equations, i.e. given $\mathbf{r}_i(t)$ and $\mathrm{d}\mathbf{r}_i(t)/\mathrm{d}t$, then

$$\mathbf{r}_i(t+\Delta) = \sum_{k=0}^{5} \frac{\Delta^k}{k!} \frac{\mathrm{d}^h\mathbf{r}_i(t)}{\mathrm{d}t^k}$$

and

$$\frac{\mathrm{d}\mathbf{r}_i(t+\Delta)}{\mathrm{d}t} = \sum_{k=1}^{5} \frac{\Delta^{k-1}}{(k-1)!} \frac{\mathrm{d}^k\mathbf{r}_i(t)}{\mathrm{d}t^k}.$$

The coefficients $(\mathrm{d}^k\mathbf{r}_i(t)/\mathrm{d}t^k\,(k>1)$ are computed using their analytical expressions, which are derived from the equations of motion.

The time step Δ is computed from

$$\Delta = \varepsilon \left[\frac{r_{ij}^{3/2}}{(Gm_j)^{1/2}}, \frac{r_{ij}}{v_{ij}} \right]_{\min}$$

that is, for all i, j the quantities in the square bracket are computed; the minimum value found is multiplied by the constant ε to give the time step Δ, ε being around 5×10^{-2}.

For the computations the following units were used: length: $R = 10^4$ AU; mass: $M = 1\ M_\odot$; time: $t = (R^3/GM_\odot)^{1/2} = 1.6 \times 10^5$ yr; velocity: $v = (GM_\odot/R)^{1/2} = 0.3$ km/s.

Our numerical experiments have the inconvenience that the energy of the test binary is not constant in time, and therefore cannot be used as a test of the accuracy of the computation. For this reason, it was necessary to compute some special cases as tests. For example, setting the masses of the field stars equal to zero ($m_i = 0$ for $i \geqslant 3$), the only perturbations suffered by the test binary will be due to numerical

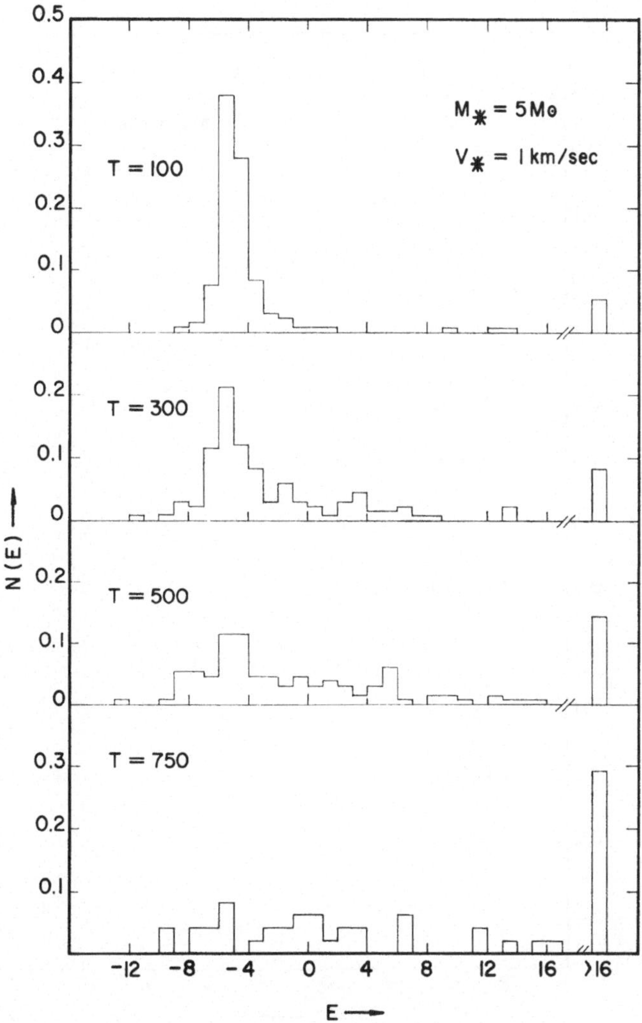

Fig. 1. Evolution in time of the distribution function of the energies of pairs in a field of stars with $M_* = 5\ M_\odot$ and $v_* = 1$ km/s. At time zero, all pairs have the same energy $E_0 = -5 \times 10^{-2}$.

errors. These special test cases were integrated for a very long time ($t \approx 10^9$ yr), giving energy errors of about 0.01%. It is therefore expected that, in the actual numerical experiments, the changes in the energy of the secondary are mainly caused by the random action of the field stars, and *not* by numerical errors.

4. Results and Discussion

Our discussion is based on numerical results for 254 cases. Each case within a series (see Table II) represents the particular history of a binary subject to the random perturbations peculiar to this binary. In all the cases the binaries have the same initial conditions, namely $a = 10^5$ AU $e = 0$. However, since each case is subject to different

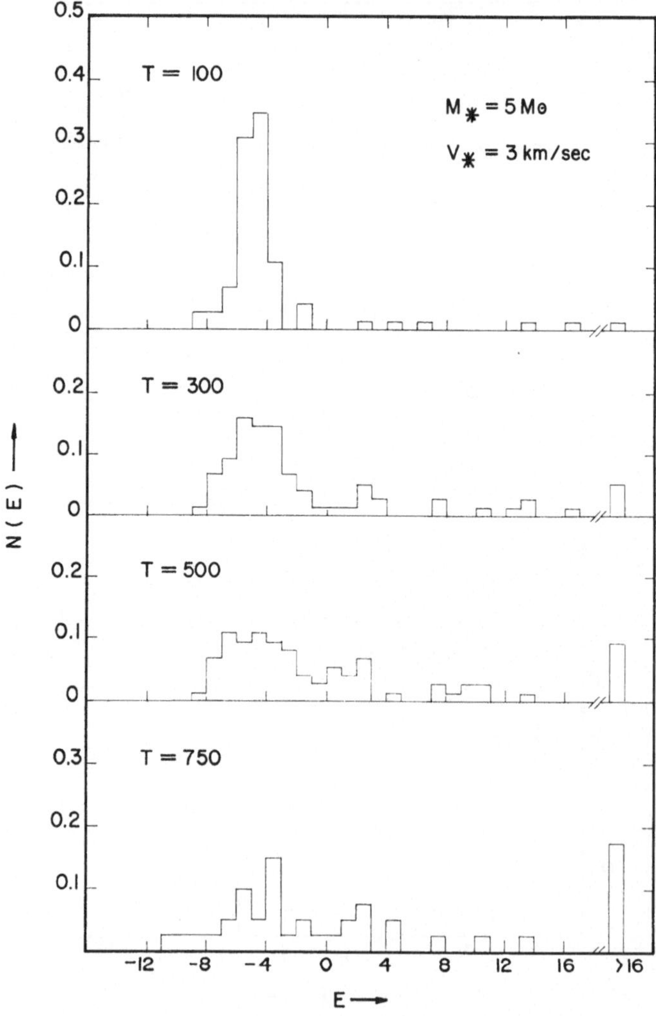

Fig. 2. Evolution in time of the distribution function of the energies of pairs in a field of stars with $M_* = 5\ M_\odot$ and $v_* = 3$ km/s. At time zero, all pairs have the same energy $E_0 = -5 \times 10^{-2}$.

random perturbations, at the end of time T_f the group of binaries of a given series will have a spread of values of their binding energies, eccentricities and semi-major axes.

The numerical results on which the present discussion is based are condensed in a number of histograms, which show the evolution in time of the energies and excentricities of the test binaries. These histograms are shown in Figures 1 to 3 and 10 to 12.

From the basic material described above we computed the first two moments of the distribution function of energies and plotted their evolution in time. Of particular interest is the evolution of the mean energy $\langle E \rangle$ of the binaries, since in the theories of Ambartsumian and Oort the dissolution time T_D of a binary is achieved when its

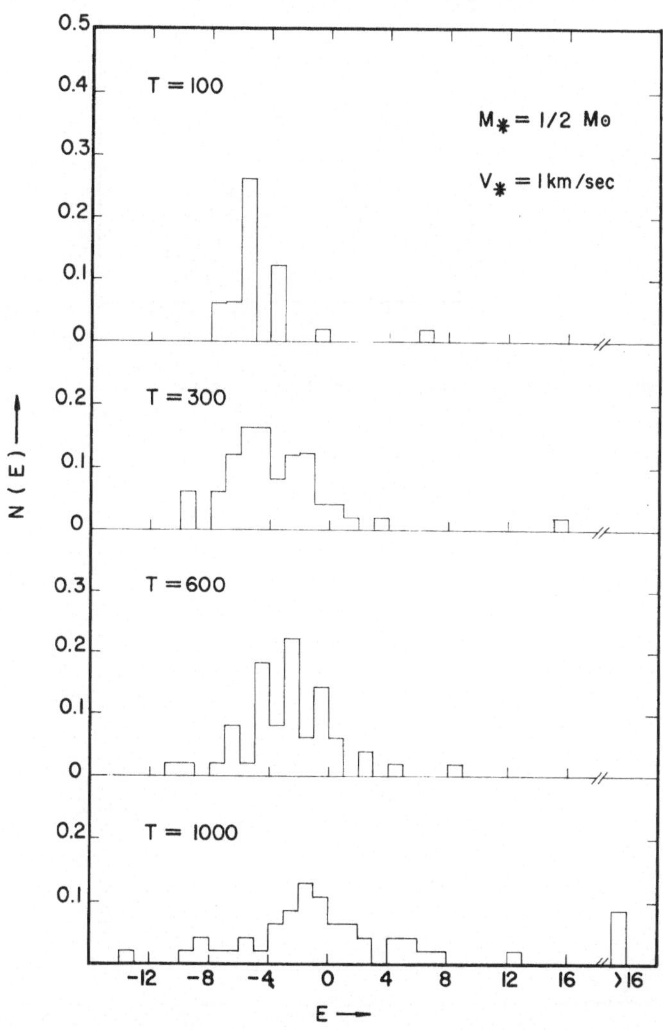

Fig. 3. Evolution in time of the distribution function of the energies of pairs in a field of stars with $M_* = \frac{1}{2} M_\odot$ and $v_* = 1$ km/s. At time zero, all pairs have the same energy $E_0 = -5 \times 10^{-2}$.

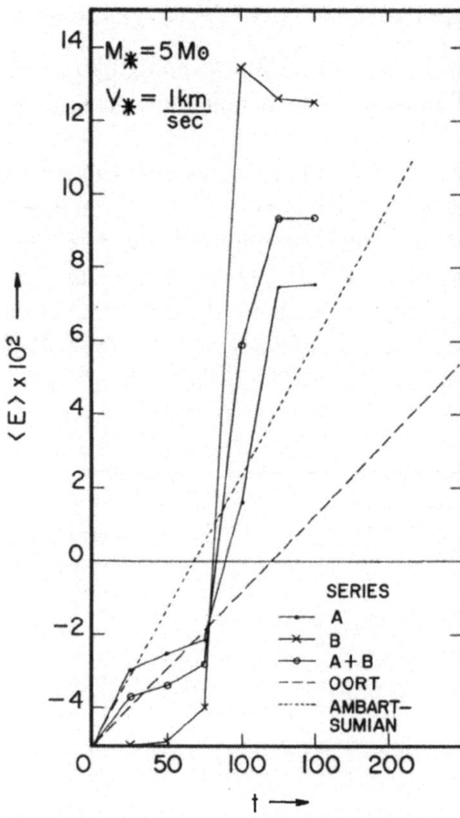

Fig. 4. Evolution of the mean energies $\langle E \rangle$ of pairs, for field stars with $M_* = 5\ M_\odot$ and $v_* = 3$ km/s.

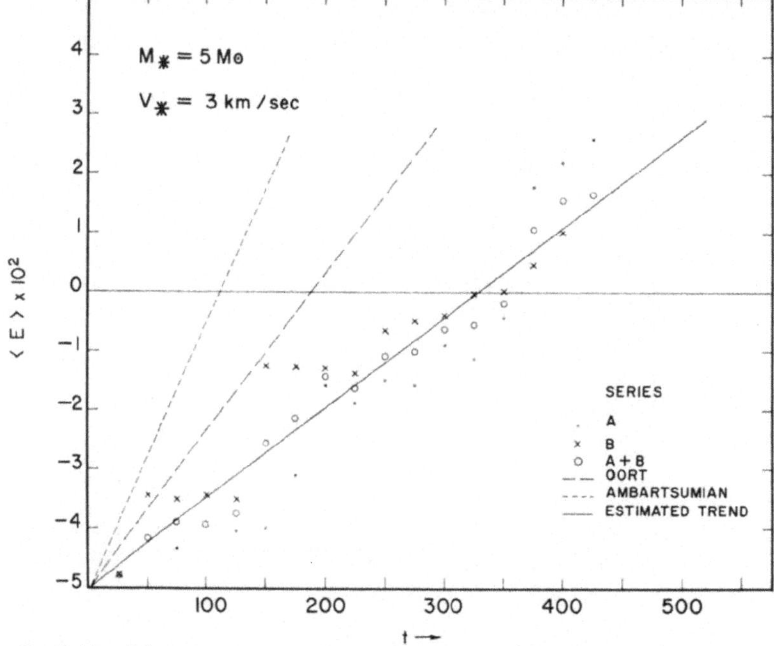

Fig. 5. Evolution of the mean energies $\langle E \rangle$ of pairs, for field stars with $M_* = 5\ M_\odot$ and $v_* = 3$ km/s.

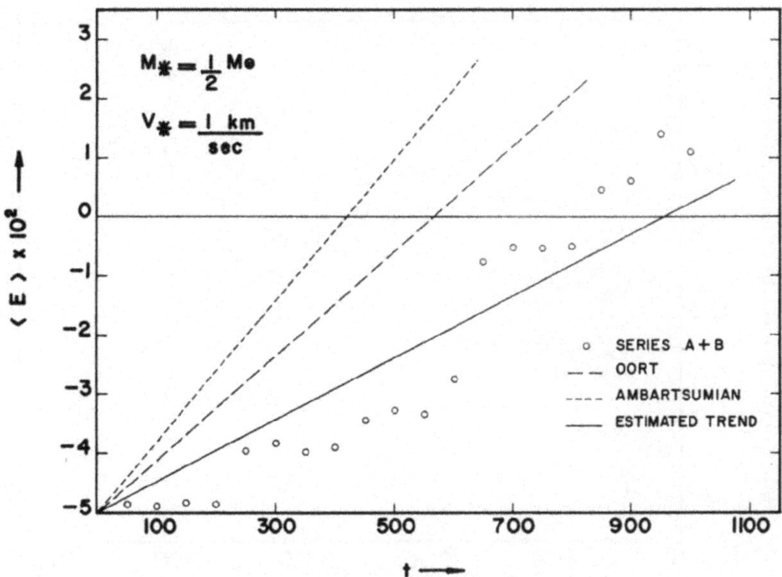

Fig. 6. Evolution of the mean energies $\langle E \rangle$ of pairs, for field stars with $M_* = \frac{1}{2} M_\odot$ and $v_* = 1$ km/s.

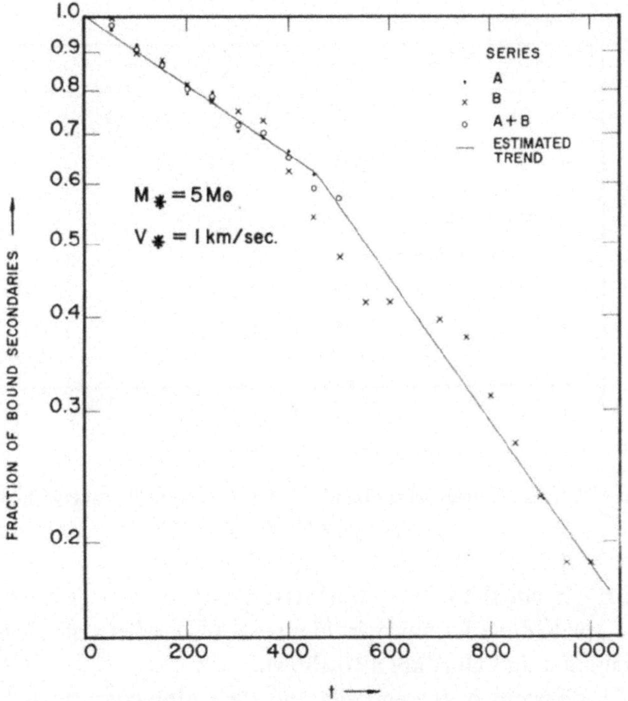

Fig. 7. Fraction of bound binaries as a function of time for field stars with $M_* = 5\, M_\odot$ and $v_* = 1$ km/s.

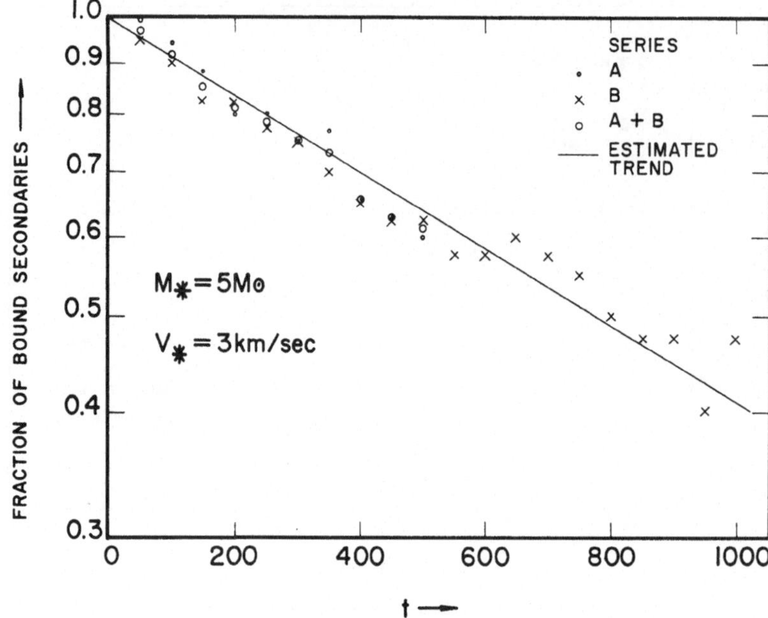

Fig. 8. Fraction of bound binaries as a function of time for field stars with $M_* = 5\ M_\odot$ and $v_* = 3$ km/s.

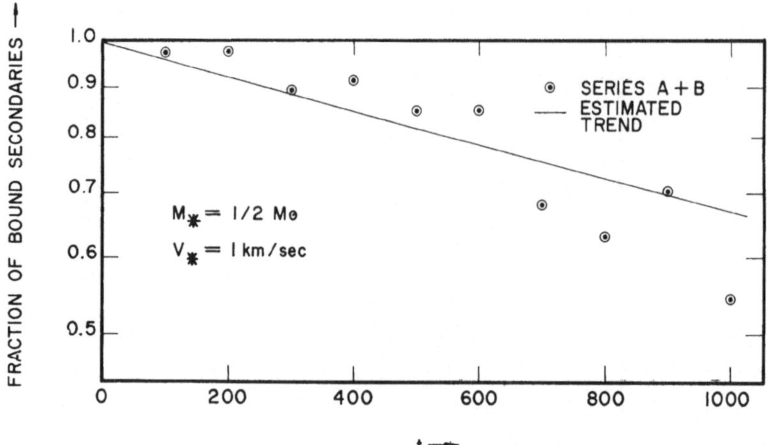

Fig. 9. Fraction of bound binaries as a function of time for field stars with $M_* = \frac{1}{2}\ M_\odot$ and $v_* = 1$ km/s.

mean energy $\langle E \rangle$ is equal to zero (Ambartsumian), or to $-E(t=0)$ (Oort). The evolution of $\langle E \rangle$ with time is shown in Figures 4 to 6, where the theoretical values due to Ambartsumian and Oort are also shown.

The study of histograms of the energies and their moments shows that the distribution function of the energies of the binaries evolves in two ways: (i) the distribution function is gradually shifted towards higher energies, and (ii) the spread of the energies

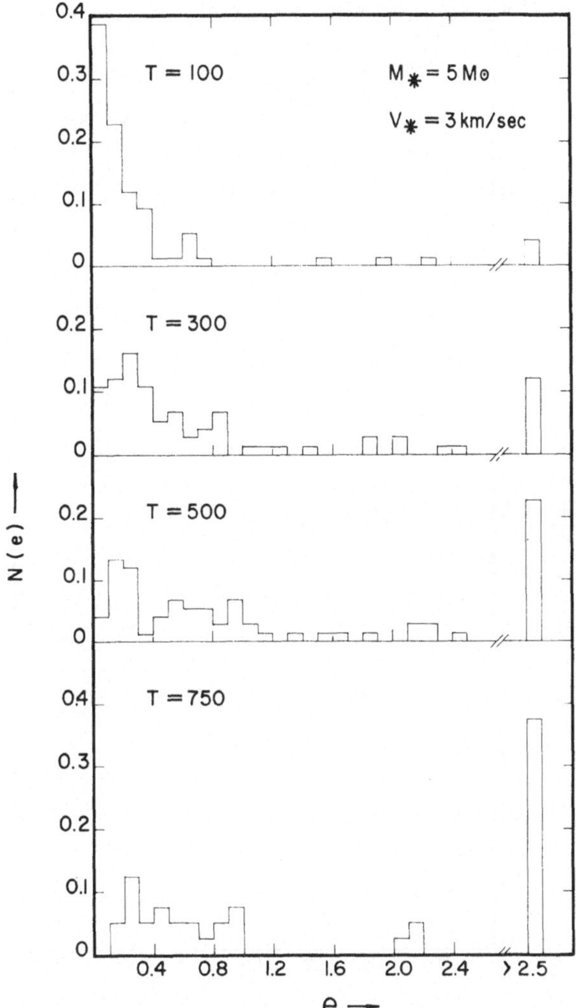

Fig. 10. Evolution in time of the distribution function of the eccentricities for field stars with $M_* = 5\ M_\odot$ and $v_* = 1$ km/s. Initially all pairs have $e = 0$.

increases steadily from an initial value of zero. Furthermore, the initial distribution function, in which all the secondaries have the same energy, evolves towards a distribution with a highly asymmetrical tail in the direction of positive energies. These results suggest that the use of the mean energy as an indicator of the degree of dissolution of a binary is a poor one. This is true for the following reasons:

(a) Shortly after the system begins to evolve, a few pairs acquire such large energies that the mean value of the energy $\langle E \rangle$ becomes equal to zero when most of the secondaries are still bound to their primaries.

(b) To define the dissolution time only in terms of the first moment of the distribution function of the energies i.e. of $\langle E \rangle$, is to ignore the role played by the con-

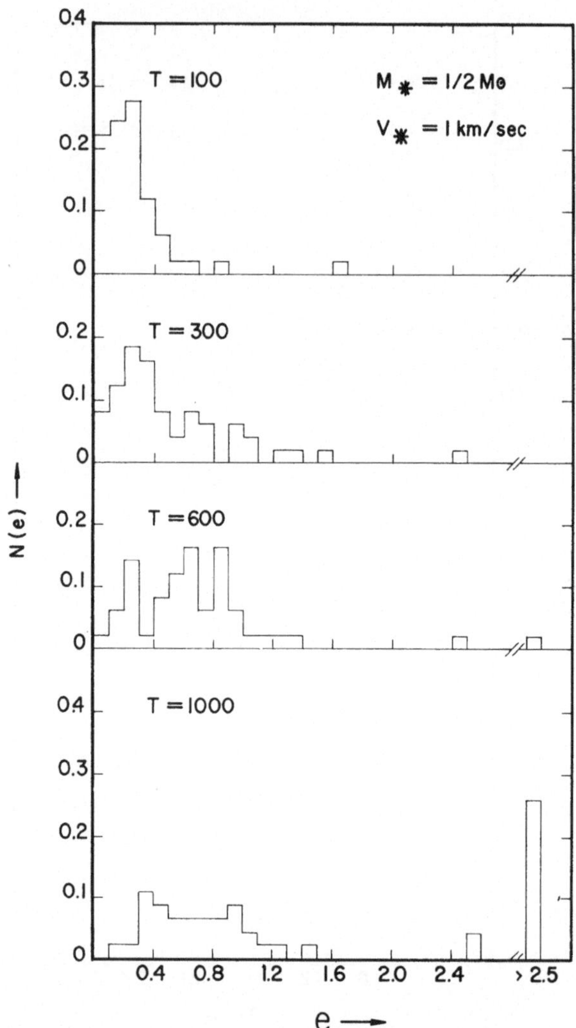

Fig. 11. Evolution in time of the distribution function of the eccentricities for field stars with $M_* = 5\ M_{\odot}$ and $v_* = 3$ km/s. Initially all pairs have $e = 0$.

tinuous increase in the dispersion of the distribution function of the energies.

(c) Since the concept of the dissolution time of a binary is a statistical one, the dissolution time should be defined as the time necessary for a given population of pairs (all with the same initial elements) to be reduced to a fraction of its original number. A particular pair in the sample is considered as dissolved when its binding energy becomes zero.

The mean energy $\langle E \rangle$ of the pairs increases linearly with time, although in some instances it does so in a very erratic way. This behavior, which is shown in Figures 4 to 6 agrees qualitatively with the predictions of the theories, but the rate of increase is smaller than the theoretical one.

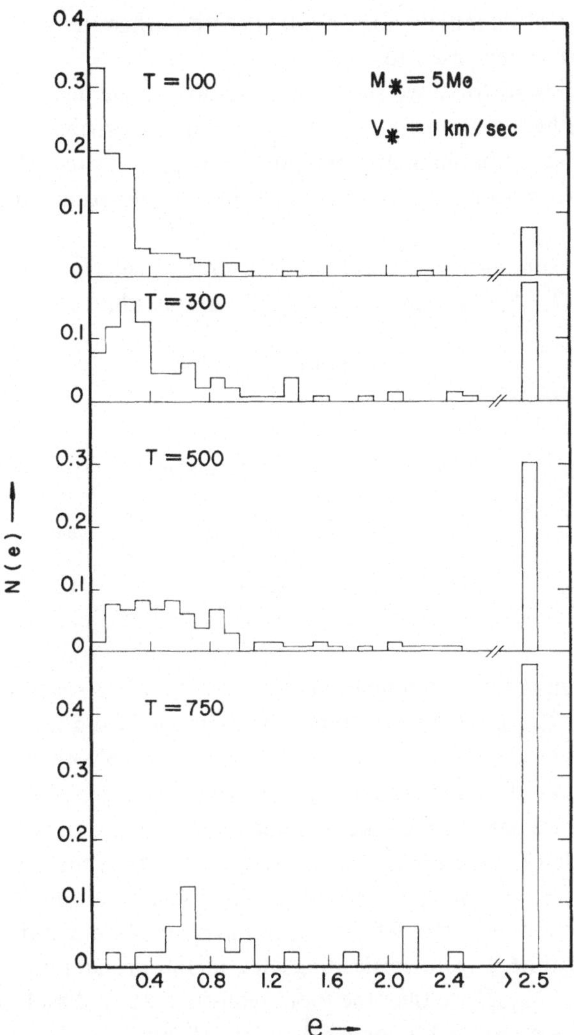

Fig. 12. Evolution in time of the distribution function of the eccentricities for field stars with $M_* = \frac{1}{2} M_\odot$ and $v_* = 1$ km/s. Initially all pairs have $e = 0$.

As the distribution function of energies broadens and drifts towards higher energies, more and more pairs cross the line of zero energy and become disrupted. As a result of this 'diffusion' the population of bound secondaries decreases steadily, as shown in Figures 7 to 9. It is interesting to note that the decrease of the population of bound secondaries follows an exponential law. In one instance, however (see Figure 7), it appears that the decay constant changed at about time 500. However, we note that from this time on, the trend of the population is more uncertain. This exponential decrease, which reminds us of the decay of radioactive nuclei, suggests that we should take as the dissolution time the time $T_{1/e}$ after which the population of bound systems has dropped to $1/e$ of its initial value, in other words, when the probability of a pair remaining bound has decreased to $1/e$.

Finally, in Figures 10 to 12 we show the evolution of the distribution function of excentricities. The histograms of e do not agree with the distribution $p(e)=2e$ found by Jeans (1919) and Ambartsumian (1937) on the hypothesis that the energies of the pairs follow a distribution $f(E)=\text{const}$. The distribution of eccentricities tends to be uniform.

In Table III we summarize the results of the present calculations. The theoretical dissolution times T_D due to Ambartsumian, Chandrasekhar and Oort are listed in

TABLE III

		T_D				
$M_*/M_\odot$	v_* [km/s]	Ambartsumian	Chandrasekhar	Oort	$T_{\langle E\rangle}$	$T_{1/e}$
5	1	66	380	244	~85	930
5	3	111	380	366	325	1065
0.5	1	400	380	1188	955	2400

non-dimensional units for 3 combinations of the masses and space velocities of field stars. The column $T_{\langle E\rangle}$ gives the dissolution times derived from the numerical experiment, following the definition used in the theories. The column $T_{1/e}$ gives the time when the population of bound secondaries has dropped to $1/e$ of its initial value.

From Table III we can see that the dissolution time $T_{1/e}$ differs significantly from the theoretically determined times. Furthermore, if we take the series characterized by $M_*=5\ M_\odot$ and $v_*=1$ km/s, we see that at the dissolution time $T_{\langle E\rangle}=85$ the fraction of bound secondaries is still 90% of the initial population. Clearly, this definition of the dissolution time gives a misleading idea of the probability of having a pair dissolve after time $T_{\langle E\rangle}$. If we take the more realistic time of dissolution $T_{1/e}$, we see that in all cases it is considerably longer than any of those given by the theories.

Another result of interest that emerges from the present calculations is that both dissolution times $T_{\langle E\rangle}$ and $T_{1/e}$ depend on the masses and velocities of the field stars; this contradicts Chandrasekhar's theory. However, it is also true that the time of dissolution $T_{1/e}$ is not as sensitive to the changes in M_* and v_* as predicted by Ambartsumian's and Oort's formulae.

The present calculations also showed that frequently one has to deal with binaries with semi-major axes larger than one parsec. For these pairs the effect of the gravitational field of the galaxy may be important and should be further investigated.

It is unfortunate that, because of insufficient computer time, we have not yet been able to simulate the velocities of the field stars in the solar vicinity; the present numerical results cannot be scaled safely from $v_*=1$ km/s to $v_*=20$ km/s. Thus, we cannot yet answer the important question as to whether the present cloud of comets around the sun is a relic from the days of formation of the planetary system.

To summarize the main results of the present paper we would like to list the following conclusions:

(1) With the passage of time $\langle E \rangle$ grows linearly and $\sigma_E \to \infty$.

(2) The distribution of eccentricities after one dissolution time does not follow Jeans-Ambartsumian's law, but tends instead towards a uniform distribution.

(3) The fraction of bound binaries decreases exponentially with time.

(4) The time of dissolution of a pair should be defined as the time $T_{1/e}$ when the probability of a pair remaining bound has decreased to $1/e$.

(5) $T_{1/e}$ does depend on M_* and v_* but not as sensitively as predicted by some of the theories.

(6) The dissolution times $T_{1/e}$ are considerably longer than the theoretical ones.

Acknowledgements

We gratefully acknowledge that the computations were done on the Burroughs 5500 of the Centro de Cálculo Electrónico, U.N.A.M.

It is a pleasure to thank Dr. Ivan King and Miss Christine Allen for a number of useful discussions.

References

Ambartsumian, V. A.: 1937, *Russ. Astron. J.* **14**, 207.

Chandrasekhar, S.: 1944, *Astrophys. J.* **99**, 54.

Cruz-González, C. and Lecar, M.: 1968, in *Colloque sur le Problème des N Corps*, p. 209 (éditions du CNRS, Paris); *Bull. Astron., 3rd Serie* **3**, 209.

Jeans, J. H.: 1919, *Monthly Notices Roy. Astron. Soc.* **79**, 408.

Oort, J. H.: 1950, *Bull. Astron. Neth.* **11**, 91.

Strömgren, E.: 1914, *Publ. Obs. Copenhagen* **19**, 193.

ON THE REPRODUCIBILITY OF RUN-AWAY STARS FORMED IN COLLAPSING CLUSTERS*

CHRISTINE ALLEN

Instituto de Astronomía, Universidad Nacional Autónoma de México

and

ARCADIO POVEDA

Instituto de Astronomía, Universidad Nacional Autónoma de México, and Observatorio Astrofísico Nacional, S.E.P.

Abstract. To test the stability of the trajectories of run-away stars we present the results of a comparative study of 26 star clusters involving very strong encounters. Each one of these clusters was computed with different time steps, different techniques of integration and on different machines. We confirmed that, in general, the values of the energies, velocities and positions for the individual stars are not reproducible from run to run. We found, however, that in all the cases that have been tested, the run-away star preserved its energy, velocity and position from one run to another. It was also found that escapers are more reproducible than stars with negative energy.

1. Introduction

Run-away stars form an important class of objects, discovered by Blaauw (1961) when he found that the space velocities of some O and B stars were very large compared with the typical velocities of stars of the same spectral type. If one draws the distribution of velocities of early type stars (O B stars) one finds that they tend to cluster at velocities close to zero, with a mean value of 6.8 km/s in one direction; their numbers decrease very fast towards higher velocities. However, at velocities larger than 30–35 km/s a second group of O B stars emerges, whose velocities spread all the way from the lower limit of 30-35 km/s up to 200 km/s.

Run-away stars are interesting because they cannot be considered simply as the tail of the velocity distribution of slowly moving O B stars. Furthermore, about 20% of the O stars are high velocity stars, while only 2% of the B stars fall in this group.

To understand the characteristics of run-away stars, Blaauw advanced the hypothesis that the explosion of the primary star in a massive binary will release the secondary with a high space velocity; when the secondary star happens to be an O B star, a run-away star will be produced.

In an alternate model (Poveda *et al.*, 1967), the high velocities of run-away stars are the result of strong dynamical interactions. These occur during the early stages of formation of small compact clusters of O B stars. To test this hypothesis, we integrated numerically more than 150 examples of parent clusters. The various initial conditions were inspired by our present knowledge of the physics of star formation. We assume, in fact, that a compact cloud fragments into a few proto-stars of equal

* Part of this work was done at the Department of Aerospace Engineering and Engineering Mechanics of the University of Texas at Austin during a summer stay of C.A.

M. Lecar (ed.), Gravitational N-Body Problem, 114–123. All Rights Reserved

masses. Since any two proto-stars cannot occupy the same volume of space, their positions resemble those of cannon balls closely packed in a sphere. We take, therefore, initial positions corresponding to the vertices of regular polyhedra, plus random deviations from the latter of the order of 10% of the initial radius of the cluster. We computed also a number of cases without random deviations. We take initial velocities at random in direction and magnitude, limited to no more than $\frac{1}{2}$ to 1 km/s. We find that, on the average, each cluster containing 5 or 6 stars produces somewhat more than one run-away star.

The success of reproducing by *N*-body computations the observed properties of run-away stars was uncertain on account of the mounting evidence that results of individual stars in *N*-body calculations are not reproducible; this became particularly evident after the comparative study of integration of the standard 25-body case (Lecar, 1968). In the latter, for instance, the escapers vary considerably from one method of integration to the next. Thus, it becomes necessary to check to what extent the run-away stars produced in our computations are the result of numerical instabilities and, in particular, what is the degree of reproducibility that run-away stars show relative to the usual escapers. In the remaining part of this paper we describe the results of numerical simulations performed to answer the previous questions.

2. The Numerical Simulation

To study the reproducibility of run-away stars we made a number of comparisons of detailed results obtained when computing the same cluster with different steps of integration, different numerical techniques, and on different machines. The basic material was the pool of more than 150 cases considered in our study on the origin of the run-away stars (Poveda *et al.*, 1967; Allen and Poveda, 1968). These clusters were integrated by von Hoerner's technique, as used by him in his pioneer paper on *N*-body calculations (von Hoerner, 1960). The comparisons were made for the following series of computations:

(a) A group of 15 cases selected at random from the pool of the basic material was computed anew with the same program and on the same machine as the one used for the basic material, but with smaller steps of integration.

(b) A group of 11 cases from the same pool was selected on account of having very close encounters (10^{-5} to 10^{-6} of the original radius of the cluster) and large errors in the total energy of the cluster. Each one of the clusters in this group was computed again with the program developed by the Department of Aerospace Engineering and Engineering Mechanics of The University of Texas at Austin using a 7th-order Runge-Kutta-Fehlberg technique.

(c) The same group of 11 cases was computed again using the same technique of integration as in (b), but regularizing the equations of motion of the closest pair by means of the Kustaanheimo-Stiefel transformation.

The computations of the basic material as well as that of Series (a) were done at the Computing Center of the National University of Mexico. The computations of

Series (b) and (c) were done at the Computing Center of The University of Texas at Austin.

To study the reproducibility of the numerical computations we give in Tables I and II a summary of our results.

TABLE I

Reproducibility of run-away stars with different time steps

15 Single precision

*	E	v	r
1	5.8	5.4	0.8
2	*65.6*	*11.5*	*4.0*
3	14.1	6.7	1.1
4	−148.0	1.3	0.9
5	1.5	4.5	0.8
6	−119.0	7.8	0.9

15 Double precision

*	E	v	r
1	6.1	5.6	1.1
2	*65.6*	*11.5*	*4.0*
3	−151.1	6.6	1.3
4	− 58.7	2.9	1.0
5	9.9	4.9	1.2
6	−126.2	9.6	1.3

$E = -17.28$

Single precision: $dE = 2.62 \times 10^{-2}$, $t = 0.715$

Double precision: $dE = 2.58 \times 10^{-2}$, $t = 0.715$

27 Single precision

*	E	v	r
1	0.7	2.2	1.6
2	−52.8	5.1	1.5
3	1.5	2.4	2.8
4	8.8	4.6	2.7
5	*18.9*	*6.3*	*6.0*
6	−62.8	2.4	1.5

27 Quadruple precision

*	E	v	r
1	−45.5	1.0	1.0
2	−18.9	8.3	1.2
3	1.2	2.3	2.8
4	−28.4	4.6	1.2
5	*18.9*	*6.3*	*6.0*
6	−33.0	2.4	1.2

$E = -17.25$

Single precision: $dE = 6.75 \times 10^{-3}$, $t = 1.298$

Quadruple precision: $dE = -1.99 \times 10^{-5}$, $t = 1.297$

28 Single precision

*	E	v	r
1	− 96.3	16.8	0.6
2	−101.0	2.7	0.6
3	14.5	6.3	0.6
4	*49.3*	*10.0*	*4.0*
5	1.3	2.8	1.4
6	12.3	5.4	2.4
7	−203.0	8.2	0.6

28 Quadruple precision

*	E	v	r
1	−126.1	12.0	0.4
2	−171.3	7.9	0.4
3	− 39.2	9.8	0.4
4	*49.4*	*10.0*	*4.0*
5	1.1	2.8	1.4
6	12.1	5.4	2.4
7	− 44.9	1.7	0.9

$E = -29.65$

Single precision: $dE = 5.10 \times 10^{-2}$, $t = 0.675$

Quadruple precision: $dE = -6.11 \times 10^{-4}$, $t = 0.674$

Table I (continued)

30 Single precision

*	E	v	r
1	− 29.1	3.3	0.7
2	24.2	7.3	2.1
3	*63.4*	*11.4*	*4.0*
4	− 88.5	9.0	0.9
5	− 32.7	2.0	0.7
6	− 120.5	4.3	0.9

$dE = 4.10 \times 10^{-4}$
$t = 0.732$

30 Quadruple precision

*	E	v	r
1	− 21.8	3.6	0.7
2	1.0	2.9	0.9
3	*63.4*	*11.3*	*4.0*
4	− 200.0	20.1	1.0
5	− 5.9	2.3	1.2
6	− 286.7	15.1	1.0

$E = -17.26$
$dE = 4.72 \times 10^{-2}$
$t = 0.733$

TABLE II

Reproducibility of run-away stars with and without regularization

82 *R*

*	E	v	r
1	− 333.478	15.244	1.839
2	*17.999*	*12.784*	*4.000*
3	88.793	13.644	2.391
4	3.539	5.419	1.989
5	19.716	12.917	3.993
6	− 334.984	15.148	1.840

$t = 0.7336$
$dE = -4.5983 \times 10^{-6}$

82 *N*

*	E	v	r
1	9.480	5.822	2.589
2	*17.986*	*12.783*	*3.999*
3	25.526	8.168	2.406
4	− 195.237	13.751	1.501
5	19.702	12.917	3.993
6	− 255.894	8.232	1.501

$E = -17.2928$
$dE = -1.8106 \times 10^{-4}$

86 *R*

*	E	v	r
1	− 15.239	16.051	0.820
2	− 3.238	4.712	0.632
3	− 162.532	6.695	0.871
4	*122.580*	*15.724*	*4.000*
5	− 180.150	3.216	0.873
6	− 115.949	7.590	0.826

$t = 0.6300$
$dE = -5.5176 \times 10^{-7}$

86 *N*

*	E	v	r
1	14.728	6.196	1.310
2	− 766.420	31.209	0.734
3	− 13.104	2.820	0.675
4	*122.583*	*15.724*	*4.000*
5	− 630.224	35.306	0.735
6	− 22.880	2.706	0.824

$E = -17.2928$
$dE = 1.0559 \times 10^{-4}$

41 *R*

*	E	v	r
1	− 152.102	3.999	1.119
2	− 141.549	6.198	1.118
3	*107.223*	*14.699*	*3.987*
4	− 21.505	5.020	1.057
5	4.339	4.841	0.700

$t = 0.4362$
$dE = -4.0363 \times 10^{-7}$

41 *N*

*	E	v	r
1	− 152.021	3.971	1.119
2	− 141.437	6.185	1.118
3	*107.223*	*14.699*	*3.987*
4	− 21.505	5.020	1.057
5	4.339	4.840	0.700

$E = -22.0249$
$dE = 1.3539 \times 10^{-3}$

Table II (continued)

164 *R*				164 *N*			
*	E	v	r	*	E	v	r
1	247.161	22.338	1.953	1	− 0.809	2.905	0.520
2	−923.301	10.249	1.022	2	−486.886	12.464	1.180
3	*294.624*	*24.317*	*4.000*	3	*294.597*	*24.316*	*4.000*
4	−696.318	23.643	1.023	4	137.717	16.775	1.653
5	56.556	10.902	0.949	5	− 3.799	1.677	0.277
6	22.876	7.377	0.947	6	−529.792	8.338	1.181
	$t=$ 0.5412				$E=-17.2947$		
	$dE=-6.7166\times10^{-5}$				$dE=$ 4.9600×10^{-3}		

In Table I we list for a representative sample of the stars belonging to the 15 cases of Series (a) the values of their energies E, velocities v, and distances r to the center of mass of each cluster. The numbers listed correspond to the time when the run-away has reached distance 4 or larger (4 times the value of the initial radius of the cluster). In non-dimensional units a run-away star has a velocity larger than $v=5.5$. In Table I we list, for each case, two or three sets of final values. The first one, called single precision, corresponds to the computations carried with the time step used in the basic material (Poveda *et al.*, 1967). The runs named double or quadruple precision were made with time steps one half or one fourth as small as the ones used in the single precision runs.

Following the same format of Table I, we list in Table II the final results of a sample of the stars of the eleven cases of Series (b) and (c). The letter R after the case identi-

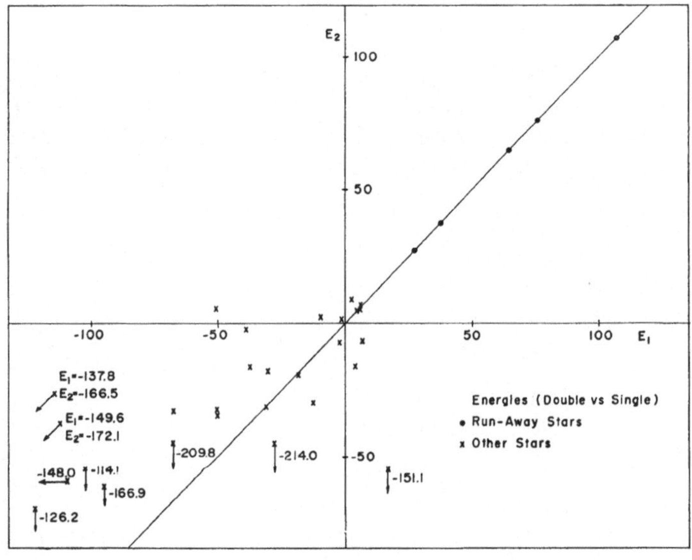

Fig. 1.

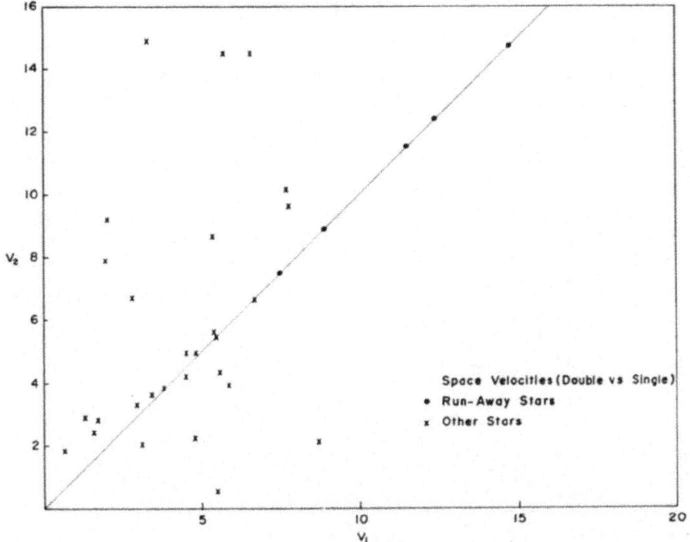
Space Velocities (Double vs Single)
Run-Away Stars
Other Stars
V_2
V_1

Fig. 2.

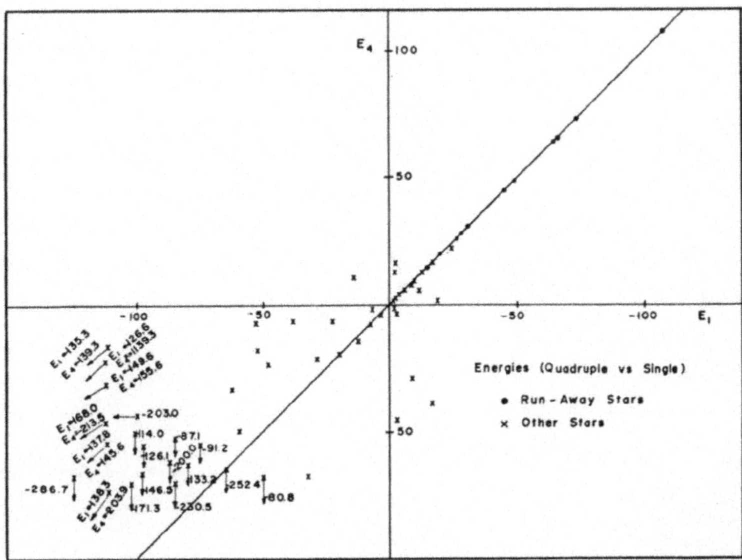
Energies (Quadruple vs Single)
Run-Away Stars
Other Stars
E_4
E_1

Fig. 3.

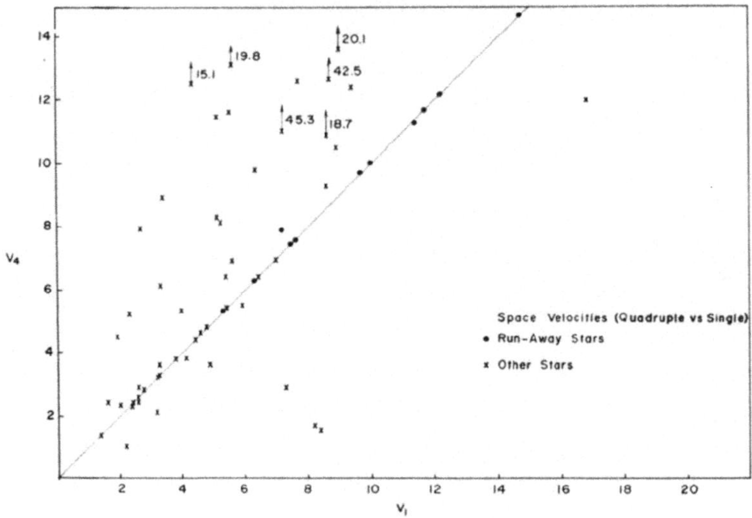

Space Velocities (Quadruple vs Single)
• Run-Away Stars
x Other Stars
V_4
V_1

Fig. 4.

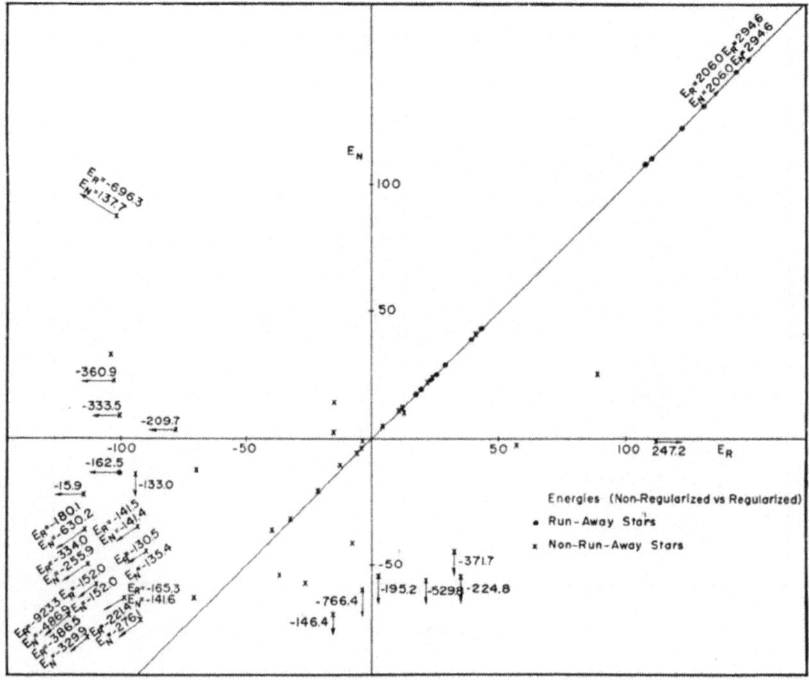

Energies (Non-Regularized vs Regularized)
• Run-Away Stars
x Non-Run-Away Stars
E_N
E_R

Fig. 5.

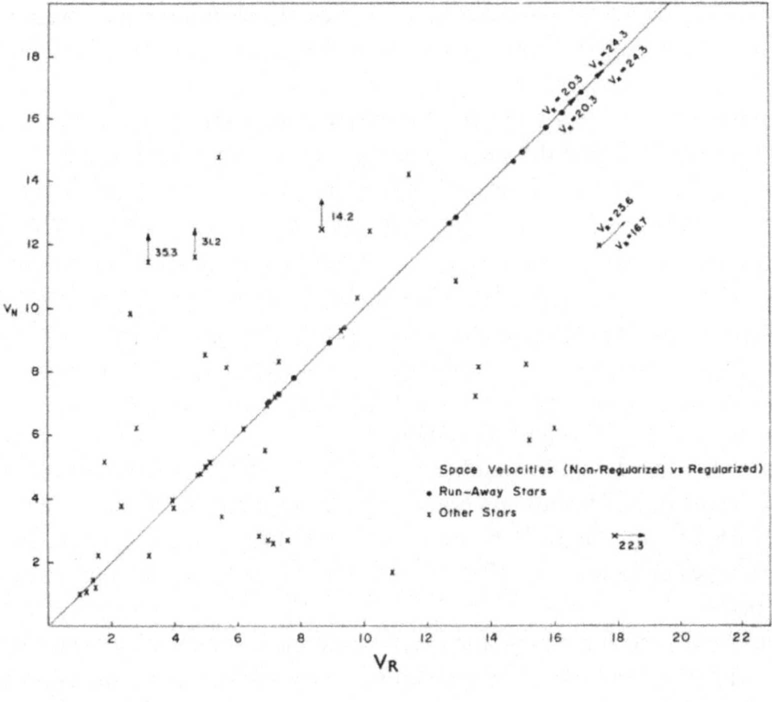

Fig. 6.

fication means that it was run with regularization; the letter N stands for a run computed without regularization.*

The material of Tables I and II is shown in graphical form in Figures 1 through 6. For each star, a point is determined by pairs of values of the energy, or velocity, as obtained in the different runs of Series (a), (b) and (c). In all of these graphs, run-away stars are marked with filled circles; when more than one run-away per cluster is formed, we mark with a filled circle only the one formed first. The remaining stars are indicated with crosses.

3. Results and Discussion

An examination of Tables I and II shows that, in general, the individual values are not reproducible when the same case is computed with different steps of integration or with different methods. This confirms the well-known experience of workers in this field, and is consistent with the general trend found in the comparative study of the 25-body case (Lecar, 1968).

The present results differ, however, from previous ones in two important aspects:

(1) Some cases – like number 41 and others – show a high degree of reproducibility, to the point that all runs lead very closely to the same detailed values of E, v and r for every star in the cluster. These cases were run backwards in time and the initial

* The authors will be glad to supply, on request, the initial conditions for all cases.

positions and velocities were recovered. In particular, when these cases were run with regularization, the initial values were recovered with errors in the third decimal or better.

(2) Run-away stars show a remarkable degree of reproducibility when their parent clusters are computed with different time steps and different methods. We found this property in all of the 26 different cases that have been run independently at least three times. When more than one run-away star is produced, however, the one that is always recovered is the one that is formed first. In some cases – like numbers 86, 164, etc. – the only star whose properties are reproduced from run to run is the run-away star. Interestingly, even the tight pair which is always formed in these cases is not reproduced. This is curious, because it is the negative energy of the close pair that compensates for the large positive energy of the run-away star.

The reproducibility of run-away stars is so remarkable that even in one particular case, run with a very low precision*, the error in the energy of the run-away star was only of a factor of 2. This low precision run had an error in the total energy of the cluster of $\Delta E/E = 10^9$; the various stars at the end of the computation showed errors in their energies of factors of 10^3 to 10^9, yet the run-away star had an error of only a factor of 2.

We found one further result of interest in the 26 cases presented here: slow escapers, i.e. stars with $E>0$ but $v<5.5$, show statistically a higher degree of reproducibility than the stars left behind with negative energy.

The properties described above can be easily seen in Figures 1 through 6, where the filled circles which correspond to the run-away stars fall on the 45° line of reproducibility. A number of other stars also fall on this line, but the scatter for non-run-away stars is considerable.

At the present moment, it is hard to find a convincing explanation for the reproducibility of run-away stars. It is also hard to see why some initial conditions lead to a relatively stable dynamical evolution. We may conclude, however, that a significant range of initial conditions exists, which, through very strong triple and quadruple encounters, leads to the formation of high velocity stars (run-away stars). The latter exhibit a degree of reproducibility unknown in previous N-body calculations.

Acknowledgements

We wish to express our gratitude to Professors Victor Szebehely and Dale G. Bettis of the Department of Aerospace Engineering and Engineering Mechanics of The University of Texas at Austin for the use of their programs and facilities, and for the computing time necessary to run Series (b) and (c). Our thanks are due also to Mr. Otis Graf for his valuable help. Christine Allen is particularly grateful for the hospitality granted to her during her summer stay at the Department of Aerospace Engineering and Engineering Mechanics.

* We are indebted to Mr. C. Cruz for running this case.

References

Allen, C. and Poveda, A.: 1968, *Astron. J.* **73**, 86.
Blaauw, A.: 1961, *Bull. Astron. Inst. Neth.* **15**, 265.
Lecar, M.: 1968, 'A Comparison of Eleven Numerical Integration of the Same Gravitational 25-Body Problem', in *Colloque sur le Problème des N Corps* (ed. by J. Delhaye), Paris, France, p. 91.
Poveda, A., Ruiz, J., and Allen, C.: 1967, *Bol. Obs. Tonantzintla y Tacubaya* **4**, 86.
von Hoerner, S.: 1960, *Z. Astrophys.* **50**, 184.

NUMERICAL EXPERIMENTS ON PAIR CORRELATIONS AND ON 'THERMODYNAMICS'

R. H. MILLER

University of Chicago

Abstract. A 'conventional' *n*-body calculation was constructed to carry out a number of numerical experiments on small stellar systems. Two of these experiments will be reported here, although something of the others will creep in. All these experiments refer to a 32-body system, in which the equations of motion were handled 'exactly', with the usual checks on constancy of the ten first integrals of motion. Because the emphasis was on the experiments, the calculation was constructed without particular attention to running speed or minimum storage requirements. Otherwise, the calculation was quite conventional.

1. 'Thermodynamics'

An *n*-body calculation provides one way of attempting the thermodynamic 'gedanken-experimente' of using a stellar system as the thermodynamic medium in a Carnot engine. These experiments have not gotten as far as putting the system into a Carnot engine, although routines have been constructed to do that. The work to be reported was undertaken to explore the practical problems that might be encountered. How long would a system have to sit to reach an equilibrium state? How fast could a cylinder wall move and still carry the system through a sequence of quasi-equilibrium states? Or is an equilibrium state attainable at all? The interaction of a stellar system with a 'thermodynamic enclosure' – even one without moving walls – can provide a good starting-point.

The system was placed in a box. Whenever a star struck the box, it was caught and thrown back into the box. It was reprojected randomly directed inward, with a randomly selected speed such that its new kinetic energy would be exponentially distributed with a mean value that was called the 'temperature' of the box. The star was re-projected from the same point, so there was no change of potential energy. Changes of momentum and of kinetic energy were tallied. Finally, the box was endowed with a 'heat capacity', such that the difference of kinetic energy received and given up could cause a change in the 'temperature' of the box. The exponential distribution of kinetic energy is appropriate to a Maxwellian velocity distribution. The random velocities of re-projection are satisfactory for all parts of the Carnot cycle, although a specular reflection might be easier to apply to the adiabatic part.

Two sequences of experiments were run. For the first sequence, the box was 'cold' – its temperature was always zero. The second sequence used non-zero temperatures and various heat capacities. The final equilibrium state of the second sequence should define a thermodynamic temperature for the stellar system by the classical definition of temperatures from the equilibrium of two systems in contact, if indeed such an equilibrium can be reached. It is not clear that an equilibrium can be reached, even by ascribing a negative heat capacity to the box.

M. Lecar (ed.), Gravitational N-Body Problem, 124–130.

The system should quickly readjust itself to something it likes from any initial condition. In fact, initially the particles randomly filled a rectangular parallelopiped in the phase space whose projection onto the configuration space and onto the momentum space was a cube in each space. The cube edge in configuration space was adjustable, but always within the box, while the cube edge in momentum space was chosen to make the ratio of kinetic to potential energy correspond to the virial theorem.

The cold box was typical of most experiments. Since there is an infinite energy store available, arbitrarily large amounts of energy can be transferred to the enclosure. The history of energy transfer to the enclosure in the longest run is shown in Figure 1 (solid curve; the dashed curve shows a shorter run with a different starting condition to illustrate the reproducibility of the experiments). The rate of energy transfer to the enclosure is nearly constant over the duration of the experiment – it does not diminish

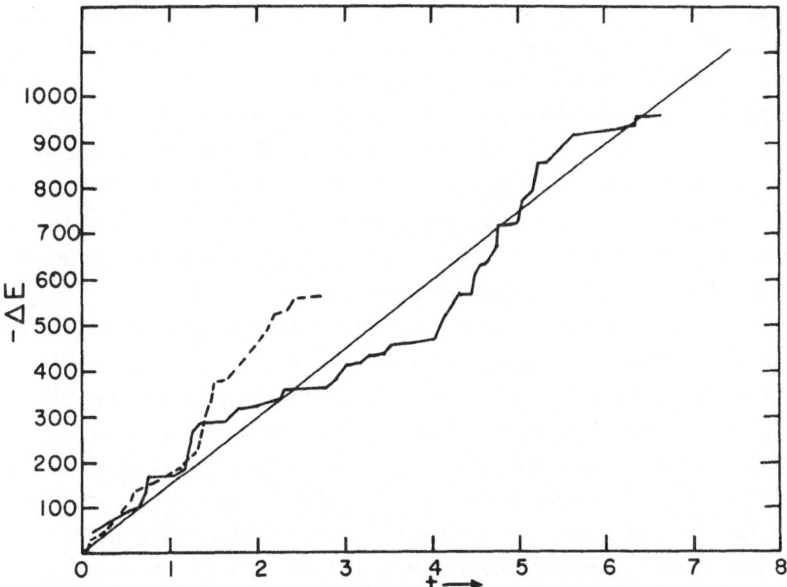

Fig. 1. Energy transferred to a 'cold' enclosure by a 32-particle stellar system. The solid line is one calculation, the dashed line another from similar initial conditions. The initial kinetic energies were about 250 in the units of the ordinate. The time-units of the abscissa are approximately crossing-times of the initial state.

or accelerate appreciably over the duration of the run – even though the total energy transferred to the enclosure was about 4 times the initial kinetic energy present in the system. (The initial kinetic energy was about 250 in the units of the ordinate). It was characteristic of these runs that the virial theorem was approximately satisfied without allowing for 'surface pressure' terms. Thus, near the end of the run, the stellar system was very small and compact, somewhere within the box, with very high kinetic energy.

The time-scale in Figure 1 is the time-variable of the differential equations, which is the time-scale of an observer looking at the cluster from outside. In time-units appropriate to the cluster (crossing-times, for example), the rate of energy transfer decreases. The crossing-time (estimated from the virial theorem) was about 1 time-unit at the beginning of the calculation, but was about $\frac{1}{10}$ time-unit at the end.

There is no preference for any one star to fly out – the identity of stars that strike the box seems random. Occasionally, it seems that the same one comes out repeatedly, but the effect is never significant. At the end of the long run, the selection of stars that have struck the enclosure seems to be drawn from the appropriate multinomial distribution.

There was very little change in the average amount of energy carried by a star that struck the enclosure during the calculation; a possible slight trend toward increasing the amount of energy carried is heavily masked in the fluctuations. The average energy delivered to the box was about the same as the average initial kinetic energy. Similarly, the rate at which stars struck the enclosure did not show a detectable trend. The distribution of intervals between times when stars struck the box seemed random – it looked like a Poisson process with a fixed rate at about 10–12 per initial crossing time. There was no detectable serial correlation of intervals.

The zero-temperature case is interesting because the stars flying out are caught, their kinetic energy removed, and they are merely released where caught. They start falling toward the center of the remaining cluster – and have very little chance to avoid passing through the most dense part. This arrangement is very efficient at extracting energy from the cluster. It is almost a Maxwell demon.

The rate of energy transfer to the box was about half the total initial kinetic energy in the system per initial crossing time.

The second series of experiments, that with nonzero box temperature, did not tend toward an equilibrium. It tended to have the star system give up energy to the box. With the box temperature colder than the stellar system, this is what would be expected. But with the box hotter than the system, the system still gave up energy to the box. This is not in the direction expected of a system with a definable 'specific heat'. (See the discussion by Lynden-Bell and Wood, 1968, with which a good deal of this work can be compared). The general trends are: (1) with the box temperature up to about twice that of the average star, the energy transfer to the box occurred at nearly the rate observed with a cold box. (2) With the box temperature about four times that of the average star, the rate was appreciably diminished. (3) With the hotter box, the fluctuations in the energy transferred were quite large – fluctuations as great as half the initial kinetic energy of the system were not unusual, in spite of the bound set on the potential energy by placing the system in a box. A general trend of transferring energy to the box was superposed on the large fluctuations.

Qualitatively, the results of these experiments were anticipated. It was expected that fluctuations in the complexion of a stellar system would lead to localities that were hot, and in which a 'heat imbalance' would exist compared to the rest of the system. The questions to be answered by experiment had to do with whether these fluctuations would last long enough to lead to a locally unstable situation. The usual continuum theories (see, e.g., Lynden-Bell and Wood, 1968; Larson, 1970) tend to minimize or to ignore these fluctuations. (Formally, they are there, hidden in some of the higher moments of the distribution functions; when the sequence of moment equations is closed by relating high-order moments to lower order moments, the fluctuations are ignored at that order of the moment equations). Fluctuations should become less important as the number of particles increases, but the amplifying effect of the localized 'runaway' makes fluctuations have a much larger effect than expected – their importance does not diminish like the square root of the particle number, for example. A naive interpretation of these experiments would lead to the conjecture that a stellar system might *always* be unstable against such 'runaways'. The importance of localized fluctuations does not seem to have been stressed in discussions applied to stellar systems.

Similarly, with the cold box, the unanticipated result is quantitative. The *rate* of energy transfer to the box is surprising. It goes much faster than Larson's (1970) discussion would lead you to expect.

2. Pair Correlations

Pair correlations are elusive quantities that are expected in stellar systems, but which have not even been conclusively demonstrated to exist, much less measured. It has been widely accepted, from the earliest days of n-body calculations, that binaries form in the calculations. This implies the existence of pair correlations. But that statement has not been turned into anything quantitative.

In the BBGKY hierarchy, the two-body distribution function can be expressed as the product of two one-body terms plus a pair correlation term: $f^{(2)} = f^{(1)} \cdot f^{(1)} + g^{(2)}$. The term $g^{(2)}$ is the pair correlation. In any real situation, the pair correlation is heavily masked by the product of one-body distribution functions. These experiments were undertaken in hopes of making some statements about the form of $g^{(2)}$; however, it seems that about all that can be done is to show that there is probably something there.

A 32-body system was started from an initial condition in which the angular momentum was accurately zero, so it should reduce to something spherically symmetrical in a rather short time. The initial-state angular momentum was about 10^{-10} of what it would have been had simple random loadings been used as they were for the other experiments described in this paper. Under spherical symmetry, the two-body terms can depend on 8 invariant quantities and the single-particle terms on 3 each. These quantities won't be listed because the important point is to find convenient physically meaningful combinations of them rather than just to provide a simple enumeration. The game is to find combinations that should behave differently with a nonzero $g^{(2)}$ from what would be expected if it were zero-ideally, combinations whose behaviour should be radically different. This is not particularly simple, and I am not at all sure that I have used anything like the best discriminant.

Taking the binaries as a model, it might be expected that the $g^{(2)}$ term would manifest itself through a tendency in configuration space for stars to be closer than a random distribution and in momentum space for them to be farther apart, but farther apart in a certain specified way. Of course, there could be a kind of 'anti-binary' pairing that was the other way around – the tests used so far would probably not find such a term. The tests that I have been able to devise are reasonably specific as to a model; they are certainly not uniformly powerful against all alternatives.

In configuration space, the two stars of a pair and the centroid of the system form a triangle. The orientation of that triangle is unimportant because of the assumed spherical symmetry, but the lengths of its three sides or the lengths of the two sides that intersect at the centroid together with their included angle are invariants. In the absence of pair correlations, the cosine of that angle should be uniformly distributed. The test used so far consists in seeing whether the cosine is uniformly distributed.

Over some 15 different experiments, a weak tendency toward an excess of cosines above about 0.9 has shown up. In any one case, there is less than a 1% chance that the distribution is uniform – in some cases it is very much less than that. A typical case is probably at about the 0.1% level. There is no experiment that doesn't show an effect at the 1% level. Thus, it seems quite likely that something is there. Further statements are not possible at this time; in particular quantitative measures or possible functional forms for $g^{(2)}$ can't be obtained from the data as yet.

Some further observations on these experiments:

(1) There is considerable variation from experiment to experiment. The effect shows up clearly in spite of this variation.

(2) A modification of the calculation (described below) to cause an iterative refinement of the first ten integrals of the motion made no noticeable difference so far as this experiment is concerned.

(3) All experiments showed weak correlation through the cosine of the included angle criterion. Particular cases were:

(a) Initially uncorrelated systems developed a correlation.

(b) Systems with initial correlation retain some correlation, if the initial correlation is moderate.

(c) Systems with strong initial correlation lose some of the initial correlation, and relax toward about the same degree of correlation as the initially uncorrelated systems achieve.

(d) Some associated pairings of the strongly correlated systems remain associated for a long time. These initial correlations looked something like binaries.

(4) Very few bound binaries formed, according to a test based on the ratio of kinetic to potential energy for the pair, treated as if it were isolated. This ratio is too large for binding in all but a very few cases. Even pairings that remained associated were not genuine binaries by this test.

(5) Two cases run with the near forces cut off showed the same kind of correlation as the other normal experiments.

(6) One case started with very tight correlation in the initial state weakened the

correlation and extended the range of the correlation terms farther and farther as the calculation proceeded.

(7) With 32 stars, there are 496 pairs. The effect shows up with 25–50 pairs having the cosine of the included angle greater than about 0.93, where the expected number is about 17. The distribution would be binomial in the absence of an effect.

(8) An experiment, started with much too tight a correlation in the initial state, developed like a stellar system consisting of binaries – the binaries remained bound as the system evolved. This, of course, is just what real stellar systems do.

(9) The correlation developed typically within about a crossing-time and persisted thereafter to the end of the calculation. The calculations usually were carried out only to 2 or 2.5 crossing-times.

3. Other Experiments

Two other experiments were run which will be mentioned only briefly.

A. ITERATIVE REFINEMENT OF THE ENERGY AND ANGULAR MOMENTUM

The first ten integrals of the motion can be conserved to arbitrary precision by iterative refinement. The representative point in the phase space is shifted back onto the hypersurface in which these integrals are conserved after it has drifted off the hypersurface because of inexact integration. The problem, of course, is that the exact point on the hypersurface to which the computed point should be moved is not known. Ambiguity is avoided by using a least-squares criterion. The local gradient of each of the first ten integrals consists of quantities that are readily available: constants, coordinates or velocity of a particle, or components of the force acting on a particle. The least-squares condition can be met by restricting the movement of the computed point to the subspace spanned by the gradients of the integrals. A vector displacement should be a linear combination of these gradients. The tool that accomplishes this is the generalized inverse (since the matrix is not square).

This procedure has been incorporated into the calculation. It turns out to be unnecessary to refine the configuration and velocity centroids. The remaining matrix is $4(6n)$; the computation of the generalized inverse is dominated by the 4 and not by the $6n$, so the amount of computation required to compute the corrections is not prohibitive. One iteration will usually hold the energy to ± 0.01 (with total energy around 200) and angular momentum components to about $\pm 10^{-4}$. Two iterations typically hold about 10^{-10} in energy and 10^{-13} in angular momentum. Occasionally, extra iterations are required. The calculation was run with a tolerance of 10^{-15} for the sum of squares of the deviations of energy and of the three components of angular momentum. Normally, two iterations sufficed, occasionally three were required. Once, in an extremely close encounter, the calculation was terminated on failure to reach the tolerance in 6 iterations. This happened too infrequently to require remedies.

Experimentally, it is not clear that the refinement improves the calculation appreciably. I have the *impression* that there are fewer close encounters in the calculation with the integrals better conserved. However, two calculations started from the same initial values, one with and the other without the iterative refinement, soon become two distinct calculations which cannot really be compared with one another. It would be very difficult to strengthen the impression into something like a reasonable experimental result.

B. PARALLEL CALCULATIONS COMPARING TWO SYSTEMS

This sequence of experiments was undertaken as an extension of my earlier work along these lines, as reported at the Thessaloniki conference, for example (Miller, 1966). This time, a perturbation calculation is carried out in which the equation of motion for the difference-vector (in the phase-space) is integrated along with the original system. The advantage is that several calculations can be carried along simultaneously at little extra cost.

Three kinds of initial displacements were used: arbitrary, along the trajectory, and perpendicular to the trajectory but in the integral hypersurface. Experimental results show: (1) The projection of the difference-vector is preferentially along the trajectory through the phase space, but the direction-cosine of the difference-vector onto the trajectory swings from positive to negative. (2) The projection of the difference-vector onto the gradient of the energy is essentially random. It drifts about slowly, but never gets very large. (3) Displacements along the trajectory can develop difference-vectors with components not along the trajectory only through numerical errors. They develop such components.

The iterative refinement of the first integrals was developed primarily to use with this series of experiments. The magnitude of the difference-vector grows at about the same rate whether the integrals are conserved or not. The effects of numerical imprecision noted earlier cannot be avoided merely by using computational methods that conserve the first integrals better.

Acknowledgements

The calculations reported in this paper were carried out at the Kitt Peak National Observatory. Their hospitality and generosity in making the computer available for this work is gratefully acknowledged.

References

Larson, R. B.: 1970, *Monthly Notices Roy. Astron. Soc.* **147**, 323.
Lynden-Bell, D. and Wood, R.: 1968, *Monthly Notices Roy. Astron. Soc.* **138**, 495.
Miller, R. H.: 1966, in G. Contopoulos (ed.), 'Theory of Orbits in the Solar System and in Stellar Systems', *IAU Symp.* **25**, 137.

A NUMERICAL EXPERIMENT ON RELAXATION TIMES IN STELLAR DYNAMICS

MYRON LECAR and CARLOS CRUZ-GONZÁLEZ

Smithsonian Astrophysical Observatory and Harvard College Observatory, Cambridge, Mass., U.S.A.

Abstract. The deflection of the velocity vector of a massless test star in the field of 100 stars was determined by numerical integration. The deflection due to each field star independently (with the other field stars removed) was also determined. The square of the deflection caused by the combined action of the field stars agreed quantitatively with the sum of the squares of the individual deflections and also with the theoretical estimate of Williamson and Chandrasekhar.

Williamson and Chandrasekhar (1941) estimated the relaxation time of a stellar system from the point of view of deflections suffered by a test star in encounters with a given distribution of field stars. The total deflection due to the combined effects of the field stars was estimated by calculating separately the deflection due to each two-body (test star – field star) encounter and then adding the individual deflections as statistically independent events (i.e., as the individual deflections are randomly oriented, the squares of the deflections are summed). But as distant encounters predominated, the test star suffered many encounters simultaneously, and it was not clear from their discussion why the individual encounters could be treated as independent. That independence did follow, however, as a consequence of the unperturbed-orbit approximation introduced by Henon (1958) and again by Ostriker and Davidson (1968), wherein the motion of the test star was assumed known so that the perturbations due to the field stars added separately. This approximation also simplified the analysis and allowed a correction for the finite duration of the encounter to be added. But this approximation essentially amounted to a linearization of the n-body problem, and it was not clear to us whether it was valid for a long enough time to allow a measurable deflection.

The experiment consisted of randomly distributing 100 unit-mass field stars inside a sphere. The field stars were constrained to remain at rest. A massless test star was introduced along a diameter and its orbit was integrated numerically. To simplify the analysis, the acceleration of a continuous mass distribution with the same average density as that of the field stars was subtracted from the acceleration of the test star. The acceleration of the test star was

$$\ddot{\mathbf{x}} = - \sum_{k=1}^{N} \frac{\mathbf{x} - \boldsymbol{\xi}_k}{|\mathbf{x} - \boldsymbol{\xi}_k|^3} + \left(\frac{N}{R^3}\right) \mathbf{x},$$

where $\mathbf{x}$, $\boldsymbol{\xi}$ are the coordinates of the test and field stars, respectively, N is the number of field stars, and R is the radius of the sphere.

As N becomes infinite, the acceleration vanishes and the 'unperturbed orbit' is a

M. Lecar (ed.), Gravitational N-Body Problem, 131–135.

straight line. The test star feels only the 'graininess' or the fluctuations of the deviations of field stars from a uniform distribution.

In this experiment, we let the velocity vector of the test star turn through an angle χ in a time t. We calculated $[(\sin^2\chi)/t]_{\rm exp}$. We then reintroduced the test particle with one field star at a time and determined the two-body deflection χ_i. To ascertain whether the encounters could be added as independent events, we compared $\langle(\sum \sin^2\chi_i)/t\rangle_{\rm exp}$ with $[(\sin^2\chi)/t]_{\rm exp}$. Strictly, if the unperturbed-orbit approximation is valid, the vectorial sum of the individual deflections should equal the deflection due to the combined effect of all the field stars. In general, it did not, but the agreement between $\langle(\sum \sin^2\chi_i)/t\rangle$ and $[(\sin^2\chi)/t]$ was quite a bit better.

Finally, the experimental results were compared with the theoretical estimate

$$\left\langle\frac{\sin^2\chi}{t}\right\rangle_{\rm theor} = \frac{8\pi n G^2 m^2}{v^3}\left[\ln\left(\frac{R}{l_0}\right) - 0.85\right],$$

where n is the average density of the field stars, R is the radius of the sphere, G is the gravitational constant (set equal to unity in the experiment), m is the mass of a field star (unity in the experiment), v is the velocity of the test star, and $l_0 = Gm/v^2$. Aside from the constant (-0.85), this is the result of Williamson and Chandrasekhar. The constant results from taking into account the finite duration of the encounters and the variation in the maximum impact parameter for a spherical distribution of field stars. The constant is determined in the Appendix. In our experiment, $\ln(R/l_0)$ varied from 4.7 to 8.4, so the constant modified the result by 10 to 20%. In stellar systems satisfying the Virial Theorem, $v^2 = GNm/R$, so $R/l_0 = N$ and the constant can be ignored.

The results of 189 experiments at six values of the test star's velocity are presented in Table I and Figure 1.

TABLE I

The deflection of a test star as it traverses the gravitational field of 100 stationary field stars

v	$\left\langle\frac{\sin^2\chi}{t}\right\rangle_{\rm theor}$	$\left(\frac{\sin^2\chi}{t}\right)_{\rm exp}$	$\left\langle\frac{\sum \sin^2\chi_i}{t}\right\rangle_{\rm exp}$	No. cases
6.93	0.57	0.33 $\pm$0.06	0.49 $\pm$0.12	26
10.00	0.22	0.21 $\pm$0.04	0.23 $\pm$0.04	63
14.42	0.087	0.054 $\pm$0.007	0.080 $\pm$0.020	55
20.80	0.033	0.032 $\pm$0.017	0.018 $\pm$0.020	15
30.00	0.012	0.0045 $\pm$0.0020	0.0051 $\pm$0.0019	15
43.27	0.0045	0.0017 $\pm$0.0007	0.0056 $\pm$0.0036	15

The growth of the error bars with increasing velocity is accounted for by the following argument. Let $x = \sin^2\chi$. For distant encounters $(l \gg l_0)$, $x \sim (l_0/l)^2$. Since the number of encounters in dl goes as ldl, the probability of a deflection in dx is approximately $p(x)\,{\rm d}x = \varepsilon\,{\rm d}x/x^2$, where $\varepsilon = (l_0/R)^2$ and $\varepsilon \leqslant x \leqslant 1$. Rybicki (1970) has

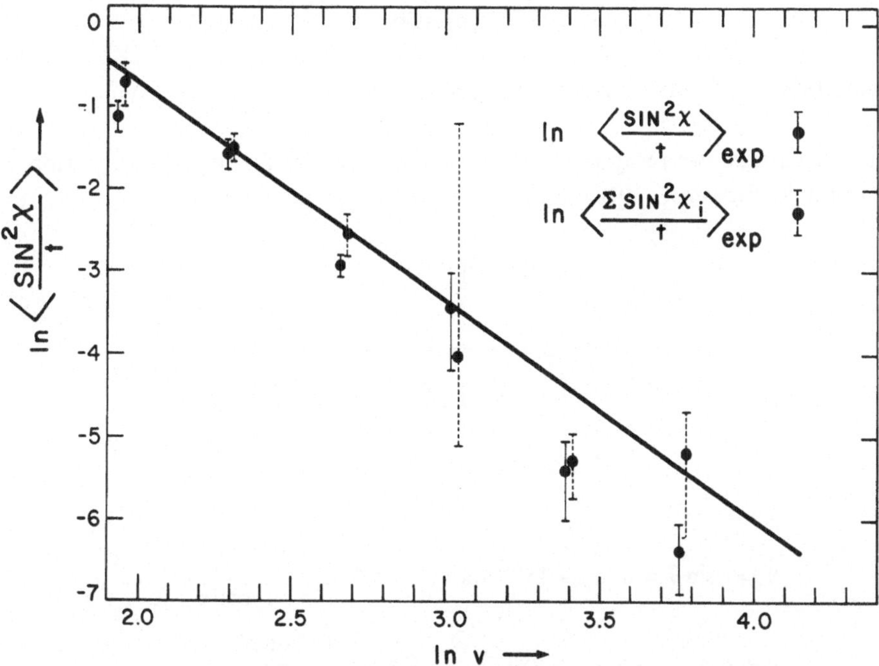

Fig. 1. The natural logarithm of the experimental determinations of $\langle(\sin^2\chi)/t\rangle$, the theoretical value of $\langle(\sin^2\chi)/t\rangle$ (solid line), plotted against the natural logarithm of the velocity.

shown that for randomly oriented vectors $\boldsymbol{\varepsilon}_i$, if $\mathbf{E}=\sum_{i=1}^{N}\boldsymbol{\varepsilon}_i$, then

$$\langle E^2\rangle = N\langle\varepsilon^2\rangle \quad\text{and}\quad \langle E^4\rangle = N\langle\varepsilon^4\rangle + 2N(N-1)\langle\varepsilon^2\rangle^2,$$

where the averaging over orientation has been performed explicitly and the remaining average ($\langle\ \rangle$) is over magnitude. Making use of this result, we find that

$$\frac{\sigma^2}{\langle x\rangle^2} = \frac{\langle x^2\rangle - \langle x\rangle^2}{\langle x\rangle^2} = \frac{1}{N}\frac{\langle x^2\rangle}{\langle x\rangle^2} + 1 - \frac{2}{N}.$$

Since $\langle x\rangle \cong \varepsilon\ln(1/\varepsilon)$ and $\langle x^2\rangle\cong\varepsilon$ and $1/\varepsilon > N^2$,

$$\frac{\sigma}{\langle x\rangle} \cong \frac{1}{N^{1/2}}\frac{(1/\varepsilon)^{1/2}}{\ln(1/\varepsilon)}.$$

Since $1/\varepsilon\sim v^4$, $\sigma\sim v^2$. If $v^2\cong GNm/R$, $1/\varepsilon\cong N^2$ and $\sigma/\langle x\rangle\cong N^{1/2}/\ln N$.

Finally, we point out that this experiment specifically excluded 'collective effects' (i.e., 'polarization' of the distribution of field stars by the test star). From the calculations of Gasiorowicz *et al.* (1956) and Gilbert (1968), it seems that for 'Jeans-stable' systems, polarization can modify the relaxation time by a factor of 2. These effects are only beginning to be understood; they are discussed in some depth in the contributions by Gilbert, Kalnajs, and Kulsrud in this issue.

Appendix

THEORETICAL DETERMINATION OF $\langle(\sin^2\chi)/t\rangle$

For convenience, we recapitulate the standard formulas for hyperbolic motion (see, for example, Plummer, 1918) in the notation used in the text:

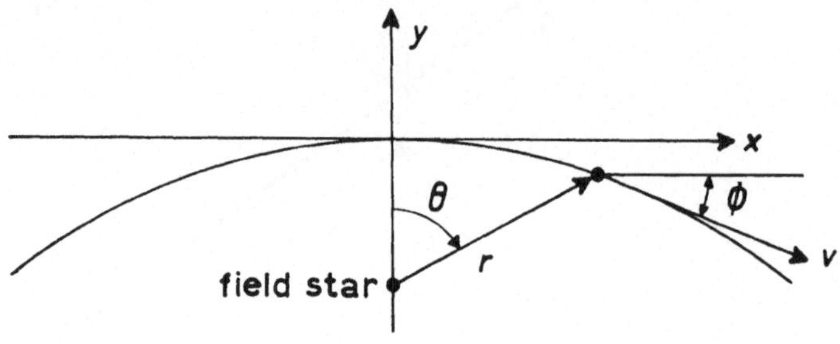

$$x = r\sin\theta = a(\varepsilon^2 - 1)^{1/2}\sinh\tau\,, \qquad \dot{x} = \frac{a(\varepsilon^2-1)^{1/2}\cosh\tau}{\varepsilon\cosh\tau - 1}\,,$$

$$y = r\cos\theta = a(\varepsilon - \cosh\tau)\,, \qquad \dot{y} = \frac{-a\varepsilon\sinh\tau}{\varepsilon\cosh\tau - 1}\,,$$

$$\omega t = \varepsilon\sinh\tau - \tau\,,$$

$$\tan\phi = |\dot{y}/\dot{x}|\,.$$

Let v be the velocity of the test star at infinity. Then

$$a = Gm/v^2 = l_0\,, \quad \omega = v/l_0\,, \quad \varepsilon^2 = 1 + (l/l_0)^2\,.$$

The energy per unit mass is

$$E = \tfrac{1}{2}(a\omega)^2 = \tfrac{1}{2}v^2\,,$$

and the angular momentum per unit mass is

$$L = lv\,;$$

ϕ_i is the angle between the velocity vector, when the test star is at x_i, and the x axis. If, during an encounter of finite duration, the test star travels from $-x_1\,(x_1>0)$ to $+x_2$,

$$\sin\chi = \sin(\phi_1 + \phi_2)\,,$$

where

$$\sin\phi_i = \frac{(l_0/l)\,(x_i/l)}{[1+(x_i/l)^2]^{1/2}}\cos\phi_i\,, \quad i = 1, 2\,,$$

$$\cos\phi_i = \frac{1}{(1+(l_0/l)^2\{(x_i/l)^2/[1+(x_i/l)^2]\})^{1/2}}\,.$$

The unperturbed orbit approximation is valid in the limit of large l/l_0, where $\cos\phi_i \to 1$.

In the experimental configuration, the test star travels a distance

$$2R = x_1 + x_2 .$$

Let

$$x_1 = R + (R - x_2), \quad x_2 = R - (R - x_2).$$

If we let

$$l = \xi l_0 , \quad R = \xi_m l_0 , \quad \text{and} \quad R - x_2 = \eta R ,$$

then

$$\sin \chi = \frac{1}{\xi} \left\{ \frac{\xi_m (1 - \eta)}{[\xi^2 + \xi_m^2 (1 - \eta)^2]^{1/2}} + \frac{\xi_m (1 + \eta)}{[\xi^2 + \xi_m^2 (1 + \eta)^2]^{1/2}} \right\}.$$

The number of encounters per second with impact parameters l to $l + \mathrm{d}l$ is

$$2\pi n v l \, \mathrm{d}l ,$$

and the maximum value of l at a given x is $R(1-\eta^2)^{1/2}$. Thus,

$$\left\langle \frac{\sin^2 \chi}{t} \right\rangle_{\text{theor}} = \frac{2\pi n G^2 m^2}{v^3} \int_0^1 \mathrm{d}\eta \int_1^{\xi_m(1-\eta^2)^{1/2}} \frac{\mathrm{d}\xi}{\xi} \times \left\{ \frac{\xi_m^2 (1 - \eta)^2}{\xi^2 + \xi_m^2 (1 - \eta)^2} + \right.$$

$$\left. + \frac{\xi_m^2 (1 + \eta)^2}{\xi^2 + \xi_m^2 (1 + \eta)^2} + \frac{2\xi_m^2 (1 - \eta^2)}{[\xi^2 + \xi_m^2 (1 - \eta)^2]^{1/2} [\xi^2 + \xi_m^2 (1 + \eta)^2]^{1/2}} \right\},$$

where the integration over η accounts for the spherical distribution of field stars. If we neglect this, for the moment, and set $\eta = 0$,

$$\left\langle \frac{\sin^2 \chi}{t} \right\rangle_{\text{theor}} \cong \frac{2\pi n G^2 m^2}{v^3} \times 4 \times \int_1^{\xi_m} \frac{\mathrm{d}\xi}{\xi} \times \frac{\xi_m^2}{\xi^2 + \xi_m^2} \cong$$

$$\cong \frac{8\pi n G^2 m^2}{v^3} (\ln \xi_m - \tfrac{1}{2} \ln 2), \quad \text{for } \xi_m \gg 1 .$$

The integration over η has been carried out, and $(\frac{1}{2}) \ln 2 \cong 0.347$ is replaced by 0.853.

References

Gasiorowicz, S., Neuman, M., and Riddel, Jr. R. J.: 1956, *Phys. Rev.* **101**, 922.
Gilbert, I. H.: 1968, *Astrophys. J.* **152**, 1043.
Henon, M.: 1958, *Ann. Astrophys.* **21**, 186.
Ostriker, J. P. and Davidson, A. F.: 1968, *Astrophys. J.* **151**, 679.
Plummer, H. C.: 1918, *An Introductory Treatise on Dynamical Astronomy*, London.
Rybicki, G.: 1970, Private communication.
Williamson, R. E. and Chandrasekhar, S.: 1941, *Astrophys. J.* **93**, 305.

RECENT DEVELOPMENTS OF INTEGRATING THE GRAVITATIONAL PROBLEM OF *N*-BODIES

VICTOR SZEBEHELY and DALE G. BETTIS

The University of Texas at Austin, U.S.A.

Abstract. This paper discusses the formulation and the numerical integration of large systems of differential equations occurring in the gravitational problem of *n*-bodies.

Different forms of the pertinent differential equations of motion are presented, and various regularizing and smoothing transformations are compared. Details regarding the effectiveness and the efficiency of the Kustaanheimo-Stiefel and of other methods are discussed. In particular, a method is described in which some of the phase variables are treated in the regularized system and others in the ordinary system. This mixed method of numerical regularization offers some advantages.

Several numerical integration techniques are compared. A high order Runge-Kutta method, Steffensen's method, and a finite difference method are investigated, especially with regard to their adaptability to regularization.

The role of integrals and integral invariants is displayed in controlling the accuracy of the numerical integration.

Numerical results are described with 5, 25 and 500 bodies participating. These examples compare the various integration techniques, several regularization methods and different logics in treating binaries.

1. Introduction

In the development of stellar systems close approaches seem to play a dominant role. Indeed, there is sufficient amount of evidence in the literature to show that binary formation as well as escape are consequences of close approaches.

On the other hand, the computational difficulties emerge at these very same close approaches, consequently numerical results not treating close approaches with sophistication may be subject to suspicion.

The ignorer of close approaches is the ignoramus of stellar dynamics, regardless of his 'physical' excuses. If meaningful comparisons between various statistical theories and numerical experiments are desired, then crude approximations during the process of numerical integration cannot be allowed. The number of binaries formed depends critically on the accuracy with which the integration is performed. To study the effects of the initial conditions on the development of the system must accommodate a systematic variation of the initial conditions. Poor and inaccurate numerical processes during the integration introduce errors at every step, effectively changing the initial conditions.

The above few examples amplify the conviction that as long as the purpose of numerical integration is to obtain results to be compared with either observations or with approximate theories, the higher the reliability and accuracy of numerical process, the more significant the comparison becomes. Without further emphasizing the above thesis, a seldom realized and often unobserved phenomenon is mentioned. The time spent on a given computer for a given number of bodies and fixed initial conditions will depend on the required accuracy of the numerical integration. It is a

M. Lecar (ed.), Gravitational N-Body Problem, 136–147. *All Rights Reserved*

most revealing fact that the expected longer computational time for higher accuracy is not always forthcoming. It has occurred during numerical experiments that a crude and inaccurate method lead to the early (and fictitious) formation of binaries because of numerical error accumulation. The computation of the development of this system, with its binaries, may take a considerable time on the computer. On the other hand, a more accurate numerical method might *not* lead to the early formation of binaries and the integration, consequently, may take less time!

This paper is dedicated to the development of accurate numerical techniques applicable to the integration of the gravitational problem of many bodies. Inasmuch as fast and accurate numerical methods require optimal formulation of the pertinent differential equations, the first part of the paper deals with this question. Once the equations are presented in their smoothed and efficient form, the appropriate method of numerical integration may be investigated. This is followed by a discussion of the role of the integrals and integral invariants as error-controlling devices.

2. Formulation of the Equations of Motion

2.1. Concept of Regularization

The purpose of this part is the application of certain smoothing transformations to the system of differential equations describing the motion of n gravitating bodies. It is recognized that the probability of collisions in the mathematical sense is zero in the problem of n bodies. Consequently, our main concern is the treatment of close approaches. The analytical methods of eliminating singularities which occur at binary collisions are well understood and the same methods may be applied to the close approach of two bodies. There is no attempt made in this paper to treat triple or more complex approaches since analytical treatments are not available for the collision of more than two bodies.

Once a generalized concept of regularization is accepted, we may formulate freely the rules to be followed to 'regularize' the equations of motion. Our main purpose in this paper is to present such a formulation of the equations of motion which allows numerical integration with high accuracy and in a short time.

The configuration which we expect to dominate during the evolution of a cluster of stars consists of an arbitrary number of close binaries existing at the same time and an arbitrary number of close approaches occurring.

The analytic problem of regularizing close approaches of two stars was treated in three dimensions successfully by Sundman (1912) by transformation of the independent variable. Levi-Civita (1903), transforming the dependent variables also, solved the problem in two dimensions. Levi-Civita's transformation was generalized to three dimensions by Kustaanheimo and Stiefel (1965). Implementations of Levi-Civita's two-dimensional regularization go back to several years when the two-dimensional motion of three gravitating point masses were treated successfully with this method by Szebehely and Peters (1967). Application of the Kustaanheimo-Stiefel (K-S) transformation as well as the Levi-Civita formulation are not immediate to the problem of

n bodies. There are several possible formulations of these transformations and it is rather critical to select the best representation for a successful numerical integration.

2.2. Transformation of the Independent Variable

The following description shows a method by which the system of differential equations might be treated. At this point it is noted that all regularizations transform the independent (time) variable, but the most successful regularizations also transform the dependent variables. We will first discuss the regularizations that transform only the independent variable, and afterwards, the K-S transformation which involves both the independent and the dependent variables.

Let the dynamical system be represented by the equations

$$\mathrm{d}^2\bar{r}_i/\mathrm{d}t^2 = f_i(\bar{r}_1, \ldots, \bar{r}_n) \tag{1}$$

where $\bar{r}_i$ is the position vector of the ith participating body, n is the number of bodies, t is the time and $i=1, \ldots, n$.

Without loss of generality, let us consider five bodies (of n) and let us assume that the first and second bodies show a close approach to each other. Also, assume that close approach occurs between the 4th and 5th bodies, while the third body is removed from both binaries. Such a configuration may be generalized but almost all the essential problems are encountered in the afore-mentioned system. If $r_{12} < r_{45} \ll r_{i3}$ and $r_{12} < r_{45} \ll r_{kl}$ where $i \neq 3$; $k=1, 2$; $l=4,5$, we speak of isolated binaries or isolated close approaches. Here $r_{ij} = |\bar{r}_i - \bar{r}_j|$.

If the transformation of the time variable $\mathrm{d}t = g_{12}(\bar{r}_1, \bar{r}_2)\,\mathrm{d}\tau_{12}$ is applied to the system of Equations (1), the first two equations become

$$\frac{\mathrm{d}^2\bar{r}_1}{\mathrm{d}\tau_{12}^2} = F_1\left(\bar{r}_1, \ldots, \bar{r}_n, \frac{\mathrm{d}\bar{r}_1}{\mathrm{d}\tau_{12}}, \frac{\mathrm{d}\bar{r}_2}{\mathrm{d}\tau_{12}}\right),$$

$$\frac{\mathrm{d}^2\bar{r}_2}{\mathrm{d}\tau_{12}^2} = F_2\left(\bar{r}_1, \ldots, \bar{r}_n, \frac{\mathrm{d}\bar{r}_1}{\mathrm{d}\tau_{12}}, \frac{\mathrm{d}\bar{r}_2}{\mathrm{d}\tau_{12}}\right).$$

The equation of the third body remains

$$\frac{\mathrm{d}^2\bar{r}_3}{\mathrm{d}t} = f_3(\bar{r}_1, \ldots, \bar{r}_n).$$

For the binary system of the fourth and fifth bodies, the transformation $\mathrm{d}t = g_{45}(\bar{r}_4, \bar{r}_5)\,\mathrm{d}\tau_{45}$ leads to

$$\frac{\mathrm{d}^2\bar{r}_4}{\mathrm{d}\tau_{45}} = F_4\left(\bar{r}_1, \ldots, \bar{r}_n, \frac{\mathrm{d}\bar{r}_4}{\mathrm{d}\tau_{45}}, \frac{\mathrm{d}\bar{r}_5}{\mathrm{d}\tau_{45}}\right),$$

$$\frac{\mathrm{d}^2\bar{r}_5}{\mathrm{d}\tau_{45}} = F_5\left(\bar{r}_1, \ldots, \bar{r}_n, \frac{\mathrm{d}\bar{r}_4}{\mathrm{d}\tau_{45}}, \frac{\mathrm{d}\bar{r}_5}{\mathrm{d}\tau_{45}}\right).$$

Since these two transformations of the independent variable depend on the coordinates of the two bodies participating in the close approach, there are three different independent 'time' variables, τ_{12}, τ_{45} and t for our configuration. Whenever two or more close approaches are treated simultaneously, the system of differential equations will contain several of these pseudo-times, all of which must be related to the actual time at every integration step. All of the equations may be integrated with respect to the same independent variable by introducing a common 'general time variable', which, in practice is usually one of the regularizing variables. In our configuration the three time variables are related by

$$\mathrm{d}t = g_{12}\,\mathrm{d}\tau_{12} = g_{45}\,\mathrm{d}\tau_{45}\,.$$

Thus, any two of these times may be represented in terms of the third. By selecting the one with the smallest factor (for example, assuming that $g_{12} < g_{45}$), we have

$$\mathrm{d}t = g_{12}\,\mathrm{d}\tau_{12} \quad \text{and} \quad \mathrm{d}\tau_{45} = \frac{g_{12}}{g_{45}}\,\mathrm{d}\tau_{12}\,,$$

as the appropriate unifying time relations.

One manner in which these relations may be applied is the following. Equations (1) may be rewritten as a system of first-order differential equations:

$$\left.\begin{aligned} \frac{\mathrm{d}\bar{r}_1}{\mathrm{d}\tau_{12}} &= \bar{V}_1\,, & \frac{\mathrm{d}\bar{V}_1}{\mathrm{d}\tau_{12}} &= \bar{F}_1\,, \\ \frac{\mathrm{d}\bar{r}_2}{\mathrm{d}\tau_{12}} &= \bar{V}_2\,, & \frac{\mathrm{d}\bar{V}_2}{\mathrm{d}\tau_{12}} &= \bar{F}_2\,, \end{aligned}\right\} \tag{2}$$

$$\frac{\mathrm{d}\bar{r}_3}{\mathrm{d}t} = \bar{V}_3\,, \quad \frac{\mathrm{d}\bar{V}_3}{\mathrm{d}t} = \bar{f}_3\,; \tag{3}$$

$$\left.\begin{aligned} \frac{\mathrm{d}\bar{r}_4}{\mathrm{d}\tau_{45}} &= \bar{V}_4\,, & \frac{\mathrm{d}\bar{V}_4}{\mathrm{d}\tau_{45}} &= \bar{F}_4\,, \\ \frac{\mathrm{d}\bar{r}_5}{\mathrm{d}\tau_{45}} &= \bar{V}_5\,, & \frac{\mathrm{d}\bar{V}_5}{\mathrm{d}\tau_{45}} &= \bar{F}_5\,. \end{aligned}\right\} \tag{4}$$

In terms of the variable τ_{12} the equations for the third body are expressed as

$$\frac{\mathrm{d}\bar{r}_3}{\mathrm{d}\tau_{12}} = \bar{V}_3 g_{12}\,, \quad \frac{\mathrm{d}\bar{V}_3}{\mathrm{d}\tau_{12}} = \bar{f}_3 g_{12}\,,$$

and similarly Equations (4) become

$$\frac{\mathrm{d}\bar{r}_4}{\mathrm{d}\tau_{12}} = \bar{V}_4 \frac{g_{12}}{g_{45}}\,, \quad \frac{\mathrm{d}\bar{V}_4}{\mathrm{d}\tau_{12}} = \bar{F}_4 \frac{g_{12}}{g_{45}}\,,$$

$$\frac{\mathrm{d}\bar{r}_5}{\mathrm{d}\tau_{12}} = \bar{V}_5 \frac{g_{12}}{g_{45}}\,, \quad \frac{\mathrm{d}\bar{V}_5}{\mathrm{d}\tau_{12}} = \bar{F}_5 \frac{g_{12}}{g_{45}}\,.$$

In general, when several binaries are treated simultaneously, we have

$$\mathrm{d}t = g_{12}\,\mathrm{d}\tau_{12} = g_{45}\,\mathrm{d}\tau_{45} = \cdots = g_{kl}\,\mathrm{d}\tau_{kl}.$$

By selecting the pseudo-time which belongs to min g_{kl}, these differential relations may be used to express all differential equations in terms of a unifying time-variable.

A simple and an effective time-transformation consists of choosing the minimum distance occurring in the system as the function g_{kl} such that

$$\mathrm{d}t = (\min R_{kl})^{\alpha}\,\mathrm{d}\tau_{kl}, \tag{5}$$

where α is a real number and where $R_{kl}=|\bar{R}_{kl}|=|\bar{R}_k-\bar{R}_l|=R$.

Considering the equations of motion of the closest kth and lth body,

$$\ddot{\bar{r}}_k = \bar{f}_k(\bar{r}_1, \ldots, \bar{r}_n),$$

$$\ddot{\bar{r}}_l = \bar{f}_l(\bar{r}_1, \ldots, \bar{r}_n),$$

where the dots denote differentiations with respect to t, the transformation (5) leads to

$$\begin{aligned} \bar{r}_k'' - \frac{\alpha R'}{R}\bar{r}_k' &= R^{2\alpha}\bar{f}_k, \\ \bar{r}_l'' - \frac{\alpha R'}{R}\bar{r}_l' &= R^{2\alpha}\bar{f}_l, \end{aligned} \tag{6}$$

where differentiation with respect to τ_{kl} is denoted by the prime. In terms of a system of first order, Equations (6) become

$$\bar{r}_k' = \bar{V}_k R^{\alpha}, \quad \bar{V}_k' = \bar{f}_k R^{\alpha},$$

$$\bar{r}_l' = \bar{V}_l R^{\alpha}, \quad \bar{V}_l' = \bar{f}_l R^{\alpha}.$$

The remaining equations may be expressed in terms of the new independent variable τ_{kl}:

$$\bar{r}_i' = \bar{V}_i R^{\alpha}, \quad \bar{V}_i' = \bar{f}_i R^{\alpha}, \quad i \neq k, l.$$

Up to this point, the value of α has not been specified. If $\bar{f}_k$ and $\bar{f}_l$ of Equations (6) contain divisors of the form r_{kl}^{β}, it is advantageous to select $\alpha=\beta/2$ in order to regularize the functions $\bar{f}$. In particular, numerical experiments show that $\alpha=\frac{3}{2}$ has distinct advantages. For the unperturbed problem of two bodies, the choice $\alpha=1$ is identical to using the eccentric anomaly as the independent variable. When $\alpha=2$, the pseudo-time becomes the true anomaly. Numerical integrations of artificial Earth satellites suggest that the true (eccentric) anomaly has advantages near perigee (apogee). The choice $\alpha=\frac{3}{2}$ may be considered a compromise.

This evidence is further corroborated by our numerical experiments with the IAU 25-body problem, described by Lecar (1968). The IAU 25-body problem was integrated for $\alpha=0$, 1, $\frac{3}{2}$, 2, 3. The error in the total energy, $\Delta\varepsilon$, and the CDC 6600 computing time at $t=10$, are listed in Table I.

TABLE I

Comparison of time transformation in the 25-body problem using $dt = R^{\alpha} d\tau_{kl}$

α	$\Delta\varepsilon/\varepsilon\ (t = 10)$	Computer time (s)
0	4×10^{-8}	662
1	3×10^{-8}	532
$\frac{3}{2}$	2×10^{-8}	485
2	5×10^{-9}	631
3	8×10^{-8}	760

2.3. The Kustaanheimo-Stiefel Regularization

Numerical results of 5, 25 and 500 bodies show the definite advantage of the use of the Kustaanheimo-Stiefel transformation over transforming only the independent variable. The K-S transformation consists of a time transformation combined with a coordinate transformation. The new time, τ_{kl}, is introduced, as before, by

$$dt = R d\tau_{kl},$$

where R is the smallest distance in the system of n bodies.

The coordinate transformation in matrix notation is

$$\bar{R} = \mathscr{L}(\bar{u})\,\bar{u},$$

with $\bar{R}$ being the position vector of one member of the binary with respect to the other, i.e.

$$\bar{R} = \bar{r}_k - \bar{r}_l.$$

The 4×4 matrix $\mathscr{L}(\bar{u})$ is given by

$$\mathscr{L}(\bar{u}) = \begin{pmatrix} u_1 & -u_2 & -u_3 & u_4 \\ u_2 & u_1 & -u_4 & -u_3 \\ u_3 & u_4 & u_1 & u_2 \\ u_4 & -u_3 & u_2 & -u_1 \end{pmatrix},$$

where u_i is the ith component of the vector $\bar{u}$.

A detailed derivation and a systematic discussion of the K-S regularization and its use in stellar dynamics is offered by us elsewhere. After transforming the equations of motion of the closest pair to their center of mass by the Jacobian relations,

$$\bar{Q} = \frac{m_k}{m_k + m_l}\,\bar{r}_k + \frac{m_l}{m_k + m_l}\,\bar{r}_l,$$

$$\bar{R} = \bar{r}_k - \bar{r}_l,$$

the equations of motion become

$$\ddot{\bar{Q}} = \bar{g}(\bar{r}_1, \ldots, \bar{r}_n),$$

$$\ddot{\bar{R}} + \frac{\bar{R}}{R^3} = \bar{F}(\bar{r}_1, \ldots, \bar{r}_n).$$

The first of these equations is regular, but the second contains the small divisor R. By applying the K-S transformation to $\bar{R}$, we have

$$\bar{u}'' + \frac{h}{2}\bar{u} = \bar{\phi}(\bar{u}, \bar{u}', \bar{r}_i), \quad i = 1, 2, \ldots, n, \quad i \neq k, l,$$

where primes denote derivatives with respect to τ_{kl}. The negative of the two-body binding energy, h, may be computed from the regular differential equation

$$h' = \theta_1(\bar{u}, \bar{u}', \bar{r}_i), \quad i \neq k, l,$$

or from the explicit equation of the form

$$h = \theta_2(\bar{u}')/R.$$

It is noted that this explicit equation is not regular.

In order to integrate the system of differential equations with the common independent variable τ_{kl}, it is only necessary to write the remaining equations as a system of first order equations and, as we did previously, multiply the differential equations by the factor R:

$$\bar{Q}' = \bar{v}_0 R, \quad \bar{v}_0' = \bar{g}R,$$

$$\bar{r}_i' = \bar{v}_i R, \quad \bar{v}_i' = \bar{f}_i R, \quad i = 1, 2, \ldots, n, \quad i \neq k, l.$$

Here $\bar{v}_0$ is the velocity of the center of mass of the closest binary.

This technique may be extended to m distinct pairs of bodies by selecting a suitable unifying time variable.

A comparison between the use of the K-S transformation for the closest pair of bodies (KS1) and the transformation of the independent variable only, is given in Table II for the IAU 25-body problem. The first entry ($\alpha = \frac{3}{2}$) in Table II refers to the results obtained by the use of Equation (5) with $\alpha = \frac{3}{2}$. The CDC 6600 computing time and the error in the constant of energy are tabulated for $t = 20$.

TABLE II

Comparison of transformations

Transformation	Computer time (min)	$\frac{\Delta\varepsilon}{\varepsilon}$
$\alpha = \frac{3}{2}$	19.5	3×10^{-8}
KS1	10.4	5×10^{-10}
KS2	20.0	2×10^{-9}

Results analogous to those in Table II were obtained for various n-body calculations using $n=5$, 10, 25, 100, 500. These indicated that regularizing the one pair (KS1) was not only more economical with respect to computing time, but was also more accurate than the results obtained by transforming the independent variable only ($n=\frac{3}{2}$). Table II shows also the results obtained by regularizing the closest *two* pairs of bodies using the K-S transformation (KS2). In these experiments the closest two pairs were *always* regularized. Thus, a considerable amount of execution time was needed to establish the initial conditions and the transformation equations associated with each new combination of the closest two pairs. Further experiments indicated the rule that the closest pair of bodies should be regularized only if the equations of motion of the pair were suffering from a loss of accuracy of the close encounter. In other words, it is *not* advisable to *always* regularize the closest pair. In order to optimize the computer time and the computational accuracy during a close encounter, criteria for regularization may be based upon the relative distance, relative binding energy, etc.

We note that regularizing the closest pair will not alleviate the loss of accuracy resulting from multiple encounters, such as triple or higher order collisions.

3. The Selection of the Method of Numerical Integration

In order to establish an accurate and efficient algorithm for numerical integration, the relative efficiency between various finite difference, Runge-Kutta, and power series methods for the solution of ordinary differential equations was examined. Recognizing the fact that the accuracy and efficiency of the methods are machine-dependent for a specific value of n, we found that using the CDC 6600, a seventh order Runge-Kutta method proposed by Fehlberg (1968) was the most efficient integration method when $n \leqslant 100$. For n larger than 100, it appears that a finite difference method or a recurrent power series may be more appropriate.

As an indication of the efficiency of the seventh-order Runge-Kutta method combined with the KS1 transformation, we list in Table III the CDC 6600 computing times and the errors in the constant of energy for the IAU 25-body problem for $t=50.0$ and $t=75.0$.

TABLE III

IAU 25-body problem with KS1 transformation

t	$\Delta\varepsilon/\varepsilon$	CDC 6600 time (min)
50.0	4.0×10^{-9}	49.5
75.0	1.1×10^{-7}	89.0

4. The Use of Integrals and Integral-Invariants

It is known that the ten integrals which exist in the gravitational problem of n bodies may be used either to check or to control the accuracy of the numerical integration.

These integrals are also available to reduce the degrees of freedom of the dynamical system by eliminating some of the variables or to act as equations of constraints, forcing the solution in the phase space to remain on the corresponding integral surfaces. This last idea is expanded in P. Nacozy's article which appears in the same volume. The accuracy-controlling role of an integral can be realized by forcing the program of the numerical integration to be more accurate whenever the integral deviates by a prescribed amount from its original (initial) value.

The use of integrals to eliminate variables and, in this way, to reduce the degrees of freedom, is well known and it is well documented in the literature. Especially the classical result by Lagrange (1772) should be mentioned who succeeded in reducing the order of the system of differential equations representing the problem of three bodies from 18 to 6 by making use of the 10 classical integrals, of the elimination of the nodes and of the elimination of time. Such a reduction of the problem of n bodies may also be performed as shown by Bennett (1904), nevertheless, the practical significance of this process might be questionable. For example, for a system of n bodies the order is $6n$ and a reduction to a system of the order $6n-12$ is of small help, indeed, especially when $n \gg 3$.

The wide use of integrals to check the accuracy of the result of a numerical integration is well known. The fact that one relation between the variables is checked means that the deviation of the solution path from an integral surface is computed in the $6n$ dimensional space. If there are large deviations observed, we should suspect the result but the reverse is not true, i.e., the solution path being close to an integral surface or close to the intersection of several (say s) such surfaces still leaves $6n-s$ possible deviations.

It has been suggested to us by Dr A. Deprit (1970) that an integral-invariant might be used for some of the aforementioned purposes also. Inasmuch as such integral-invariants cannot be used to eliminate variables and in this way to reduce the degrees of freedom of the system, one might experiment with their use as further checks of the accuracy. One anticipates that such checks are of high value because the sensitivity of integral-invariants to errors is expected to be greater than the sensitivity of, say, the energy integral. This expectation is based on the intimate relation existing between the variational equations and invariant integrals and on the well known sensitivity of the variational equations.

Indeed, it is known that the integral-invariants of a given system of differential equations are the integrals of the variational equations. We note paranthetically that Poincaré's (1890) pth order integral invariant may be defined as the p-tuple integral taken over a p-dimensional region of the phase space, if it has the same value at all times. The reader might be referred to Whittaker (1904, Chapter 10) for additional details.

Poincaré (1905), in his *Leçons*, suggests the establishment of an integral invariant for the problem of three bodies. This integral invariant we have succeeded in generalizing to the problem of n bodies. In fact, it may be shown that Poincaré's integral-invariant is intimately connected with Jacobi's equation, also known as Lagrange's identity in stellar dynamics, see e.g. Chandrasekhar (1942) and Whittaker (1904, p. 324).

We shall now show how Jacobi's equation may be used to check the accuracy of the numerical integration when it is formulated as an integral-invariant.

Defining the moment of inertia of our system by

$$I = \sum_{i=1}^{n} m_i \bar{r}_i^2 ,$$

where $\bar{r}_i$ is the position vector of the ith body with mass m_i, we have Jacobi's equation:

$$\frac{d^2 I}{dt^2} = 2(2T - V),$$

where the kinetic and potential energies are

$$T = \tfrac{1}{2} \sum_{i=1}^{n} m_i \dot{\bar{r}}_i^2$$

and

$$V = G \sum_{1 \leqslant i < j \leqslant n} \frac{m_i m_j}{r_{ij}} .$$

Integrating Jacobi's equation, we have

$$\dot{I}(t) = 2 \int_0^t (2T - V)\, dt + \dot{I}(0) .$$

The left side of the above equation is computed by the equation

$$\dot{I}(t) = 2 \sum_{i=1}^{n} m_i \dot{\bar{r}}_i \cdot \bar{r}_i ,$$

which involves both the velocities and the positions of all bodies at time t.

The integral on the right side may be written as

$$K(t) = 2 \int_0^t V(\alpha_0 - 1)\, dt ,$$

where α_0 is the virial coefficient, defined by

$$\alpha_0 = 2T/V .$$

At any instant, when the virial theorem is satisfied ($\alpha_0 = 1$), the integrand is zero. At close approaches $\alpha_0 \to 2$, which may be seen considering the expression for the total energy

$$E = T - V .$$

From this we have

$$\frac{2E}{V} = \alpha_0 - 2$$

and since at close approaches $V \to \infty$, $\alpha_0 \to 2$.

The function $K(t)$ may be evaluated by solving the differential equation

$$\frac{\mathrm{d}K}{\mathrm{d}t} = 2(2T - V),$$

where the kinetic and potential energies must be evaluated as functions of the time. Inasmuch as the total energy is computed for checking purposes in any case, the additional computation required does not seem to be too taxing.

Numerical experiments performed on systems consisting of 5, 25, and 500 gravitating bodies show that the most sensitive checks are offered by the integral of energy and by the above-described integral-invariant. Smaller variations are shown by the integrals of momentum and by the integrals of the center of mass.

It is strongly recommended that other integral-invariants be constructed and tested numerically since it is not inconceivable that some might exist with considerably higher sensitivity than the integral of energy.

The following numerical results describe the behavior of integrals and of our integral-invariant during the process of the numerical integration.

For $n=5$, 25 and 500 the changes of the values of the integral invariant $\Delta\sigma$, of the energy $\Delta\varepsilon$, of the angular momentum $\Delta\mu$, and of the center of mass $\Delta\gamma$, are presented in Table IV. The case $n=5$ is an example investigated by Poveda (1967).

TABLE IV

Variation of integrals and of the integral-invariant

n	$\Delta\sigma$	$\Delta\varepsilon/\varepsilon$	$\Delta\mu$	$\Delta\gamma$	Method
5	1×10^{-8}	1×10^{-5}	2×10^{-10}	4×10^{-10}	*
5	2×10^{-7}	4×10^{-7}	1×10^{-10}	1×10^{-12}	KS1
25 ($t=20$)	2×10^{-10}	3×10^{-8}	6×10^{-12}	3×10^{-13}	$r^{3/2}$
25 ($t=20$)	2×10^{-10}	6×10^{-10}	7×10^{-12}	2×10^{-13}	KS1
25 ($t=20$)	1×10^{-9}	2×10^{-9}	5×10^{-11}	1×10^{-13}	KS2
25 ($t=50$)	8×10^{-11}	4×10^{-9}	2×10^{-11}	4×10^{-12}	KS1
500	2×10^{-7}	2×10^{-10}	3×10^{-10}	2×10^{-11}	$r^{3/2}$
500	2×10^{-7}	5×10^{-11}	2×10^{-9}	9×10^{-13}	KS1

* No transformation of the dependent or independent variables.

In this example, the original differential equations as well as the ones obtained by the one-pair K-S regularization were integrated. The integrals are also compared for the IAU 25-body problem at $t=20$ and $t=50$, and for a 500-body problem after the first crossing time.

The change in the integral invariant is defined by

$$\Delta\sigma = \dot{I}(t) - \dot{I}(0) - K(t)$$

and the errors in the total energy, angular momentum and center of mass are defined as their deviations from the initial values. The results for the total energy are made dimensionless by dividing the deviations by the initial values of the energy. The initial values of the other integrals and that of the integral invariant are zero, consequently, the actual values of the errors (deviations from zero) are given.

Acknowledgements

Partial support of the Air Force Office of Scientific Research, Grant Number AFOSR 69-1744A is acknowledged. The cooperation of our associates at The University of Texas at Austin, Department of Aerospace Engineering and Engineering Mechanics, especially the competent and devoted assistance of Mr. Otis Graf, is gratefully acknowledged. We are grateful for Professor A. Poveda's and for Miss Christine Allen's permissions to incorporate their competent comparison studies of five-body systems. Our thanks are extended to the personnel of the Computation Center of The University of Texas at Austin for their advice and erudite cooperation.

References

Bennett, T. L.: 1904, *Mess. Math.* (2) **34**, 113.
Chandrasekhar, S.: 1942, *Principles of Stellar Dynamics*, 199, Constable and Co. Publ., London, p. 199. Also (1960) Dover, New York.
Deprit, A.: 1970, private communications.
Fehlberg, E.: 1968, NASA TR, R-287.
Jacobi, C. G. J.: 1866, *Vorlesungen über Dynamik*, Reimer Publ. Berlin, p. 22.
Kustaanheimo, P. and Stiefel, E.: 1965, *J. Math.* **218**, 204.
Lagrange, J.: 1772, *Oeuvres* **6**, Gauthier-Villars, Paris, 1873.
Lecar, M.: 1968, *Bull. Astron.* (3), **3**, 91.
Levi-Civita, T.: 1903, *Ann. Math.* (3), **9**, 1.
Poincaré, H.: 1905, *Leçons de Mécanique Céleste*, Vol. **1**, pp. 19–22, Gauthier-Villars, Paris.
Poveda, A., Ruiz, J., and Allen, C.: 1967, *Publ. Obs. Tonantzintla and Tacubaya*, Mexico.
Sundman, K. F.: 1912, *Acta Math.* **36**, 105.
Szebehely, V.: 1968, *Bull. Astron.* (3), **3**, 33.
Szebehely, V. and Peters, C. F.: 1967, *Astron. J.* **72**, 876.
Whittaker, E. T.: 1904, *Analytical Dynamics*, Cambridge University Press, Cambridge, p. 267. Also (1944), Dover, New York.

A MULTI-PARTICLE REGULARISATION TECHNIQUE*

D. C. HEGGIE

Institute of Theoretical Astronomy, Cambridge

Abstract. Certain features of a practical regularisation method, in which the potential or kinetic energy is used as a time-regularising function, are described. For two-body encounters the method is less powerful than Kustaanheimo-Stiefel regularisation, but it has wider applicability.

1. Early experience of N-body computations by many workers made clear the need for an efficient (i.e. rapid and accurate) method of treating close two-body encounters and, in particular, stable binaries. Two methods, both based on a description of such events as perturbed two-body motion, have been proposed and used: the 'binary method' of Aarseth (1970), and Kustaanheimo-Stiefel regularisation, described by Peters (1968). In practice, the binary method is unsatisfactory for moderate or heavy perturbations, and regularisation is inefficient at dealing with relatively stable three-body configurations, where the choice of the pair of particles whose relative motion is to be regularised requires repeated alteration. It is therefore desirable to devise an efficient means of treating such cases, and one possible method is the subject of this paper. Although it is not suggested that the method be used for straightforward two-body encounters, for which powerful techniques already exist, this is nevertheless the simplest situation in which to examine its properties.

In the K.-S. regularisation method, in which both time and position are transformed, the unperturbed relative motion of two point masses is described by the equations

$$u_k'' = \frac{h}{2\mu} u_k \qquad (k = 1, 2, 3, 4) \tag{1}$$

where $\mathbf{u}$ is the 4-vector of transformed coordinates, h is the total energy, μ is the reduced mass, and a prime denotes differentiation with respect to regularised time. Suppose, however, that only the time transformation is performed, that is, we define the regularised time, s, by the differential equation

$$ds = g(\mathbf{x}, \dot{\mathbf{x}})\, dt$$

where g is some well-behaved function of $\mathbf{x}$, the relative position vector, and of $\dot{\mathbf{x}}$, its derivative with respect to physical time t. Then if m_i $(i=1, 2)$ are the masses of the two particles, and we write $r \equiv |\mathbf{x}|$, the normal Newtonian equations

$$\ddot{x}_k = -(m_1 + m_2)\, x_k r^{-3} \qquad (k = 1, 2, 3) \tag{2}$$

become

$$g^2 x_k'' + g^{-1} \dot{g} \dot{x}_k = F_{ck} \qquad (k = 1, 2, 3)$$

* This work was carried out during tenure by the author of a Studentship from the U.K. Science Research Council.

M. Lecar (ed.), Gravitational N-Body Problem, 148–152. *All Rights Reserved*

where F_{ck} is the right hand side of (2). Choosing $g(\mathbf{x}, \dot{\mathbf{x}})=r^{-1}$, this equation may be written as

$$x_k'' = r^2 F_{ck} + r\dot{r}\dot{x}_k \qquad (k = 1, 2, 3). \tag{3}$$

It is this equation which will be generalised to the N-body problem ($N \geqslant 3$), but in order to study its properties, it is convenient to write it in the form

$$x_k'' = -(m_1 + m_2) x_k r^{-1} + r\dot{r}\dot{x}_k - \dot{\mathbf{x}}^2 x_k + \\ + 2x_k \left(\frac{h}{\mu} + \frac{m_1 + m_2}{r}\right) \qquad (k = 1, 2, 3)$$

where, as before,

$$h \equiv \tfrac{1}{2}\mu\dot{\mathbf{x}}^2 - \frac{m_1 m_2}{r}.$$

Hence

$$x_k'' = \frac{2h}{\mu} x_k + C_k \qquad (k = 1, 2, 3) \tag{4}$$

where

$$\mathbf{C} \equiv (m_1 + m_2)\,\mathbf{x} r^{-1} + \dot{\mathbf{x}} \times (\dot{\mathbf{x}} \times \mathbf{x}).$$

With the aid of (2) it is easy to show that $\mathbf{C}$ is a constant vector, which, incidentally, is directed along the line of apsides. It is of interest to note that, since (4) may be readily generalised to the case of perturbed two-body motion and the first derivatives of $\mathbf{C}$ and h are regular at $r=0$, these equations could be used in a two-body regularisation method; however, that is not the subject of this paper. Comparison of (4) with (1) shows that, when the coordinate transformation is performed, the period of the motion (in case $h<0$) is doubled. Therefore, we may expect that, to achieve a certain accuracy in a numerical integration, twice as many steps per physical orbit are needed to integrate (3) as to integrate (1). That (3) is nevertheless generally more efficient than the ordinary equations (2), can be appreciated in the following way. Suppose one uses a time-step criterion based on the rate of convergence of the Taylor series for the velocity, such as

$$(\Delta t)^2 = \eta^2 \frac{\eta|\dot{\mathbf{x}}| + \Delta t|\ddot{\mathbf{x}}|}{\eta|\dddot{\mathbf{x}}| + \Delta t|\mathbf{x}^{\mathrm{IV}}|}$$

and the analogue for Δs, the regularised time-step for integrating (3). In this expression, which is a modification of that used by Aarseth (1968), η is an adjustable constant. Then at pericentre one finds the approximate result

$$\left(\frac{\Delta t_3}{\Delta t_2}\right)^2 = \frac{2a}{r} \tag{5}$$

where a is the semi-major axis, and subscripts refer to the above numbering of the equations; one sees that, for close encounters, a much larger timestep can be taken for system (3) than for system (2). In practice the advantage gained is not as great as this, for reasons that will appear later.

2. In general, a close 2-body encounter in an N-body problem ($N \geqslant 3$) can be treated similarly if the total potential energy, V, is used as the regularising function, for V is dominated by any two-body encounter. The regularised force per unit mass on any particle is

$$F_k = V^{-1}(V^{-1}F_{ck} - V'x_k') \qquad (k = 1, 2, 3) \tag{6}$$

where F_{ck} is the corresponding Newtonian force. This system of equations is unsatisfactory for two reasons:

(i) A close encounter between two bodies reduces the time step for all bodies, because of the second term in the expression for the force. This is very inefficient for large N.

(ii) The calculation of V, involving about $\frac{1}{2}N^2$ full precision distance calculations per step, is very time-consuming. It could be calculated from the regularised kinetic energy, T, and the total energy, H, by the relation

$$V^2T - V - H = 0.$$

However this procedure is very inaccurate, because (Figure 1) $(dT/dV) = 0$ at $V = -2H$, i.e. when the virial ratio is unity. In most practical cases, V soon tends to fluctuate about this value, and then it is ill-determined by T.

The second objection can be met by using as regularising function the kinetic energy

$$\mathcal{T} \equiv \tfrac{1}{2} \sum_{i=1}^{N} m_i \dot{\mathbf{x}}_i^2 .$$

Because of the presence of the term in $\mathcal{T}'$ in the regularised force, $\mathbf{F}$ must be known for every particle at each time-step. An explicit calculation is too expensive if individual

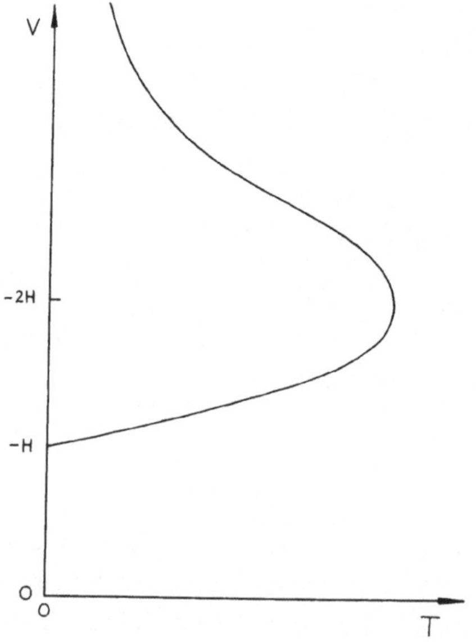

Fig. 1. The relation between V and T for $H < 0$, where T is the V-regularised kinetic energy.

time-steps are taken by each particle, and one can show that the method is unstable if **F** is extrapolated from previous, explicitly calculated values. It is therefore expedient to use the same time-step for each particle, which reinforces the first objection above.

One may meet the first objection by taking as regularising function the kinetic energy, $\mathscr{T}^*$, of a small number N_{reg} of the particles, where $N_{\text{reg}} \ll N$. In practice one selects those involved in binaries or in encounters, so that very often $N_{\text{reg}}=0$ or 2; and one's choice is reviewed from time to time. Then one still uses the unregularised equations, and individual time-steps, for all other particles. For the 'regularised' particles, analogues of (6) are used, with a common time-step: this is not inefficient if N_{reg} is small, and indeed the efficiency may be improved, because several parts of a time-step calculation are shared between all such particles. At this stage there is little to choose between $\mathscr{T}^*$ and V^* as regularising functions, and in the remainder of the paper the former is referred to.

3. It remains to point out a number of features of the equations which become apparent in practical application. Here reference is made to computations carried out on the Institute's IBM 360/44, in which variables are normally carried to a standard precision of about six significant decimal digits; for very accurate work, a facility exists whereby the precision may be extended to up to about sixteen digits.

Consider an isolated binary system. Suppressing the asterisks, we have

$$\mathbf{x}_i'' = T\,(T\mathbf{F}_{ci} - \mathscr{T}'\mathbf{x}_i') \qquad (i = 1, 2) \tag{7}$$

where $T\mathscr{T}=1$. Weighting each equation by the corresponding mass and adding, one obtains, for the motion of the centre of mass, the equation

$$\bar{\mathbf{x}}'' = -\,T\mathscr{T}'\bar{\mathbf{x}}'$$

which is unstable if $\mathscr{T}'<0$, i.e. after pericentre. In practice this instability is not serious.

It is usually sufficient to compute and carry the force in standard precision, and so it is subject to considerable rounding error which may sometimes feed back into the force via T, the regularised kinetic energy. This cannot be obviated by reducing the steplength if the rounding error in the force has non-zero expectation, which is the case with those computers in which conversion from extended to normal precision is effected by truncation rather than by rounding. Equations (7) are not easy to discuss, but the one-dimensional analogue of the equation of relative motion of the two bodies reduces, in case $m_1=m_2=2$, to

$$x'' = (x')^4 x^{-2}\,(1 + \varepsilon f(s))$$

which will be adopted as a model of the system (7). Here, $\varepsilon f(s)$ is the rounding error, so that $f(s)=\mathrm{O}(1)$ as $\varepsilon\downarrow 0$; in practice $\varepsilon\approx 10^{-6}$.

Using ε as expansion parameter we find, for $\varepsilon\downarrow 0$, the particular asymptotic solution

$$x \sim x_0 + \varepsilon x_1$$

where

$$x_0 \equiv \tfrac{1}{8}s^2$$

and

$$x_1 \equiv -\tfrac{1}{12}s \int_{s_0}^{s} f(v)\,dv + \tfrac{1}{12}s^4 \int_{s_0}^{s} v^{-3} f(v)\,dv$$

if $x_1(s_0) = x_1'(s_0) = 0$; $s_0 < s < 0$ during the approach to encounter. If a stepwise integration procedure is used, these integrals must be replaced by sums, and one finds that, as $s \uparrow 0$ (i.e. for very close encounters), the first term of x_1 dominates. If the expectation of f vanishes, this term is $O(s\sqrt{\Delta s})$ as $s \uparrow 0$, where Δs is the steplength, and if the expectation of f is non-zero it is $O(s)$; in this case, the relative error is of order $\varepsilon |x|^{-1/2}$. Experience confirms the presence, during particularly close encounters, of an instability that cannot be removed by a reduction in steplength; it can be obviated by reducing ε (e.g. by calculating the force in extended precision), and little inefficiency is introduced as this procedure is required only in rather exceptional cases. It is probable that this instability would be less serious in a computer with a more accurate standard precision. Rounding error can still be introduced in the higher terms of the integration scheme, but may be controlled by a modest reduction of steplength. Because of velocity-dependent terms in the force, significant truncation error in the velocity must be avoided, by the same reduction in the time-step. It is for reasons such as these that Equation (5) overestimates the efficiency of the method.

A computer program, constructed on these principles, has been written in FORTRAN IV and applied to some standard problems. For example, the IAU 25-body problem (Lecar, 1968) has been integrated to $t = 10$ in 16 min, the energy decrement being only 8×10^{-7}, although this results from the fortuitous near-cancellation of two decrements of 3×10^{-6}. Even so, the test compares very well with all those discussed by Lecar. In two integrations of comparable accuracy up to $t = 5$, respectively with and without regularisation, it was found that regularisation effects a reduction in computing time of about 50%.

Acknowledgement

It is a pleasure to record the constant help and encouragement in this project from my supervisor, Dr. Aarseth.

References

Aarseth, S. J.: 1968, *Bull. Astron.* **3**, 124.
Aarseth, S. J.: 1970, *Astron. Astrophys.* **9**, 64.
Lecar, M.: 1968, *Bull. Astron.* **3**, 91.
Peters, C. F.: 1968, *Bull. Astron.* **3**, 167.

THE USE OF INTEGRALS IN NUMERICAL INTEGRATIONS OF THE *N*-BODY PROBLEM

PAUL E. NACOZY

The University of Texas at Austin

Abstract. The numerical integration of systems of differential equations that possess integrals is often approached by using the integrals to reduce the number of degrees of freedom or by using the integrals as a partial check on the resulting solution, retaining the original number of degrees of freedom.

Another use of the integrals is presented here. If the integrals have not been used to reduce the system, the solution of a numerical integration may be constrained to remain on the integral surfaces by a method that applies corrections to the solution at each integration step. The corrections are determined by using linearized forms of the integrals in a least-squares procedure.

The results of an application of the method to numerical integrations of a gravitational system of 25-bodies are given. It is shown that by using the method to satisfy exactly the integrals of energy, angular momentum, and center of mass, a solution is obtained that is more accurate while using less time of calculation than if the integrals are not satisfied exactly. The relative accuracy is ascertained by forward and backward integrations of both the corrected and uncorrected solutions and by comparison with more accurate integrations using reduced step-sizes.

1. Introduction

This paper presents a method to efficiently utilize the integrals in the numerical integration of gravitational systems. The method yields solutions of a higher accuracy while using less time of calculation than conventional procedures of numerical integration that do not use the integrals directly.

A gravitational system of n-bodies that has p integrals may be described uniquely by $(6n-p)$ position and velocity variables in the phase space. The integrals, such as energy, angular momentum, or center of mass, may be regarded as conditions of constraint imposed upon the $6n$ variables. The p integrals constrain the variables of a solution to remain on the intersection of p hypersurfaces, each of $(6n-1)$ dimensions. The intersection is a hypersurface of $(6n-p)$ dimensions.

It is common practice to integrate numerically the full, $6n$th order system of the equations of motion and to employ the integrals only as partial checks on the accuracy of the calculations. But the errors indicated by the integrals are often somewhat misleading. Since the satisfaction of the integrals is only necessary but not sufficient to guarantee accuracy of the solution, a computed solution of a gravitational system frequently has a larger error in the solution than in the integrals.

Moreover, if the error introduced by not satisfying the integrals remains in the solution during a process of numerical integration, and if the system is unstable, the solution with the error will diverge from a solution without the error. Since gravitational systems are very often unstable in the Liapunov sense, the small errors introduced by not satisfying the integrals will often become unbounded with time. Deeper and more

M. Lecar (ed.), Gravitational N-Body Problem, 153–164.

extensive discourses on this concept have been given by Miller (1964) and Szebehely (1968).

The question is whether or not the additional time of calculation necessary to satisfy the integrals is worth the resulting increase in accuracy. That is, could the increased accuracy be obtained in less time of calculation by merely decreasing the truncation error of the numerical integration and not satisfying the integrals exactly?

The answer appears to depend upon which method is used to satisfy the integrals. The integrals may be used to reduce the order of the system to $(6n-p)$, but the integration of the reduced system is often expensive in calculation time. The resulting equations of motion of the reduced system may be much more complex than the original system of order $6n$ due, for example, to the non-linearity introduced by the integral of energy. Also, the equations of order $(6n-p)$ may have lost the symmetry of the original system.

In this paper, it will be shown that the integrals may be satisfied without introducing additional complexity nor losing symmetry. The method introduces constraints on the solution of a numerical integration of the full system of order $6n$. During the computations, corrections are calculated and applied to the $6n$ variables to satisfy the integrals. The corrections are determined by the method of least-squares such that the sum of the squares of the corrections is minimized. The corrections generally are small and hence the integrals may be linearized, eliminating much of the complexity and reducing significantly the calculation time. The corrections, determined in this manner, modify those variables that are most in error so as to greatly increase the effectiveness of the method. This point will be discussed later.

The idea of using corrections of least-squares to satisfy the integrals has a geometrical interpretation. During an integration, errors in the calculation may cause the solution to leave the hypersurface of $(6n-p)$ dimensions defined by the integrals. The least-squares corrections to the $6n$ variables return the solution to the surface *along the normal* to the surface. By continually correcting the variables, the solution remains on the original integral hypersurface during the numerical integration.

Constraining the computed solutions of gravitational systems of n-bodies to remain on the proper integral surfaces has been performed previously by Aarseth (1966) and Miller (presented elsewhere in this volume). Aarseth corrects the integrals, the positions, and the velocities of the computed solution to account for the removal of escaping bodies from the system. Miller compares a corrected solution of the system with a similar, but uncorrected, solution. He finds that the two solutions diverge from each other – indicating the instability of the gravitational system. Neither of the two studies proposes to satisfy the integrals in order to produce a more accurate and efficient integration procedure.

In this paper, the method of satisfying the integrals will be derived and discussed, and its application outlined. The results of applying the method to several dynamical systems are presented. It is shown that solutions of gravitational systems that are corrected by the method are considerably more accurate and require less time of calculation than uncorrected solutions.

2. The Equations of Constraint

The equations that constrain the solution to remain on the original integral hypersurface are derived by the use of Lagrangian multipliers. A presentation of the method of Lagrangian multipliers with an excellent motivation for the present discussion is given by Lanczos (1949). A geometrical approach is given by Forsyth (1930).

To find extrema of a function of two variables, $f(x, y)$, subject to the constraining condition

$$g(x, y) = \text{constant}, \tag{1}$$

the two equations

$$\frac{\partial f}{\partial x} - \lambda \frac{\partial g}{\partial x} = 0, \quad \frac{\partial f}{\partial y} - \lambda \frac{\partial g}{\partial y} = 0, \tag{2}$$

are to be solved with Equation (1), to determine the quantities x, y, and λ. Here, λ is the Lagrangian multiplier.

For a dynamical system with two degrees of freedom, let $x = [x_1, x_2, x_3, x_4]$ be the state vector in the phase space, where x_1 and x_2 are the coordinates and x_3 and x_4 are the corresponding velocity components. Let

$$g(x) = 0, \tag{3}$$

be an integral of the system. Equation (3) defines a hypersurface of three dimensions imbedded in the phase space of four dimensions.

During a process of numerical integration of the system, a computed solution is obtained at time t:

$$\eta = \eta(t) = [\eta_1, \eta_2, \eta_3, \eta_4],$$

where η_1 and η_2 are the computed position components and η_3 and η_4 are the computed velocity components. Due to errors in the computational procedure, the integral may not be satisfied exactly but

$$g(\eta) = \varepsilon, \tag{4}$$

where ε is a small quantity. The solution has left the integral surface defined by Equation (3) and is on the surface defined by Equation (4). It is desired to make corrections $\Delta\eta = [\Delta\eta_1, \Delta\eta_2, \Delta\eta_3, \Delta\eta_4]$ to the computed vector η to obtain the vector

$$x = \eta + \Delta\eta,$$

such that

$$g(x) = 0. \tag{5}$$

The square of the magnitude of the correction vector $\Delta\eta$ may be written as

$$f(\Delta\eta) = \sum_{i=1}^{4} (\Delta\eta_i)^2. \tag{6}$$

The corrections are uniquely chosen so that the function f of Equation (6) is minimized, subject to the constraint of Equation (5). The solution may be obtained by extending Equations (2) to four dimensions, yielding

$$\Delta\eta_i - \lambda \frac{\partial g}{\partial \eta_i} = 0, \quad i = 1, 2, 3, 4. \tag{7}$$

Equations (5) and (7) may be solved for the five unknowns λ, and $\Delta\eta_i$, $i=1, 2, 3$, and 4. Unless the integral given by Equation (5) is a simple function of the variables (for instance linear), the solution of the system may be complex (or perhaps not obtainable). This is the case when the integral is the integral of energy of a gravitational system. The solution may be simplified by an expansion of the integral in powers of the corrections. The expansion becomes

$$g(x) = g(\eta) + \sum_{i=1}^{4} \frac{\partial g}{\partial \eta_i} \Delta\eta_i + \cdots. \tag{8}$$

Since the errors of the computation and hence the necessary corrections, $\Delta\eta_i$, are generally small, second and higher-order terms may be neglected.

Solving Equations (7) and (8) for the corrections $\Delta\eta_i$, with $g(x)=0$ and $g(\eta)=\varepsilon$, yields

$$\Delta\eta_i = \frac{-\varepsilon \dfrac{\partial g}{\partial \eta_i}}{\displaystyle\sum_{j=1}^{4} \left(\frac{\partial g}{\partial \eta_j}\right)^2}, \quad i = 1, 2, 3, 4. \tag{9}$$

The correction vector $\Delta\eta$ is added to the computed state vector η to obtain a new state vector x which satisfies the integral $g(x)=0$, with an error of order $|\Delta\eta|^2$. Geometrically, minimizing Equation (6) subject to Equation (8) causes the vector $\Delta\eta$ of Equation (9) to be normal to a three-dimensional plane which is approximately tangent to the surface $g=0$ at the point x. The equation of the plane is given by Equation (8), neglecting the second and higher-order terms.

The result of Equation (9) may be generalized to a dynamical system of order $6n$ having p integrals. Denote the state vector of the system by x, where x is a column vector in the phase space with components $x_j, j=1, 2, \ldots, 6n$. Denote the configuration or position vector of the system by R and the velocity vector by V. The state vector x may be written as

$$x = \begin{pmatrix} x_1 \\ x_3 \\ \vdots \\ x_{6n} \end{pmatrix} = \begin{pmatrix} R \\ V \end{pmatrix}. \tag{10}$$

The equations of motion of the system are

$$\dot{x} = \begin{pmatrix} \dot{R} \\ \dot{V} \end{pmatrix} = \begin{pmatrix} V \\ F \end{pmatrix}. \tag{11}$$

where the vector F is, in general, a function of the vector x and the time and will be defined explicitly later. The p integrals of the system may be written as

$$e_j(x) = 0, \quad j = 1, 2, ..., p. \tag{12}$$

The functions $e_j(x)$ are the components of a column vector E, so that $E(x)=0$, where 0 is a null vector.

Equation (11) may be solved by numerical integration yielding a computed solution with a column state vector η at time t. The partial derivatives of the integrals of Equation (12) with respect to the components of the computed state vector η are the elements of a matrix E', having p rows and $6n$ columns. That is,

$$E' = \begin{pmatrix} \frac{\partial e_1}{\partial \eta_1} \frac{\partial e_1}{\partial \eta_2} \cdots \frac{\partial e_1}{\partial \eta_{6n}} \\ \frac{\partial e_2}{\partial \eta_1} & \vdots \\ \vdots & \vdots \\ \frac{\partial e_p}{\partial \eta_1} \cdots & \frac{\partial e_p}{\partial \eta_{6n}} \end{pmatrix}.$$

At time t, due to errors in the computation, some or all of the p components of the vector E may be nonzero. That is,

$$E(\eta) = \varepsilon \neq 0,$$

where ε is an error vector whose elements are small quantities.

It is desired to compute a correction vector $\Delta\eta$ so that the vector

$$x = \eta + \Delta\eta,$$

will satisfy the equation

$$E(x) = 0.$$

The vector $\Delta\eta$ is chosen so that the quantity

$$\Delta\eta^T W \Delta\eta \tag{13}$$

is minimized. Here, W is a weighting matrix and the T superscript indicates matrix transpose.

As in Equation (8), each element of the vector E is expanded in powers of the vector $\Delta\eta$. The expansion becomes

$$E(x) = E(\eta) + E'\Delta\eta + \cdots.$$

Neglecting second and higher-order terms, with $E(x)=0$ and $E(\eta)=\varepsilon$, the expansion reduces to:

$$0 = \varepsilon + E'\Delta\eta\,. \tag{14}$$

Extending the solution given by Equation (7) yields

$$W\Delta\eta\, E - {}'^{T}\lambda = 0\,, \tag{15}$$

where λ is a column vector whose p components are the Lagrangian multipliers. Equations (14) and (15) are $(6n+p)$ equations to be solved for $(6n+p)$ unknowns: the components of the two vectors $\Delta\eta$ and λ. Solving Equation (15) for $\Delta\eta$ and substituting the result into Equation (14) gives

$$\varepsilon + E'W^{-1}E'^{T}\lambda = 0\,.$$

Solving for λ and substituting the result into Equation (15), the solution for the correction vector $\Delta\eta$, is

$$\Delta\eta = -\,W^{-1}E'^{T}(E'W^{-1}E'^{T})^{-1}\varepsilon\,. \tag{16}$$

The matrix $(E'\,W^{-1}\,E'^{T})$ is a $p \times p$, symmetrix matrix. If the matrix E' has rank p, the matrix $(E'\,W^{-1}\,E'^{T})$ is positive definite and non-singular.

For gravitational systems, the vector F of Equation (11) is given by

$$F = \nabla_R U\,, \tag{17}$$

where $\nabla_R U$ denotes a column vector whose $3n$ components are $\partial U/\partial x_i$, $i=1, 2, \ldots, 3n$. The quantity x_i is a component of the vector R defined earlier. The function U is the negative of the potential energy of the system and is defined as

$$U = k^2 \sum_{1 \leqq i \leqq j \leqq n} \frac{m_i m_j}{|r_{ij}|}\,,$$

where k is the Gaussian gravitational constant, m_i is the mass of the ith body and,

$$|r_{ij}| = \left(\sum_{p=0}^{2} (x_{3i-p} - x_{3j-p})^2\right)^{1/2}.$$

The energy integral of the system may be written in terms of U and the state vector x as

$$\tfrac{1}{2} \sum_{i=1}^{n} m_i \sum_{k=0}^{2} (x_{3n+3i-k})^2 - U - C = 0\,, \tag{18}$$

where C is the value of the energy for a set of initial conditions. Denote the energy integral as $e_1(x)=0$; the angular momentum integrals as $e_j(x)=0, j=2, 3, 4$; and the center of mass integrals as $e_j(x)=0, j=5, 6, \ldots, 10$. The functions $e_j(x)$ are the components of the column vector E, where

$$E(x) = 0\,.$$

A numerical integration of the system of Equation (11) with F defined by Equation (17), yields the solution vector η at time t. The errors in the computation may cause some or all of the components of E to be nonzero:

$$E(\eta) = \varepsilon,$$

where the components of the vector ε are the errors of the corresponding integrals. The correction vector $\Delta\eta$ may be calculated by using Equation (16), so that $E(\eta + \Delta\eta) = 0$. For the calculation, the quantities ε, E' and W^{-1} are needed.

During many procedures of numerical integration, the errors of the integrals, ε, are computed at various times as a check on the accuracy of the calculation. Since the correction vector $\Delta\eta$ is calculated and applied only at various times, as will be discussed later, little extra calculation is required to obtain ε. Moreover, in the calculation of the force vector F, the individual terms of the potential energy may be calculated as an intermediate step. One may save these calculated terms and compute the energy with a minimum of added effort.

The quantity E' is the partial derivative of E with respect to x and is easily computed. The partial derivative of e_1 with respect to V is equal to V multiplied by a mass. The partial derivative of e_1 with respect to R is simply minus the force F, pre-computed during the integration step. The partial derivative of e_j with respect to x is linear in x for $j = 2, 3, 4$, and is equal to a mass or to zero for $j = 5, 6, \ldots, 10$.

The quantity W is a diagonal weighting matrix to be included in the least-squares solution of Equation (16). In most variable stepsize numerical integration techniques, an error vector is computed at each integration step. The elements of the error vector correspond to an estimated truncation error in the calculation of each of the elements of the state vector x. The elements of the error vector might be the last differences of a finite difference integration method or the differences between the predictor and corrector of an integration step. The error vector is placed along the diagonal of the matrix W^{-1} in Equation (16). The weighting matrix allows the solution to correct some state variables proportionately more if there is a larger truncation error associated with the calculation of those state variables than of other variables. The concept of corrections that are weighted by the estimated truncation error is due to Gottlieb (1970).

The matrix multiplications of Equation (16) may be grouped efficiently as follows. The multiplication of the quantity $W^{-1}\,E'^T$ is performed and stored in E'^T. The multiplication $E'\,E'^T$ is performed and the product inverted. The multiplication and inversion are simplified since $(E'\,W^{-1}\,E'^T)$ is symmetric. The inverted matrix is pre- and postmultiplied by E'^T and -ε, respectively, to form $\Delta\eta$, which is then added to the computed state vector.

Finally, one advantageous property of the equation of corrections (Equation (16)) should be noted. In the numerical integration of a gravitational system, the largest errors in the computations often arise in the coordinates and velocities of the bodies that are closest to one another – the binaries. This fact may easily be seen if one looks at the estimated truncation error vector during a numerical integration. Hence, one would like to correct the coordinates and velocities of a binary or a triple system more than those of the other bodies. This is precisely the result achieved by Equation (16) (regardless of whether or not the weighting matrix is used). To verify this point, note that for the integral of energy, the top row of E' or $\partial e_1/\partial x$ contains the two vectors F and V. The largest elements of both of these vectors are those corresponding to the closest bodies. If one examines Equation (16), with only the energy integral present, he will notice that the state vectors of the bodies closest to one another are recipients of the largest corrections while the other bodies receive smaller corrections. A similar analysis for the other integrals yields the same result. In other words, the corrections to the state vector x are proportional to the partial derivatives of the integrals with respect to x. And the partials are larger the closer two bodies become.

3. Evaluation of the Method

The method presented here was applied to the numerical integration of several dynamical systems to determine its practical value. The systems considered were the harmonic oscillator, the gravitational system of two-bodies, and the gravitational system of 25-bodies.

The method was applied to the harmonic oscillator and to the system of two-bodies by the following procedure. Two sets of solutions were obtained by numerical integration with various initial conditions. One set of solutions did not utilize the integrals while the other set introduced corrections determined by the method. The corrections were applied to the state vector after each integration step. All of the integrations were compared with the true solutions of the systems to determine the relative accuracies of the uncorrected and corrected solutions. The solutions of the harmonic oscillator were obtained using a fourth-order Runga-Kutta integration routine with constant step size. Both uncorrected and corrected solutions of the harmonic oscillator used the same step-size and the same number of integration steps. The solutions of the system of two-bodies were obtained using a fourth-order, predicator-corrector integration routine with variable step-size. Both uncorrected and corrected solutions of the system of two-bodies were integrated simultaneously with the same step-sizes and with the same number of integration steps.

The application of the method to the harmonic oscillator in a phase space of two dimensions showed no noticeable differences in accuracy between the corrected and uncorrected solutions. The reason for this negative result will be discussed later.

The application of the method to the system of two-bodies in a phase space of four dimensions over a range of initial conditions showed a large difference in accuracy between corrected and uncorrected solutions. The corrected solutions were about three orders of magnitude more accurate than the uncorrected solutions. Some results are given in Table I.

In the table, two solutions are presented: one for eccentricity $e=0.1$ and the other for $e=0.6$. Both solutions have semi-major axes of $a=2.0$ and were integrated for a duration of 55 orbital periods. The column denoted by $|\Delta R|$ gives the magnitude of

TABLE I

Two-body system

Solution		$\|\Delta R\|$	$\|\Delta V\|$
$a = 2.0$	Uncorrected	2.2×10^{-2}	7.5×10^{-3}
$e = 0.1$ $T = 55$ rev.	Corrected	3.1×10^{-5}	9.4×10^{-6}
Solution		$\|\Delta R\|$	$\|\Delta V\|$
$a = 2.0$	Uncorrected	2.4×10^{-1}	7.9×10^{-2}
$e = 0.6$ $T = 55$ rev.	Corrected	1.4×10^{-4}	2.2×10^{-5}

the error in position and the column denoted by $|\Delta V|$ gives the magnitude of the error in velocity. The row denoted by 'Uncorrected' gives the errors at the final time for the numerical integration without corrections. The row denoted by 'Corrected' gives the errors at the final time for the numerical integration that performs corrections after each integration step to satisfy the integrals of energy and of angular momentum. The results of solutions with other initial conditions were similar to the results shown in Table I. The results of the application given in Table I indicate that numerical integrations of the gravitational system of two-bodies that identically satisfy the integrals are more accurate than integrations that do not. The demonstration of the overall accuracy and efficiency of the method by comparison of times of calculation as well as accuracy is given later in the application of the method to the system of 25-bodies.

The different results obtained for the harmonic oscillator and the system of two-bodies offers an explanation of when and why the method appears to be of value. Two points were mentioned in the introduction of this paper: (1) the errors in the integrals are generally less than the errors in the computed solution; and (2) if a dynamical system is unstable, the solution with an error will diverge from a solution without the error. With these points, the following conclusions may be given. The errors in the integral of the harmonic oscillator are small compared to the error in the state variables of the solution. Since the harmonic oscillator is a stable system, a solution with a small error will not diverge from a system without the error. This would explain the result indicating no difference between the corrected and the uncorrected solutions for the harmonic oscillator. The errors in the integrals of the system of two-bodies are also small relative to the state variables of the solution. But the system of two-bodies is unstable in the Liapunov sense and hence the system with the errors will diverge from the system without the errors. And, as seen in Table I, the corrected solution lies several orders of magnitude closer to the true solution than the uncorrected solution.

The method was applied to a gravitational system of 25-bodies using the standard initial conditions as given by Lecar (1968). A highly-accurate, uncorrected numerical integration of the system was performed at the outset. The integration technique employed methods of regularization and was developed by Szebehely and Bettis (described by them elsewhere in this volume). The truncation error of the integration was lowered to the limit of the computer capability. The system was integrated forward and backward in time, and the accuracy verified. This solution was taken as an accurate, standard solution with which other, less accurate solutions were compared.

The numerical integration routine that was used to evaluate the correction method presented here is a 7th-order, Runga-Kutta-Fehlberg, variable-step method (Fehlberg, 1966), applied to the problem of 25-bodies. Two sets of solutions were obtained with the numerical integration. First the system of order $6n$ was integrated without using the integrals. Then the system of order $6n$ was integrated and all or various combinations of the ten integrals of the system were satisfied. Various truncation error tolerances were allowed and all integrations were extended to time equal five units. This time is after a very close encounter of two-bodies and just before the total collapse of the system. All solutions were compared with the more accurate standard solution described above and also were integrated in reverse, from time equal five units to time equal zero. The comparisons and the reversals showed similar accuracies, hence only comparisons of the various solutions with the standard solution are presented here.

Some results of the comparisons are shown in Tables II and III.

TABLE II

25-Body problem – accuracy comparison

	Mean error	Time of calculation (seconds)
Uncorrected	1.6×10^{-1}	178
Corrected (1)	5.1×10^{-3}	179
Corrected (2)	1.0×10^{-3}	178

TABLE III

25-Body problem – Time of calculation comparison

	Mean error	Time of calculation (seconds)
Uncorrected	5.0×10^{-3}	228
Corrected (1)	5.1×10^{-3}	179
Corrected (2)	5.3×10^{-3}	166

The first column of Tables II and III gives the various solutions that were obtained. The first solution, denoted as 'Uncorrected', is a numerical integration of the system of 25-bodies without corrections to satisfy the integrals. The second solution, denoted as 'Corrected (1)', is the integration of the system with corrections, satisfying all ten integrals: energy, angular momentum, and center of mass. The third solution, denoted as 'Corrected (2)', is the integration with corrections, satisfying only the energy integral. Both solutions, Corrected (1) and Corrected (2), performed unweighted, least-squares corrections determined by Equation (16) given above. The second column in Tables II and III gives the error of each solution. The state vector of each solution at the final time of five units, containing 150 components of the positions and the velocities of the 25 bodies, was compared with the final state vector of the accurate, standard solution described above. Denoting the 150 differences between the standard and less accurate solutions by ε_i, $i=1, 2, \ldots, 150$; the mean error is computed by

$$\sigma = \left(\frac{1}{150} \sum_{i=1}^{150} \varepsilon_i^2\right)^{1/2}.$$

The quantity σ is given in the second column of the Tables II and III, denoted by 'Mean Error'. In addition to the calculation of the σ's, all numerical integrations were reversed. The errors indicated by the forward and backward integrations for all of the various solutions are consistent with the errors ε_i. Also, the dispersion of the quantities ε_i about the mean error σ is similar for all solutions. Hence, only the mean error σ is given in the tables. The third column of the tables, denoted by 'Time of Calculation', gives the execution time that a CDC 6600 computer required to generate each solution to a time equal to five units and with the accuracy given in the second column of the tables.

The results shown in Tables II and III indicate that the method presented here yields a more efficient numerical integration process. In Table II, a greater accuracy is obtained with the method while using the same time of calculation. And in Table III, the same accuracy is obtained with the method while using less time of calculation.

It may be seen from Tables II and III that satisfying only the energy integral ('Corrected (2)') produces more efficient numerical integrations than the integrations satisfying all ten integrals ('Corrected (1)'). The reasons for this are: (1) The energy error was about 10^5 times larger than the errors in the angular momentum and center of mass integrals; (2) The satisfaction of all ten integrals requires the inversion of a 10×10 matrix of Equation (16) as well as various matrix multiplication operations with 10×150 matrices. Whereas satisfaction of only the energy integral requires just a scalar division for the inversion and dot products of vectors of 150 dimensions. Hence, the results of Tables II and III show that, for the gravitational system and initial conditions of this application, and probably for many other systems and initial conditions, the added improvement of the corrections due to the inclusion of angular momentum and center of mass is small at a large computational expense.

Several time-saving techniques have been incorporated into the correction procedure. The most important is to calculate corrections not at every integration step but only when the corrections become significant. In the application presented here, if the error of the integral of energy increased to a value of approximately 100 times less than the desired or requested truncation error, only then were corrections calculated to satisfy the integrals.

The results shown in Tables I, II, and III, were not obtained using a weighting matrix in the calculation for the corrections. Some numerical integrations were obtained using weighted, least-squares corrections, where the weights were the truncation error vectors described in the preceding section. No appreciable difference in accuracy was noticed between the solutions using weighted corrections and solutions using unweighted corrections. This result is preliminary since too few solutions have been obtained using weighted, least-squares corrections to form a definite conclusion.

The method presented here may be applied to the numerical solution of any system of differential equations that possesses integrals. For gravitational systems, this could include the equations of motion of the restricted problem of three bodies and the equations of motion of a particle under the attraction of a non-spherical solid body. The equations of motion of the system may also be formulated in a set of regularized variables, as long as the variables are constrained by one or more integral relations.

Acknowledgements

The motivation for this study came from a paper by Szebehely (1968). For the many illuminating discussions during the preparation of this study, I am most grateful to Professor Szebehely. To D. Bettis and R. Gottlieb goes my warmest thanks for many discussions and contributions. In particular, I thank Mr. Gottlieb for his contribution of the weighting matrix so that the estimated truncation errors may be included into the corrections. I thank R. Trevino for his diligence in performing many of the numerical comparisons, C. Weiss for his study of the magnitude of the corrections, and A. Rios-Neto for applying the method to the harmonic oscillator. The financial support of the National Science Foundation, Grant No. GP-27369, and the Office of Naval Research, Grant No. N00014-67-A-0126-0007, is gratefully acknowledged.

References

Aarseth, S. J.: 1966, *Monthly Notices Roy. Astron. Soc.* **132**, 35.
Fehlberg, E.: 1966, *Z. Angew. Math. Mech.* **46**, 1.
Forsyth, A. R.: 1930, *Geometry of Four Dimensions*, I, Cambridge, p. 80.
Gottlieb, R.: 1970, private communication.
Lanczos, C.: 1949, *The Variational Principles of Mechanics*, Toronto, p. 43.
Lecar, M.: 1968, *Bull. Astron.* **3**, 91.
Miller, R. H.: 1964, *Astrophys. J.* **140**, 250.
Szebehely, V. G.: 1968, *Bull. Astron.* **3**, 33.

PART II

COLLISIONLESS SYSTEMS

A. ANALYTIC TREATMENTS

COLLISIONLESS STELLAR DYNAMICS*

G. CONTOPOULOS
University of Thessaloniki, Greece

1. Introduction

Collisionless Stellar Dynamics deals with Stellar Systems for times small with respect to the relaxation time. Thus it disregards completely individual encounters between stars. Our basic equation is the collisionless Boltzmann equation

$$\frac{\partial f}{\partial t} + \bar{v} \cdot \frac{\partial f}{\partial \bar{x}} - \frac{\partial V}{\partial \bar{x}} \cdot \frac{\partial f}{\partial \bar{v}} = 0. \tag{1}$$

This equation has been used in Stellar Dynamics ever since Jeans and Eddington. However, most of the recent progress in Collisionless Dynamics comes from Plasma Physics. This branch of Physics has had a spectacular development in the last two decades. Its practical applications and in particular the prospect of harnassing thermonuclear power have proved much more attractive than the theoretical delights of the dynamics of Stellar Systems. This is why the effort put into Plasma Physics much exceeds that put into Stellar Dynamics over the past decades. Thus Stellar Dynamics had to learn new methods and approaches from Plasma Physics, and that has changed its character accordingly. We will see some of these new approaches presently.

The basic Equation (1) of Collisionless Dynamics can be considered from two main points of view.

(a) The 'classical' point of view is to consider the potential V as known and solve Equation (1) as a linear partial differential equation for the distribution function f. This is the problem of the integrals of motion. In fact f is an integral of the motion of test particles in the potential V.

In this area belong questions about 'third' integrals and adiabatic invariants, the isolating and non-isolating character of the integrals, etc. All these problems belong essentially to Orbit Theory.

(b) The second point of view is to consider V connected to f by Poisson's equation

$$\nabla^2 V = 4\pi G \varrho = 4\pi G \int f \, \mathrm{d}\bar{v}. \tag{2}$$

This is the self-consistent or self-gravitating problem. In that case Equation (1) is essentially non-linear. We can divide the self-consistent problems in three main categories:

* Invited paper presented at the IAU Colloquium No. 10, in Cambridge, England.

M. Lecar (ed.), Gravitational N-Body Problem, 169–178. *All Rights Reserved*

(1) Zero order solutions, e.g. spherical, axisymmetric, or plane parallel solutions (usually stationary); (2) Small perturbations of the above zero-order solutions; and (3) Very large perturbations.

We will concentrate here on recent developments connected with the second problem.

2. Small Perturbations

We distinguish again two types of problems, (A) linear and (B) non-linear. Most of the progress in Collisionless Stellar Dynamics deals with the linear theory, while non-linear effects have drawn attention only quite recently.

Among the particular problems of the linear theory are (i) The problem of modes; (ii) The initial value problem; and (iii) Variational approaches in stability analysis.

One of the first to introduce Plasma Physics methods into Stellar Dynamics was Lynden-Bell. In his 1962 paper on 'The Stability and Vibrations of a Gas of Stars' he followed Landau's treatment of the initial value problem in the case of an infinite system with axial symmetry, by using Fourier and Laplace transforms in space and time.

Many papers using Plasma Physics methods in Collisionless Stellar Dynamics have appeared since 1962. A useful introduction to these problems is found in Lynden-Bell's article on 'Cooperative Phenomena in Stellar Dynamics', in the Cornell lectures on Relativity Theory and Astrophysics (1967).

One area of current interest is the problem of modes.

Stellar systems have not the variety of modes appearing in Plasma Physics, due to the existence of both magnetic and electric fields and of more than one kind of particles. However some of the particular modes found in plasmas may have their counterparts in Stellar Dynamics. Recent studies of inhomogeneous plasmas, or plasmas with particular distributions of velocities, have disclosed a variety of new modes that are important in particular cases. There are also modes due to the spatial variation and curvature of the magnetic field, which could not be accounted for by previously known theory.

The basic equation in these studies is an integral equation, which is linear in the (perturbed) density but highly nonlinear in the frequency.

I will write down the *form* of such an equation in the case of a flat spiral galaxy, that has a special interest for us. We assume that the perturbed potential V_1, the corresponding surface density σ_1, and the distribution function f_1 depend on the time and azimuth through $\exp[(i\omega t - m\theta)]$, where m is the number of spiral arms, and $\Omega_s = \omega/m$ is the angular velocity of the spiral pattern. Namely we have

$$V_1 = V_1^*(r) \exp[i(\omega t - m\theta)] \tag{3}$$

and similar expressions for σ_1 and f_1.

The collisionless Boltzmann Equation (1) gives f_1 in terms of V_1, while Poisson's equation V_1 in terms of σ_1. Then if we integrate f_1 with respect to all the velocities, to find $\sigma_1 = \int f_1 \, \mathrm{d}\bar{v}$, we derive an integral equation of the form

$$r\sigma_1^*(r) = \int K_{m,\omega}(r, r')\, r'\sigma_1^*(r')\, \mathrm{d}r' . \tag{4}$$

This equation is similar to a homogeneous Fredholm equation of the second kind, except that its dependence on ω is non-linear. In fact the kernel $K_{m,\omega}(r, r')$ is a complicated function of r, r', m and ω. This equation is the basic equation of the problem of modes. It was first formulated by Kalnajs (1965) and later by Shu (1968, 1970).

The solution of this equation gives the eigenvalues ω and the eigenfunctions $r\sigma_1^*(r)$ at the same time (except for an arbitrary factor). However one cannot find an explicit solution of this equation in general. If $K_{m,\omega}(r, r')$ is known one can find solutions by numerical methods. This was done by Kalnajs (1970), in the case of a model of the Andromeda galaxy. He found a trailing growing wave with angular velocity $\Omega_s = 30$ km s^{-1} kpc^{-1} and rather open spiral arms.

In particular cases the integral Equation (4) can be simplified considerably. It is common in Plasma Physics to derive particular *dispersion relations* from the basic integral equation, by considering particular limiting cases. In the case of a flat galaxy the best known dispersion relation is the one derived by Lin (1966) and his associates.

The basic assumption of Lin's theory is that the radial wavelength λ of the spiral waves is small. In an 'asymptotic' approximation we omit all higher order terms in λ except the lowest, and we derive from the integral Equation (4) a dispersion relation of the form

$$D(\omega, k, r) = 0 . \tag{5}$$

This relation connects $\omega = 2\Omega_s$ (where Ω_s is the angular velocity of the spiral pattern consisting of two arms), with the wavenumber $k = 2\pi/\lambda$ for every r.

Similar dispersion relations for a flat system of two populations and for a cylindrical system were given by Marochnik and Suchkov (1969a, b). According to Marochnik and Suchkov the origin of the spiral arms is a two-stream instability arising from the interaction of the two populations of stars, that have different mean velocities. However most people working in this field assume a velocity distribution with only one maximum.

A dispersion relation gives a continuum of eigenvalues ω unless we specify some boundary conditions. Namely for every given ω Equation (5) gives the wavenumber k as a function of r, and by integration we find the spiral arms

$$\theta' = \theta - \Omega_s t = \tfrac{1}{2} \int k(r)\, \mathrm{d}r + \mathrm{const}\,(+\,\pi) . \tag{6}$$

It is rather difficult to specify boundary conditions for the spiral arms (6) without some arbitrariness. We may even assume that non-linear effects at the boundaries (e.g. the inner Lindblad resonance and the particle resonance) make the boundaries loose, in such a way that all solutions (6) are in principle permissible.

At this point Lin's theory appeals to observations. If we ask what spiral of the form (6) can best represent what we know about our Galaxy we find a value of $\Omega_s=\omega/2$ near 13 km s^{-1} kpc^{-1}.

On the other hand Kalnajs' modes need only one quite general restriction, namely that the energy of the perturbation is finite.

Kalnajs found that growing modes are isolated, thus we cannot expect many nearby modes ω. Of course in principle the integral equation may give more closely spaced modes (nearby values of ω) for small wavelengths ($|k|$ large). If the ω's are real they *may* even form a continuum.

Lin does not deny the existence of Kalnajs' modes. He stresses the possibility of coexistence of various modes (which is obvious in a linear theory) but he relies on the observations to stress the importance of his particular solution for ω.

At this point the question of the origin of spiral waves becomes important. The mechanism of generation of spiral waves will specify which modes will be excited. The answer can be formulated in a general way as follows: Any initial perturbation, unless devised artificially, will excite all the modes, both Kalnajs' and Lin's. After a transition period all the 'noise' will be eliminated and the modes will stand out. We can even state further that k in Lin's dispersion relation can be either negative (trailing spirals) or positive (leading spirals); thus both leading and trailing waves are excited, although some of them more than the others. The situation is roughly like hitting a bell; whatever the hitting all the modes of the bell are excited more or less.

The question now arises which modes will survive for very long times. We can be sure of Kalnajs' modes, because they are growing (until non-linear effects will stop this growth).

On the other hand tight spirals, like those observed in our Galaxy, whether they form a mode or not, go through a transition period of growth if they are to survive for a certain time. During such a period we found (Contopoulos, 1971) that initially leading spirals are transformed into trailing ones near the inner Lindblad resonance. This effect explains why trailing spirals are usually preferred. It is thus expected that Lin's trailing modes have the greatest probability of survival.

We must distinguish here between Lin's modes, which satisfy the dispersion relation (4), and Lin's spiral model of the Galaxy, which is a particular mode, corresponding to a particular fixed value of ω. The probability of having a fixed ω is essentially zero. Therefore in general ω varies in space (and time) and a differential rotation of the pattern results. At this point comes Toomre (1969) with his analysis of the behavior of groups of spiral waves. Such waves have a group velocity directed inwards, such that it brings the whole observed spiral waves near the inner Lindblad resonance after a little more than 10^9 yr. This is the main difficulty of Lin's original theory at the present moment. This theory has explained away the winding difficulty of the spiral arms due to differential rotation of the stars, by considering the spiral arms as waves. However it does not seem easy to explain away also the differential rotation of the pattern.

Kalnajs' modes do not suffer from such a difficulty. The eigenvalues of such modes are fixed by definition, except for possible slight changes due to non-linear effects. Further Kalnajs' modes are growing, thus they are developing spontaneously. However these modes do not seem to be compatible with observations in our Galaxy.

We can summarize the situation as follows.

The linear theory predicts two kinds of modes for a flat axisymmetric galaxy like our own: Kalnajs' open modes, which are trailing and growing, and Lin's tight modes, which are neither growing nor damping. Observations in our Galaxy favor Lin's modes. However such modes form groups of waves that are, presumably, eliminated in a rather short time scale.

An escape from the difficulty may come either from the non-linear theory of spiral structure, as suggested by Lin, or by reinterpreting the observations to fit Kalnajs' mode, as suggested by Kalnajs. Of course one may also try to introduce elements from other theories besides the gravitational theory. However, before we embark into such an adventure, we should first explore all the possibilities offered by the gravitational theory itself.

Here we will discuss the new effects introduced by the non-linear theory of spiral structure.

3. Non-Linear Theory

In many cases non-linear effects in a galaxy are rather small. In a few cases, however, they play a rather crucial role.

Among the non-linear effects let us mention (i) The appearance of harmonics; (ii) The 'saturation' of growing waves; (iii) Resonances; and (iv) Mode Interactions.

The first effect is easy to describe. If we write f and V in successive orders of approximation,

$$\begin{aligned} f &= f_0 + f_1 + f_2 + \cdots, \\ V &= V_0 + V_1 + V_2 + \cdots, \end{aligned} \tag{7}$$

we can split the collisionless Boltzmann equation, written in the form

$$(f, V) = 0, \tag{8}$$

into a set of equations

$$(f_0, V_0) = 0, \tag{9}$$

$$(f_1, V_0) + (f_0, V_1) = 0, \tag{10}$$

$$(f_2, V_0) + (f_1, V_1) + (f_0, V_2) = 0, \tag{11}$$

etc.

The first equation gives the zero-order solution and the second the linearized solution. Then (f_1, V_1) contains the factor

$$\exp[2i(\omega t - 2\theta)], \tag{12}$$

therefore the third equation gives f_2, and V_2 (through Poisson's equation) with a quadruple symmetry, i.e. four spiral arms. These arms will be much weaker than the two basic arms, except near the Lindblad resonances.

The second effect is usually described in Plasma Physics by a quasi-linear theory. For example in the case of a two-stream instability the quasi-linear theory accounts for the formation of a plateau in the distribution function, so that one of the streams is, in fact, eliminated and the growth stops. Such an effect has been described, in the case of a galaxy composed of two populations with different mean velocities, by Marochnik and Suchkov.

In both these effects the linear theory gives the basic solution, while nonlinear effects are considered small.

On the other hand near the main resonances in a galaxy, namely the Lindblad resonances and the particle resonance, the linear theory is not applicable. Near the Lindblad resonances the linearized Equation (10) gives a term f_1, which is larger than the zero-order term f_0, therefore the whole approximation scheme, underlying the linearization, fails.

At exact resonance the linearized equation gives a secular term. Thus we can apply its results for short times, while f_1 is still small. However, at later times the particles are trapped near two 'resonant' periodic orbits, and deviate considerably from the unperturbed epicyclic orbits of the axisymmetric case.

The resonant periodic orbits are like two perpendicular ellipses. Orbits of this form were considered by Lindblad, who called them dispersion orbits. The main difference is that our resonant periodic orbits have a fixed orientation in the rotating system. We can say that, because of non-linear effects, only two dispersion orbits are, in fact, realized. All the particles near resonance follow orbits near the one or the other resonant periodic orbit. Such orbits are called 'tube' orbits; they are of the same nature as the tube orbits found in other galactic problems by Ollongren (1965) and explained by Contopoulos (1965).

It is interesting to follow the evolution of the orbits in the case of a spiral field growing from zero. The particles start moving along the usual epicycles, but as they experience more and more the growing spiral field they approach tube orbits of the above type. This is a trapping process that differentiates the regions near resonances from the rest of the galaxy.

The distribution function at the Lindblad resonances depends on three integrals of motion of the axisymmetric background, namely the energy, the angular momentum, and the phase difference between the rotational and the epicyclic motion. The last integral is usually discarded as non-isolating. However at resonances it is isolating, and essential in avoiding secular terms in higher approximations (Contopoulos, 1970).

The density distribution near the inner Lindblad resonance can be considered as a superposition of two elliptical distributions at right angles to each other. The importance of these distributions depends on the initial distribution of matter. If they are roughly equal we have a rough quadruple symmetry in zero order; this case is the most probable one if the initial distribution of matter was roughly axisymmetric. If, however, one of the two elliptical distributions predominates, the density distribution has a bar-like structure near the center. In the case of our Galaxy such phenomena take place up to a distance of 5 kpc from the center.

Such configurations are not necessarily self consistent. However one can find self consistent solutions of this form and the only remaining question is how an initial perturbation settles into such a configuration.

The particle resonance presents a different problem. The lowest order solution does not give large terms there. However, this solution is wrong, because the underlying assumption that the orbits are approximately epicycles is not correct. The particles *near* the particle resonance do not circulate around the center of the galaxy, but they librate around the Lagrangian points L_4, L_5, at right angles from the minimum of potential (Barbanis, 1970). Thus in the case of a growing field particles move away from the spiral arms and their maximum of density is 90° away from the minima of potential at this distance. The situation is similar to the concentration of the Trojan asteroids, except that the place of Jupiter is taken by the whole spiral arms and the deviation is 90°, rather than 60°.

This effect produces a breaking of the spiral arms at the co-rotation distance.

4. Mode Interactions

The last non-linear effect mentioned above is the interaction of waves. The interaction between various modes in a plasma has been studied in recent years (see, e.g. the book by Kadomtsev on Plasma Turbulence (1965) and more recent papers). In the best known case, of 'weak turbulence', we have interactions of only a few modes. The simplest and most important interaction is that of three modes. Namely if we have three modes with eigenvalues ω_1, ω_2, ω_3 and wavenumbers $\bar{k}_1$, $\bar{k}_2$, $\bar{k}_3$ satisfying the relations

$$\begin{aligned} \omega_1 + \omega_2 &= \omega_3 , \\ \bar{k}_1 + \bar{k}_2 &= \bar{k}_3 , \end{aligned} \tag{13}$$

then the first two modes excite the third.

In the case of a flat galaxy we can understand such an interaction as follows.

In the linear approximation the various modes are independent. If we have two modes we can write

$$f_1 = f_{11} + f_{12} , \quad V_1 = V_{11} + V_{12} . \tag{14}$$

In the next approximation (f_1, V_1) contains a term with the factor

$$\exp\left[i\left(\Phi_1 + \Phi_2 + (\omega_1 + \omega_2)\, t - 4\theta\right)\right] , \tag{15}$$

where $\Phi_i' = k_i$.

Thus f_2 and V_2 contain terms of the same form. Such a term in V_2 corresponds to a forcing of the galaxy by an 'imposed' potential of quadruple symmetry with angular velocity $(\omega_1 + \omega_2)/4$. If this 'imposed' potential coincides with a mode it will be amplified, otherwise it will be damped by phase mixing. The phenomenon involves a 'resonance' between the imposed field and an eigenmode of the system. This 'resonance' is of a quite different kind from the resonances considered above (Lindblad resonances, etc.). It is like the resonant response of an organ pipe to a tune by a similar instrument.

The above imposed four-armed wave can be considered as the harmonic of a two-armed wave of the form

$$\exp\left[i\left(\tfrac{1}{2}\Phi_3 + \tfrac{1}{2}\omega_3 t - 2\theta\right)\right]. \tag{16}$$

Therefore if we count in terms of two-armed waves (waves with -2θ azimuth dependence), we have to give a multiplicity 2 to a harmonic of the above type and consider the interaction as a four-wave interaction.

In the case of two-armed spirals all wave interactions involve at least four waves. In fact the Plasma Physics condition for the addition of wavenumbers corresponds here to a condition of addition of the radial wavenumbers k_i *and* of the azimuthal angles, and the latter can match only in pairs.

A slightly more general interaction is that of four waves, where

$$\begin{aligned} \omega_1 + \omega_2 &= \omega_3 + \omega_4, \\ k_1 + k_2 &= k_3 + k_4. \end{aligned} \tag{17}$$

We may have ω_2 negative, but then ω_3 or ω_4 must be also negative, so that the azimuthal angles match.

Wave interactions in a galaxy are much more complicated than in a homogeneous plasma. The differences are

(a) the fact that k_i are functions of r, besides ω_i, and

(b) in the case of waves with amplitude varying fast with r, the relations (17) are not sufficient; one has to match also the amplitudes of the waves.

However, if we disregard this last difficulty, we can consider the possible interactions between Lin's modes (ω small, $|k|$ large) and Kalnajs' mode (ω_K large and k_K negative and absolutely small). For the present purpose we assume that ω_K is fixed and k_K is a known function of r. Then we may have

(i) Interactions between various Lin's modes.

Lin's dispersion relation gives, for any given ω and r, two values of $|k|$. The first is large and corresponds to tight spirals, while the second is small and corresponds to open spirals. The wavelength λ (distance between successive arms) of the second solution is about 40 kpc for $r=10$ kpc and larger for smaller r.

It is doubtful whether we can apply, even qualitatively, a dispersion relation derived under the assumption that λ is small to this second kind of waves. However let us consider, for the present, both kinds of waves for the same value of ω.

If all four k_i in the second relation (17) are absolutely large we must have equal numbers of positive k_i (leading waves) and negative k_i (trailing waves) in the two members of this equation. The same is true if all four k_i are small. Thus nothing particular comes out this way.

If three k_i are large and one small the relation (17) is not satisfied. We can only satisfy it if three k_i are small and one large, e.g. if $k_1 = k_2 = -k_3$, $k_4 = 3k_1 < 0$. This case represents an interaction between two open trailing waves and one open leading wave to produce a tight trailing wave. Then Equation (17) is satisfied for a given ω for every r. In fact if we eliminate k_1 between the relations

$$D(\omega, k_1, r) = 0, \quad D(\omega, 3k_1, r) = 0, \tag{18}$$

we find ω as a function of r. Thus ω is not constant in general.

(ii) Interaction between three Lin's modes and a Kalnajs' mode.

Let us take $\omega_1 = \omega_2$, $k_1 = k_2 < 0$, and $\omega_3 = \omega_K$, $k_3 = k_K(r)$. Then

$$D(\omega_1, k_1, r) = 0, \tag{19}$$

and if Equations (17) are satisfied we have also

$$D(\omega_4, k_4, r) \equiv D(2\omega_1 - \omega_K, 2k_1 - k_K(r), r) = 0. \tag{20}$$

By eliminating k_1 between these two equations we find ω_1 as a function of r; then ω_4 is also a function of r.

It may be that in one or the other of the above cases we have roughly constant ω. If this is the case, then such waves can interact for long times.

We have checked that if the eigenvalue $\omega_1 = \omega_2$ is ~ 40 km s^{-1} kpc^{-1} (Lin's mode), while ω_3 is ~ 55 km s^{-1} kpc^{-1} (Kalnajs' mode), then ω_4 is ~ 25 km s^{-1} kpc^{-1} (Lin's mode) and the second relation (17) is also approximately satisfied.

We can consider this case as an interaction between a harmonic wave of Lin's type with a Kalnajs' wave to produce a 'classical' Lin wave (the one used by Lin and his associates). Such an interaction would be stronger near the inner Lindblad resonance, where a harmonic wave with quadruple symmetry appears already in zero order in the distribution function.

Of course this particular interaction is suggested here in a highly tentative way. Its applicability is based mainly on the assumption that near the inner Lindblad resonance a stationary wave with a rough quadruple symmetry is established, which has a different angular velocity than the usual Lin wave, and which extends outwards with small amplitude.

Such a suggestion has the following advantages.

(1) It does not need the introduction of leading waves (in fact, leading waves are distorted into trailing ones near the inner Lindblad resonance).

(2) It does not need the second branch of Lin's dispersion relation.

(3) It introduces quite naturally both Kalnajs' and Lin's modes. Kalnajs' mode, which is growing, can interact with the quadruple wave and excite the usual Lin's mode.

(4) This mechanism may give a solution of the group velocity difficulty of Lin's waves. Although Lin's waves as a whole move inwards, and are lost near the inner Lindblad resonance they may be continuously excited by the above interaction. The energy in such a case can be provided by Kalnajs' mode, which is growing, therefore it does not need replenishment.

Although my suggestion is highly tentative there is no doubt that wave interactions are important in many cases. They open many possibilities that have not yet been explored. Thus it would be useful to continue further the study of wave interactions in our Galaxy, to see if they can solve some of the main difficulties of the present day spiral wave theory.

References

Barbanis, B.: 1970, in W. Becker and G. Contopoulos (eds.), 'The Spiral Structure of our Galaxy', *IAU Symp.* **38**, 343.

Contopoulos, G.: 1965, *Astron. J.* **70**, 526.

Contopoulos, G.: 1970, *Astrophys. J.* **160**, 113.

Contopoulos, G.: 1971, *Astrophys. J.* **163**, 181.

Kadomtsev, B. B.: 1965, *Plasma Turbulence*, Academic Press.

Kalnajs, A.: 1965, Ph.D. Thesis, Harvard University.

Kalnajs, A.: 1970, in W. Becker and G. Contopoulos (eds.), 'The Spiral Structure of our Galaxy', *IAU Symp.* **38**, 318.

Lin, C. C.: 1966, *SIAM J. Appl. Math.* **14**, 876.

Lynden-Bell, D.: 1962, *Monthly Notices Roy. Astron. Soc.* **124**, 279.

Lynden-Bell, D.: 1967, in *Relativity Theory and Astrophysics 2. Galactic Structure* (ed. by J. Ehlers), Amer. Math. Soc., Providence, p. 131.

Marochnik, L. S. and Suchkov, A. A.: 1969a, *Astron. Zh.* **46**, 319; 524 = *Soviet Astron.* **13**, 252; 411.

Marochnik, L. S. and Suchkov, A. A.: 1969b, *Astrophys. Space Sci.* **4**, 317.

Ollongren, A.: 1965, *Ann. Rev. Astron. Astrophys.* **3**, 113.

Shu, F. H.: 1968, Ph.D. Thesis, Harvard University.

Shu, F. H.: 1970, *Astrophys. J.* **160**, 89.

Toomre, A.: 1969, *Astrophys. J.* **158**, 899.

ON THE ORIGIN AND PERMANENCE OF GALACTIC SPIRALS

C. C. LIN

Massachusetts Institute of Technology, Cambridge, Mass., U.S.A.

Abstract*. The origin of galactic spirals is attributed to Jeans instability in the outer parts of disk shaped galaxies where the gas is predominant. The material spiral arms thus produced with the aid of differential rotation initiate a group of waves propagating towards the galactic center, creating a spiral pattern which *corotates* with the material in the outer parts of the galactic disk. This point of view is substantiated by Frank H. Shu's calculations of the spiral patterns of the Milky Way System and of the external galaxies M 33, M 51 and M 81. Corotation, as described, is found in all cases.

The permanence of a quasi-stationary spiral pattern, however, depends on mechanisms for re-enforcing these waves by 'reflection' from the central regions. A three-step process is envisaged:

(1) propagation of a group of 'short waves' towards the center of the galaxy, accompanied by an increase of the amplitude of the gravitational field;

(2) distortion of the galaxy by these waves, resulting in the formation of a bar-like structure near the center or of an oval structure at the ring of inner Lindblad resonance; and

(3) re-enforcement of the 'short waves' by the 'long waves' at the radius of co-rotation where the two families of waves have the same wavelength.

These concepts are supported by agreement with observations of (1) the 'long wave' pattern calculated by Frank H. Shu for the external galaxy M 51, and (2) the distribution of ionized hydrogen in the Milky Way System calculated by Stuart Feldman.

* A more detailed presentation of these ideas may be found in the Invited Discourse delivered by the writer at the General Assembly of the International Astronomical Union (Aug. 24, 1970). The text has been published in *Highlights of Astronomy, 1970* (D. Reidel, Dordrecht, 1971). The work reported was partially supported by a grant from the National Science Foundation.

THE HOSE-PIPE INSTABILITY IN STELLAR SYSTEMS

R. M. KULSRUD and JAMES W. K. MARK*
Plasma Physics Laboratory, Princeton University, Princeton, New Jersey 08540
and
A. CARUSO
Laboratori Gas Ionizzati, Frascati, Italy

1. Introduction

Indications are that instabilities play an important role in many of the phenomena of stellar dynamics. Examples of such phenomena are: the formation of spiral arms, and the evolution of stellar clusters at a rate faster than one would expect from normal two-body collisions. The analogy with the situation in plasma physics, where similar phenomena are known to be dominated by instabilities, is very suggestive that one might seek for instabilities in stellar dynamics that correspond to similar ones in plasma physics.

However, there is a difficulty. Most plasma instabilities only occur for $\lambda > D$, where D is the Debye length. (For shorter lengths, collective forces are insufficient to hold particles bunched against dispersive thermal forces.) On the other hand, plasma instabilities are usually treated under the simplifying assumption of an infinite uniform medium. This assumption is valid for many plasma systems whose size L is many times larger than D. In gravitating systems, it cannot be strictly valid, since one always has the approximate relation $L \sim D$, where D here refers to the Jeans length. Thus, in treating collective instabilities in stellar systems one is forced to study them on the basis of an inhomogeneous theory. This prevents a bodily carrying over of all plasma instabilities into stellar dynamics.

Therefore, in searching for collisionless instabilities one cannot rely on those already discovered in plasma theory, but must start from scratch, working with inhomogeneous theory. One must take as simple an equilibrium as possible and treat its possible instabilities exactly, by the full Poisson-Vlasov equations in order to be certain as to the correctness of a decision about stability or instability. Two such simple equilibria are the spherically symmetric one, characteristic of stellar clusters, and the one-dimensional slab equilibrium closely related to the galactic disk.

The former type has been rather extensively investigated (Antonov, 1960. 1962, 1969; Lynden Bell and Sanitt, 1969; Ispser and Thorne, 1968), but as yet no instability has been found. The nonrotating slab, although not so realistic since it is only a true equilibrium in one direction, has been investigated with more success. In this paper we wish to discuss an instability in this equilibrium.

* Present address: Mathematics Department, Massachusetts Institute of Technology, Cambridge, Mass.

M. Lecar (ed.), Gravitational N-Body Problem, 180–183.

2. The Slab Equilibrium

By a slab equilibrium we mean a one-dimensional, self-consistent equilibrium with the distribution function $F=F(\varepsilon, v_y, v_z)$, $\varepsilon \equiv v_x^2/2+\Phi$, and potential $\Phi=\Phi(x)$. (We restrict ourselves to those equilibria for which $\partial F/\partial\varepsilon \equiv F_\varepsilon<0$.) One-dimensional perturbations, depending only on x, are stable (Milder, 1966). However, perturbations ϕ in Φ, f in F (which depend on y as $\phi \sim \exp iky$, and which are symmetric in x) are unstable for sufficiently small k (Kulsrud and Mark, 1970) – see Figure 1a. These are of the same character as normal Jeans instabilities for fluid systems, and have the same growth

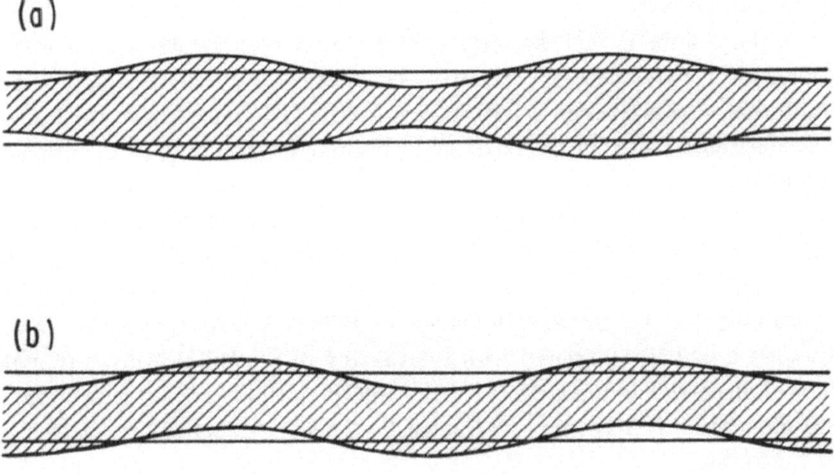

Fig. 1. (a) Jeans instability; (b) Hose-pipe instability.

rates and critical wave number. However, before one can be sure that such instabilities are realistic, one must worry about rotational forces or whatever gives the slab equilibrium in the y and z directions. If one agrees that rotational forces are negligible if the growth rate ω exceeds the rotation rate Ω, one can conclude that these instabilities are real.

In this paper we wish to report on the other possibility for instability; namely, the buckling instability of the slab for which ϕ is not symmetric in x (Figure 1b). This instability occurs when $\sigma_y>\sigma_x$, where $\sigma_y \equiv \langle v_y^2\rangle$ and $\sigma_x \equiv \langle v_x^2\rangle$, where $\langle v_x^2\rangle$ and $\langle v_y^2\rangle$ are the dispersions of the velocity in the x and y directions, respectively. It will grow when the centrifugal force arising from stars passing over the buckling exceeds the gravitational restoring force. This instability is analogous to the hose-pipe instability in plasma physics, but there are differences. For long horizontal wavelengths the gravitational restoring force dominates and the buckling is stable. Thus, the instability occurs for *shorter wavelengths* only, but not, however, for wavelengths short enough to be comparable to the thickness of the slab. On the other hand, the plasma instability exists for all wave numbers. This instability is thus complementary to the Jeans instability, in which the longer wavelengths are unstable.

There are two reasons for discussing this instability. First, we admit that the model is not quite realistic, since the horizontal forces are not self-consistent because the slab extends infinitely in the y and z directions. Still, it is possible to give a precise treatment of the Vlasov-Poisson equations for the linearized small-amplitude instabilities. By employing one small parameter kx_0, where $2x_0$ is the slab thickness, a simple closed solution can be obtained to lowest significant order. This solution would ordinarily be that for simple gravitational waves of the slab, but if we order $\sigma_y/\sigma_x \sim (kx_0)^{-1}$ the wave is converted to an instability, while the perturbation theory remains correct.

Second, the instability has the interesting property that wavelengths *shorter* than a certain critical wavelength are unstable. This critical wavelength is approximately the length that an equilibrium would have if it were self-consistent in the horizontal y and z directions. Thus, the approximation of homogeneity in these directions may not be so bad.

As a possible illustration of this instability we may consider the galactic disk with the y and z directions in the plane of the disk. The large velocity difference between population I and population II gives a large value for σ_y. If we assume $\sigma_y/\sigma_x \approx 10$, one finds instability for $1 < \lambda/2\pi x_0 < 10$ or $1 \ll \lambda < 10$ kpc, where $x_0 = 150$ pc, $\lambda = 2\pi/k$. These values might be compared with the spacing of the spiral arms, although the direction of dispersion is not correct for this interpretation. (Also, it is problematical whether population II stars may be treated by the slab model as they have a much greater thickness.)

3. The Dispersion Relation

The actual calculation of the growth rate of the instability is as follows: the equilibrium equation is

$$(\mathrm{d}^2\Phi/\mathrm{d}x^2) = 4\pi G \int F_0\left[(v_x^2/2) + \Phi, v_y, v_z\right] \mathrm{d}^3v\,. \tag{1}$$

Assume, for definiteness, that F_0 vanishes for $|x| > x_0$. The linearized equation for f_1 can be solved and substituted into the equation for ϕ_1 to get

$$\begin{aligned}(\mathrm{d}^2\phi/\mathrm{d}x^2) - k^2\phi &= 4\pi G n \\ &= 4\pi G \int \mathrm{d}^3v\left[(\partial F/\partial\varepsilon)\,\phi - J\left(\bar{\omega}\,(\partial F/\partial\varepsilon) + k\,(\partial F/\partial v_y)\,x\right)\right],\end{aligned} \tag{2}$$

where $\bar{\omega} = \omega - kv_y$,

$$J(\varepsilon, x) = \int_0^{\tau} \mathrm{d}\tau'\phi' \cos\bar{\omega}\,(\tau - \tau') - (\sin\bar{\omega}\tau/\cos\bar{\omega}\tau_0) \int_0^{\tau_0} \mathrm{d}\tau'\phi' \cos\bar{\omega}\,(\tau - \tau'), \tag{3}$$

$$\tau(\varepsilon, x) = \int_0^{x} \mathrm{d}x/v_x(x, \varepsilon), \tag{4}$$

and $\tau_0 = \tau(\varepsilon, x_0)$. For ϕ even in x, replace $\sin\bar{\omega}\tau/\cos\bar{\omega}\tau_0$ by $-\cos\bar{\omega}\tau/\sin\bar{\omega}\tau_0$. The boundary condition on ϕ is

$$(\mathrm{d}\phi/\mathrm{d}x) = -|k|\,\phi, \qquad x = \pm x_0. \tag{5}$$

Now, expand in k, but assume $v_y^2/v_x^2 \sim O(k^{-1})$. Then, to lowest order, the solution is $\omega=0$ and $\phi = \delta\,\mathrm{d}\Phi_0/\mathrm{d}x$, corresponding to a grid displacement δ in the x direction. For this value one finds (if $\langle v_y \rangle = 0$)

$$n = \phi(\mathrm{d}N/\mathrm{d}\Phi) + \int \mathrm{d}^3 v F_\varepsilon \delta x(\omega^2 + k^2 \langle v_y^2 \rangle). \tag{6}$$

In order to find a solution which in first order satisfies the boundary condition (5), we must choose $\omega \sim O(k^{1/2})$. Then, substituting the result into (2) and demanding a solution in next order which satisfies (5) ,we get [by integrating Equation (2)]

$$\begin{aligned} \omega^2 &= 2\pi k G N - k^2 \langle v_y^2 \rangle \\ &= 4\pi G \langle \rho \rangle \left[k x_0 - \alpha (k x_0)^2 (\langle v_y^2 \rangle / \langle v_x^2 \rangle) \right]. \end{aligned} \tag{7}$$

Here, $\langle v_y^2 \rangle$ is the mean value of v_x^2 averaged over all particles, $N = 2x_0 \langle \varrho \rangle$. N is the number of particles per unit area and α a constant of order unity defined by

$$\alpha \langle v_x^2 \rangle = 4\pi G \langle \varrho \rangle x_0. \tag{8}$$

From Equation (7) we see that the system is indeed unstable for

$$1 \gg k x_0 > \langle v_x^2 \rangle / \alpha \langle v_y^2 \rangle. \tag{9}$$

This instability was discovered independently by Toomre (1967). He has found in a numerical investigation that the instability only occurs for $\sigma_x/\sigma_y < 0.1$, so it would appear difficult for such an instability to occur.

Acknowledgement

One of the authors (R.M.K.) was supported by U.S. Air Force Office of Scientific Research Contract AF-44620-70-C-0033.

We should like to thank Alar Toomre for many helpful ideas in connection with this work.

References

Antonov, V. A.: 1960, *Soviet Astron.* **4**, 859.

Antonov, V. A.: 1962, *J. Leningrad Univ.* 7, 135 (available in translation Princeton Plasma Physics Laboratory Report PPL-Trans-1 (1969) 34 pp.).

Ipser, J. R. and Thorne, K. S.: 1968, *Astrophys. J.* **154**, 251.

Kulsrud, R. M. and Mark, J. W. K.: 1960, *Astrophys. J.* **160**, 471.

Lynden Bell, D. and Sanitt, N.: 1969, *Monthly Notices Roy. Astron. Soc.* **143**, 167.

Milder, D. M.: 1969, 'Rotating Stellar Systems', unpubl. Ph.D. Thesis, Harvard Univ.

Toomre, A.: 1966, 'A Kelvin-Helmholtz Instability', Notes from the Geophysical Fluid Dynamics Summer Program, Woods Hole Oceanographic Inst., pp. 111–114.

ON THE STABILITY OF AN ENCOUNTERLESS SELF-GRAVITATING CONSTANT DENSITY SYSTEM

S. GOLDSTEIN

Dept. of Physics and Astronomy, Tel-Aviv University, Tel-Aviv, Israel

Abstract. The stability of a constant density, self-gravitating system is investigated.

The system considered is one-dimensional, collisionless and described by the sheet model.

The equilibrium distribution function $F(E)$, E being the energy, is such that the system has constant density in real space over a finite region.

An analytical treatment as well as computer experiment show stability for symmetric disturbances.

1. Introduction

The general equilibrium distribution function in one-dimension for collisionless systems is given by $F(E)$, where F is any positive or zero function and E being the energy per unit mass of a particle.

Using Poisson's equation, one can determine the density from $F(E)$ and vice-versa. For self-gravitating systems it is possible to construct density profiles that decrease, increase or are constant over a finite region. We shall limit ourselves to one case with constant density for $|x|<L$ and zero outside. This case may be of interest because it has not been treated in self-consistent way for a collisionless system, and the question of stability for such a system is still open.

The distribution function for the system just mentioned is:

$$F(E) = \begin{cases} \dfrac{K}{[E_m - E]^{\frac{1}{2}}} & E < E_m \\ 0 & E \geqslant 0 \end{cases}$$

where $E=\frac{1}{2}v^2+\psi(x)$, $\psi(x)$ being the self-consistent gravitational potential. K and E_m are parameters that can be calculated by the total mass and energy of the system.

Note that the function $F(E)$ is increasing and has an integrable singularity. The spatial density is constant as shown in Appendix A.

In recent years some stability criteria have been formulated (Antonov, 1961; Lynden-Bell, 1969) for self-gravitating systems but they refer only to monotonic decreasing functions $F(E)$ and do not apply to the case we treat.

In stability analysis one usually investigates perturbations with time dependence of the form $e^{\omega t}$ and obtains a solution for ω. If $Re\,\omega>0$ then the system is unstable.

In addition to analytical treatment, one can perform a computer experiment. If the system is unstable it will show up in its phase space pictures.

In the following, we first present the results of the computer experiment (Section 2) and next, the analytical treatment (Sections 3 and 4).

M. Lecar (ed.), Gravitational N-Body Problem, 184–193.

2. Computer Experiment

The system was simulated by the sheet model using 900 sheets thrown initially in phase space according to the function $F(x, v)$ (see Appendix A).

The accuracy of the total energy is $\Delta E/E \simeq 10^{-5}$ per period.

The development in time of the system indicates stability and is shown as pictures in phase space (Figure 1). The collective kinetic period is 3.14, that is; Figure 1c shows the situation after more than 6 periods.

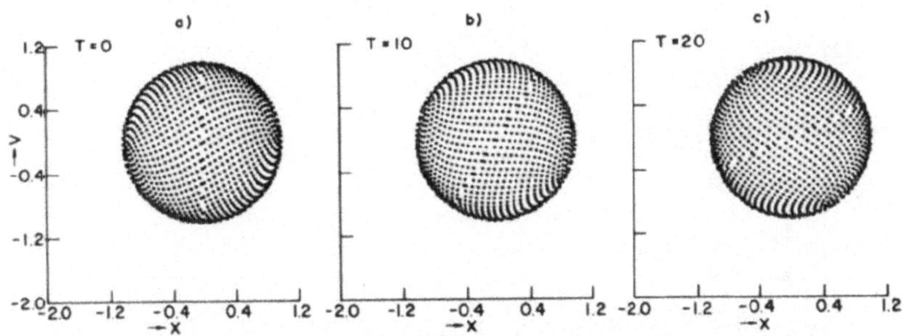

Fig. 1. Phase space picture of the evolution in time. The kinetic energy period is $T=3.14$.

It seems from the pictures that the particles are too well ordered in phase space with very small perturbations. This is not the case. In Table I the density as a function of x is described. There are fluctuations in the density up to 10% and their level does not change with time.

TABLE I

Density as function of x at two different times. In equilibrium the density is constant $\varrho=0.50$

X/L	0.05	0.15	0.25	0.35	0.45	0.55	0.65	0.75	0.85	0.95
$T=1$	0.48	0.49	0.52	0.47	0.53	0.45	0.48	0.52	0.47	0.54
$T=20$	0.50	0.49	0.47	0.51	0.50	0.51	0.47	0.55	0.50	0.48

We may conclude that if there are any instabilities at all, their time scale is much larger than the collective period.

3. Statement of the Problem

In this chapter we shall formulate an analytical approach to investigate stability of a general one-dimensional system and use the results for one case in the next chapter.

Starting with a given distribution function $F_0(E)$ for a self-gravitating system, one can solve for the potential $\psi_0(x)$ by using Poisson equation (with the units described in Appendix A):

$$\frac{d^2\psi_0}{dx^2} = 2\int F(\tfrac{1}{2}v^2 + \psi_0(x))\,dv\,. \tag{3.1}$$

The contour in phase space given by

$$E = \tfrac{1}{2}v^2 + \psi_0(x) \tag{3.2}$$

defines the particles trajectories.

As the contours are closed the motion of every particle is periodic with period $T(E)$. Define:

$$\Omega(E) = \frac{2\pi}{T(E)} \quad \text{and} \quad \Omega(E)\,t = \alpha \tag{3.3}$$

then, after solving (3.2), one gets:

$$x = x(E, \alpha) \tag{3.4}$$

$$v = \frac{dx}{dt} = \frac{\partial x}{\partial \alpha}\cdot\Omega(E)\,. \tag{3.5}$$

Equations (3.4) and (3.5) define the transformation $x, v \leftrightarrow E, \alpha$.

Using (3.3) and the relation $dv/dt = -(d\psi_0/dx)$ one obtains:

$$\frac{\partial v}{\partial \alpha} = \frac{dv}{dt}\cdot\frac{1}{\Omega(E)} = -\frac{d\psi_0}{dx}\cdot\frac{1}{\Omega(E)}\,. \tag{3.6}$$

Define the operator

$$\Omega\cdot\frac{\partial}{\partial \alpha} = \Omega\left[\frac{\partial x}{\partial \alpha}\cdot\frac{\partial}{\partial x} + \frac{\partial v}{\partial \alpha}\cdot\frac{\partial}{\partial v}\right] = v\cdot\frac{\partial}{\partial x} - \frac{d\psi_0}{dx}\cdot\frac{\partial}{\partial v}\,. \tag{3.7}$$

In order to investigate stability we treat the total distribution function $f_T = F_0 + f$, where f is the perturbation and the total potential $\psi_T = \psi_0 + \psi$.

Poisson's equation gives:

$$\frac{d^2\psi_T}{dx^2} = 2\int f_T\,dv = 2\int F_0\,dv + 2\int f\,dv\,. \tag{3.8}$$

Then, using the definition of ψ_T and relation (3.1) the equation for ψ is:

$$\frac{d^2\psi}{dx^2} = 2\int f\,dv\,. \tag{3.9}$$

Vlasov equation for f_T is:

$$\frac{\partial f_T}{\partial t} + \left(v\frac{\partial}{\partial x} - \frac{\partial \psi_T}{\partial x}\frac{\partial}{\partial v}\right) f_T = 0\,. \tag{3.10}$$

Assuming that f is small, Equation (3.10) may be linearized and one obtains

$$\frac{\partial f}{\partial t} + \left(v\frac{\partial}{\partial x} - \frac{d\psi_0}{dx}\frac{\partial}{\partial v}\right) f - \frac{\partial \psi}{\partial x}\frac{\partial F_0}{\partial v} = 0\,. \tag{3.11}$$

The second term is the operator of (3.7) acting on f, and for the last term:

$$\frac{\partial F_0}{\partial v} = \frac{\mathrm{d}F_0}{\mathrm{d}E} \cdot \frac{\partial E}{\partial v} = v \cdot \frac{\mathrm{d}F_0}{\mathrm{d}E} \tag{3.12}$$

and by Equation (3.7)

$$\Omega \frac{\partial \psi}{\partial \alpha} = v \frac{\partial \psi}{\partial x}. \tag{3.13}$$

Equation (3.11) gets the simple form:

$$\frac{\partial f}{\partial t} + \Omega(E) \frac{\partial f}{\partial \alpha} - \Omega(E) \frac{\partial \psi}{\partial \alpha} \frac{\mathrm{d}F_0}{\mathrm{d}E} = 0. \tag{3.14}$$

Assume that

$$f = f(x, v)\, e^{\omega t} \tag{3.15}$$

then, the potential has the same time-dependence and Equation (3.9) is unchanged. Equation (3.14) will have the form:

$$\omega f + \Omega \frac{\partial f}{\partial \alpha} - \Omega \frac{\partial \psi}{\partial \alpha} \frac{\mathrm{d}F_0}{\mathrm{d}E} = 0. \tag{3.16}$$

The perturbation can be written as the sum:

$$f = f^+(x, v) + f^-(x, v)$$

where

$$f^+(x, v) = f^+(-x, -v)$$

$$f^-(x, v) = -f^-(-x, -v).$$

Equations (3.9) and (3.16) separate for f^+ and f^-, as is shown in Appendix B.

In the remainder of this paper we shall restrict ourselves to the stability analysis of the symmetric part only, i.e., we shall consider $f = f^+(x, v)$.

4. Solution for Constant Density

The system described in the introduction has a constant density ϱ_0 and the potential is (see (A.10) in Appendix A):

$$\psi_0(x) = \varrho_0 x^2. \tag{4.1}$$

The period for every particle is independent of energy and $\Omega = 2\pi/T$ is given by:

$$\Omega = \sqrt{2\varrho_0} \tag{4.2}$$

and the transformations described in (3.4), (3.5) are:

$$x = \sqrt{\frac{E}{\varrho_0}} \sin\alpha \tag{4.3}$$

$$v = \sqrt{2E} \cos\alpha. \tag{4.4}$$

f is subject to the conditions that:

$$f(E, \alpha) = \begin{cases} \neq 0 & E < E_m \\ = 0 & E > E_m . \end{cases} \tag{4.5}$$

Because $f \neq 0$ for $E > E_m$, Equation (3.16) where $\mathrm{d}F_0/\mathrm{d}E = 0$ is:

$$\omega f + \Omega \frac{\partial f}{\partial \alpha} = 0 \tag{4.6}$$

with the solution $f = f(E) \cdot e^{(-\omega/\Omega\alpha)}$; but f is periodic in α, that is:

$$\frac{\omega}{\Omega} = \pm\, in$$

n being integer. This results in oscillatory constant amplitudes for f which are non-interesting, proving (4.5).

The statement in (4.5) includes another one: the density of the disturbance is limited in the region $|x| < L$ as if the system had reflecting walls. We may expand the density and potential in Fourier series with basic period $2L$ taking into account only even functions.

$$\psi = \sum_k \psi_k \cos kx$$
$$k = \frac{\pi m}{L}; \qquad m = 1, 2, \ldots \tag{4.8}$$

Using (4.3) and the relation

$$\cos(z \sin\alpha) = \sum_{l=-\infty}^{\infty} J_{2l}(z)\, e^{i2l\alpha} \tag{4.9}$$

where J are Bessel functions, the potential can be written as a function of E and α.

$$\psi = \sum_l \sum_k \psi_k J_{2l}\left(k \sqrt{\frac{E}{\varrho_0}}\right) e^{i2l\alpha}. \tag{4.10}$$

With the aid of ψ we shall solve for f by (3.16), and by using (3.9) we shall get an equation for ψ. The non trivial solution of ψ results in an equation for ω that can be solved by numerical methods. We proceed with this scheme. The function $f(E, \alpha)$ has a period π in α because $f(x, v) = f(-x, -v)$ and that amounts to adding π to α in (4.3), (4.4). The expansion in α is:

$$f(E, \alpha) = \sum_{n=-\infty}^{\infty} \phi_{2n}(E)\, e^{i2n\alpha}. \tag{4.11}$$

Solving for $\phi_{2n}(E)$ by using Equation (3.16)

$$\phi_{2n}(E) = \frac{\mathrm{d}F_0}{\mathrm{d}E} \cdot \sum_k J_{2n}\left(k \sqrt{\frac{E}{\varrho_0}}\right) \cdot \psi_k \frac{i2n}{i2n + \dfrac{\omega}{\Omega}}. \tag{4.12}$$

The equation for ψ is (3.9) that has the form

$$\frac{d^2\psi}{dx^2} = \sum_k - k^2\psi_k \cos kx = 2 \int dv \sum_n \phi_{2n}(E)\, e^{i2n\alpha} . \tag{4.13}$$

We multiply both sides by $\cos k_1 x$, for convenience, and integrate over $|x| \leqslant L$

$$- k_1^2 \psi_{k_1} \cdot L = 2 \iint dx\, dv \cos k_1 x \cdot \sum_n \phi_{2n}(E)\, e^{i2n\alpha} . \tag{4.14}$$

The evaluation of the integral is done in the following steps: (a) changing variables from x, v to E, α; (b) using formula (4.3) and (4.9) for $\cos k_1 x$; (c) using (4.12) for $\phi_{2n}(E)$.

The result is:

$$- k_1^2 \psi_{k_1} L = \sum_k \psi_k A(k, k_1) \tag{4.15}$$

where

$$A(k, k_1) = \frac{4\pi}{\Omega} \sum_{l=1}^{\infty} \int_0^{E_m} dE \frac{dF_0}{dE} J_{2l}\left(k_1 \sqrt{\frac{E}{\varrho_0}}\right) J_{2l}\left(k \sqrt{\frac{E}{\varrho_0}}\right) \frac{8l^2}{4l^2 + \left(\frac{\omega}{\Omega}\right)^2} . \tag{4.16}$$

Equation (4.15) stands for an infinite set of equations for all k_1. In order to have a non-trivial solution for ψ_k, the determinant must equal zero. The matrix of the homogeneous equations is $C(k, k_1)$:

$$C(k, k_1) = A(k, k_1) + \delta(k, k_1)\, k_1^2 \cdot L . \tag{4.17}$$

We require $|C(k, k_1)| = 0$ which is the desired equation for ω.

We are left with a matrix of infinite rank. The complete solution of the mathematical problem seems very difficult.

We have treated the problem by starting with a matrix of first rank and solving it for ω. Next we increased gradually the rank of the matrix and repeated the calculation for ω. The procedure that we adopted was to search for solutions subject to the restriction

$$\left|\frac{\omega}{\Omega}\right| < 2l - 1 \tag{4.18}$$

l being an integer. The reason for this is the way ω appears in the infinite series in (4.16). The l-th term may dominate the whole series if

$$\left(\frac{\omega}{\Omega}\right)^2 \simeq - 4l^2 . \tag{4.19}$$

In order to truncate the series we assumed $|\omega/\Omega| < 3$ which means that for $l \geqslant 3$

$$\left|\frac{8l^2}{4l^2 + \left(\frac{\omega}{\Omega}\right)^2}\right| = \left|2 \cdot \frac{1}{1 + \left(\frac{\omega}{\Omega \cdot 2l}\right)^2}\right| \simeq 2 . \tag{4.20}$$

The coefficients that multiply this term in (4.16) contain Bessel functions which diminish rapidly with increasing l (as calculated by the computer). The truncation of the series in (4.16) is thus justified and we took only the terms for $l=1, 2$.

The next step – not carried out yet – is to search for solutions satisfying (4.18) for higher number of l.

The results obtained for $|\omega/\Omega|<3$ are summarized in Table II. As seen, the solutions do not change much when increasing the rank of the matrix, they pile up near $\omega\simeq 2i$.

TABLE II

Solutions of ω for $|\omega/\Omega|<3$ and increasing the rank of the matrix considered

Rank of matrix	Solutions		
1	$\omega=\pm i1.86$		
2	$\omega_1=\pm i1.81$	$\omega_2=\pm i2.02$	
3	$\omega_1=\pm i1.81$	$\omega_2=\pm i2.04$	$\omega_3=\pm i2.004$

All the solutions for ω are imaginary. We have assumed a time dependence $e^{\omega t}$ and ω imaginary means that the system is stable, and the disturbances are oscillatory.

One class of the disturbances may oscillate with a frequency 1.81 (note that $\Omega=1$).

We plan to calculate this disturbance and to simulate it in a computer experiment to check if it will indeed oscillate with the specified frequency 1.81.

We intend to treat the case of antisymmetric disturbances in the same way as we treated the symmetric ones.

The conclusion from the results presented indicate stability of a constant density self-gravitating collisionless system.

Acknowledgements

I wish to thank Prof. M. Lecar and Drs. A. Kalnajs and S. Cuperman for very helpful discussions that encouraged me to work on this strange, difficult subject. This work has been supported by grant SFC-8-7006 from the Smithsonian Institution.

Appendix A

DESCRIPTION OF THE SYSTEM INVESTIGATED

The units we use are

$$2\pi G = 1 \tag{A.1}$$

where G is the gravitational constant.

$$M = 1 \tag{A.2}$$

where M is the mass per unit area of the total system.

The distribution function $F(E)$

$$F(E) = \frac{K}{(E_m - E)^{1/2}} \tag{A.3}$$

where $E=\frac{1}{2}v^2+\psi(x)$, gives a constant density $\varrho(x)$ for $\psi(x)<E_m$

$$\varrho(x) = \int \mathrm{d}v\, F(x, v) = \int_{-\sqrt{2(E_m-\psi(x))}}^{+\sqrt{2(E_m-\psi(x))}} \mathrm{d}v \frac{K}{[E_m - \frac{1}{2}v^2 - \psi(x)]^{1/2}} \tag{A.4}$$

and $\varrho(x)=0$ for $\psi(x)>E_m$ where $F(E)\equiv 0$.

In order to see the constant density from (A.4) one introduces a new variable:

$$\frac{\frac{1}{2}v^2}{E_m - \psi} = y^2 \tag{A.5}$$

and solves for ϱ as a function of $\psi(x)$

$$\varrho(\psi) = \frac{4K}{\sqrt{2}} \int_0^1 \frac{\mathrm{d}y}{(1 - y^2)^{1/2}}. \tag{A.6}$$

The result shows that ϱ is independent of ψ. Let us denote that density as ϱ_0. Using (A.6) one can calculate

$$K = \frac{\sqrt{2}}{2\pi} \varrho_0. \tag{A.7}$$

Assume that the center of the system is at $x=0$ and the density is ϱ_0 for $-L<x<L$ then from (A.2):

$$\varrho_0 \cdot 2L = 1. \tag{A.8}$$

Poisson equation for one-dimensional system is:

$$\frac{\mathrm{d}^2\psi}{\mathrm{d}x^2} = 2\varrho(x) \tag{A.9}$$

and its solution for our case is simply:

$$\psi = \varrho_0 x^2 \qquad x \leqslant L \tag{A.10}$$

but, $E_m=\psi(L)$ that is

$$E_m = \varrho_0 L^2 = L/2. \tag{A.11}$$

The period of a particle is $T=2\pi/\Omega$ where for the potential in (A.10)

$$\Omega = \sqrt{2\varrho_0} = 1/L \tag{A.12}$$

and the period is independent of energy. We choose $L=1$, then: $T=2\pi$, $E_m=\frac{1}{2}$ and the distribution function in x, v is:

$$F(x, v) = \frac{1}{2\pi} \frac{1}{(1 - x^2 - v^2)^{1/2}}. \tag{A.13}$$

Appendix B

PROOF OF SEPARATION OF f^+ AND f^-

The linearized Vlasov equation for $f=f^+ + f^-$

$$\frac{\partial (f^+ + f^-)}{\partial t} + v \frac{\partial (f^+ + f^-)}{\partial x} - \frac{d\psi_0}{dx} \frac{(f^+ + f^-)}{\partial v} - \frac{(\psi^+ + \psi^-)}{\partial x} \cdot \frac{\partial F_0}{\partial v} = 0 \tag{B.1}$$

where

$$\frac{d^2\psi^+}{dx^2} = 2 \int dv\, f^+ \tag{B.2}$$

and

$$\frac{d^2\psi^-}{dx^2} = 2 \int dv\, f^-. \tag{B.3}$$

From Poisson equation:

$$\frac{d^2\psi^+}{dx^2}(-x) = 2 \int_{-\infty}^{\infty} dv\, f^+(-x, v) = 2 \int_{\infty}^{-\infty} (-)\, dv\, f^+(-x, -v)$$
$$= 2 \int_{-\infty}^{\infty} dv\, f^+(x, v) = \frac{d^2\psi^+}{dx^2}(x). \tag{B.4}$$

The initial conditions

$$\left.\frac{d\psi^+}{dx}\right|_{x=0} = \left.\psi^+\right|_{x=0} = 0 \tag{B.5}$$

imply that ψ^+ is an even function and in the same way ψ^- is odd.

Let us investigate the parity of the operators acting on f, with respect to $x, v \to -x, -v$

$\dfrac{\partial}{\partial t}$ is even and unaffected

$v \dfrac{\partial}{\partial x}$ is even.

$d\psi_0/dx$ is odd because $\psi_0(x)=\psi_0(-x)$, that is, the operator $d\psi_0/dx \cdot \partial/\partial v$ is even.

The function $\partial F_0/\partial v = v \cdot \mathrm{d}F_0/\mathrm{d}E$ is odd because $\mathrm{d}F_0/\mathrm{d}E$ is even. The operator $v(\mathrm{d}F_0/\mathrm{d}E) \cdot \partial/\partial x$ is thus even. Formula (B.1) can be written as the sum of two functions, even part for f^+ and odd for f^-, each one equal to zero. The equations for f^+ or f^- are similar:

$$\frac{\partial f^+}{\partial t} + v \frac{\partial f^+}{\partial x} - \frac{\mathrm{d}\psi_0}{\mathrm{d}x} \frac{\partial f^+}{\partial v} - \frac{\partial \psi^+}{\partial x} \frac{\partial F_0}{\partial v} = 0. \tag{B.6}$$

References

Antonov: 1961, *Sov. Astron. – AJ* **4**, 859.

Lynden Bell, D.: 1969, *Monthly Notices Roy. Astron. Soc.* **144**, 189.

EXACT STATISTICAL MECHANICS OF A ONE-DIMENSIONAL SELF-GRAVITATING SYSTEM

GEORGE B. RYBICKI

Smithsonian Astrophysical Observatory, Cambridge, Mass. 02138, U.S.A.

Abstract. The statistical mechanics of an isolated self-gravitating system consisting of N uniform mass sheets is considered using both canonical and microcanonical ensembles. The one-particle distribution function is found in closed form. The limit for large numbers of sheets with fixed total mass and energy is taken and is shown to yield the isothermal solution of the Vlasov equation. The order of magnitude of the approach to Vlasov theory is found to be $0(1/N)$. Numerical results for spatial density and velocity distributions are given.

1. Introduction

Consider the one-dimensional model of an N-body self-gravitating system described by the Hamiltonian

$$\begin{aligned} H(\mathbf{p}, \mathbf{x}) &= T(\mathbf{p}) + V(\mathbf{x}) \\ &= \sum_{n=1}^{N} p_n^2/(2\sigma_n) + 2\pi G \sum_{n>m} \sigma_n \sigma_m |x_n - x_m| \, . \end{aligned} \tag{1.1}$$

Physically this represents a system of N parallel, uniform mass sheets with positions $\mathbf{x}=(x_1, x_2,..., x_N)$ and momenta $\mathbf{p}=(p_1, p_2,..., p_N)$ that move along the x-axis under the influence of their mutual gravitation. Each sheet has a certain surface mass density σ_n, and it may pass freely through any other sheet, since there is no hard-core included in the interaction potential. G is the gravitational constant.

This model has some direct astrophysical relevance to the problem of the mass distribution normal to a highly flattened galaxy. More indirectly it has been used to investigate such questions as: (1) Relaxation by particle effects and the validity of Vlasov theory for large N; (2) Relaxation by mean field effects and the validity of Lynden-Bell's theory of violent relaxation; (3) Mass segregation effects.

A self-gravitating system of this type, if left to develop for a sufficiently long time, will reach the ultimate relaxed state of thermodynamic equilibrium. The precise description of this state is of considerable interest. For example, the degree of relaxation of any system must be defined relative to this state as a standard. Also the degree of validity of Vlasov theory in the thermodynamic equilibrium state is a useful indicator of the general validity of Vlasov theory for any system. Finally, by comparison to numerical experiments that have reached quasi-equilibrium, one may find evidence for approximate integrals of motion.

It should be made clear, however, that the state of thermodynamic equilibrium for this one-dimensional self-gravitating system sheds no light whatsoever on the general problem, since there is no state of thermodynamic equilibrium for three-dimensional systems. This is easily seen from the virial theorem $E=-T$, which predicts a negative heat capacity for the system, in direct violation of general statistical mechanical

M. Lecar (ed.), Gravitational N-Body Problem, 194–210. All Rights Reserved

theorems. The nonexistence of thermodynamic equilibrium for three-dimensional systems is also manifest from the divergence of the relevant partition functions.

For the case of a one-dimensional self-gravitating system with equal masses, $\sigma_n = \sigma$, confined to a 'box' of length L, Salzburg (1965) found the thermodynamic properties of the system. This was done using an extension of the methods of Lenard (1961) and Prager (1961), which were developed for treating the analogous one-dimensional plasma. Nonextensive thermodynamic properties were found when the usual thermodynamic limit of $N \to \infty$, keeping N/L fixed, was taken.

The calculations of Salzburg are not directly applicable to the self-gravitating systems of interest in stellar dynamics, which are generally regarded as existing in free space, not enclosed by external walls. Also, the chemical thermodynamics approach is not sufficiently detailed to answer questions about the intrinsic structure of the system, such as the density distribution relative to the center of mass. Furthermore the appropriate large N limit for such a system is not the usual thermodynamic limit, but rather one in which the total mass and energy are kept fixed. While interesting in its own right, the calculation of Salzburg needs to be extended for the present purposes.

In this paper, by suitable modification of the methods of Lenard, Prager, and Salzburg, the single particle distribution function will be found in closed form for an isolated, equal mass model on the basis of both canonical and microcanonical ensembles. Specific account is taken of the integrals of motion due to the separability of the center of mass motion. This is done by performing the integration over the phase space subject to the constraints $\bar{x} = 0$ and $\bar{p} = 0$, where

$$\bar{x}(\mathbf{x}) \equiv N^{-1} \sum_n x_n \tag{1.2}$$

$$\bar{p}(\mathbf{p}) \equiv \sum_n p_n \tag{1.3}$$

are the coordinate and momentum of the center of mass. This is equivalent to choosing a frame of reference in which the center of mass is at rest at the origin. The proper investigation of the intrinsic structure of the system requires such a procedure; otherwise certain average properties such as density would be uniform in space due to the uniform motion of the center of mass.

For the microcanonical ensemble there is the additional constraint of fixed total energy E:

$$E = H(\mathbf{p}, \mathbf{x}). \tag{1.4}$$

For the canonical ensemble there is no such constraint, but the integrations are done with the weighting function $\exp(-\beta H)$, where $\beta = (kT)^{-1}$; T is the temperature and k is Boltzmann's constant.

The physical nature of the canonical ensemble is admittedly somewhat obscure in the present instance, since the system has been assumed to be in momentum isolation, and it is difficult to see how it then can also be in energy contact with a heat bath. However, it will be assumed that such a physical arrangement can be made and that the canonical ensemble has some meaning. In any case the canonical results are necessary mathematical preliminaries to finding the more physically realistic microcanonical results.

The possibility of doing the necessary configurational integrations in closed form rests first of all on the fact that it is always possible to reduce any such integration to one over a particular ordering of the coordinates, say $x_1 \leqslant x_2 \leqslant \cdots \leqslant x_N$. Then the potential takes the simple form

$$V(\mathbf{x}) = 2\pi G\sigma^2 \sum_{n>m} (x_n - x_m) \tag{1.5}$$

without absolute value bars. Letting $\lambda = 2\pi G\sigma^2$ this may be written, after some manipulation,

$$V(\mathbf{x}) = \lambda \sum_{l=1}^{N-1} l(N-l)\cdot(x_{l+1} - x_l). \tag{1.6}$$

This equation has a simple physical interpretation. Consider the work done in reducing an interval $x_{l+1} - x_l$ to zero while keeping rigid connections between the members of each group of particles to the left and to the right, so that all other intervals remain constant. Since the force is independent of distance, this is equivalent to the work done in moving just two sheets of mass $l\sigma$ and $(N-l)\,\sigma$ into coincidence over a distance $x_{l+1} - x_l$, that is, $-\lambda l(N-l)\cdot(x_{l+1} - x_l)$. Each interval may in turn be reduced to zero by a similar process, and eventually all particles will be coincident, so that $V=0$. Thus the potential energy of the original configuration is given by the sum (1.6). The importance of Equation (1.6) is that it expresses the potential as a sum of independent contributions, which makes it possible to do the configurational integrations, after a simple change of variables. This is the essential analytical trick employed by Lenard and Prager in the plasma case.

The combinatorial difficulties of the plasma case fortunately do not arise here in performing the necessary phase averages, because all sheets have the same mass. On the other hand the analytical difficulties in the present calculation exceed those of the plasma case, because of the more detailed information sought, and it is somewhat remarkable that the results come out in such simple form. By treating a system with various masses, one would encounter the combinatorial difficulties as well.

In the limit of large N, with fixed total mass and energy, it is shown that the single particle distribution function approaches the isothermal solution of the Vlasov equation. The order of this approach is $1/N$, which implies something about the order of the two-particle distribution function.

Numerical results are presented for the spatial density of an N-particle system for both the canonical and the microcanonical results. Comparison with the numerical experiments of Hohl (1968) for $N=3$ indicates that there is an approximate integral of motion in this case.

2. The Canonical Ensemble

The canonical one-particle distribution function $f_c(p, x)$ may be defined as the phase space average of the quantity

$$N^{-1} \sum_{n} \delta(p - p_n)\, \delta(x - x_n) \tag{2.1}$$

with weighting function $\exp(-\beta H)$ and with the constraints (1.2) and (1.3). Thus,

$$f_c(p, x) = (zN!)^{-1} \iint \mathbf{dp}\, \mathbf{dx}\, \delta(\bar{x})\, \delta(\bar{p}) \exp(-\beta H)\, N^{-1} \sum_{n} \delta(p - p_n)\, \delta(x - x_n) \tag{2.2}$$

where

$$z = (N!)^{-1} \iint \mathbf{dp}\, \mathbf{dx}\, \delta(\bar{x})\, \delta(\bar{p}) \exp(-\beta H). \tag{2.3}$$

Planck's constant has been omitted here, since it cancels in all relevant formulae. Because of the identity of particles the average may be taken of $\delta(x-x_N)\,\delta(p-p_N)$ instead of the symmetrized form (2.1). With the separability of the Hamiltonian $H=T+V$, this implies the factorization $f_c = \varrho_c(x)\,\theta_c(p)$, where

$$\begin{aligned}
\varrho_c(x) &= (QN!)^{-1} \int \mathbf{dx}\, \delta(\bar{x}) \exp(-\beta V)\, \delta(x - x_N) \\
Q &= (N!)^{-1} \int \mathbf{dx}\, \delta(\bar{x}) \exp(-\beta V) \\
\theta_c(p) &= (R)^{-1} \int \mathbf{dp}\, \delta(\bar{p}) \exp(-\beta T)\, \delta(p - p_N) \\
R &= \int \mathbf{dp}\, \delta(\bar{p}) \exp(-\beta T).
\end{aligned} \tag{2.4}$$

Because the kinetic energy is simply a sum of quadratic terms, $\theta_c(p)$ may be easily found. In the following development integrals are taken over all space, unless otherwise indicated. Using the Fourier representation of the δ-function,

$$\delta(\bar{p}) = (2\pi)^{-1} \int dk \exp\Big(ik \sum_{n} p_n\Big) \tag{2.5}$$

we have

$$\begin{aligned}
R\theta_c(p) &= (2\pi)^{-1} \int dk \int dp_1 \ldots dp_N \exp\Big[ik \sum_{n} p_n - \beta \sum_{n} p_n^2/(2\sigma)\Big]\, \delta(p - p_N) \\
&= (2\pi)^{-1} \exp[-\beta p^2/(2\sigma)] \int dk\, \times \\
&\quad \times e^{ikp} \int dp_1 \ldots dp_{N-1} \prod_{n=1}^{N-1} \exp[ikp_n - \beta p_n^2/(2\sigma)] \\
&= (2\pi)^{-1} \exp[-\beta p^2/(2\sigma)] \int dk\, e^{ikp} \exp[-(N-1)\,\sigma k^2/(2\beta)] \\
&= (N-1)^{-1/2} (2\pi\sigma\beta^{-1})^{(N/2)-1} \exp[-\beta N p^2/(2\sigma(N-1))]
\end{aligned} \tag{2.6}$$

where the well-known formula for the Fourier transform of a Gaussian has been used twice. Since θ_c has unit normalization,

$$\int \theta_c(p)\,\mathrm{d}p = 1 \tag{2.7}$$

integration of Equation (2.6) yields

$$R = N^{-1/2}(2\pi\sigma\beta^{-1})^{(N-1)/2} \tag{2.8}$$

so that

$$\theta_c(p) = \left[\frac{\beta N}{2\pi\sigma(N-1)}\right]^{1/2} \exp\left[-\frac{\beta N p^2}{2\sigma(N-1)}\right]. \tag{2.9}$$

Thus the momentum part of the one-particle distribution is Maxwellian, as could have been predicted. The unfamiliar factors $N/(N-1)$ are explained by the fact that N particles must share the thermal kinetic energy of $N-1$ degrees of freedom, since one degree of freedom has been lost by virtue of the center of mass constraint.

In order to find the density ϱ_M, first a symmetrized form of Equation (2.4) is written:

$$Q\varrho_c(x) = (N!)^{-1} \int \mathrm{d}\mathbf{x}\, \delta(\bar{x}) \exp(-\beta V)\, N^{-1} \sum_n \delta(x - x_n). \tag{2.10}$$

Because of the symmetry the integrations may be taken over any ordering of the variables and the result multiplied by $N!$. The ordering chosen here is $x_1 \leqslant x_2 \leqslant \cdots \leqslant x_N$. Defining the Fourier transform of the density ϱ_c by

$$\bar{\varrho}_c(k) = \int \mathrm{d}x \exp(ikx)\, \varrho_c(x) \tag{2.11}$$

Equation (2.10) may be written:

$$Q\bar{\varrho}_c(k) = N^{-1} \sum_n F_n(k) \tag{2.12}$$

where

$$F_n(k) = \int_{-\infty}^{\infty} \mathrm{d}x_1 \int_{x_1}^{\infty} \mathrm{d}x_2 \int_{x_2}^{\infty} \mathrm{d}x_3 \ldots \int_{x_{N-1}}^{\infty} \mathrm{d}x_N\, \delta(\bar{x}) \exp(-\beta V) \exp(ikx_n). \tag{2.13}$$

Advantage is now taken of the result (1.6) by introduction of the variables

$$\begin{aligned} u_l &= x_{l+1} - x_l, \quad 1 \leqslant l \leqslant N-1 \\ u_N &= N^{-1} \sum_m x_m = \bar{x}. \end{aligned} \tag{2.14}$$

The Jacobian of this transformation is

$$J = \frac{1}{N} \begin{vmatrix} -1 & 1 & 0 & \ldots & 0 \\ 0 & -1 & 1 & \ldots & 0 \\ 0 & 0 & -1 & \ldots & 0 \\ \vdots & & & & \vdots \\ 1 & 1 & 1 & \ldots & 1 \end{vmatrix} = \frac{1}{N} \begin{vmatrix} -1 & 1 & 0 & \ldots & 0 \\ 0 & -1 & 1 & \ldots & 0 \\ 0 & 0 & -1 & \ldots & 0 \\ \vdots & & & & \vdots \\ 0 & 0 & 0 & \ldots & N \end{vmatrix}. \tag{2.15}$$

The determinant has been transformed by addition of n times the n-th row to the last row for $n=1, 2,\ldots, N-1$. This is now in tridiagonal form with the determinant equal to the product of the diagonal elements, so that $|J|=1$.

The inverse of this transformation is easily found. By writing out the summations and regrouping terms, the results

$$\begin{aligned} \sum_{l=n}^{N-1} u_l &= x_N - x_n \\ \sum_{l=1}^{N-1} l u_l &= N(x_N - u_N) \end{aligned} \tag{2.16}$$

are obtained. Therefore

$$\begin{aligned} x_n &= u_N + N^{-1} \sum_{l=1}^{N-1} l u_l - \sum_{l=n}^{N} u_l \\ &= u_N - N^{-1} \sum_{l=n}^{N=1} D_{nl} u_l \end{aligned} \tag{2.17}$$

where

$$\begin{aligned} D_{nl} &= -l, \qquad n > l \\ &= N - l, \quad n \leqslant l. \end{aligned} \tag{2.18}$$

With the notation

$$C_l = l(N - l) \tag{2.19}$$

the potential takes the form

$$V(\mathbf{x}) = \lambda \sum_{l=1}^{N-1} C_l u_l . \tag{2.20}$$

Transforming to the new variables in Equation (2.13) yields

$$\begin{aligned} F_n(k) = & \int_{-\infty}^{\infty} \mathrm{d}u_N \int_0^{\infty} \mathrm{d}u_1 \int_0^{\infty} \mathrm{d}u_2 \ldots \int_0^{\infty} \mathrm{d}u_{N-1}\, \delta(u_N) \times \\ & \times \exp\left[- \beta\lambda \sum_{l-1}^{N-1} C_l u_l - ikN^{-1} \sum_{l=1}^{N-1} D_{nl} u_l + iku_N\right] \\ = & \prod_{l=1}^{N-1} \int_0^{\infty} \mathrm{d}u_l \exp\{-(\lambda\beta C_l + ikN^{-1} D_{nl})\, u_l\} \\ = & \prod_{l=1}^{N-1} [\lambda\beta C_l + ikN^{-1} D_{nl}]^{-1} . \end{aligned} \tag{2.21}$$

The value of Q follows by setting $k=0$:

$$Q = F_n(0) = \prod_{l=1}^{N-1} [\lambda\beta l(N-l)]^{-1} = (\lambda\beta)^{-(N-1)} [(N-1)!]^{-2} . \tag{2.22}$$

The parameter $\alpha = k/(N\beta\lambda)$ is introduced, so that one has

$$\prod_{l=1}^{N-1} [C_l + i\alpha D_{nl}] = \prod_{l=1}^{n-1} [l(N-l) - i\alpha l] \prod_{l=n}^{N-1} [l(N-l) + i\alpha(N-l)]$$
$$= \prod_{l=1}^{n-1} [l(N-l-i\alpha)] \prod_{l=n}^{N-1} [(N-l)(l+i\alpha)]$$
$$= (n-1)! \frac{\Gamma(N-i\alpha)}{\Gamma(N-n-i\alpha+1)} (N-n)! \frac{\Gamma(N+i\alpha)}{\Gamma(n+i\alpha)}. \tag{2.23}$$

Therefore Equation (2.12) becomes

$$\bar{\varrho}_c(k) = \frac{[(N-1)!]^2}{\Gamma(N+i\alpha)\Gamma(N-i\alpha)} \frac{1}{N} \sum_{n=1}^{N} \frac{\Gamma(n+i\alpha)\Gamma(N-n+1-i\alpha)}{(n-1)!(N-n)!}. \tag{2.24}$$

This expression may be reduced considerably. First note the general result of the Beta-function theory

$$\int_0^1 z^{\mu-1}(1-z)^{\nu-1}\,dz = \frac{\Gamma(\mu)\Gamma(\nu)}{\Gamma(\mu+\nu)}, \quad Re(\mu) > 0, \quad Re(\nu) > 0. \tag{2.25}$$

Then with use of the binomial theorem,

$$\frac{1}{N}\sum_{n=1}^{N} \frac{\Gamma(n+i\alpha)\Gamma(N-n+1-i\alpha)}{(n-1)!(N-n)!} = \frac{1}{N}\sum_{n=1}^{N} \frac{N!}{(n-1)!(N-n)!} \times$$
$$\times \int_0^1 z^{n+i\alpha-1}(1-z)^{N-n+i\alpha}\,dz = \int_0^1 z^{i\alpha}(1-z)^{-i\alpha}[z+(1-z)]^{N-1}\,dz$$
$$= \frac{\Gamma(1+i\alpha)\Gamma(1-i\alpha)}{\Gamma(2)}. \tag{2.26}$$

Thus

$$\bar{\varrho}_c(k) = [(N-1)!]^2 \frac{\Gamma(1+i\alpha)\ \Gamma(1-i\alpha)}{\Gamma(N+i\alpha)\Gamma(N-i\alpha)}$$
$$= \prod_{l=1}^{N-1} [l^2(l+i\alpha)^{-1}(l-i\alpha)^{-1}] \tag{2.27}$$

and we obtain the remarkably simple form for the Fourier transform of the density,

$$\bar{\varrho}_c(k) = \prod_{l=1}^{N-1} \frac{l^2}{l^2+\alpha^2} = \prod_{l=1}^{N-1} \frac{l^2}{l^2 + k^2/(N\beta\lambda)^2}. \tag{2.28}$$

This Fourier transform may be inverted in closed form. As a function of the complex variable k, $\bar{\varrho}_c$ is seen to have simple poles equally spaced along the imaginary axis from $-i\,N(N-1)\,\beta\lambda$ to $+i\,N(N-1)\,\beta\lambda$, except at $k=0$. The contour for inversion lies along the real axis:

$$\varrho_c(x) = (2\pi)^{-1} \int_{-\infty}^{\infty} \bar{\varrho}_c(k)\, e^{-ikx}\, \mathrm{d}k \tag{2.29}$$

and may be deformed upward for $x<0$, enclosing the poles on the positive portion of the imaginary axis. The residue at the pole at $+iNn\beta\lambda$ is

$$\lim_{k\to iNn\beta\lambda} (k - iNn\beta\lambda)\, \bar{\varrho}_c(k) = \frac{n}{2i} \prod_{l=1}^{N-1}{}' \frac{l^2}{l^2 - n^2} \tag{2.30}$$

where the prime on the product means to omit the term $l=n$. Now

$$\begin{aligned}\prod_{l=1}^{N-1}{}' \frac{l^2 - n^2}{l^2} &= \left[\frac{n}{(N-1)!}\right]^2 \prod_{l=1}^{n-1} (l-n) \prod_{l=n+1}^{N-1} (l-n) \prod_{l=1}^{N-1}{}' (l+n) \\ &= \left[\frac{n}{(N-1)!}\right]^2 (-1)^{n-1} (n-1)!\,(N-1-n)!\, \frac{(N+n-1)!}{(2n)\, n!}.\end{aligned} \tag{2.31}$$

With the definition

$$A_l^N = \frac{[(N-1)!]^2\, (-1)^{l-1}\, l}{(N-1-l)!\, (N-1+l)!} \tag{2.32}$$

the density may be written

$$\varrho_c(x) = N\beta\lambda \sum_{l=1}^{N-1} A_l^N e^{-N\beta\lambda l|x|}. \tag{2.33}$$

The combination of Equations (2.9) and (2.33) finally gives the canonical one-particle distribution function in closed form:

$$f_c(p, x) = \frac{(N\beta)^{3/2}\, \lambda}{[2\pi\sigma(N-1)]^{1/2}} \sum_{l=1}^{N-1} A_l^N \exp\left[-\frac{\beta N}{2\sigma(N-1)} p^2 - N\beta\lambda l|x|\right]. \tag{2.34}$$

There is a quite useful integral representation for ϱ_c. Note that the upper limit on the summation in Equation (2.33) may be extended to ∞, since $A_l^N = 0$ for $l \geqslant N$ due to the factor $(N-1-l)!$ in the denominator. Using the result

$$\frac{[(N-1)!]^2}{(N-1-l)!\,(N-1+l)!} = \int_{-\pi/2}^{\pi/2} e^{-2ilt} \cos^{2(N-1)} t\, \mathrm{d}t \Bigg/ \int_{-(\pi/2)}^{\pi/2} \cos^{2(N-1)} t\, \mathrm{d}t \tag{2.35}$$

which follows from well-known integrals and performing the elementary summation yields

$$\varrho_c(x) = \frac{N\beta\lambda}{4} \int_{-(\pi/2)}^{\pi/2} \operatorname{sech}^2\left(\frac{N\beta\lambda|x|}{2} + it\right) \cos^{2(N-1)} t\, \mathrm{d}t \Bigg/ \int_{-(\pi/2)}^{\pi/2} \cos^{2(N-1)} t\, \mathrm{d}t. \tag{2.36}$$

One immediate application of this formula is to compute the central density. Setting $x=0$, we have

$$\varrho_c(0) = \frac{N\beta\lambda}{2} \frac{N-1}{2N-3} \tag{2.37}$$

since $\operatorname{secht}(it) = (\cos t)^{-1}$. This also implies the sum rule for the coefficients A_l^N,

$$\sum_{l=1}^{N-1} A_l^N = \frac{1}{2} \frac{N-1}{2N-3}. \tag{2.38}$$

The partition function

$$z = QR = (\lambda\beta)^{-(N-1)} [(N-1)!]^{-2} N^{-\frac{1}{2}} (2\pi\sigma\beta^{-1})^{(N-1)/2} \tag{2.39}$$

is proportional to $\beta^{-(3/2)(N-1)}$, so the average energy is

$$\langle E \rangle = -\frac{\partial}{\partial\beta} \log z = \tfrac{3}{2}(N-1)\,\beta^{-1}. \tag{2.40}$$

3. The Microcanonical Ensemble

The canonical results just obtained apply to a system in contact with a heat bath that keeps the system at a temperature T. The energy of such a system is not well defined, and it undergoes thermal fluctuations of the order of kT. It is more realistic, from the point of view of stellar dynamics, to assume that the total energy of the system is fixed at the value E. This requires the use of the microcanonical ensemble in the statistical mechanical treatment. Rather than the weighting function $\exp(-\beta H)$, the integrations over the phase space are then to be done with the constraint (1.4) which may be conveniently handled by the inclusion of the weighting function $\delta(E-H)$ in the phase integrals.

In this section it will be shown that it is possible to obtain the microcanonical results from the canonical results just obtained. The microcanonical one-particle distribution function is defined as the average of the quantity (2.1) over the phase space with constraints (1.2), (1.3), and (1.4). Thus

$$f_{MC}(p,x) = (\Omega N!)^{-1} \iint \mathbf{dp}\,\mathbf{dx}\,\delta(\bar{x})\,\delta(\bar{p})\,\delta(E-H)\,N^{-1} \sum_n \delta(p-p_n)\,\delta(x-x_n) \tag{3.1}$$

where

$$\Omega = (N!)^{-1} \iint \mathbf{dp}\,\mathbf{dx}\,\delta(\bar{x})\,\delta(\bar{p})\,\delta(E-H). \tag{3.2}$$

Note that these quantities are related to the corresponding canonical ones (2.2) and (2.3) by Laplace transformation:

$$z f_c = \int_0^\infty e^{-\beta E} (\Omega f_{MC})\,\mathrm{d}E \tag{3.3}$$

$$z = \int_0^\infty e^{-\beta E} \Omega \, \mathrm{d}E. \tag{3.4}$$

Therefore, by inversion,

$$\Omega f_{MC} = (2\pi i)^{-1} \int_C e^{\beta E} (z f_C) \, \mathrm{d}\beta \tag{3.5}$$

$$\Omega = (2\pi i)^{-1} \int_C e^{\beta E} z \, \mathrm{d}\beta \tag{3.6}$$

where the contour C extends from $-i\infty$ to $i\infty$ to the right of all singularities. First note the general result

$$(u)_+^{\gamma-1}/\Gamma(\gamma) = (2\pi i)^{-1} \int_C e^{\beta u} \beta^{-\gamma} \, \mathrm{d}\beta. \tag{3.7}$$

The notation $(\ \ )_+$ is defined by

$$\begin{aligned} (u)_+ &= u, \quad u \geqslant 0 \\ &= 0, \quad u < 0. \end{aligned} \tag{3.8}$$

The above inversions may now be done in closed form, using equations (2.33), (2.40), and (3.7). The results are:

$$f_{MC}(p, x) = \frac{N\lambda}{E} \left(\frac{N}{2\pi\sigma(N-1)E} \right)^{1/2} \frac{\Gamma\left(\frac{3N}{2} - \frac{3}{2}\right)}{\Gamma\left(\frac{3N}{2} - 3\right)} \times$$

$$\times \sum_{l=1}^{N-1} A_l^N \left(1 - \frac{Np^2}{2\sigma(N-1)E} - \frac{N\lambda |x| \, l}{E} \right)_+^{(3N/2)-4} \tag{3.9}$$

$$\Omega = \frac{(2\pi\sigma)^{(N-1)/2} \, E^{(3N-5)/2}}{\lambda^{N-1} [(N-1)!]^2 \, N^{1/2} \Gamma\left(\frac{3}{2}(N-1)\right)}. \tag{3.10}$$

Integrating this over p and x yields the density and the momentum distributions,

$$\begin{aligned} \varrho_{MC}(x) &= \int f_{MC}(p, x) \, \mathrm{d}p \\ &= \frac{N\lambda}{E} \tfrac{1}{2}(3N-5) \sum_{l=1}^{N-1} A_l^N \left(1 - \frac{N\lambda}{E} l |x| \right)_+^{\frac{3}{2}N - \frac{7}{2}} \end{aligned} \tag{3.11}$$

$$\theta_{MC}(p) = \int f_{MC}(p, x) \, \mathrm{d}x$$

$$= \frac{\Gamma\left(\frac{3N}{2} - \frac{3}{2}\right)}{\Gamma\left(\frac{3N}{2} - 2\right)} \left(\frac{N}{2\pi\sigma(N-1)E} \right)^{1/2} \left(1 - \frac{Np^2}{2\sigma(N-1)E} \right)_+^{(3N/2)-3}. \tag{3.12}$$

To obtain Equation (3.11) the formula

$$\int_{-1}^{+1} (1 - t^2)^{\mu - 1} \, \mathrm{d}t = \pi^{1/2} \frac{\Gamma(\mu)}{\Gamma(\mu + \frac{1}{2})} \tag{3.13}$$

was used, which follows from the transformation $t = 2z - 1$ in Equation (2.25) with $\mu = \nu$ and use of the duplication formula for Γ-functions. The normalization of ϱ_{MC} was used to evaluate

$$\sum_{l=1}^{N-1} \frac{1}{l} A_l^N = \tfrac{1}{2} \tag{3.14}$$

which was then used to obtain Equation (3.12).

The central density in the microcanonical case follows from Equations (3.11) and (2.38):

$$\varrho_{MC}(0) = \frac{N\lambda}{E} \frac{1}{4} \frac{(3N - 5)(N - 1)}{(2N - 3)}. \tag{3.15}$$

4. The Limit of Large N

The large N limit of most interest in stellar dynamics is the one in which the total energy E and total mass $M = N\sigma$ are fixed. This will be called the *Vlasov limit*, since this is the limit that is expected to lead to the usual Vlasov results. Since there is no well-defined total energy for the canonical ensemble, in this case the limit will be taken for fixed *average* total energy $\langle E \rangle$, given by Equation (2.40). In this section the notations E and $\langle E \rangle$ will be used interchangeably.

The first step in investigating the Vlasov limit is to introduce scaled variables that depend only on the fixed quantities E and M and on the number N. It is also necessary to adopt velocity rather than momentum as the basic variable in the distribution functions. To do this the dimensionless velocity, position, and Fourier transform variables

$$\eta = \frac{p}{\sigma V}$$
$$\xi = \frac{x}{L} \tag{4.1}$$
$$K = kL$$

are defined, where the characteristic velocity V and length L are given by

$$L = \frac{2E}{3\pi G M^2}, \quad V^2 = \frac{4E}{3M}. \tag{4.2}$$

The particular numerical factors used here have been chosen to simplify the final results. Scaled distribution functions are then defined by

$$\begin{aligned} f^*(\eta, \xi) &= \sigma V L f(\sigma V \eta, L\xi) \\ \varrho^*(\xi) &= L\varrho(L\xi) \\ \theta^*(\eta) &= \sigma V \theta(\sigma V \eta) \\ \bar{\varrho}^*(K) &= \bar{\varrho}\left(\frac{K}{L}\right) = \int e^{iK\xi} \varrho^*(\xi)\,\mathrm{d}\xi\,. \end{aligned} \tag{4.3}$$

These are normalized in the following manner:

$$\iint f^*(\eta, \xi)\,\mathrm{d}\eta\,\mathrm{d}\xi = \int \varrho^*(\xi)\,\mathrm{d}\xi = \int \theta^*(\eta)\,\mathrm{d}\eta = 1\,. \tag{4.4}$$

These definitions lead to the canonical ensemble results

$$f_c^*(\eta, \xi) = 2\pi^{-1/2}\left(1 - \frac{1}{N}\right) \sum_{l=1}^{N-1} A_l^N e^{-2(1-(1/N))l|\xi| - \eta^2} \tag{4.5}$$

$$\theta_c^*(\eta) = \pi^{-1/2} e^{-\eta^2} \tag{4.6}$$

$$\varrho_c^*(\xi) = 2\left(1 - \frac{1}{N}\right) \sum_{l=1}^{N-1} A_l^N e^{-2(1-(1/N))l|\xi|} \tag{4.7}$$

$$= \frac{1}{2}\left(1 - \frac{1}{N}\right) \cdot \frac{\int_{-(\pi/2)}^{\pi/2} \operatorname{sech}^2\left(\left(1 - \frac{1}{N}\right)\xi + it\right) \cos^{2(N-1)} t\,dt}{\int_{-(\pi/2)}^{\pi/2} \cos^{2(N-1)} t\,dt} \tag{4.8}$$

$$\varrho_c^*(0) = \frac{(N-1)^2}{N(2N-3)} \tag{4.9}$$

$$\bar{\varrho}_c^*(K) = \prod_{l=1}^{N-1} \frac{l^2}{l^2 + \left(\frac{NK}{2(N-1)}\right)^2} \tag{4.10}$$

which follow from Equations (2.34), (2.9), (2.33), (2.36), (2.37), and (2.28), respectively. For the microcanonical ensemble the results are:

$$f_{MC}^*(\eta, \xi) = \frac{4}{3N}\left(\frac{2}{3\pi(N-1)}\right)^{1/2} \frac{\Gamma\left(\frac{3N}{2} - \frac{3}{2}\right)}{\Gamma\left(\frac{3N}{2} - 3\right)} \sum_{l=1}^{N-1} A_l^N \left(1 - \frac{2\eta^2}{3(N-1)} - \frac{4l|\xi|}{3N}\right)_+^{(3N/2)-4} \tag{4.11}$$

$$\theta^*_{MC}(\eta) = \frac{\Gamma\left(\frac{3N}{2} - \frac{3}{2}\right)}{\Gamma\left(\frac{3N}{2} - 2\right)} \left(\frac{2}{3\pi(N-1)}\right)^{1/2} \left(1 - \frac{2\eta^2}{3(N-1)}\right)_+^{(3N/2)-3} \tag{4.12}$$

$$\varrho^*_{MC}(\xi) = 2\left(1 - \frac{5}{3N}\right) \sum_{l=1}^{N-1} A_l^N \left(1 - \frac{4l|\xi|}{3N}\right)_+^{\frac{3}{2}N - \frac{7}{2}} \tag{4.13}$$

$$\varrho^*_{MC}(0) = \frac{(N-1)(3N-5)}{3N(2N-3)} \tag{4.14}$$

which follow from Equations (3.9), (3.12), (3.11), and (3.15), respectively.

In Figures 1 and 2 the canonical and microcanonical ensemble densities $\varrho^*(\xi)$ are plotted for various values of N. It can be seen that as $N \to \infty$ these curves appear to approach a limit. This limit is easily found in the canonical case from Equation (4.8). Apart from the factors $(1-1/N)$, which approach unity, this equation expresses $\varrho^*_c(\xi)$ as an average of $\operatorname{sech}^2(\xi + it)$ over the range of t, $-(\pi/2) \leqslant t \leqslant (\pi/2)$, with weighting function $\cos^{2(N-1)} t$. For large N this weighting function becomes highly peaked in the neighborhood of $t=0$, so that as $N \to \infty$ it picks out the single value at

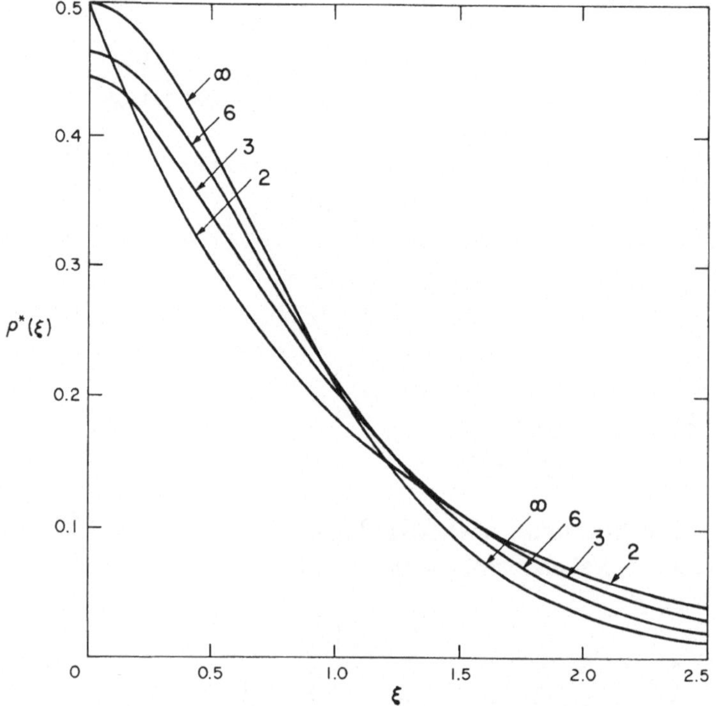

Fig. 1. Canonical density vs ξ for various values of N.

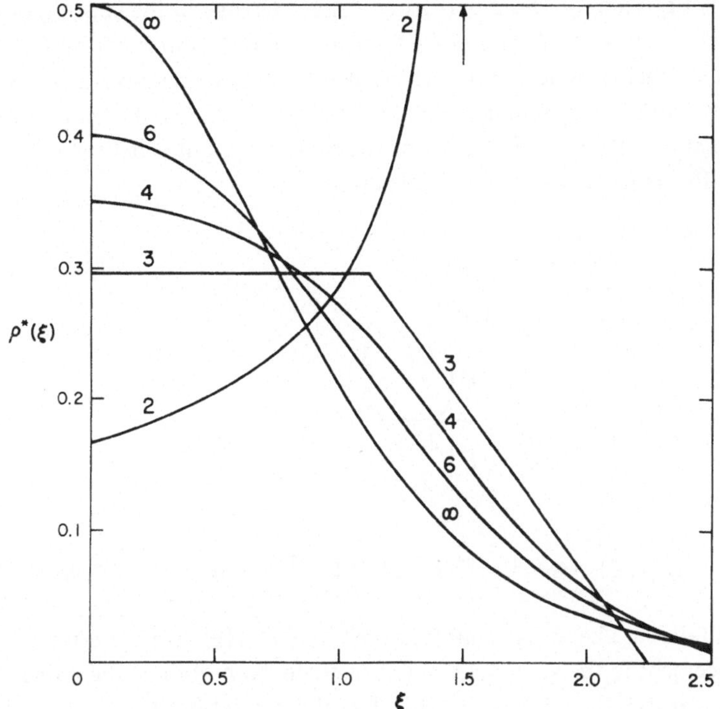

Fig. 2. Microcanonical density vs ξ for various values of N. The arrow locates the vertical asymptote for the case $N=2$.

$t=0$. Therefore,

$$\varrho_c^*(\xi) \to \tfrac{1}{2}\,\mathrm{sech}^2\xi \tag{4.15}$$

and with Equation (4.6) this implies

$$f_c^*(\eta, \xi) \to f_V^*(\eta, \xi) \tag{4.16}$$

where

$$f_V^*(\eta, \xi) = \frac{\pi^{-1/2}}{2}\, e^{-\eta^2}\,\mathrm{sech}^2\,\xi\,. \tag{4.17}$$

This may be recognized as the isothermal solution of the Vlasov equation, found by Camm (1950). The Vlasov results will be denoted by a subscript V.

This limiting form of the density appears in a quite picturesque way in terms of the Fourier transform (4.10). It is clear that as $N \to \infty$,

$$\bar{\varrho}_c^*(K) \to \prod_{l=1}^{\infty} \frac{l^2}{l^2 + \left(\dfrac{K}{2}\right)^2} = \frac{\pi K}{2}\,\mathrm{csch}\,\frac{\pi K}{2} \tag{4.18}$$

using the infinite product representation of $\mathrm{csch}(\pi K/2)$. The inverse transform of this again yields Equation (4.15). Apart from a slight change in the spacing of the poles, the main effect of finite N is simply to eliminate from the function (4.18), having an infinite number of poles, all but the $2(N-1)$ poles closest to the origin.

The microcanonical ensemble results also approach the isothermal solution of the Vlasov equation. This is difficult to prove rigorously, but a heuristic proof can be indicated: In the limit $N\to\infty$, one notes that

$$\begin{aligned} &A_l^N \to (-1)^{l-1}\, l \\ &\left(1-\frac{2\eta^2}{3(N-1)}-\frac{4l|\xi|}{3N}\right)_+^{(3N/2)-4} \to e^{-\eta^2-2l|\xi|} \\ &\frac{4}{3N}\left(\frac{2}{3\pi(N-1)}\right)^{1/2}\frac{\Gamma\left(\frac{3N}{2}-\frac{3}{2}\right)}{\Gamma\left(\frac{3N}{2}-3\right)} \to 2\pi^{-1/2} \end{aligned} \tag{4.19}$$

so that

$$f_{MC}^*(\eta,\xi)\to 2\pi^{-1/2}\sum_{l=1}^{\infty}(-1)^{l-1}\, l e^{-\eta^2-2l|\xi|}=\frac{\pi^{-1/2}}{2}e^{-\eta^2}\operatorname{sech}^2\xi\,. \tag{4.20}$$

This proof fails to take into account that the limits (4.19) are not uniform in l, so that their replacement in the sum (4.13) is not justified. Nonetheless, the numerical results obtained seem to indicate that this is in fact the correct limit.

The approach to the Vlasov theory may be estimated from the central densities:

$$\frac{\varrho_c^*(0)}{\varrho_V^*(0)}=\frac{\left(1-\frac{1}{N}\right)^2}{\left(1-\frac{3}{2N}\right)}=1-\frac{1}{2N}+0\left(\frac{1}{N^2}\right) \tag{4.21}$$

$$\frac{\varrho_{MC}^*(0)}{\varrho_V^*(0)}=\frac{\left(1-\frac{1}{N}\right)\left(1-\frac{5}{3N}\right)}{\left(1-\frac{3}{2N}\right)}=1-\frac{7}{6N}+0\left(\frac{1}{N^2}\right). \tag{4.22}$$

This approach is seen to be of order $0(1/N)$. The canonical results approach Vlasov theory faster than the microcanonical results. This is to be expected, since the velocity part of the canonical distribution is already precisely in Vlasov form.

The microcanonical ensemble for $N=2$ consists of the single system of two particles moving symmetrically about $x=0\,(\xi=0)$, along with all other systems derived from this one by a phase difference only. Each particle moves in a constant force field with a turning point occurring at a certain distance $|\xi|=\frac{3}{2}$. The density is proportional to the inverse of the velocity so that it approaches infinity at the turning point; this is indicated in Figure 2 by the arrow. The velocity distribution (see Figure 3) is uniform between certain maximum limits $|\eta|=(\frac{3}{2})^{1/2}$; this is because each particle spends the same time in each velocity range due to the constant acceleration.

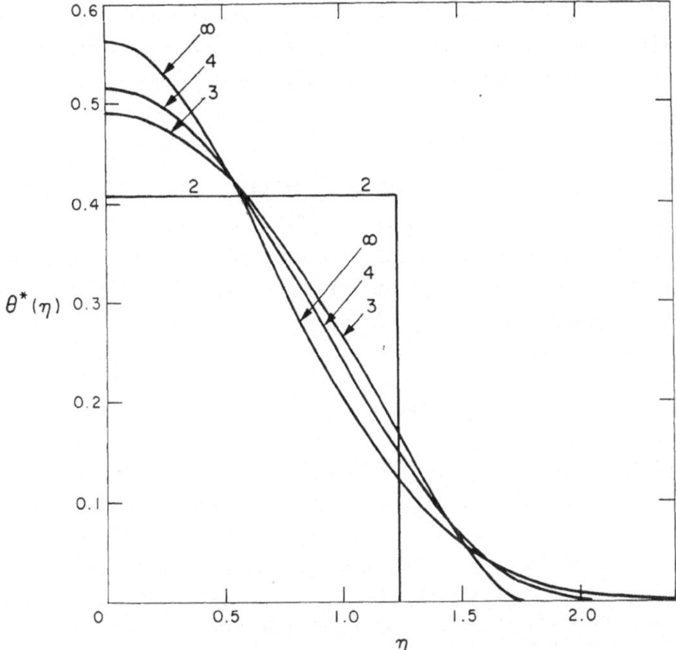

Fig. 3. Microcanonical velocity distribution vs η for various values of N. The case $N=\infty$ is also the canonical velocity distribution for all values of N.

For $N=3$ the microcanonical density is uniform over a certain range with limits $|\xi|=\frac{9}{8}$, and then falls linearly to zero at $|\xi|=\frac{9}{4}$. These points represent the respective limits of the regions where at most two or at most one particle can contribute to the density. Note that the velocity distribution in this case already approximates the Maxwellian Vlasov results quite well.

Numerical experiments with $N=3$ have been performed by Hohl (1968). The experimental velocity distribution found seems to fit the theoretical curve quite well, but the density curve shows a number of features not found in the theoretical curve. From this one may conclude that there must be an approximate integral of motion for the particular case treated by Hohl, which prevented complete ergodic behavior. It would be interesting to repeat the $N=3$ experiment with other initial conditions to see how strict these integrals may be.

For values of N larger than 3 the microcanonical density distributions appear fairly smooth, although all of these possess discontinuities in high order derivatives at certain points. The velocity distribution rapidly approaches the Vlasov result.

5. Final Remarks

Similar derivations for the two-particle distribution function have been attempted by the author. The purpose is to rigorously estimate the two-particle correlation function, which would give a definitive answer to the question of the order of magnitude of approach to Vlasov theory. This program has not as yet been successful, because of the considerable algebraic complexities involved, which make the necessary order of magnitude estimates difficult.

One possible generalization of this theory is to treat cases having two (or more) types of masses. It would be interesting to see whether such a mass spectrum affects the approach to Vlasov theory. Another possible study is suggested by the fact that the density curves found for finite N seem to differ from the Vlasov result primarily by a scale error. It is possible that the introduction of a simple N-scaling could improve the approach to Vlasov theory to, say, $0(1/N^2)$.

Acknowledgement

The author would like to thank Dr. Myron Lecar for introducing him to problems of self-gravitating systems and for helpful advice and encouragement during this study.

References

Camm, G. L.: 1950, *Monthly Notices Roy. Astron. Soc.* **110**, 305.
Hohl, F.: 1968, NASA TR R-289.
Lenard, A.: 1961, *J. Math. Phys.* **2**, 682.
Prager, S.: 1961, *Advan. Chem. Phys.* **4**, 201.
Salzburg, A. M.: 1965, *J. Math. Phys.* **6**, 158.

B. NUMERICAL EXPERIMENTS

NUMERICAL EXPERIMENTS IN COLLISIONLESS SYSTEMS

R. H. MILLER

University of Chicago, Dept. of Astronomy and Astrophysics

Gravitational *n*-body calculations handling 10^5 particles have become routine in the last three or four years. The plasma physicists have also run calculations with comparable numbers of particles – you will hear of these from Dawson this afternoon. In the stellar dynamics area, two groups have been making these calculations – our own group with Kevin Prendergast and William Quirk, and the competition, the group that formed when Roger Hockney and Frank Hohl merged (and later parted ways). Our calculation is described in Miller and Prendergast (1968); theirs is best described by Hohl and Hockney (1969), and some very thorough discussions of computational speed of force calculations by Hockney (1970).

The interpretation of these calculations is beset by the same kinds of difficulties that trouble any numerical esperiments – the astronomical value results from the questions asked and the kinds of experiments performed. And, all too often, the experimenter's prior prejudices come through unaffected by any experimental evidence. The astronomical interpretations of the results have been published elsewhere (Hockney and Hohl, 1969; Miller *et al.*, 1970; Quirk, 1970).

In this report, I want to take a different tack: this is supposed to be a meeting of experts in *n*-body calculations, so I want to concentrate on difficulties with the calculations. In particular, the emphasis will be on attempts to convince you that the bits running around inside those nice, big computers bear some relationship to the physics of stellar systems. It is not *ipso facto* evident that they do: the mere fact that the experimenter intends his calculation to relate to some kind of system in the sky does not assure any similarity.

These calculations deal with particles moving on a plane under $1/r^2$ forces. The ability to handle large numbers of particles comes from the fact that forces are computed only at a restricted set of points – if needed, forces are obtained for intermediate locations by interpolation rules. This avoids the need to compute forces between particle pairs, and makes the amount of computation necessary to obtain the forces independent of the number of particles. Of course, the main computational difficulty in gravitational *n*-body problems comes from the need to handle close encounters – or the possible formation of binaries. In these calculations, the forces are cut off for close encounters, thus sidestepping that problem. The justification ultimately lies in the observation that we know how to correct stellar dynamical calculations for the divergences at close encounters, but the long-range effects are the feature that defies theoretical treatment.

There is nothing magic about two dimensions. Three dimensional calculations have been made (Miller and Alton, 1968). Storage requirements and the amount of computation increase explosively in going from two to three dimensions, and the problem

M. Lecar (ed.), Gravitational N-Body Problem, 213–230. *All Rights Reserved*

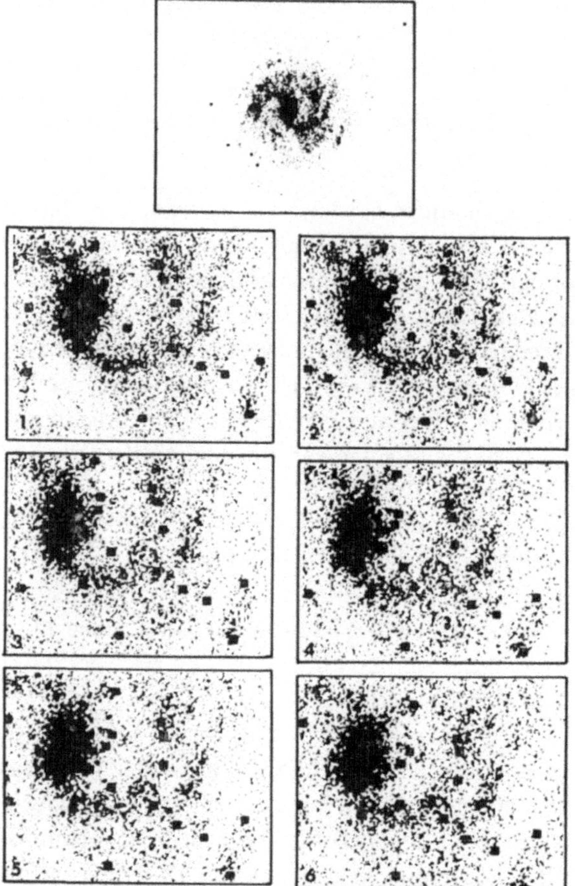

Fig. 1. Particles moving through spiral feature, from Miller *et al.* (1970). Complete pattern at beginning is shown. Time between frames is about 1/150 pattern rotation time.

of displaying the results in a comprehensible manner is much more difficult in three dimensions. There are so many challenging problems in two dimensions that there is little urge to go on to three. We just haven't gotten to three dimensions yet.

The early models used periodic configuration spaces, but both groups now have gotten rid of the periodic replications. A rather coarse grid for computing the forces remains.

The astronomically interesting problems so far attacked using these computer models are (1) the 'Jeans instability' – gravitational collapse of an initially uniform plane system, (2) attempts at static self-consistent models, and (3) spiral patterns. If you concentrate on the spiral patterns, you are led to the other two, and the spirals are most interesting, so we'll follow that route today.

The spiral patterns obtained do some of the things that you think spirals should do.

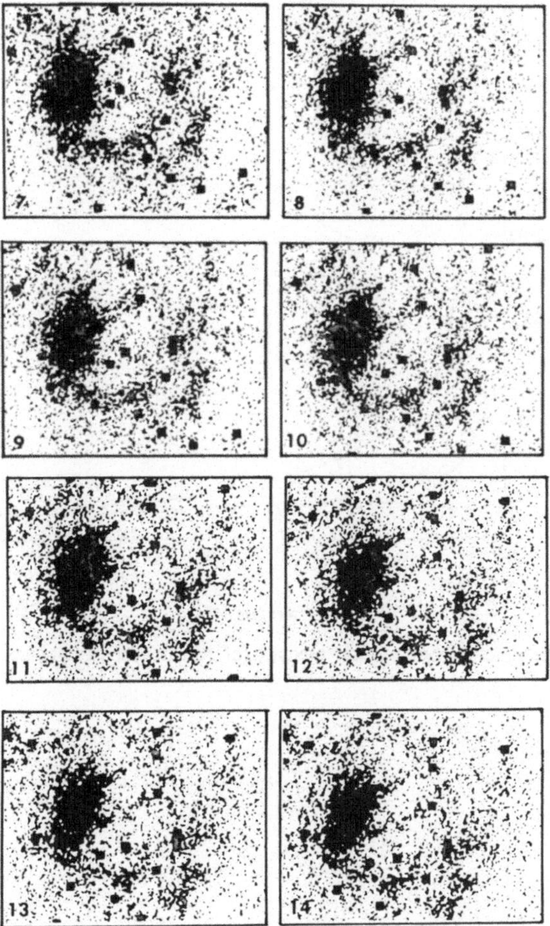

Fig. 2. Continuation of sequence of Figure 1.

For example, individual mass points ('gas' particles) can be seen to move through the spiral features (Figures 1 and 2). The usual galactic parameters can be computed – the density of 'stars' and of 'gas' separately (Figure 3), the epicyclic frequencies and the angular velocity of the pattern (Figure 4), and the mean circular and radial velocity components (Figure 5). So far everyting looks very nice and reasonable – but a closer look at Figure 5 shows that the 'stars' have half the local circular velocity. Aside from complicating the definitions of epicyclic frequencies, forcing care to use gradients of the forces rather than 'Oort constants' determined from the mean velocities, this means that an equilibrium model must be pressure-supported, or must have very large velocity dispersions. This is the first clear sign of trouble. While some real galaxies have significant differences between gas velocities and star velocities (Code, 1967), our own galaxy, for example, shows very little difference between velocities determined from 21-cm observations and those determined optically form the stellar motions.

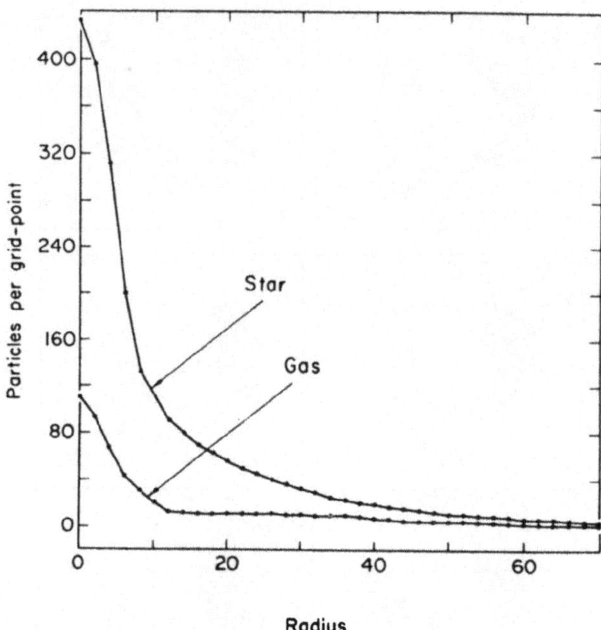

Fig. 3. Galactic parameters from model. Density of 'Gas' and of 'Stars'. From Quirk (1970).

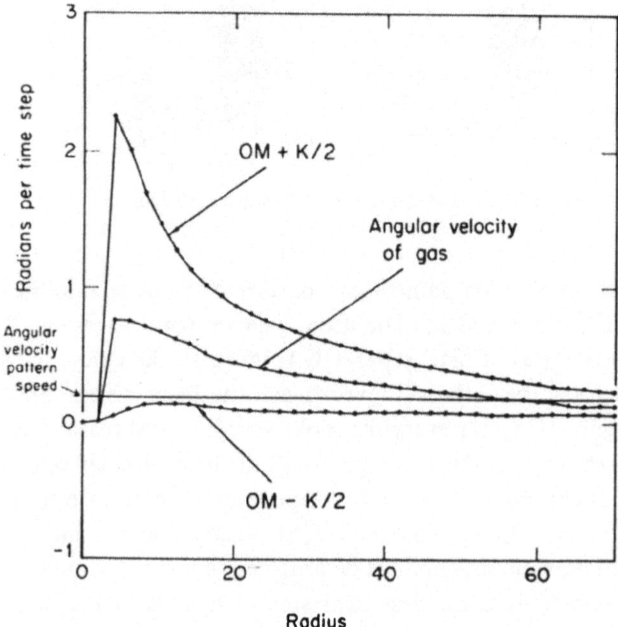

Fig. 4. Galactic parameters from model. Epicyclic frequencies and the angular velocity of gas. The pattern angular velocity is shown. From Quirk (1970). It is conceivable that the feature setting barred spirals apart from normal spirals might be the absence of an inner Lindblad resonance, as in this figure.

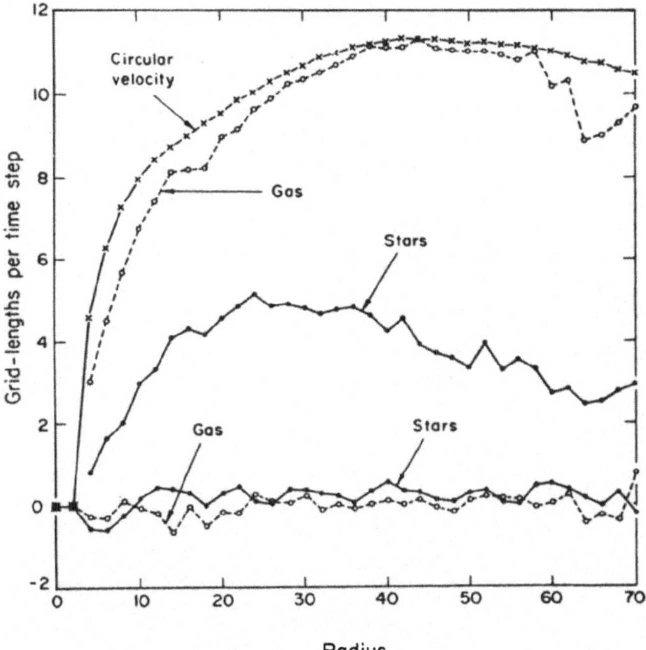

Fig. 5. Galactic parameters from model. Local circular velocity and local standards of rest for 'gas' and for 'stars'. From Quirk (1970). The circumferential mean velocity of the 'stars' is only half the local circular velocity, implying predominance of pressure-support.

Scaled to our own galaxy, the pressure support required to make up this difference would require velocity dispersions in excess of 100 km/s, quite different from the 30–40 observed.

Both groups have obtained spiral patterns. Neither has obtained spirals in a pure stellar case, and the artifices used are interesting in themselves. Hockney and Hohl have placed their system in a background potential – essentially a Schmidt model. They then get spirals that live as long as the dollars last to explore them. Our group used a two-component system in which the component that looked like 'gas' showed a spiral pattern. With both groups, the reason that spirals could not be obtained in a pure stellar case was that the pure stellar cases got too 'hot' – the velocity dispersions or pressure became very large. Figure 6 shows a curve of mean velocities determined for a purely stellar system back in the days when we were first trying to produce spirals in stellar systems by cooling them to make them move with the circular velocity. Figure 6 gave the condition just before that kind of 'cooling' was attempted. Again, the local standard of rest has half the local circular velocity. This condition of 'hot' populations is present in Hohl's calculations, too. Figure 7, which appeared in an unpublished report Hohl was kind enough to send, shows Q, the ratio of the actual velocity dispersion in the computer model to Toomre's (1964) velocity dispersion required to stabilize the system against axisymmetric modes. Figure 7 is one of Hohl's

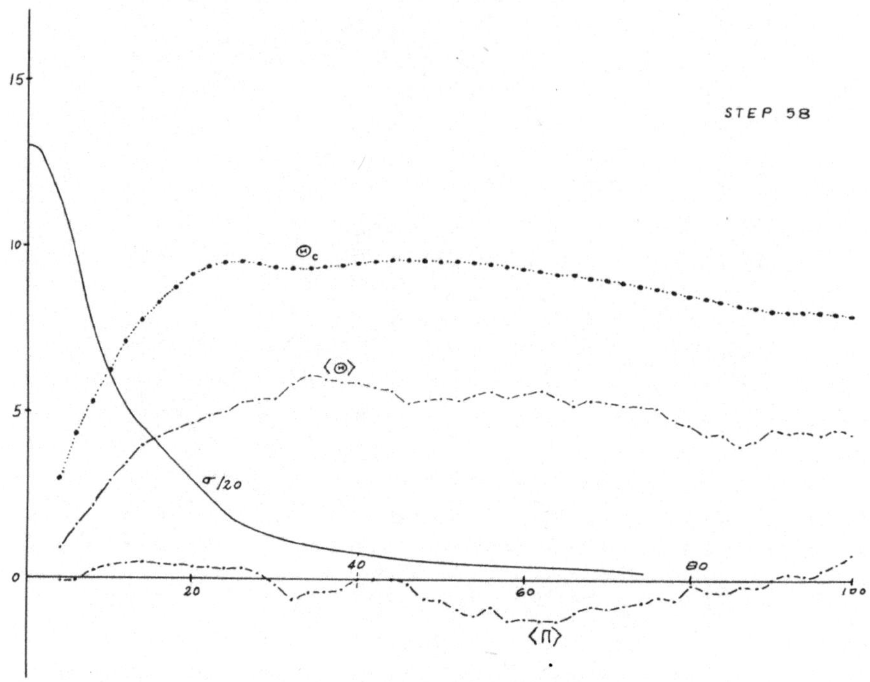

Fig. 6. Galactic parameters from a different model. 'Star' velocities compared to local circular velocity from a pure 'star' case run some time ago.

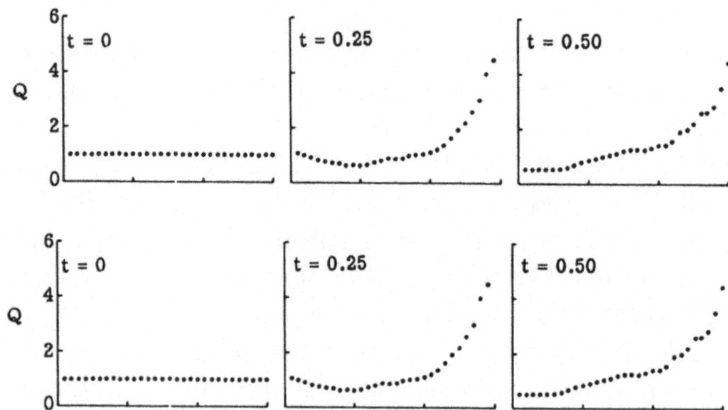

Fig. 7. Q, the ratio of the actual velocity dispersion in the model to that required for stability by the Toomre (1964) criterion. Hohl's data, courtesy of Frank Hohl.

more extreme cases, but it shows that these models can get very 'hot'. Toomre (1964) calculated that the velocity dispersions observed in our galaxy are, if anything, just a little bit too small for comfort under the stability criterion.

It is tempting to speculate that we may have constructed models that were a little bit too 'cool' and underwent a gravitational collapse. But is the price a stellar system must pay for having violated the stability criterion (somewhere, not even everywhere) a subsequent heating to Q of a least 2, and more likely 3? If so, it is difficult to see how any galaxy could be as cool as our own.

Here, then, is the major problem that I want to stress. Why are these systems so hot? No one has yet succeeded in building a 'cool' stellar model. And no one has seen a reasonable looking spiral in a 'hot' population. We have gotten around this by introducing a second population that is strongly 'cooled' all the way through the calculation; a spiral pattern appeared in one of these experiments. Hohl has placed a stellar system in a background potential – a Schmidt (1965) model of the galaxy – and then finds that spirals can form and persist. But even then, Hohl finds Q to be $1\frac{1}{2}$ to 2 or more – so even these models are 'hot'. Hohl is now undertaking experiments at 'weak cooling' in which the velocity dispersion is very gently reduced taking care not to get too close to the Toomre stability criterion.

Hohl's experiments using a fixed background potential, and some of our own along the same lines, leave the impression that the systems do not build up this high velocity dispersion merely to have a way of providing a potential field that will not collapse under its own weight. Toomre's models (1963, 1964) can be stabilized without such a large velocity dispersion, and should provide a noncollapsing potential. But are nonaxisymmetric models harder to stabilize? Ng's (1967) axisymmetric models are just as hot, but they were constructed in such a way that it was practically assured at the outset that they would turn out to be hot.

The 'gas' population seems to do more than merely to provide a population with small velocity dispersion that could feel irregularities in the potential to show up a spiral pattern. There is a cooperation between the two populations to make the spiral pattern (Miller *et al.*, 1970; Quirk, 1970). So the dissipative character may be necessary; but Hohl's model in a background is not dissipative.

This rather aimless discussion is included to point out the difficulty of really coming to grips with the essential problem: Why are the stars so hot? Almost any discussion of these models comes back to that fundamental question. Sometimes the point comes up in discussion that we shouldn't worry about needing a two-population system of 'gas' and 'stars' in order to build a model that shows a spiral pattern, since real galaxies show their spiral patterns in gas and in the hot young stars that have been created out of the gas but cannot live long enough to leave the neighborhood of the gas. That is fine for rationalizing a need for two populations; but real galaxies do not have stellar populations that are anywhere nearly this hot. So, even with two populations, we don't have a good model of a real galaxy. With one of the populations replaced by an 'equivalent gravitational potential', the remaining population is uncomfortably hot.

It is not clear where the difficulty lies – but as experimentalists, we must adopt the viewpoint that it most likely lies in our experiments. Unfortunately, computer experiments cannot be used to prove that systems must reach this 'hot' condition – the result need not follow from all initial conditions. If the argument is made probabilistically, that we have somehow sampled the parameter space of possible initial states, a careful look will convince you that very few initial conditions have actually been tried. And even then, the search stopped once spirals were found. It was much more exciting to explore the properties of those spirals. The only way to prove that 'hot' systems are not necessarily the only final state for a computer model is to build a 'cool' model. But no one has yet built a 'cool' static self-consistent model.

The answer may lie in studies of the stability of purely 'stellar' systems. There are, of course, some features of the computer models that make the experiments seem strained or artificial. In a sense, neither the usual theoretical model nor the computer model properly represents a real stellar system. Each is an approximation that overlooks certain essential features. The approximations are complementary. The theoretical (continuum) models overlook the grainy structure of a stellar system, while the computer models are much grainier than actual stellar systems. It appears that a 'grainy' fluctuation in the force field tends to destabilize a stellar system, so continuum models may underestimate the difficulty of achieving stable systems, while computer models may exaggerate it. Computer models have other features that stellar systems lack. The discrete number representation in a computer makes it impossible to define an infinitesimal perturbation – any change is finite (although it may be quite small). This restricts the methods available for studying stability. Distinctions (such as between instability and failure to be static-self-consistent), which are clearly defined for the continuum theoretical models may not be meaningful in computer experiments. The distinction might be meaningful for a stellar system and still be difficult or impossible to apply operationally because an instability cannot be recognized.

The importance of the question, quite aside from a natural desire to make our models be as close to nature as possible. lies in the interpretations that are otherwise made of things like Toomre's stability criterion and the observed velocity dispersions. They figure heavily in the arguments of Lin *et al.* (1969), for example, in discussions of whether spiral galaxies that we observe are stable configurations.

Some attempts to construct static self-consistent models will be taken up next, then a discussion of some numerical properties of the Miller-Prendergast (1968) model as a preliminary to a different way of undertaking the construction of self-consistent models. Along the way, the advertised features of our model as providing an exact handling of the collision-free Boltzmann, or Vlasov, equation will be pointed out.

1. Self-Consistent Model

The models that we have used as initial conditions for various calculations have not been designed to be static self-consistent models. Invariably, the models have evolved through spectacular collapses into systems that were largely pressure-supported rather than being 'cool' in the sense of having small velocity dispersions.

The 'hot' condition does not seem to be necessary for disk galaxies, but the velocity dispersions of our evolved computer models are substantially greater (perhaps twice or more) than required for stability, and are disconcertingly large for comparison with real stellar systems. A static self-consistent model that can be maintained in the computer with smaller velocity dispersions is essential for studying actual galaxies, particularly if the galaxies are to be built without resorting to several populations.

This work was undertaken to see whether a 'cool' static self-consistent model could be maintained in the computer. A successful attempt would show that there is nothing intrinsic to plane stellar systems that requires unusually large velocity dispersions in order to remain stable. The model was tested for its stability limits and to see what kinds of instabilities might develop. A complication in such a study is that the onset and early growth of an instability cannot be readily detected. Rather, the instability must grow to finite amplitude to be detected – unless we can devise a more sensitive indicator. A genuine instability is difficult to distinguish from mere failure to construct a truly static self-consistent model. Three-dimensional modes, such as bending, are not allowed.

A. THE MODEL

The model chosen was essentially 'Model I' of Toomre (1963). All of the standard galactic parameters can be easily worked out for this model, and the stability criterion of Toomre (1964) can be applied as well. This model has, in obvious notation:

$$\varrho(\varpi, z) = \mu(\varpi)\,\delta(z) = \frac{Mb}{2\pi}(\varpi^2 + b^2)^{-3/2}\,\delta(z), \tag{1}$$

for which the corresponding potential is

$$V(\varpi, z) = -\,GM\left[\varpi^2 + (b + |z|)^2\right]^{-1/2}. \tag{2}$$

Here b is a scale parameter that sets the dimensions of the system. If all particles have unit mass, M is just the number of particles.

In the initial conditions generated for a machine calculation, a number of particles was chosen (125000), and a scale parameter b. With $b=13$, the number of particles per location drops below $\frac{1}{2}$ at a radius $\varpi=80$. Velocities were chosen for the particles at each configuration space location to make a Gaussian (Schwarzschild) velocity distribution whose mean is the local circular velocity and with a disperion whose axis ratios conformed to the usual ones computed from the Oort constants (see e.g. Chandrasekhar, 1960, Sec. 4.3) with the actual root-mean-square velocities being a multiple, T, of a convenient form that is large at the center and decreases outward. The actual velocity of each particle was generated by a pseudorandom number generator, to meet these conditions.

The velocity dispersion that results with Toomre's stability criterion is large enough to require allowance for pressure-support. This is done by reducing the local circular velocity according to the usual hydrodynamical equations of stellar dynamics (see, e.g. Chandrasekhar, 1960. Sec. 4.8 (iv)). It can happen (at T about 1.5 that of the Toomre condition) that the system becomes wholly pressure supported in the center – there is then no rotation in those parts of the system that are pressure-supported.

B. MODIFICATION TO THE ACTUAL FORCE LAW

The force between pairs of particles, used in the calculation, is derivable from a potential

$$\varphi(x, y) = \text{const.}\,[(x - x')^2 + (y - y')^2 + a^2]^{-1/2}. \tag{3}$$

This modification was used to avoid computational troubles from near encounters. It is convenient to think of this potential in a 3-dimensional space. The particles move in a plane at z = constant, and the force on a test particle is measured a units above that plane, at $z = a$. Only those force components that lie in the plane $z = a$ enter the problem – the test particle is not permitted to leave the plane. For most calculations $a = 3$. This change, which seems small, produces a surprisingly large change in the actual force field near the center of a configuration like that of Equation (1). Corrected expressions for the parameters involving the force (rotational velocity, Oort constants, epicyclic frequency, and so on) can be obtained from the usual expressions by substituting $c = b + a$ in place of b wherever it appears. Thus all the model parameters can be exactly corrected for the revised force law.

A remarkable feature of this force law is that the same notion can be used to make the appropriate modification to the stability calculation. Toomre has also noticed that this force law permits the stability calculation to be carried through. The result is interesting by itself because it separates long-range from short-range contributions to the predicted stability conditions. Toomre's calculation proceeds from a linearized collision-free Boltzmann equation (Equation (43) of Toomre, 1964) in which the only term affected by the modified force law (assuming that the model representing the unperturbed system is properly self-consistent with the revised force-law) is that containing the force due to the perturbed density distribution. The force for each Fourier mode is obtainable from the usual solution to the Poisson equation,

$$V'(x, z) = \text{const.}\,\frac{2\pi G}{k}\,e^{ikx}e^{-k|z|}, \tag{4}$$

which indicates that the replacement $G \to Ge^{-ka}$ wherever G appears will properly account for the revised force law. Toomre's calculation now follows, yielding a modified dispersion relation

$$\frac{\kappa^2 a}{G\mu} = 2\pi\beta\,\frac{1 - e^{-y}I_0(y)}{\sqrt{y}}\,e^{-\beta y}, \tag{5}$$

where

$$y = \frac{\sigma_u^2 k_n^2}{\kappa^2}, \quad \beta = \frac{\kappa a}{\sigma_u}, \tag{6}$$

k_n is the wavenumber of the neutrally stable mode, σ_u is the velocity dispersion in the radial direction, and κ is the epicyclic frequency. The system is stable if $\kappa^2 a/G\mu$ is greater than the righthand side of Equation (5). Let $K(\beta)$ be the maximum value attained by the righthand side of Equation (5) for any value of $y(\geqslant 0)$ and the given value of β. Since the lefthand side contains (local) quantities that are completely determined by the model, this relation can be solved for β anywhere in the system. An interesting feature is that $K(\beta)$ attains an asymptotic value $2\pi/e = 2.31145...$ as $\beta \to \infty$ and any value of $\kappa^2 a/G\mu$ greater than this can be stabilized with any value of σ_u including $\sigma_u = 0$, by taking β infinite. Thus, with this modification to the force law (near cutoff in the forces), there are situations in which rotation alone is sufficient to stabilize the system – a result that appears quite different from Toomre's. (The distinction from the results of Lynden-Bell (1962) and Lee (1967) concerning rotational stabilization should be noted: these both refer to three-dimensional mass-distributions.)

In the model designed for the machine calculation, with $(GM)^{1/2} = 3.0155\ c$ (a value that sets the maximum force to 3.5), the rotation should have been sufficient to stabilize the system inside $\varpi \sim 26$. The actual velocity dispersion (in the radial direction) that is required for stability is shown, in units appropriate to the model, in Figure 8.

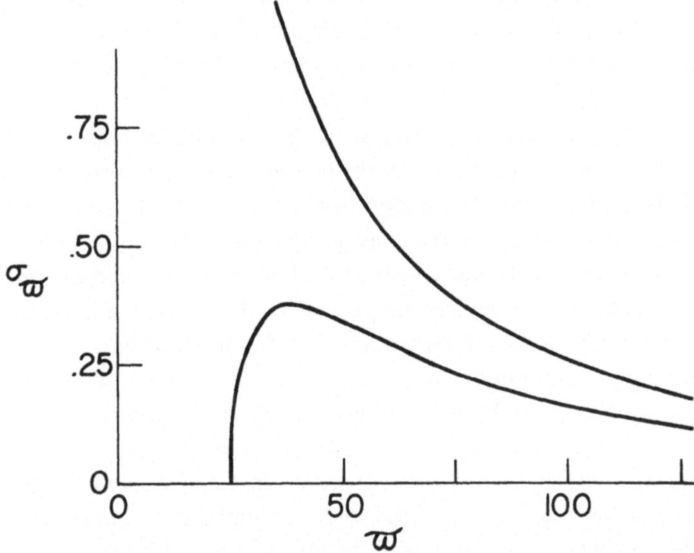

Fig. 8. Velocity dispersion in the radial direction required to stabilize configuration against axisymmetric modes. The lower curve is for the modified force law, according to Equation (5); the upper curve is for the same configuration with $1/r^2$ forces. The upper curve would reach 2.91 at the center.

With reasonable values for the velocity dispersions, the greatest instability should occur around $\varpi = 70$ in these models – fairly far out where the density is already quite low. Furthermore, the velocity dispersions required to stabilize are reduced by at least a factor of two below those expected without allowing for the near cutoff of the forces.

C. EXPERIMENTAL RESULTS

In the machine representation, all physical quantities are integers. To avoid sharp discontinuities in the initial density and velocity fields, we rounded by adding a pseudorandom number, uniform in (0, 1) to the 'real' value before truncating. The rounding is unbiased, but has twice the variance of the usual rounding rule. The variance acts like a velocity dispersion (with a circular velocity ellipse), and can contribute to the stabilization of the system through random motions, In fact, this term alone should be sufficient to stabilize the systems with the modified force law.

Roundoff 'noise' in the force-values is expected to be destabilizing, in contrast to the stabilizing influence of velocity dispersions. This can be seen by multiplying the linearized Boltzmann equations appropriate to different phase space points, and assuming that cross terms correlating forces to velocities, velocities to positions, and forces to coordinates can be neglected. The term with force covariances enters with the opposite sign to those with velocity variances. In physical terms, fluctuations in the force seem directly to induce fluctuations in the part of the distribution function that depends upon the velocities, broadens the velocity dependence. The effect of a velocity spread behaves differently, since it is already included in the distribution function. Quantitative estimates of the effect are complicated by correlations in the 'fluctuations' of force-values at different spatial locations. The force rounding remained with the 'add one-half' rule, to preserve the reversibility that we feel is an important feature of our model.

Two series of experiments were run with these configurations. The experiments started from an initial load produced by the methods just described, then followed the evolution of the system for a few integration steps. Normally, ten integration steps were sufficient to tell how the system was going to behave.

Each series of experiments was a sequence of runs with different values of T, the parameter that controls the velocity dispersion in the initial load. The two series of experiments were run with different numbers of configuration space locations per periodic cell length. The number of lattice points per periodic length was increased by the square of the factor by which the timestep is reduced, so force-values were unaltered.

The results of each experiment were observed in step-by-step density plots of the configuration and through values of the force tabulated for certain control points in the system. Some runs, duplicated with different sequences of pseudorandom numbers used in the initial loading routine, gave results that were qualitatively similar. The extent of quantitative difference indicates the reliability of the numerical results. The similarities of two systems starting with different pseudorandom number sequences were comparable with the retention of gross symmetry within a given system. Thus, these results are not peculiar to the particular set of pseudorandom numbers used.

The experimental results may be qualitatively described as follows. First, the systems run with our usual 256×256 grid appeared not to be static self-consistent models. They tended to rearrange themselves rather quickly, although those with larger velocity dispersions did so less rapidly. This probably means that discrete models are not adequately approximated by models designed according to continuous density distributions, although the departure is less troublesome for 'hotter' systems. There were no instabilities that were recognizable as such.

Second, the experiments with 'reduced time-step' (more points per configuration space lattice), were more nearly static. Runs were made with $\frac{1}{5}$ to $\frac{1}{31}$ of the normal time-step, with T about 1.3 times the value required for stabilization in the absence of 'quantization noise'. All these runs behaved about the same way, so $\frac{1}{5}$ was used thereafter. A run with $T=0$ showed a clumping similar to that seen earlier with the plane gravitational instability – very irregular patterns leading to large density variations characterized by fairly short length scales. These seem to be genuine instabilities, as distinguished from mere failure to have achieved a static self-consistent model, although the distinction is not clear-cut. The instability was strongest near the center, where the system should have been rotationally stabilized.

A sequence of runs was made with $T=0, 0.1, 0.2, 0.3, 0.4$ and 0.5. Stabilization should be achieved (without 'quantization noise') at $T=0.267$ for the form of velocity dispersion used. The velocity dispersion was largest at the center and fell off uniformly toward the outside; the precise analytic form is not important. The system becomes completely pressure-supported near the center at $T=0.457$. In general, larger values of T lead to systems that are more nearly static, self-consistent, and stable, as expected. At $T=0.1$, an instability is still present but it grows less rapidly. Values of T greater than 0.2 do not produce recognizable instabilities, although the systems still rearrange themselves in search of an equilibrium configuration. The rearrangement is slowed with larger values of T. There does not seem to be a recognizable boundary between genuine instabilities (if that is what they are) and rearrangements. There is rather a uniform trend toward slower rearrangement and longer characteristic lengths with larger values of T (larger velocity dispersions).

Hockney and Hohl (1969) have carried out similar experiments with a different model. Their low-velocity-dispersion configurations form into patterns characterized by short length scales; configurations with increased velocity dispersion change more slowly and form into patterns with larger length scales. Their treatment of the forces effectively introduces a near-cutoff that must influence the analytic stability criterion in some way like that of subsection *b*; their treatment of the forces also introduced local departures from the analytic form that are equivalent to the stochastic contribution noted above. It is difficult to distinguish an instability from a failure of self-consistency in their models, just as it is in ours. The picture that emerges from their published results seems similar.

D. DISCUSSION

These experimental results seem to disagree with expectations on two counts. (1) Something happens, even though 'quantization noise' resulting from the discrete allowed velocities alone should be sufficient to stabilize the system. As noted earlier, 'quantization noise' in the forces tends to destabilize the configuration; it works against the effects of the velocity dispersion, and could make the system unstable. There is not a clear-cut stability threshold to compare with theory. (2) That 'something' is strongest near the center, a place where rotation alone should be adequate for stabilization.

There are several points at which these experiments fail to meet the conditions of the stability calculations. First, the system is discrete, so the continuous model is an approximation. It is difficult to estimate the effect of the approximation and even more difficult to do the calculation for a discrete system (but see Section 3.) Second, the onset of the instability is not detected. The instability (if indeed that is what it is) may well have started where theory indicated and later have spread to the center, where we were able to observe it. Third, the configuration is no longer axisymmetric at the time that it is clear that something has happened. We cannot say whether the unstable mode(s) that started the whole process were axisymmetric; the final state that we observe is not. Fourth, the initial state is not strictly axisymmetric. Here again, we do not know how important the 'perturbations' of the computer model are.

Our earlier experiences in trying to produce 'cool' models led to the conjecture that the 'hot' condition resulted from a rearrangement following a gravitational instability which developed at some place that was too 'cool'. The response to this instability could be a local 'heating'. The fact that our self-consistent equilibrium systems were so far beyond the stability threshold is disturbing. It suggests that the penalty for violating the stability criterion might be rather severe; that it could lead to 'heating' far in excess of that just sufficient to lift the instability, and that this 'heating' could extend over the entire system even if it arose in response to a localized instability.

The experiments reported here neither confirm nor disprove this conjecture. The cause cannot easily be tied in with instabilities because it is operationally difficult to identify instabilities as such. The picture evidently can be tied to a failure to be self-consistent as well as to instabilities. Systems that started with low velocity dispersions (small values of T) rather quickly developed 'clumps' of 'hot' stars. Those with higher velocity dispersions did not change so rapidly, but did rearrange themselves, doubtless with some resultant 'heating'. None of these models was run long enough to tell how 'hot' the system would have looked if it had reached an equilibrium state, but it seems likely that any one of them would have become as 'hot' as our other typical models. A 'cool' stable equilibrium was not reached, but this does not prove that none exists.

2. Numerical Considerations

The tendencies of computed collisionless systems to get hotter than the real galaxies that we see in the sky might result from the approximations necessary to represent the systems in a computer. The entire set of numerical problems is of utmost importance in n-body calculations in general. Because some of these matters, especially as applied to the large n-body calculations, have been discussed in detail elsewhere, we merely refer to these discussions (Miller, 1970).

3. A Thought-Experiment

Static self-consistent models should be straightforward to construct in the discrete phase space if they are undertaken directly rather than by analogy to continuous models as described in Section 1. These discrete static self-consistent models are instructive, although they may not be very interesting to construct in the computer.

In the discrete phase space, a point moves over an integer lattice according to the rules of the 'game' described by Miller and Prendergast (1968). If the model were truly static and self-consistent, the force at any allowed configuration space point would be the same at one integration step as at any other.

A realistic 'galaxy' model should contain one isolated region of nonzero forces (periodicity is unimportant for this discussion) in an infinite expanse where the forces are zero. Orbits that are restricted to the region in configuration space where the forces are nonzero occupy a bounded portion of the phase space – there is a maximum velocity that a particle may have and still be in a bound orbit. Any allowed phase space point is either on a periodic orbit or on an escape orbit (it is on a unique orbit, and the number of points in the region containing bounded orbits is finite). This is the discrete analogue of Hopf's first theorem (see, e.g., Contopoulos, 1966). The static self-consistent model can be constructed of these periodic orbits.

Consider a periodic orbit containing one particle at each phase space point along the orbit. At any integration step, the particle occupying any one of these points will move to another point along the orbit, but the original point will be filled by a particle that has moved up from the previous point. The integration process looks like a huge game of 'musical chairs', with just as many 'players' as 'chairs'. After the integration step, the entire system looks exactly as it did before the step.

The problem of constructing static self-consistent models consists in: (1) Design a 'reasonable' force-field; (2) Find all the periodic orbits in that force field; (3) Find a subset of the periodic orbits that, when projected to configuration space, will produce the original force field. If there is no such subset, pick a new force field and start over. The whole process sounds trivial – but it is not at all trivial except in some particularly simple cases.

Such a system, once constructed would not be very interesting. It would sit forever and never change. As an example, suppose there were a ring of particles – the discrete analogue of a set of particles at the vertices of a regular polygon, rotating in the plane. In the discrete phase space, this could be mimicked by two-dimensional harmonic oscillator orbits, chosen to be periodic in one rotation around the origin (same periodicity in x and in y; two-dimensional harmonic oscillator orbits tend to make Lissajous figures with periods not 1:1 in the discrete phase space). The forces produced by the ring of particles, in the neighborhood of the ring itself, will look like harmonic oscillator forces. Adjust the force constant to balance things.

This ring would, if not disturbed, circulate forever. Yet it represents a system that is known to be unstable in continuous space. This system is stable although 'cold' in the sense of having no random velocities. It is not a very interesting galaxy model. The interest attaches to the response of the system to a perturbation – but this system cannot be perturbed by arbitrarily small amounts. The least perturbation that can be given to it is to remove one of the particles to another orbit. If this were done, the ring would probably disintegrate to a chaotic motion very quickly.

The same thought-experiment can be carried out with the kind of model attempted earlier. Suppose we had succeeded in constructing a static self-consistent model. It would circulate forever, each particle on its own periodic orbit. But then remove one particle somewhere, either leaving a hole or replacing the particle elsewhere in some other orbit – preferably periodic so that it will stay around for a while. The two anomalies will move around the phase space until they happen to reach locations at which they conspire to make the force value round differently than it would in the absence of the perturbation. More particles will be disturbed by this change. Thereafter, the system is no longer static and self-consistent; it will take off and can change drastically. The argument is exactly like the reversibility argument of Miller (1970).

Without actually doing the experiment, we cannot say whether the system that would eventually result would be 'cool' or 'hot', or whether it would be approximately axisymmetric. Some points are clear, however. First, it is simple to estimate how long, on the average, a particle or a hole would have to move around the system before it would, in a probabilistic sense, cause a computed force value to round differently. For the model tested, this is 10–15 integration steps. It is this long because the force near a single particle is much smaller in magnitude than the interval between successive allowed values of the force, and because the force field around a particle effectively extends only over a very small spatial region. The system might not ever notice that it had been perturbed. This, of course, is just what is meant, operationally, by an 'infinitesimal' perturbation. But the system response is nonlinear, and it is the nonlinear response that causes the trouble. A second point is that the model described in Section 2 is hard to make self-consistent because of a design error – it had particles in regions of zero force. These must be on escape orbits. This is not serious with periodically replicated systems.

This discussion makes it clear that there is a smallest perturbation that can be given to the system and a smallest perturbation to which the system can respond. Because of the discrete number representation in computers, any model in a computer shares these attributes. But with the very fine discrete phase space allowed with full-precision computer numbers, the number of particles in the self-consistent model must be very large. It is straightforward to estimate how the number of particles scales with the number of points allowed in each coordinate direction and in each velocity component as the spatial lattice and time-steps are separately refined. The present calculation is just crude enough to build somewhat realistic static self-consistent models out of the number of particles that it can handle. But we have not yet developed an algorithm that will generate such models at will. Initial steps in this direction have produced some very interesting results, showing that a phase space point may be on a periodic orbit while its neighbors are on escape orbits – some regions have small fractions of periodic orbits.

There are no nice continuity properties to the phase space, of course; there is not a closed surface inside which all orbits are periodic and outside which all orbits are of the escape variety. Some orbits require several thousand steps to repeat periodically; some have extraordinary shapes. The bookkeeping required to keep track of even these few periodic orbits seems prohibitive.

The notion of such self-consistent models can be extended to models that move in regular ways. A pattern that translates itself across the plane without change of shape is one such possibility – patterns that appear to rotate seem more difficult.

A system containing many 'holes' should be able to respond to smaller perturbations. This might provide a useful route to attack the self-consistency problem posed in this report.

A truly static self-consistent model seems difficult to build. An attempt to do so is very instructive, and provides considerable insight into the way the computer models behave.

4. Conclusions

The search for reasons why the computer models might be so hot has led through some interesting investigations. At the end if all these investigations, we still don't know whether it is possible to build a static self-consistent model that is cool. However, along the lines of the 'thought-experiment' of Section 3, it seems quite likely that one can be constructed. Whether it would be stable (whatever that means) to small disturbances is unclear, but the suspicion is strong that, if the system felt the distrubance at all, it might well be quite unstable. That means that its response to as small a disturbance as it could feel would be a major rearrangement that might result in its being as hot as all the other models.

A related question is why spiral patterns do not last any longer than they do. Hohl's last quite long – for reasons that are as mysterious as why ours don't last very long. The obvious reasons might be (1) that there aren't enough particles to build a long-lived spiral, or (2) that the spirals dissipate because of numerical effects. After all this, there is no clear-cut basis to choose between the two alternatives, but there is no *a priori* reason why numerical effects should work against the maintenance of spirals instead of helping to maintain them.

Acknowledgements

It is a pleasure to acknowledge the assistance of many people who have participated in this program. The bulk of the calculations were carried out at the Goddard Institute for Space Studies in New York through the courtesy of Dr Robert Jastrow, Director. My colleagues on this project, Kevin Prendergast and William J. Quirk of Columbia University have been full partners in the work at all stages. Many thanks are due to Frank Hohl for generously taking time to give me rather complete descriptions of what he and Roger Hockney have been doing – I hope that the descriptions of his work in this report do him justice. Hohl has essentially gotten as far in this work as we have, a fact that is obscured by the heavy emphasis on our work in this report.

References

Chandrasekhar, S.: 1960, *Principles of Stellar Dynamics*, Dover Publications, Inc., New York.
Code, A. D.: 1967, *Astron. J.* **72**, 789–90 (A).
Contopoulos, G.: 1966, 'Problems of Stellar Dynamics', pp. 169–257 in *Relativity Theory and Astrophysics* (J. Ehlers, ed.) Vol.1 American Mathematical Society, Providence, R.I.
Hockney, R. W.: 1970, *Methods in Computational Physics*, Vol. 9, pp. 136–212, (B. Adler, S. Fernbach, and M. Rotenberg, eds.), Academic Press New York.
Hockney, R. W. and Hohl, F.: 1969, *Astron. J.* **74**, 1102.
Hohl, F. and Hockney, R. W.: 1969, *J. Comp. Phys.* **4**, 306.
Lee, E. P.: 1967, *Astrophys. J.* **148**, 185.
Lin, C. C., Yuan, C., and Shu, F. H.: 1969, *Astrophys. J.* **155**, 721.
Lynden-Bell, D.: 1962, *Monthly Notices Roy. Astron. Soc.* **144**, 279.
Miller, R. H.: 1970, *J. Comp. Phys.* **6**, 449.
Miller, R. H. and Alton, N.: 1968, 'Three Dimensional *N*-Body Calculation', ICR Quarterly Report, No. 18 (August 1, 1968), Institute for Computer Research, University of Chicago (Unpublished).
Miller, R. H. and Prendergast, K. H.: 1968, *Astrophys. J.* **151**, 699–709.
Miller, R. H., Prendergast, K. H., and Quirk, W. J.: 1970, *Astrophys. J.* (to be published).
Ng, E.: 1967, *Astrophys. J.* **150**, 787.
Quirk, William J.: 1970, Ph.D. Thesis, Columbia University, (Unpublished).
Schmidt, M.: 1965, *Galactic Structure*, pp. 513–30. (A. Blaauw and M. Schmidt, eds.) University of Chicago Press, Chicago, Vol. V, of Stars and Stellar Systems, (B. M. Middlehurst and G. Kuiper, eds.).
Toomre, A.: 1963, *Astrophys. J.* **138**, 385.
Toomre, A.: 1964, *Astrophys. J.* **139**, 1217.

DYNAMICS OF PLANE STELLAR SYSTEMS

FRANK HOHL

NASA, Langley Research Center, Hampton, Virginia

Abstract. The evolution of initially balanced rotating disks of stars is investigated with a computer model for isolated disks of stars. An isolated, initially cold balanced disk is found to be violently unstable. Adding a sufficient amount of velocity dispersion will stabilize all small-scale disturbances. However, the disks are still unstable against slowly growing long wave-length modes and after about two rotations most disks tend to assume a bar-shaped structure. It is found that the final mass distribution over most of the disk can be closely approximated by an exponential variation, irrespective of the initial mass distribution. The gravitational two-stream instability is investigated by means of a modified computer model for infinite doubly periodic stellar systems.

1. Introduction

The only means available to us for performing gravitational experiments on the evolution of stellar systems is the use of computer models. Such numerical methods can be called 'experimental stellar dynamics' and they play a role similar to the experimental methods in physics. The type of computer model used in the present paper applies to systems that can be considered collisionless. The force acting on a particular star is determined by the mean gravitational field of the system and the effect of nearby stars can be neglected. Chandrasekhar (1960) points out that for the Galaxy a star can describe at least 100 rotations about the galactic center before the effects of encounters with other nearby stars becomes important. A galaxy can therefore be studied by using a 'collisionless' computer model in which close encounters are effectively neglected. The first such models used were simple one-dimensional models, where the stars are stratified into plane parallel layers or mass sheets (Lecar, 1966; Hohl and Feix, 1967; Hohl, 1968a). The one-dimensional model is useful as an approximation to the motion of stars normal to the galactic plane of the Galaxy. The one-dimensional model was also used to investigate 'violent relaxation' of self-gravitating systems (Hohl and Campbell, 1968, 1969; Cuperman *et al.*, 1969). Henon (1964) used a model of concentric spherical mass shells to study the dynamical mixing of spherical star clusters. A model of concentric mass rings was used by Toomre (1954). Two-dimensional models, in which the motion of infinitely long mass rods is followed, have been used by Hockney (1967) and by Hohl (1968b, 1969).

Two computer models for self-consistent disk galaxies have recently been described (Miller and Prendergast, 1968; Hohl and Hockney, 1969). The model of Miller and Prendergast is doubly periodic and the forces, star positions and velocities are allowed to attain only discrete (integer) values which are less than some given maximum value. In a recent report Miller *et al.* (1970) used their model to study the evolution of spiral structure and they find that spiral patterns persist for about three galactic rotations. This spiral pattern was obtained for the 'gas' component of systems which consisted of 'gas' and stars. The model described by Hohl and Hockney is

M. Lecar (ed.), Gravitational N-Body Problem, 231–249.

for isolated disk galaxies and the star positions and velocities are allowed to attain essentially any value. Hockney and Hohl (1969) used the model to study the effects of velocity dispersion on the stability of an initially balanced uniformly rotating disk of stars. To investigate the development of spiral structure the model was modified by Hohl (1970a, b) to include a fixed central force similar to the Schmidt model of the Galaxy. It was then found that spiral structure persisted for more than eight rotations.

In the present paper the evolution and final state of self-consistent and initially stationary disks of stars is investigated by means of the computer model for isolated disks of stars developed by Hohl and Hockney (1969). A number of initially cold (zero velocity dispersion) and balanced disk galaxies were investigated previously (Hohl, 1970) and were found to be violently unstable. The results presented here are for disks with an initial velocity dispersion equal to that required for stabilization of all axisymmetric instabilities as calculated by Toomre (1964). Such an initial velocity dispersion will stabilize all small-scale disturbances but the disk is still unstable against slowly growing nonaxisymmetric disturbances which cause the disk to assume a bar-shaped structure in less than two rotations. The development of a bar-like structure appears to be a general feature for disks of stars and has been found to occur for a large variety of initial conditions (Hohl, 1970). The later stages of the calculations by Miller *et al.* (1970) also give an indication of the development of a bar-like structure.

2. The Model

The model used for the present calculations is described in detail by Hohl and Hockney (1969) and a listing of the computer program used for the potential calculation is given elsewhere (Hohl, 1970). We include here only a summary of the model.

The model for the disk galaxy consists of a large number of representative stars (here 50000 or 100000) that are confined to move in the galactic disk. An N by N (here 64 by 64 or 128 by 128) array of cells is superposed over the plane of the disk for the purpose of calculating the gravitational potential. At the center of each cell a mass density is defined which is given by the number of stars in that cell. The mass density distribution is used to obtain the gravitational potential at the center of each cell. From the gravitational potential the force acting at the position of a star is calculated by means of a bilinear interpolation among four surrounding cell centers. Newton's equations of motion are then used to advance the position and velocity of each star by a small time step. Usually there are 200 or 400 time steps per 'galactic rotation'. One complete cycle for advancing the motion of the system by a small time interval consists of the following steps:

(i) The coordinates of each star are examined to determine the mass density over the N by N array of cells.

(ii) The gravitational potential corresponding to the mass density distribution is determined by using fast Fourier transform methods.

(iii) The gravitational potential is used to advance the motion of all the stars by the use of time centered finite difference equations. The cycle repeats at step (i). If a star leaves the N by N array of cells it is still included in the calculations by approximating the force acting on the star. For the doubly periodic system a star leaving the array at one side will enter again at the opposite side with the same velocity components.

The effects of varying the number of stars and various other discretization parameters was investigated by Hohl and Hockney (1969). For a 50000 star system the effect of binary collisions for the model has been estimated (Hohl, 1970) to be such that the collision time equals about 100 'galactic rotations'.

3. Gaussian Mass Distribution

A. INITIAL CONDITIONS

The first system to be investigated is a disk with a Gaussian surface mass density distribution given by

$$\mu(r) = \mu(0) \exp\left\{-\pi \frac{\mu(0)}{M} r^2\right\} \tag{1}$$

where M is the total mass of the disk. For the present calculation the value of $1/(\pi\mu(0)/M)$ was taken to be 50 kpc^2. The rotational velocity required to balance a cold disk with a Gaussian surface density was calculated by Toomre (1963) as

$$V_\theta^2 = \pi^2 G\mu(0) \sqrt{\frac{\sigma(0)}{M}}\, {}_1F_1\left(1.5; 2; \frac{\pi\mu(0)}{M} r^2\right) \tag{2}$$

where ${}_1F_1$ (1.5; 2; x) is the confluent hypergeometric function. Toomre (1964) found the root-mean-square radial velocity dispersion required to stabilize all axisymmetric disturbances anywhere in the disk to be

$$\sigma_{r,\,\min} = 3.36\, G\mu/\kappa \tag{3}$$

where μ and κ are the local values of the density and epicyclic frequency respectively. To determine whether Toomre's minimum velocity dispersion will stabilize a disk with a Gaussian mass distribution the initial velocity dispersion is taken as

$$\sigma_r(r) = \sigma_{r,\,\min} \tag{4}$$

for the radial component and

$$\sigma_\theta(r) = \frac{\kappa(r)}{2\omega_0(r)} \sigma_{r,\,\min} \tag{5}$$

for the azimuthal component. The angular velocity $\omega_0(r)$ is defined by

$$\omega_0^2 = \frac{1}{r}\frac{\partial\phi}{\partial r}$$

where ϕ is the gravitational potential. Because of the effective pressure caused by the added velocity dispersion the initial angular velocity of the stars, $\omega(r)$, for a balanced disk is lower than $\omega_0(r)$ and is given by

$$\omega^2 = \omega_0^2 + \frac{1}{r\mu(r)}\frac{\partial}{\partial r}(\mu(r)\,\sigma_r^2(r)) + \frac{1}{r^2}[\sigma_r^2(r) - \sigma_\theta^2(r)]. \tag{6}$$

Figure 1 shows the variation of ω_0, ω and κ as a function of r for the disk with a Gaussian mass distribution.

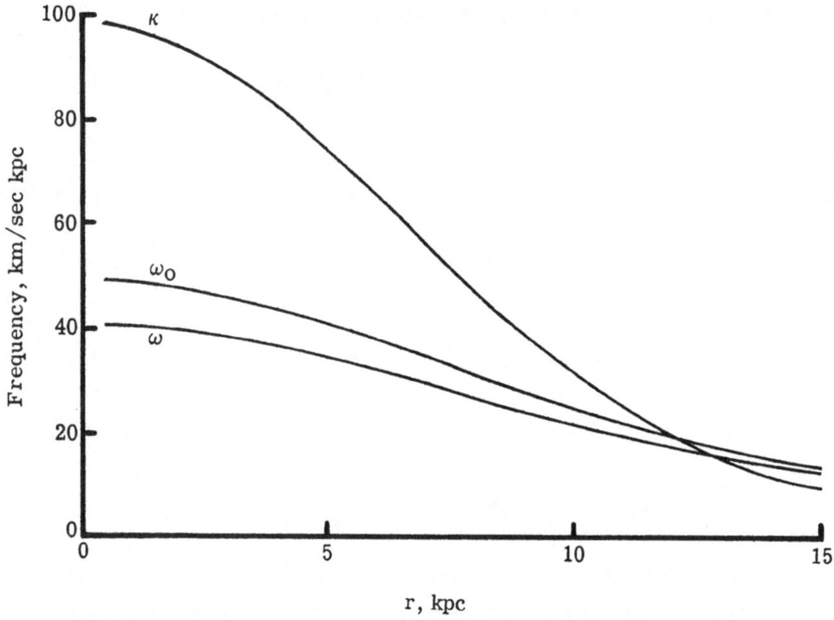

Fig. 1. Variation of ω_0, ω, and κ as a function of r for a disk of stars with a Gaussian mass density. The angular velocity ω_0 is that for a cold balanced disk, ω is given by Equation (6), and κ is the epicyclic frequency.

B. EVOLUTION OF THE DISK

The evolution of the disk is shown in Figure 2. There are 50000 stars each of mass $Gm = 77.4$ in the disk and 200 time steps per rotation are used in the calculation. The time is given in units of the rotational period of the cold balanced disk

$$\tau_0 = 2\pi/\omega_0(r)$$

at $r = 10$ kpc. The rectangular border enclosing the disk is at $x = \pm 19$ kpc and $y = \pm 19$ kpc. The evolution displayed in Figure 2 shows that even though the disk is stabilized against all small-scale disturbances and against axisymmetric disturbances according to Toomre's formula, it is still unstable against relatively slowly growing nonaxisymmetric disturbances. After two rotations the system has assumed a bar-shaped structure which changes very little during the following five rotations. At

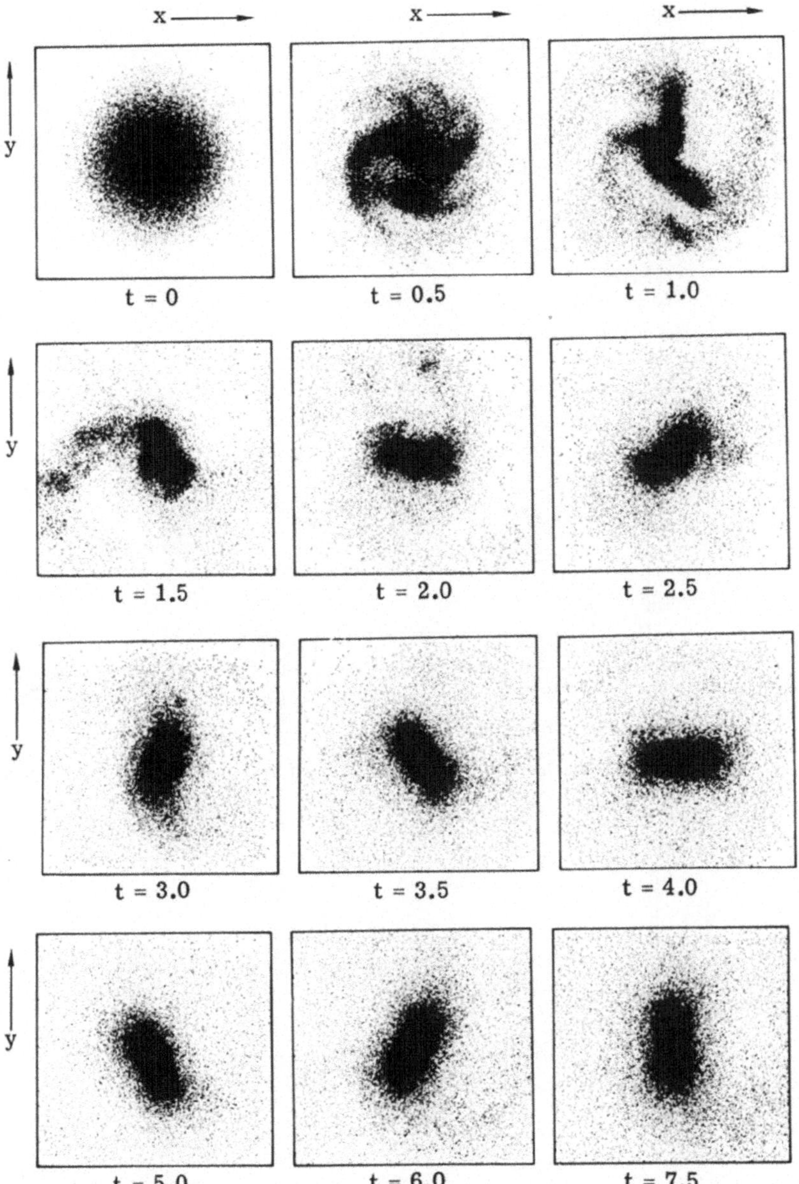

Fig. 2. Evolution of an initially stationary disk of stars with a Gaussian density distribution and with an initial velocity dispersion equal to $\sigma_{r,\min}$ and $\sigma_{\theta,\min}$ as calculated by Toomre.

$t=2.0$ the bar is rotating with a period of about 1.33, at $t=3.5$ it rotates with a period of about 1.45. After $t=5.0$ the rotational period of the bar appears to remain constant at about 1.6. Note that these times are in units of τ_0.

The change (error) in the total energy of the system is less than 1% for the first three rotations and the change in the angular momentum is about 0.1%. It should

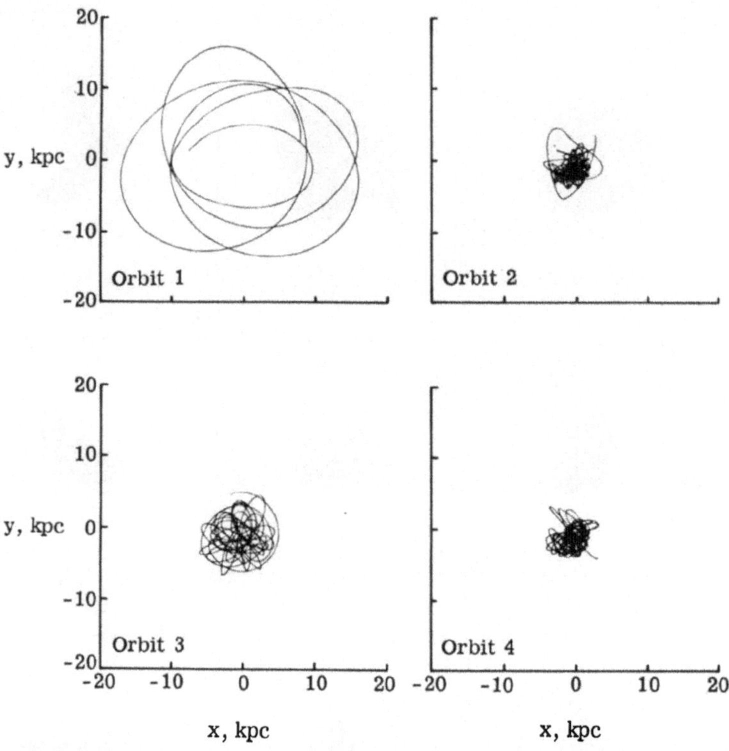

Fig. 3. Four individual star orbits for the disk with an initially Gaussian mass distribution.

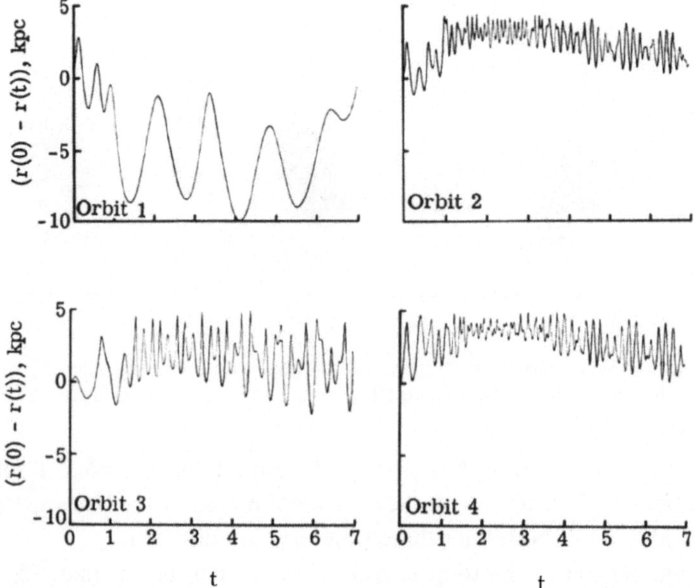

Fig. 4. Deviation from circular motion for the four star orbits shown in Figure 3.

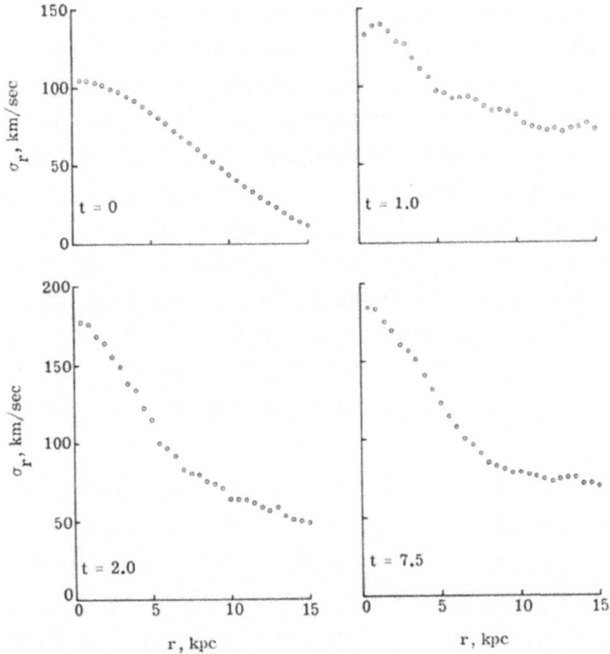

Fig. 5. Evolution of the radial velocity dispersion for the disk with a Gaussian mass distribution shown in Figure 2.

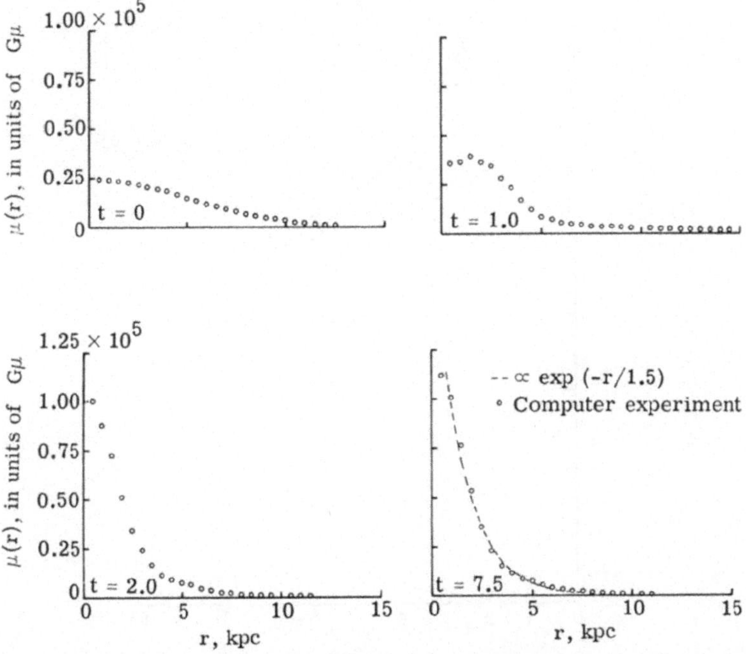

Fig. 6. Evolution of the mass density for the disk with a Gaussian mass distribution. The mass density is plotted in units of $G\mu$ per kpc^2.

also be noted that at $t=0$ the virial theorem is satisfied; that is, the negative of the potential energy equals twice the kinetic energy of the disk.

Four individual star orbits taken at random from the 50000 stars in the disk are shown in Figure 3. The orbits are plotted by simply connecting the position of a star at each time step by straight line segments. The orbits indicate that stars initially near the center of the disk have a tendency to become trapped in even tighter orbits as the central mass density increases. Stars further out have a tendency to escape from the system. The orbits are plotted for the interval $t=0$ to $t=7$. The deviation of the star motion from circular orbits is displayed in Figure 4 which shows the difference between the initial star radius and the star radius at time t.

The frequency of the oscillations in the star orbits is that expected from epicyclic theory. In order to obtain more quantitative information than can be obtained from Figure 2, the disk is divided into a number of concentric rings each of $\frac{1}{2}$ kpc width.

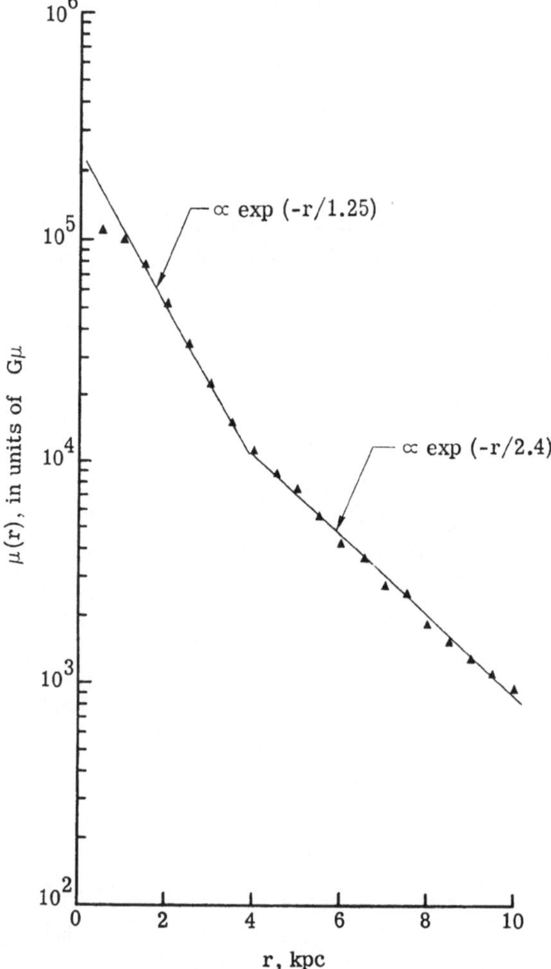

Fig. 7. Logarithmic plot of the density distribution at $t=7.5$ for the Gaussian disk.

The radial dependence of various parameters averaged azimuthally over each ring is then obtained. Figure 5 shows the evolution of the radial velocity dispersion obtained in the above manner for the disk shown in Figure 2. Initially, the velocity dispersion for large r increases rapidly as can be seen from a comparison of the results at $t=0$ and $t=1.0$. Again, the time shown is in units of $2\pi/\omega_0$ at $r=10$ kpc. After two rotations, the radial velocity dispersion in the central part of the disk has increased to about 175 km/s from the initial value of 105 km/s. During the following five and one half rotations the velocity dispersion only increases slightly, indicating that at $t=2$ the system has nearly reached a steady state.

In interpreting the results shown in Figure 5 and in subsequent figures, the monaxisymmetric shape of the disk should be kept in mind. Figure 6 shows the evolution of the azimuthally averaged mass density as a function of radius. The figure shows that the central mass density of the disk increases by a factor of four during the first two rotations. Again there is little further change in the distribution during the following five and one half rotations. The dashed line shown at $t=7.5$ corresponds to an exponential density distribution with a scale length of 1.5 kpc,

$$\mu(r) \propto \exp[-r/1.5].$$

It can be seen that the exponential variation closely fits the mass distribution of the

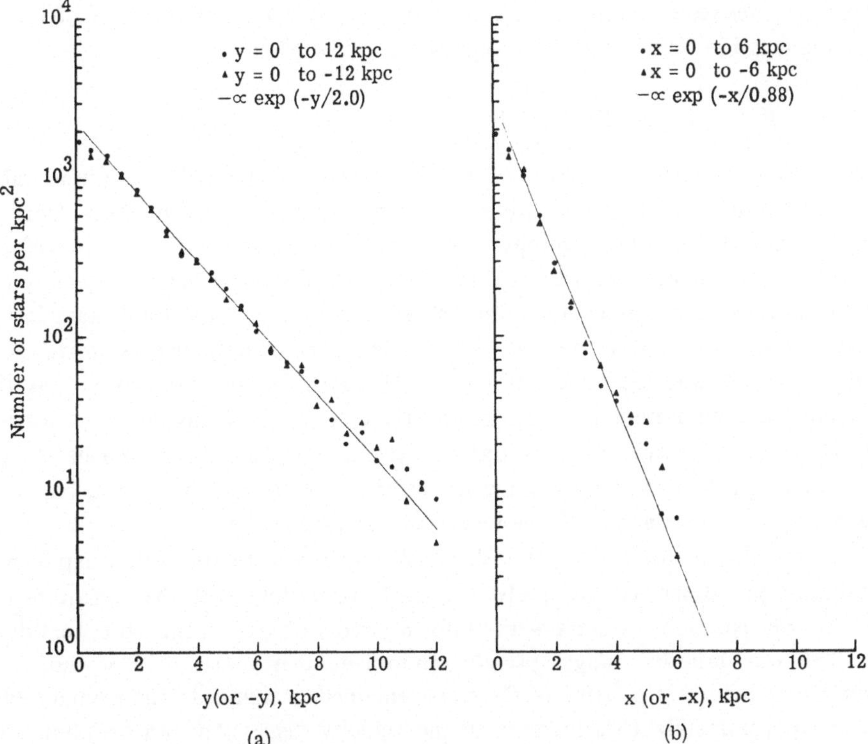

Fig. 8. Logarithmic plots of the density variation along the bar (a), and across the bar (b).

disk. A better indication of how closely the mass density fits an exponential distribution can be obtained by plotting the azimuthally averaged density on a logarithmic scale. The result is shown in Figure 7 for $t=7.5$. Up to a radius of about four kpc the density varies as exp $(-r/1.25)$ and for r larger than 4 it varies like exp $(-r/2.4)$. The variation shown in Figures 6 and 7 is the azimuthally averaged density variation and it is of interest to get an indication of the variation in any particular radial direction. At $t=7.5$ the bar is in a vertical position and from the number of stars in the cells along the y-direction for $x=0$ and along the x-direction for $y=0$ we obtain the variation along and across the bar. Note that $x=y=0$ represents the center of the system. The results are shown in Figure 8 where the number of stars per kpc^2 is plotted on a logarithmic scale along the bar (a), and across the bar (b). Along the bar the variation is closely approximated by an exponential variation exp $(-y/2.0)$ and across the bar it approximates exp $(-x/0.88)$. The variation shown in Figure 7 appears to be a combination of the exponential variation across and along the bar.

4. Exponential Mass Distribution

A. INITIAL CONDITIONS

For the disk with an initial Gaussian mass distribution it was found that the final mass density was closely described by an exponential distribution. It is therefore of interest to investigate whether a disk with an initially exponential mass distribution will remain stable. The initial mass density for a disk given by

$$\mu(r) = \mu(0) \exp(-r/r_0)$$

where $\sigma(0)$ is the central density and $r_0=3$ kpc was investigated. The initial values of ω_0 and ω as a function of r are obtained by finite difference methods from the computer model. The disk contains 50000 stars. The mass per star is $Gm=60$ and the initial mass density was cut off at 15 kpc (or five-scale lengths). This cutoff should not effect the dynamics of the disk since only 4% of the total mass for the exponential distribution lies outside 15 kpc. The initial velocity dispersion was given by Equations (4) and (5). It was found (Hohl, 1970) for the exponential disk that the evolution is note quite as violent as was the case for the Gaussian mass distribution. However, after only 1.5 rotations the system had again assumed a bar-shaped structure with a halo of stars moving in larger orbits around the central bar. Only slow changes in the shape of the system occurred after $t=1.5$.

In an attempt to force a more stable disk the disk with the initial exponential distribution just described was evolved up to five rotations with the constraint that every quarter rotation the stars were uniformly redistributed in the azimuthal direction. This was done by using a pseudo random number generator. The radial and azimuthal velocity components of the stars remained unchanged. The primary effect was an approximately 15% increase in the velocity dispersion and an about 20% increase in the central density. The evolution of the disk starting from this slightly hotter and more centrally condensed exponential disk is discussed in the next section.

B. EVOLUTION OF THE DISK

Figure 9 displays the evolution of the exponential disk. The evolution is not quite as violent as was the case for the Gaussian mass distribution. In addition the final

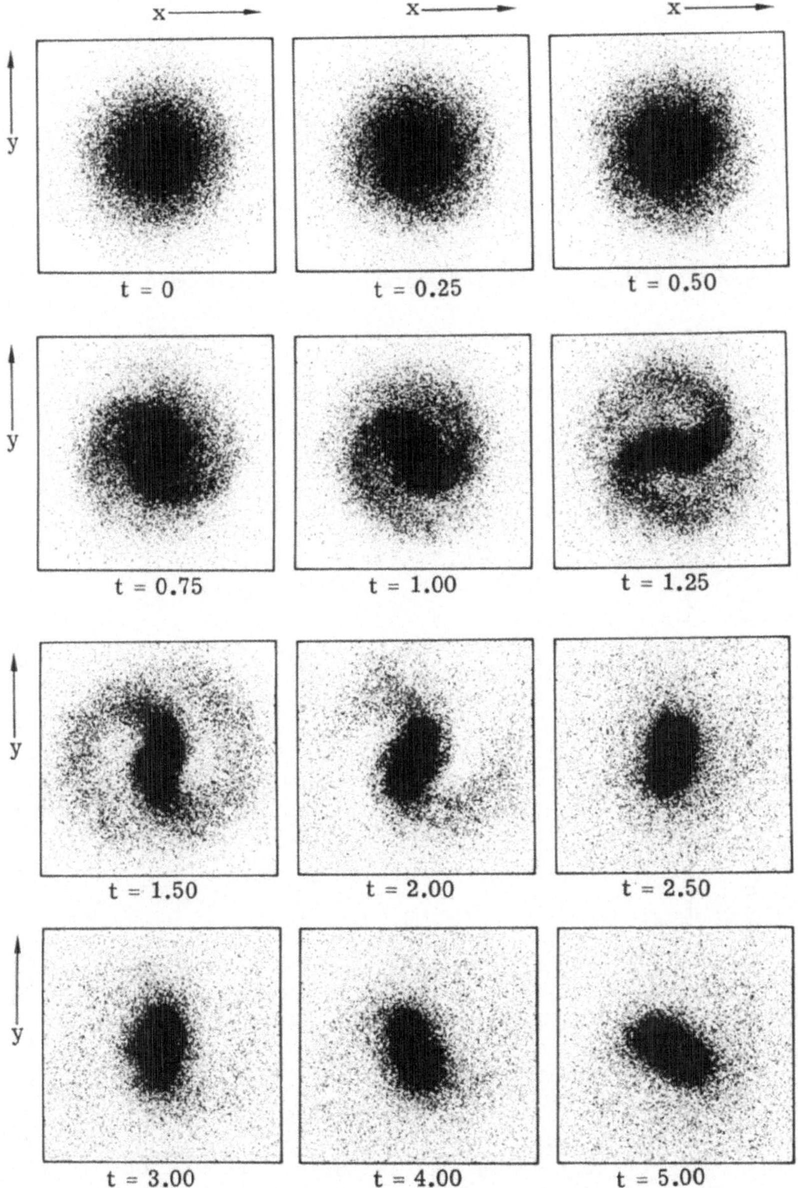

Fig. 9. Evolution of a disk of stars with an initially exponential mass distribution.

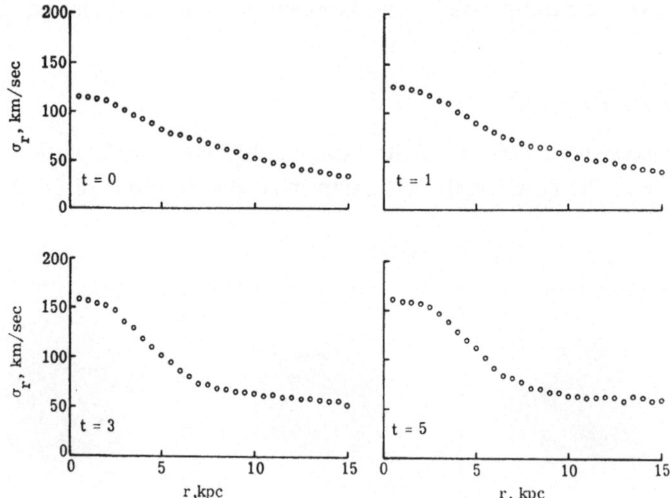

Fig. 10. Evolution of the azimuthally averaged radial velocity dispersion for the exponential disk.

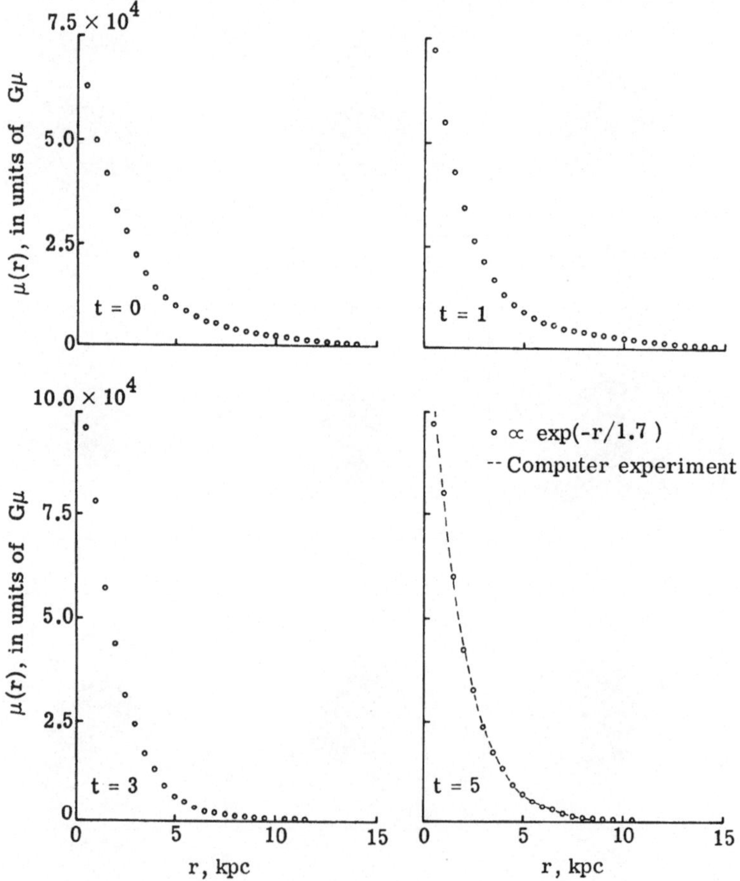

Fig. 11. Evolution of the azimuthally averaged density for the exponential disk.

bar structure is shorter and 'fatter' than that shown in Figure 2. Even though the 'initial' velocity dispersion is about 15% larger than that calculated by Toomre for the suppression of axisymmetric instability, the disk quickly assumes a bar-shaped structure, in less than two rotations. Note that the time shown is in rotational periods at $r=10$ kpc and that the rectangle enclosing the disk is at y or $x=\pm 19$ kpc.

The evolution of the azimuthally averaged radial velocity dispersion for the exponential disk is shown in Figure 10. The velocity dispersion in the central region of the disk increases from about 115 km/s to 160 km/s during the first three rotations. There is little change in the velocity dispersions during the next two rotations. Figure 11 shows the evolution of the azimuthally averaged density. The central density increases by about 50% during the evolution and changes from an initial distribution given by $\exp(-r/3)$ to a final distribution approximated by $\exp(-r/1.7)$. Again there is little change in the distribution during the last two rotations. A logarithmic plot of the density distribution at $t=5$ is shown in Figure 12. For r up to 8 kpc the density is closely approximated by a variation $\exp(-r/1.7)$. For the Gaussian case shown in Figure 7 the exponential fit in the central part of the disk extends only up to about 4 kpc. The reason for the difference in the two exponential density plots is that the final bar structure for the Gaussian disk shown at $t=7.5$ in Figure 2 is more elongated than that for the exponential disk shown at $t=5$ in Figure 9.

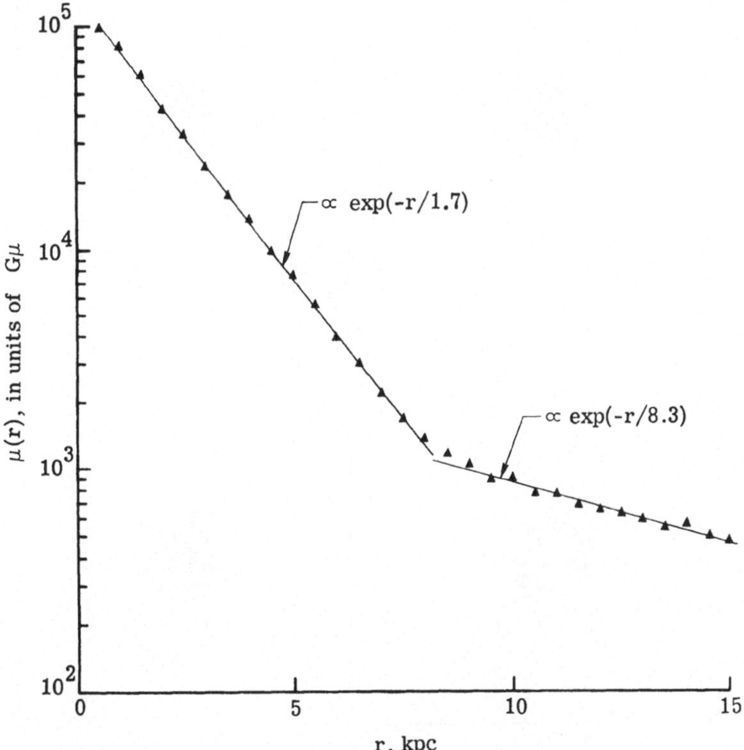

Fig. 12. Logarithmic plot of the density distribution at $t=5$ for the exponential disk.

5. Stable Axisymmetric Disk

Disks of stars with initial conditions generated according to analytical expressions such as Equations (3) to (6) were found to be unstable and to finally assume a steady state with a central bar-shaped structure. The evolution of the warm disks is found to be very similar irrespective of whether the initial mass distribution is Gaussian, exponential or some other distribution (Hohl, 1970). Also, the stationary state finally reached by the disk results in a rather hot population of stars in the outer portions of the disk. In order to generate an axisymmetric stable disk the final distribution of stars of an evolved disk like that shown at $t=5$ in Figure 9 was used as an initial condition after symmetrizing out the bar structure. All stars keep the same velocity components and radius, except that they are now axisymmetrically distributed by means of a random number generator. The evolution of such a disk is shown in Figure 13 for six rotations. The rectangular border surrounding the disk is at $x=\pm 32$ kpc and $y=\pm 32$ kpc. Also, the disk contains 100000 stars each of mass $Gm=30$. The disk is stable and all parameters remain constant during the evolution. For

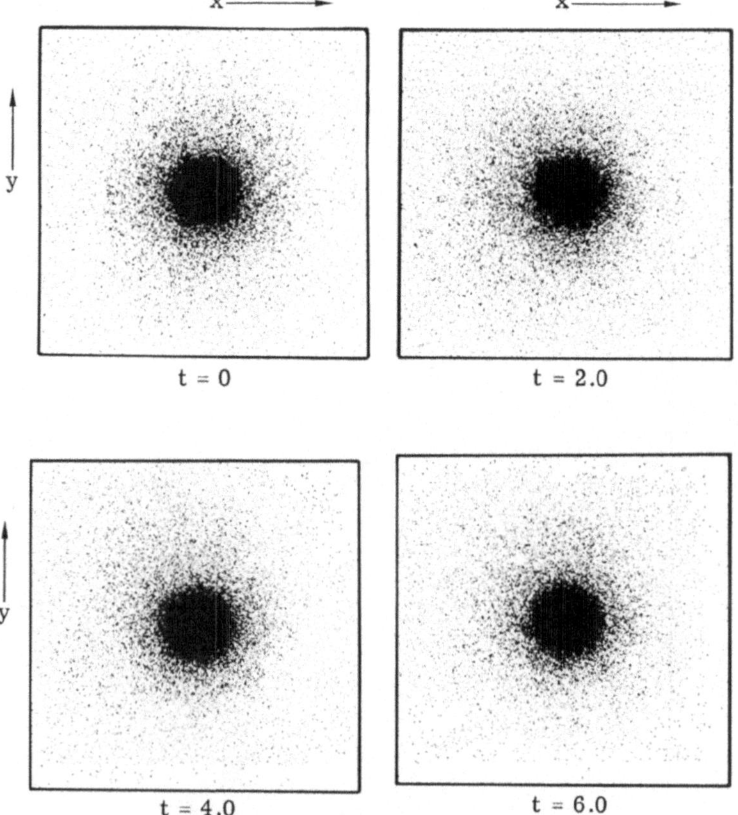

Fig. 13. Evolution of a stable axisymmetric disk of 100000 stars.

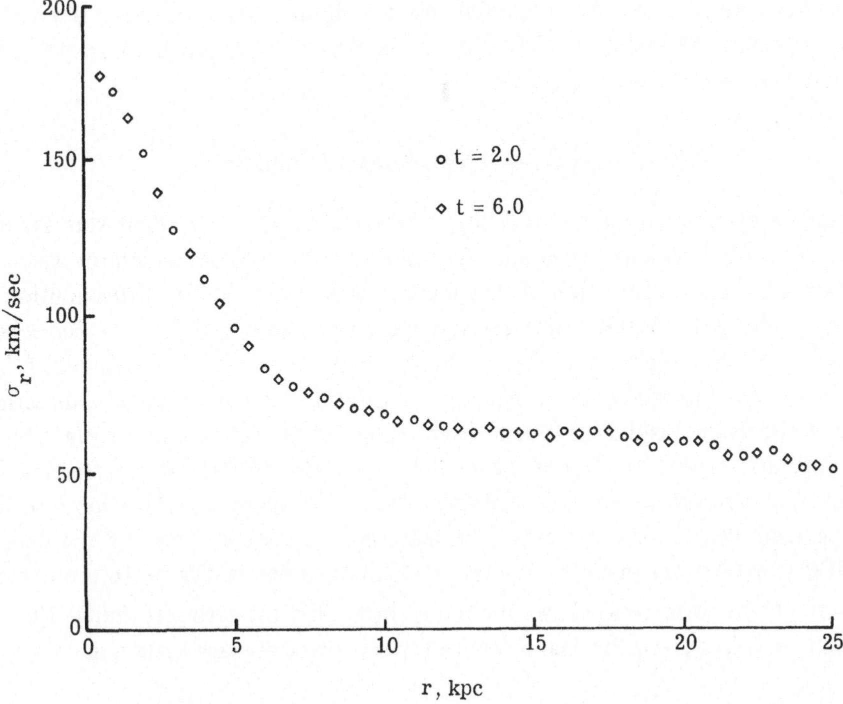

Fig. 14. Variation of the radial velocity dispersion for the stable disk shown in Figure 13.

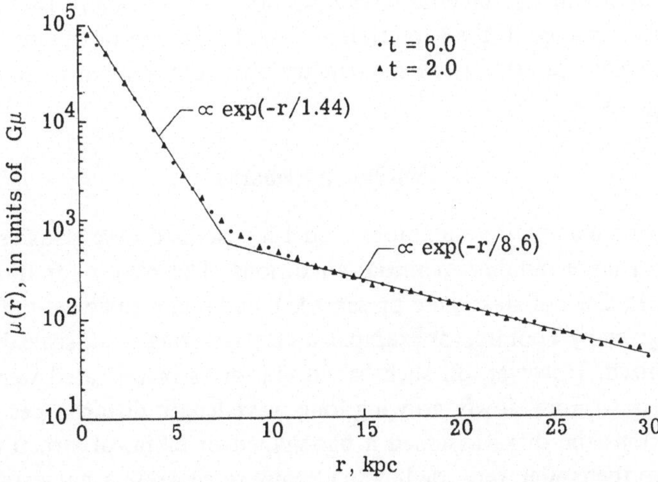

Fig. 15. Variation of the density for the stable disk shown in Figure 13.

example, the radial velocity dispersion shown in Figure 14 shows no change for the results after 2 and 6 rotations. A similar result is shown in Figure 15 for the density variation. Again the variation of the density is closely described by an exponential variation over most of the disk.

6. Gravitational Two-Stream Instability

The gravitational two-stream instability was investigated for 100000 star systems with a given initial velocity dispersion. A pseudo random number generator was used to obtain a nearly uniform initial distribution of stars. A uniform distribution of one half of the stars (50000 stars) has superposed on the initial velocity dispersion, σ, a constant streaming velocity, V, in the positive x-direction. The remaining half of the stars has superposed a streaming velocity, V, in the negative x-direction. Figure 16 shows the results for four different values of the ratio of streaming velocity to velocity dispersion, V/σ. Since the model is now doubly periodic, a star leaving the rectangular region at one side will enter the region again from the opposite side with the same velocity components. The mass per star is $Gm=0.5$ and the density per cell (64 by 64 array of cells) is $G\mu=12.5$. Times shown in Figure 16 are in units of $\tau_0=1/\sqrt{G\mu}$. Figure 16(a) shows the Jeans' instability for zero streaming, $V/\sigma=0$. Toomre (1964) has given the Jeans' length for this plane stellary system as

$$\lambda_J = \frac{\sigma^2}{G\mu}.$$

For the present system σ is 5 (5 cell dimensions per τ_0). The measured diameter of the condensations shown in Figure 16(a) is $2\lambda_J$, indicating that disturbances with wavelength $2\lambda_J$ have the fastest growth rate. Figure 16(b) and (c) show that the dynamics of the two-stream instability changes very little up to streaming velocities of $V/\sigma=1.0$. For streaming velocities greater than $V/\sigma=1.0$, Figure 16(d) shows that there is a marked change in the evolution and final structure of the instability. The system now displays an elongated filamentary structure similar to that found in certain spiral galaxies.

7. Concluding Remarks

Numerical experiments with a computer model for isolated disk galaxies have been performed for a large number of initial conditions. The results for three cases of initially balanced disks of stars were presented. By adding a sufficient initial velocity dispersion (as given by Toomre, for example) all fast-growing small-scale disturbances could be stabilized. However, all such 'stabilized' disks investigated were found to be unstable against more slowly growing long wave-length disturbances and in less than two rotations the disks assumed a bar-shaped or elliptical structure. Various modifications to the model were tried in an attempt to generate a stable axisymmetric disk. For example, a fraction of the mass in the central portion of the disk was held

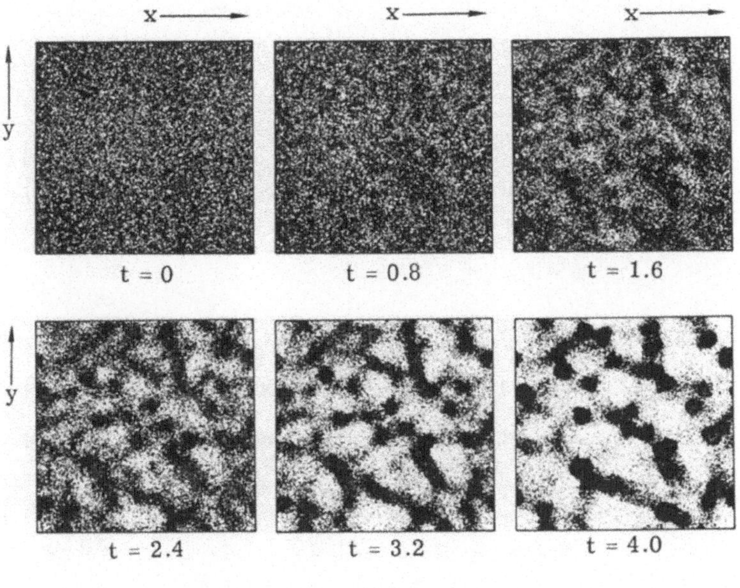

(a) $V/\sigma = 0$.

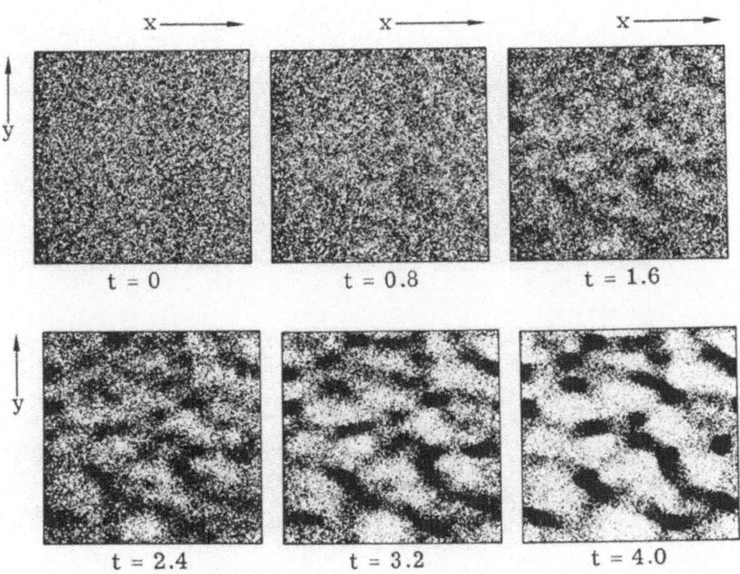

(b) $V/\sigma = 0..5$.

Fig. 16a–b. Evolution of the two-stream instability for four different streaming velocity (a) $V/\sigma = 0$, (b) $V/\sigma = 0.5$, (c) $V/\sigma = 1.0$, and $V/\sigma = 2.0$.

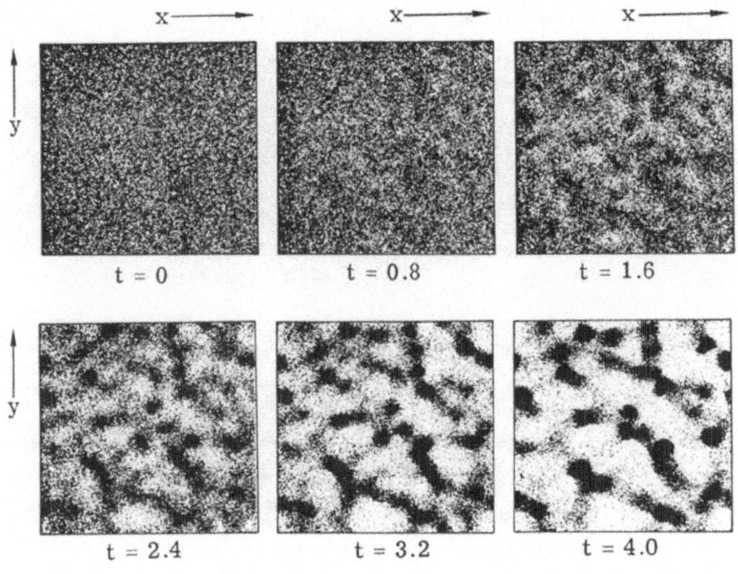

(c) V/σ = 1.0.

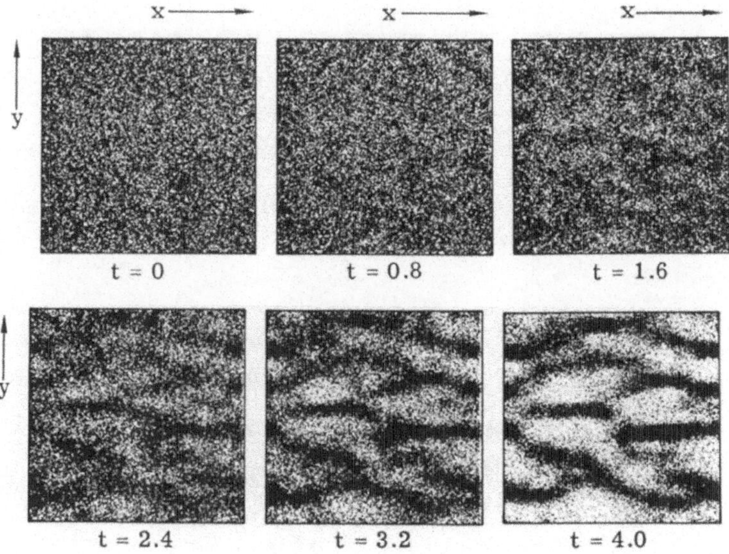

(d) V/σ = 2.0.

Fig. 16c–d.

fixed. Of course, if the fraction of mass held fixed becomes large (say 70 or 80%) then axisymmetric disks can easily be obtained. The final velocity dispersion for the disks is generally found to be from 100 to 200% larger than $\sigma_{r,\min}$ as calculated by Toomre. By evolving a disk of stars on the computer until it reached a steady state and by then symmetrizing out the central bar structure it was possible to generate a stable axisymmetric disk.

An interesting result of the calculations is that for the disks of stars so far investigated the final mass distribution in the radial direction can be closely approximated by an exponential variation irrespective of the initial conditions. This result is in agreement with observational evidence (Freeman, 1970) which indicates that the luminosity distribution of many SO and spiral galaxies can be approximated by an exponential distribution.

In investigating the gravitational instability it was found that the condensations which form have dimensions equal to about two Jeans lengths. Increasing the streaming velocity from zero to a value equal to the velocity dispersion of the system had little effect on the dynamics of the instability. For streaming velocities larger than the velocity dispersion, the system developed elongated filamentary structure.

Acknowledgement

I wish to thank Professor Alar Toomre for his many suggestions and contributions to this work.

References

Chandrasekhar, S.: 1960, *Principles of Stellar Dynamics*, Dover Publ., Inc., New York.
Cuperman, S., Goldstein, S., and Lecar, M.: 1969, *Monthly Notices Roy. Astron. Soc.* **146**, 161.
Freeman, K. C.: 1970, *Astrophys. J.* **160**, 811.
Hénon, M.: 1964, *Ann. Astrophys.* **27**, 83.
Hockney, R. W.: 1967, *Astrophys. J.* **150**, 797.
Hockney, R. W. and Hohl, F.: 1969, *Astron. J.* **74**, 1102.
Hohl, F.: 1968a, NASA TR R-289.
Hohl, F.: 1968b, *Bull. Astron.* **3**, 227.
Hohl, F.: 1969, NASA TN D-5200.
Hohl, F.: 1970a, in W. Becker and G. Contopoulos (eds.), 'The Spiral Structure of our Galaxy', *IAU Symp.* **38**, 368.
Hohl, F.: 1970b, NASA TR R-343.
Hohl, F. and Campbell, J. W.: 1968, *Astron. J.* **73**, 611.
Hohl, F. and Campbell, J. W.: 1969, NASA TN D-5540.
Hohl, F. and Feix, M. R.: 1967, *Astrophys. J.* **147**, 1164.
Hohl, F. and Hockney, R. W.: 1969, *J. Comput. Phys.* **4**, 306.
Lecar, M.: 1966, in G. Contopoulos (ed.), 'The Theory of Orbits in the Solar System and in Stellar Systems', *IAU Symp.* **25**, 46.
Miller, R. H. and Prendergast, K. H.: 1968, *Astrophys. J.* **151**, 699.
Miller, R. H., Prendergast, K. H., and Quirk, W. J.: 1970, 'Numerical Experiments on Spiral Structure', to be published.
Toomre, A.: 1963, *Astrophys. J.* **138**, 385.
Toomre, A.: 1964, *Astrophys. J.* **139**, 1217.

NUMERICAL EXPERIMENTS IN SPIRAL STRUCTURE

WILLIAM J. QUIRK*
Department of Astronomy, Columbia University

Abstract. The formation of spiral density waves has been observed in some computer simulations of the evolutions of galaxies. Analysis of these computer experiments has revealed that although both 'stars' and 'gas' participate in the spiral structure, a greater percentage of 'gas' participates than of stars. Experiments have been performed in which the evolution of these computer galaxies has been altered. These experiments have shown that the azimuthal gravitational forces and the azimuthal dependence of the radial forces produced by the spiral arms are necessary to maintain the density wave structure of the galaxy.

1. Introduction

In two previous papers (Miller and Prendergast, 1968; Miller *et al.*, 1970; hereafter referred to as papers I and II) results of computer *n*-body calculations that simulated galactic evolution were reported. In these calculations *n* was large, roughly 100000; and relaxation from two body gravitational interactions was unimportant.

The bodies were of two types – 'stars' and 'gas clouds'. Both 'stars' and 'gas clouds' interacted gravitationally, but 'gas clouds' also interacted with each other through inelastic collisions. These collisions cooled the 'gas' population, that is they reduced the random velocities. The 'gas' component of the *n*-body system resembled a real gas in two ways; it was dissipative and its particles had a finite mean free path. Using these two components three types of systems were simulated: pure 'gas', pure 'star', and pure 'gas' evolving into mixed 'star' and 'gas'.

The motions of 'stars' and 'gas cloud' was determined by a difference scheme for Newton's laws, which was second order in time for the positions and first order in time for the velocities. The method used in these computations is discussed in detail in papers I and II.

The effects of magnetic fields, cosmic rays, supernovae, ionization of hydrogen regions by young stars and many other phenomena were ignored in evolving these model galaxies. The aim of this program was to find what phenomena observed in galaxies occur solely as a result of gravitational and hydrodynamic forces and to study these phenomena.

In one of the computer experiments that simulated the evolution of a pure 'gas' system into a mixed 'gas' and 'star' system, named case 9 in paper II, a two-armed spiral structure developed which persisted for three rotations of the spiral pattern (see Figures 1a, 1b and 1c). These arms were found to be density waves. The spiral pattern rotated more or less rigidly about the center of the galaxy even though the particles in the model galaxy participated in differential rotation. In a movie made to

* Present address Department of Astronomy, California Institute of Technology.

M. Lecar (ed.), Gravitational N-Body Problem, 250–253.

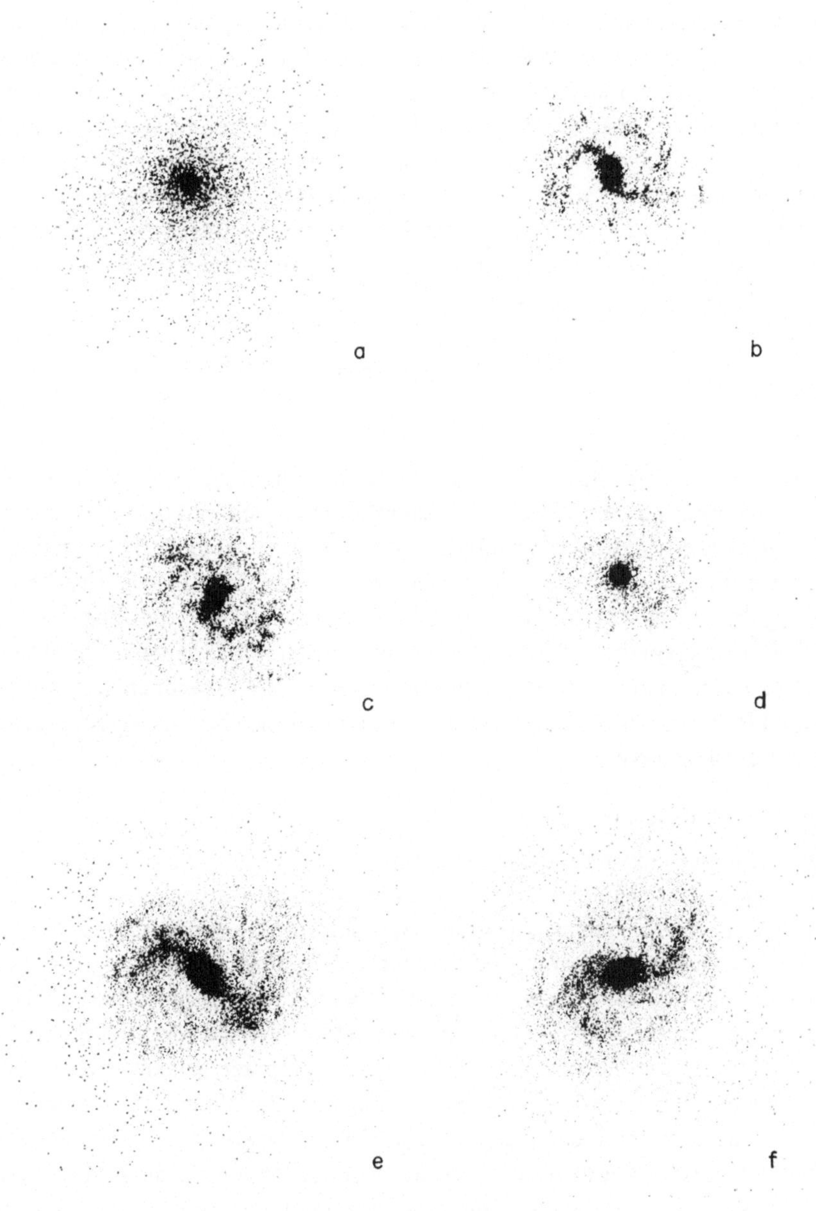

Fig. 1. Plots of positions of star and gas particles for various stages of several computer galaxies. Figure 1(a) is of stars; every eleventh particle is plotted. Figures 1(b) through 1(f) are of gas; every other particle is plotted. (a) Stars at step 101 of case 9; (b) gas at step 101 of case 9; (c) gas at step 105 of case 9; (d) result of not allowing outer part of spiral to exert forces (step 105); (e) result of not allowing inner part of spiral to exert forces (step 105); (f) same as (e), but at step 119.

show the evolution of this computer galaxy, 'gas clouds' can be seen to pass through the spiral density wave.

The existence of spirals of this type has been discussed by Linblad (1963, and earlier references) and Lin *et al.* (1969, and earlier references). The spontaneous appearance of persistent spirals in the computation is particularly noteworthy, because the numerical scheme was not specifically designed to treat problems in spiral structure. For example Newton's laws were applied in a Cartesian coordinate system rather than some coordinate system which would be more natural to a rotating circular galaxy. This paper reports on some experiments designed to investigate the maintenance and origin of spiral structure. The experiments confirm and extend the findings of similar experiments reported in paper II.

2. Experiments

A. METHOD

The experiments described in this section used a modified version of their actual density in computing their forces. For example, in order to find the effect of the forces generated by the spiral on the evolution of the galaxy a symmetrized version of the density can be used in computing forces. By symmetrizing the density (that is, by replacing the density at each point with the average density at that distance from the center of the galaxy) the azimuthal forces and the azimuthal dependence of the radial forces can be eliminated. By comparing the results of two experiments, in which spiral generated forces are present and absent, one can find the effect of these forces on some structure of the galaxy.

B. MAINTENANCE OF SPIRALS

In paper II a mechanism for causing the persistence of the spiral of case 9 was suggested by computer experiments, i.e. asymmetries in the density give rise to forces which in turn reinforced the tendency for the material to form a spiral pattern. Historically there has been a question as to whether a spiral maintained itself through gravitational self interaction of its outer parts of whether a bar in the inner part of the galaxy drove the spiral (see Linblad, 1959). To find out if either of these reinforcing mechanisms were responsible for the spiral structure in our computer galaxies two experiments were designed. In these experiments we investigated the influence of the gravitational forces produced by the inner region of a galaxy on the structure of the outer region, and the influence of the field produced by the outer region on the inner region.

The division between the inner and outer region of a computer galaxy is somewhat arbitrary. The 'gas' at step 101 of case 9 is shown in Figure 1b. The model galaxy appears to have a somewhat bar-shaped inner region surrounded by a less dense outer region clearly showing the spiral structure. This inner region extends to a radius of approximately 15 grid lengths, which we now take to be the boundary between the inner and outer regions.

The initial condition for both experiments was a stage of case 9 in which the spiral had become well developed. In the first experiment the computation of the forces was carried out using the actual density of the inner region and the symmetrized version of the outer density. Thus any azimuthal forces or azimuthal dependence of the radial forces that would ordinarily have arisen from the outer part of the spiral were not allowed to have any effect. This experiment was run for what would have been two rotations of the spiral pattern; not only did the spiral in the outer region wind up and disappear in less than a rotation of the spiral pattern, but any suggestion of a bar shape in the center also disappeared (see Figure 1d).

The second experiment used the actual density of the outer region and the symmetrized density of the inner region for computing forces. In this experiment, the spiral structure of the system persisted for as long as the experiment was carried out. The spiral arms showed a marked increase in strength over the spiral structure of case 9 at similar stages of evolution (see Figures 1e and 1f). The speed of the spiral pattern was slower than in case 9 by some fifteen percent, and the nucleus looked more bar shaped than in case 9.

These two experiments confirmed the conclusion of paper II that the persistence of the spiral was due to the forces it generated. In addition, since in both experiments the form of the spiral was markedly changed, the experiments show that it is difficult to consider the structure of just part of the galaxy; the structure of one region of a galaxy depends very much on other regions of a galaxy.

Acknowledgements

I would like to thank Dr Kevin Prendergast, my thesis advisor, for many helpful suggestions and comments. I would also like to thank Dr Richard Miller for extensive consultations and Drs Frank Hohl, Alar Toomre and J. Ostriker for several comments and suggestions. This work could not have been accomplished without the kind hospitality of the Goddard Institute for Space Studies and of its director, Dr Robert Jastrow, through whose courtesy the extensive computations reported in this paper were carried out. Dr David Stonehill of Computer Associated, Inc. was especially helpful. Programming assistance was provided by Mr Charles Koh of C. A. I.

This work was supported in part by NSF Grant GP-7010. During two of the years this work was underway I held a NASA traineeship.

References

Lin, C. C., Yuan, C., and Shu, F. H.: 1969, *Astrophys. J.* **155**, 721.
Linblad, B.: 1963, *Stockholm Obs. Ann.* **21**, 3.
Linblad, B.: 1959, in *Handbuch der Physik*, Vol. **53**, Springer-Verlag, Berlin.
Miller, R. H. and Prendergast, K. H.: 1968, *Astrophys. J.* **151**, 699.
Miller, R. H., Prendergast, K. H., and Quirk, W. J.: 1970, *Astrophys. J.* **161**, 903.
Quirk, W. J.: 1970, unpublished Ph.D. Thesis, Columbia University (available from University Microfilms, Inc., Ann Arbor, Michigan).
Quirk, W. J.: 1971, *Astrophys. J.* **167**, 7.

ON THE NUMBER OF ISOLATING INTEGRALS IN SYSTEMS WITH THREE DEGREES OF FREEDOM

CLAUDE FROESCHLE

Observatoire de Nice, Le Mont-Gros, 06 Nice (France)

Abstract. Dynamical systems with three degrees of freedom can be reduced to the study of a four-dimensional mapping. We consider here, as a model problem, the mapping given by the following equations:

$$\begin{cases} x_1 = x_0 + a_1 \sin(x_0 + y_0) + b \sin(x_0 + y_0 + z_0 + t_0) \\ y_1 = x_0 + y_0 \\ z_1 = z_0 + a_2 \sin(z_0 + t_0) + b \sin(x_0 + y_0 + z_0 + t_0) \\ t_1 = z_0 + t_0 \end{cases} \pmod{2\pi}$$

We have found that as soon as $b \neq 0$, i.e. even for a very weak coupling, a dynamical system with three degrees of freedom has in general either two or zero isolating integrals (besides the usual energy integral).

1. Introduction

Many studies have been made in the last few years of the motion of a star in an axisymmetric galaxy or in the plane of symmetry of a spiral galaxy (see Contopoulos, 1970).

It is easy to show that this problem is completely equivalent to the study of a dynamical system with two degrees of freedom. The present paper deals with the motion of a star in a galaxy without any symmetry, i.e. the study of a dynamical system with three degrees of freedom.

One of the most fruitful methods in the case of two degrees of freedom has been the method of 'Surface of Section'. This method dates back to Poincaré (1892) and consists essentially in considering not a complete trajectory in the phase space but only its successive intersections with a certain 'Surface of Section'.

Let us consider, quite generally, a conservative system with n degrees of freedom. The corresponding phase space has $2n$ dimensions. However, a given trajectory must lie on a manifold with $2n-1$ dimensions corresponding to a given value of the energy. In this manifold we define a 'surface of section' which is a given sub-manifold with $2n-2$ dimensions and we consider the successive intersections of the trajectory with this sub-manifold.

For $n=2$ the 'surface of section' has two dimensions and the corresponding mapping T which maps an intersecting point into the next one is an area preserving mapping. In particular it is easy to know whether an isolating integral does exist. In this case the set of points obtained by repeated application of the mapping seems to lie exactly on a closed curve. On the other hand it has been found that direct study of given area preserving mappings displays the usual features of dynamical systems with two degrees of freedom.

For $n=3$ the surface of section has four dimensions and the problem of finding

M. Lecar (ed.), Gravitational N-Body Problem, 254–261.

whether one or two other isolating integrals exist in addition to the integral of the energy is more difficult than for $n=2$. Nevertheless, using various numerical methods we have studied such systems taking the three-dimensional restricted problem as an example. We have found (Froeschlé, 1969, 1970) that for orbits close to one of the primaries the points appear to lie on a two-dimensional manifold and therefore two isolating integrals seem to exist besides the Jacobi integral. More distant orbits appear to fill a manifold with four dimensions and therefore the two isolating integrals have disappeared. In fact they appear to vanish at the same time.

In order to study the number of isolating integrals in dynamical systems with three degrees of freedom or more exactly to find whether cases of transition with only one isolating integral besides the integral of the energy do exist, we have taken a four-dimensional mapping T of R^4 into itself as a model problem. We study the set of points obtained by repeated applications of the mapping T, i.e.

$$P_0, P_1 = T(P_0), \ldots, P_n = T^n(P_0) = T(P_{n-1}).$$

In Section 2 we give some properties of the mapping T and its features when the coupling is equal to zero.

In Section 3 we study the number of isolating integrals for the coupled case using slice-cutting methods.

2. The Mapping

We take a mapping T of the $(xyzt)$ space over itself defined by

$$T \begin{cases} x_1 = x_0 + a_1 \sin(x_0 + y_0) + b \sin(x_0 + y_0 + z_0 + t_0) \\ y_1 = x_0 + y_0 \\ z_1 = z_0 + a_2 \sin(z_0 + t_0) + b \sin(x_0 + y_0 + z_0 + t_0) \\ t_1 = z_0 + t_0 \end{cases} \pmod{2\pi}$$

The determinant of the Jacobian matrix is equal to 1. This mapping has been suggested by Arnold (1965).

If $b=0$ then the mapping T is the product of two area-preserving mappings T_1 of (x, y) on itself and T_2 of (z, t) on itself. These mappings are inverse mappings of those given by Taylor (1969). Each of these mappings displays the well-known features of problems with two degrees of freedom (Hénon, 1969).

Figure 1 shows typical sets of points for the transformation T_1, i.e.

$$T_1 \begin{cases} x_1 = x_0 + a_1 \sin(x_0 + y_0) \\ y_1 = x_0 + y_0 \end{cases} \pmod{2\pi}$$

for some initial conditions of x_0 and y_0 given in Table I and for $a_1 = -1.3$, N being the total number of points plotted for each set of points.

For certain values of the initial conditions x_0 and y_0 the points (x_n, y_n) generated by T_1 seem to lie on an invariant curve; this indicates the invariance of one isolating integral.

For other values of x_0 and y_0 the points are scattered and fill a two-dimensional region; this indicates the disappearance of the isolating integral. We also get some islands and as the mapping is defined (mod 2π) we obtain the same island on the four corners of the picture. Of course, the same applies to T_2.

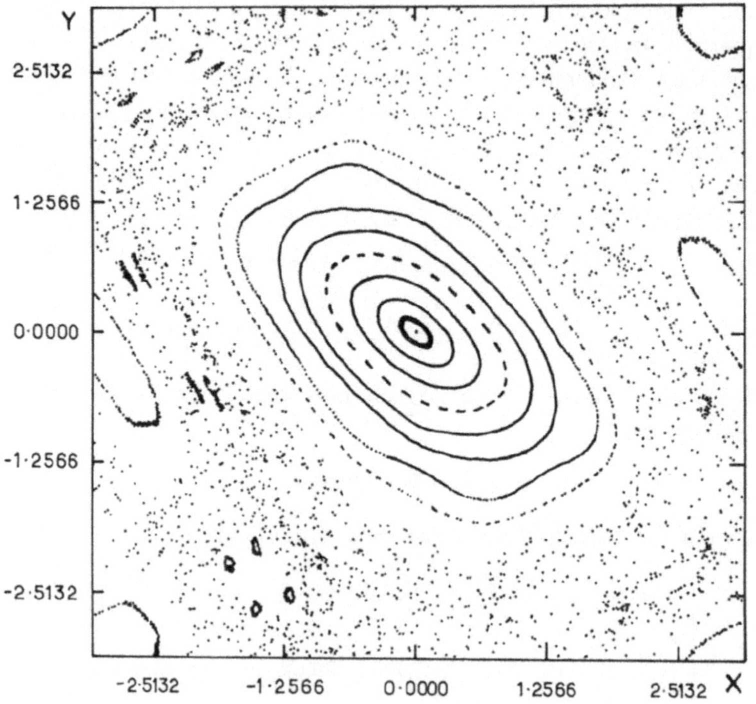

Fig. 1. The mapping T_1 for $a_1 = -1.3$.

TABLE I

x_0	y_0	N
0.1	0	500
0.3	0	400
0.5	0	500
0.7	0	600
0.9	0	800
1.1	0	800
1.3	0	800
1.5	0	1000
−2.51	−1.8	700
−2.5	−1.9	800
−1.25	−2.51	600

3. Study of the Number of Isolating Integrals for the Coupled Case

If $b \neq 0$ the term $b \sin(x_0 + y_0 + z_0 + t_0)$ introduces a coupling between (xy) and (zt). If $x_0 y_0 z_0 t_0$ are such that invariant curves exist for T_1 and T_2 we have found that, as long as b is not too large, the set of points $P_n(x_n y_n z_n t_n)$ still lies on a two-dimensional manifold. This indicates that there still exists two isolating integrals

$$f_1(x, y, z, t) = C_1$$
$$f_2(x, y, z, t) = C_2.$$

The existence of the two-dimensional manifold can be shown in the following way. Let $g(x, y, z) = C$ be the equation of this two-dimensional manifold obtained after elimination of t in $f_1 = C_1$ and $f_2 = C_2$. If $g(x, y, z) = C$ does exist then the points $((x_n, y_n)\ n = 0, 1, \ldots, N-1)$ for $|z_n - z_0| < \varepsilon_1$ should lie on a curve in the (x, y) plane, i.e. a section of the two-dimensional manifold $g(x, y, z) = C$ by $z = z_0$.

Figure 2 shows such a section when the initial conditions are $x_0 = 0.5$, $y_0 = 0.5$, $z_0 = 0.4$, $t_0 = 0.4$, $b = 0.01$, $\varepsilon_1 = 0.006$, $a_1 = a_2 = -1.3$, and $N = 80000$.

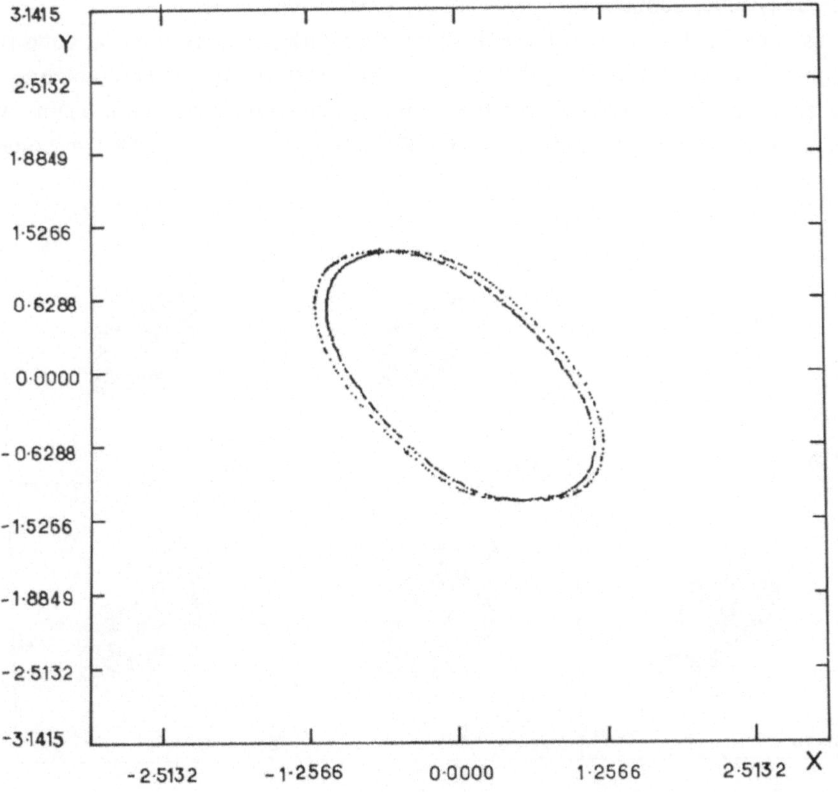

Fig. 2. Section by $z = 0.4$ of the mapping T for the initial conditions $x_0 = 0.5$, $y_0 = 0.5$, $z_0 = 0.4$, $t_0 = 0.4$ and $a_1 = -1.3$, $a_2 = -1.3$, $b = 0.01$.

The figure shows clearly two curves. This seems to indicate that the points P_n lie on a two-dimensional manifold with two sheets.

If $x_0y_0z_0t_0$ are such that an invariant curve exists for T_1 but not for T_2, then one isolating integral exists in the uncoupled case $b=0$. What can one expect in this case for $b\neq 0$? In order to study this case of transition the points P_n for which $|z_n - z_0| < \varepsilon_2$ have been plotted for various values of N, the total number of points. The condition $|z_n - z_0| < \varepsilon_2$ is no longer a section but it reduces the number of plotted points to manageable propositions.

Figure 3 summarizes the results of these experiments for $N=20000$ (20000) 120000. For each case we have the initial conditions $x_0=0.5$, $y_0=0.5$, $z_0=0.5$, $t_0=3$ and $b=0.01$, $\varepsilon_2=0.1$.

For N less than 80000, the points lie on strips including the invariant curve which corresponds to the zero coupling case. The width of these strips increases with N. For N greater than 80000 the points are scattered and ergodicity appears.

In order to follow the phenomenon in more detail we have used slice cuttings on n only (same initial conditions and same parameters). Hence we have plotted the projections on the (x, y) plane of the points P_n for which $N<n<N+100$ with $N=10000$ (10000) 200000.

Let us come back to Figure 1 which shows that with some given initial conditions the points lie either on an invariant curve or are scattered in the ergodic zone. On the other hand, Figure 4 shows the two types of behaviour. The curve begins wandering in a quasi-random fashion; it changes into islands ($N=70000$), becomes a

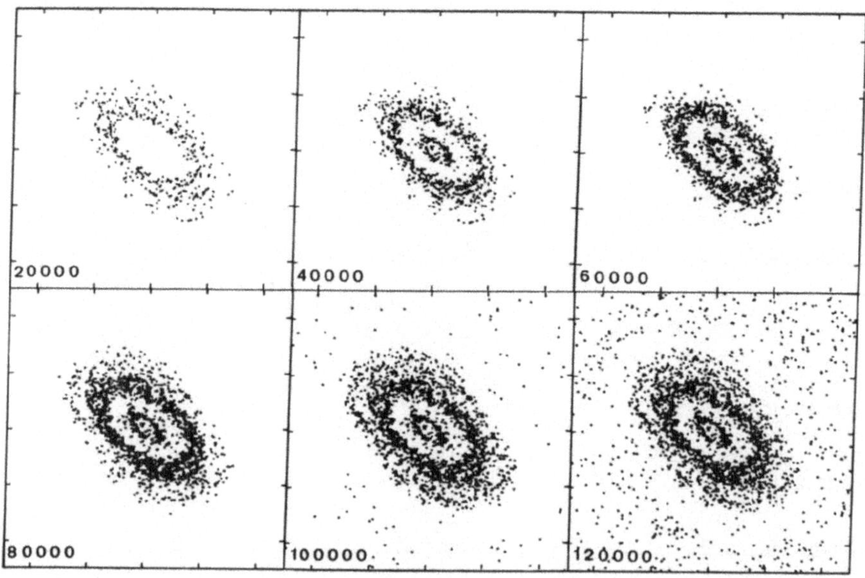

Fig. 3. Projections on the (x, y) plane of the set of points $P_n = T(P_{n-1})$, $n=0, N-1$ for the initial conditions $x_0=0.5$, $y_0=0.5$, $z=0.5$, $t_0=3$ and $a_1=-1.3$, $a_2=-1.3$, $b=0.01$ when $z_n - z_0 = 0.1$ and $N=20000$ (20000) 120000.

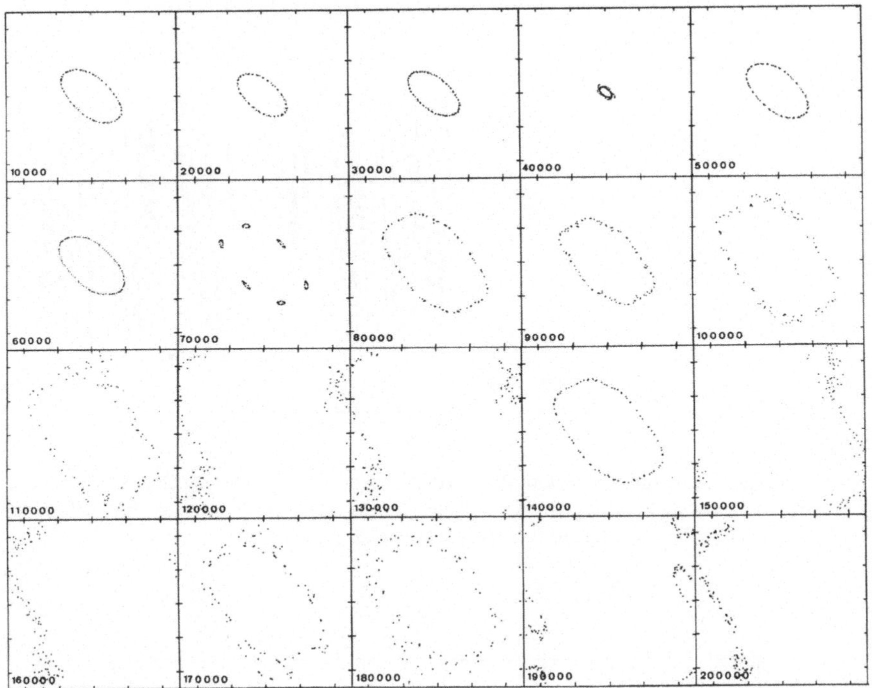

Fig. 4. Projections on the (x, y) plane of the set of points $P_n = T(P_{n-1}) n = 0, -, 110000$ for the initial conditions $x_0 = 0.5$, $y_0 = 0.5$, $z_0 = 0.5$, $t_0 = 3$, and $a_1 = -1.3$, $a_2 = -1.3$, $b = 0.01$ when $N < n \leqslant N + 100$ with $N = 10000\ (10000)\ 200000$.

curve again, then becomes ergodic ($N = 110000$ to 130000) and reverts to a curve ($N = 140000$), becomes ergodic again ($N = 150000$ to 200000). The following explanation suggests itself: since T_2 has no isolating integral, the points $(z_n t_n)$ behave in an ergodic, quasi-random fashion. Therefore the coupling term $b \sin(x_0 + y_0 + z_0 + t_0)$ produces a quasi-random perturbation of the points $(x_n y_n)$. As a result the value of the former isolating integral of T_1 is subjected to a kind of random walk.

In order to have some quantitative information about this random walk, a measure D_j of the dimension of the curve has been computed where

$$D_j = \left(\sum_{n=(j-1)\,100}^{n=j \times 100} \left(x_n^2 - a_1 \left(y_n^2 + x_n y_n\right)\right) \right) \Big/ 100 .$$

The quadratic term in this expression is constant in the vicinity of the origin in the linear approximation.

Figure 5 shows the variations of D_j with j for different cases. The lowest line shows the results for the integrable case (two isolating integrals exist) for which D_j is approximately constant. On the other hand the upper line shows the variations of D_j when only one isolating integral exists initially. The variations of D_j are quite

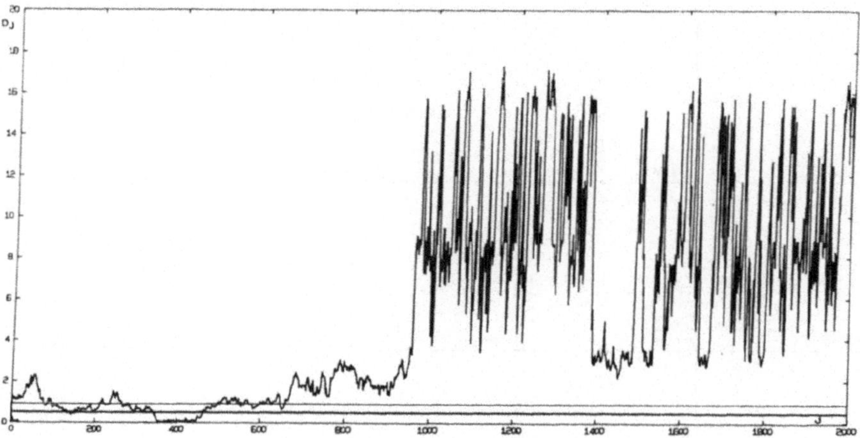

Fig. 5. D_J, a measure of the dimension of the curve, against J for different characteristic cases. – upper line for the case of Figure 4. – intermediate line for the case of Figure 4 but with $b=0$. – lowest line for the case of Figure 2.

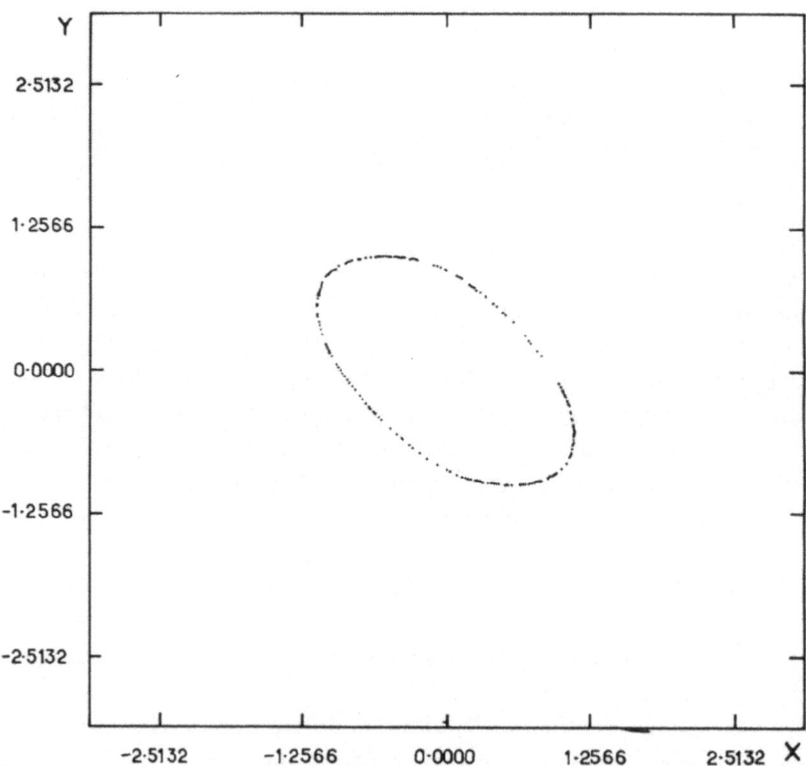

Fig. 6. Projection on the (x, y) plane of the different sections, for the case of Figure 4 but with $b=0$.

large and are in good agreement with the graphical results given by Figure 3 and Figure 4.

Figure 6 shows the same results as Figure 4 for $b=0$, where all the sections in n have been plotted on the same figure. It shows clearly that the effects of the rounding errors of the computer are negligible although they could have produced the same effects as the coupling. This is also shown in Figure 5 where the intermediate line, shows clearly that D_j remains constant with j.

4. Conclusion

The results found with the help of a four-dimensional mapping suggest that, apart from particular cases (such as $b=0$ in the present model), a dynamical system with three degrees of freedom has in general either two or zero isolating integrals (besides the usual energy integral). The disappearance of one of the two isolating integrals entails the disappearance of the other.

This agrees with results obtained earlier for a particular case of the three-dimensional restricted problem (Froeschle, 1970). A similar effect probably exists in systems with more than three degrees of freedom, i.e. a dynamical system with n degrees of freedom has in general either $n-1$ or zero isolating integrals (besides the usual energy integral).

Acknowledgement

It is pleasure to thank Dr M. Hénon for his constant guidance and encouragement, throughout this research.

References

Arnold, V. I.: private communication to Dr M. Hénon.
Contopoulos, G.: 1970, Cont. from the Ast. Dept. Univ. of Thessaloniki, No. 53.
Froeschle, C.: 1970, *Astron. Astrophys.* **4**, 115.
Froeschle, C.: 1970, *Astron. Astrophys.* **5**, 177.
Hénon, M.: 1969, *Quart. Appl. Math.* **27**, 291.
Poincaré, H.: 1892, *Les Méthodes Nouvelles de la Mécanique Céleste*, Gauthier-Villars, Paris.
Taylor, J. B.: 1969, private communication; see Culham Laboratory Progress Report CLM-PR-12.

NUMERICAL EXPERIMENTS ON LYNDEN-BELL'S STATISTICS

MYRON LECAR

Smithsonian Astrophysical Observatory and Harvard College Observatory, Cambridge, Mass., U.S.A.

and

LEON COHEN

Hunter College of the City University of New York, New York, N.Y., U.S.A.

Abstract. We performed computer experiments on 13 different initial configurations of one-dimensional self-gravitating systems. The three most and the three least violently relaxed systems were compared with the predictions of Lynden-Bell's statistical mechanics. The agreement between the experimental results and the theoretical predictions became worse as the relaxation became more violent. While all six systems were theoretically nondegenerate, the violent systems invariably flung out a halo that took most of the energy, leaving behind a low-energy degenerate core.

1. Introduction

Since Lynden-Bell (1967) proposed his statistical mechanics of violently relaxed collisionless self-gravitating systems, a considerable number of computer experiments have been done to check his predictions. (See, for example, Cuperman *et al.* (1969) and the references contained therein.) Systems with energy not much larger than the energy of a completely degenerate configuration (with the same mass and phase density) invariably satisfied the predictions closely, but they had no choice, as the completely degenerate configuration occupied most of their available phase space. For systems with higher energies, the experimental results were not uniform. Sometimes the fit was quite good (e.g., case A of Hohl and Campbell (1968) and case A of Cuperman *et al.* (1969)), but more often, a high-energy halo was formed that took most of the energy of the system, leaving the core (which contained most of the mass) degenerate. So that, compared with the predictions, the phase density was too high at very low energies, was too low at intermediate energies, and oscillated at high energies.

We were unsatisfied by the previous experiments because, in the case of Hohl and his collaborators, there was no independent measure of the violence of the relaxation and because the predicted distribution was not evaluated – instead, a curve of the same functional form as a Lynden-Bell distribution was fitted to the data. In addition, we felt that Cuperman *et al.* ran too few cases (four).

The results presented herein include a measure of relaxation; they are representative of a wider variety of initial conditions (13) and are compared with an accurate evaluation of the predicted distribution function.

2. The Experiment

We performed computer experiments on 13 different initial configurations of single-

M. Lecar (ed.), Gravitational N-Body Problem, 262–275.

phase density, one-dimensional systems. The various cases were evolved for 20 to 40 periods (40 to 80 oscillations of the kinetic energy) until they reached a steady state. In addition to the amplitude of the oscillations of the kinetic energy, the following quantities were used to estimate the violence of the relaxation:

$$\left\langle \frac{|\Delta v^2|}{v^2} \right\rangle_0 = \frac{\tau}{T_R(0)} = \frac{\Sigma_i |v_i^2 \,(\text{2nd crossing}) - v_i^2 \,(\text{1st crossing})|}{2\varepsilon/3},$$

$$\left\langle \frac{|\Delta v^2|}{v^2} \right\rangle_f = \frac{\Sigma_i |v_i^2 \,(\text{final crossing}) - v_i^2 \,(\text{1st crossing})|}{2\varepsilon/3},$$

where v_i is the velocity of the ith particle as it crosses $x=0$ with positive velocity. The change in v^2 is twice the work done on the particle in one complete oscillation; it vanishes for a particle oscillating in phase with the bulk of the system and is positive for a lagging particle (reflecting a transfer of energy from the collective oscillation to the particle).

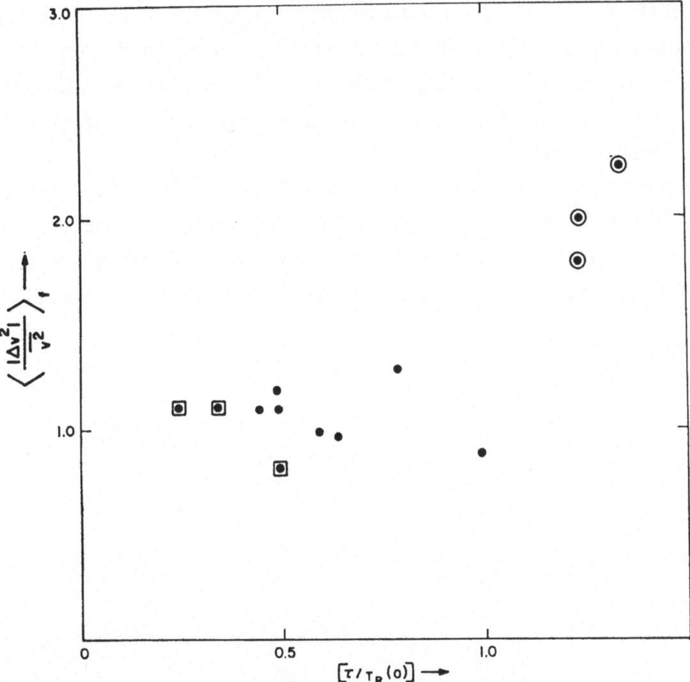

Fig. 1. Relaxation time for the 13 cases.

The values of these two measures for the 13 runs are plotted in Figure 1. The squared and circled points were chosen for comparison with the theoretical predictions to represent mildly and violently relaxed systems, respectively.

For these six cases, we also list the following quantities in Table I:

$Q = 3T/E =$ the initial value of the kinetic energy divided by its equilibrium (Virial Theorem) value.

$\varepsilon = E/E_0 =$ the total energy divided by the total energy of a completely degenerate configuration with the same mass and phase density.

TABLE I

Parameters of the three least and the three most violent cases

Case	Q	$\langle \lvert \Delta v^2 \rvert / v^2 \rangle_f$	$\tau/T_R(0)$	ε
M	1.2	0.9	0.5	11.3
G	1.2	1.2	0.3	5.7
E	1.4	1.2	0.4	5.4
I	2.0	1.8	1.3	6.5
N	2.0	2.3	1.4	2.9
P	2.2	2.0	1.3	3.3

We see that the violence of the relaxation depends primarily on the initial departure from hydrostatic equilibrium (Q) and not at all on how much larger the energy is than that of its degenerate counterpart. In fact, all the cases were theoretically non-degenerate. As can be seen from the following figures, their theoretical distributions show no low-energy plateau.

Figures 2 through 7 show, for the six cases, the initial and final phase-space configurations, the oscillations of the kinetic energy, and a comparison between the predicted and the experimental distribution functions. For convenience, the number of particles per unit energy, $N(\varepsilon)$, rather than the number of particles per unit phase-

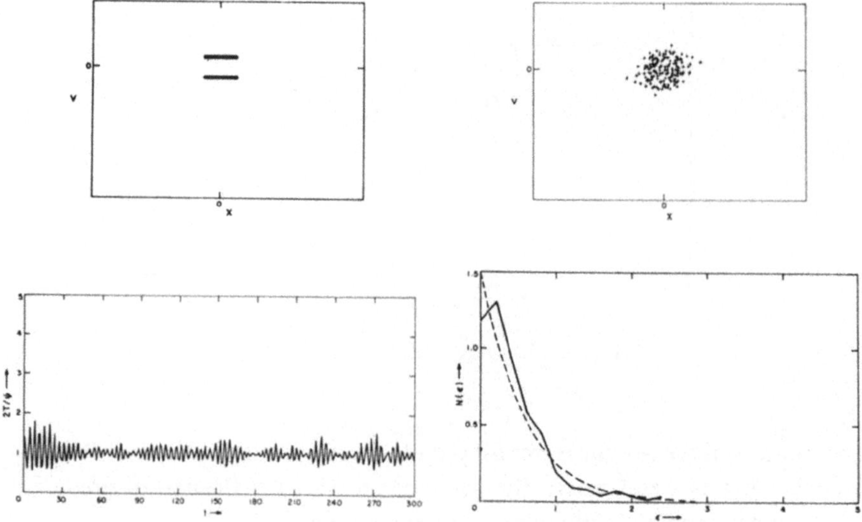

Fig. 2. Case M: $\varepsilon = 11.3$, $\eta = 5.04$, $\beta = 0.134$, $\mu = -21.0$, $\tau/T_R(0) = 0.5$.

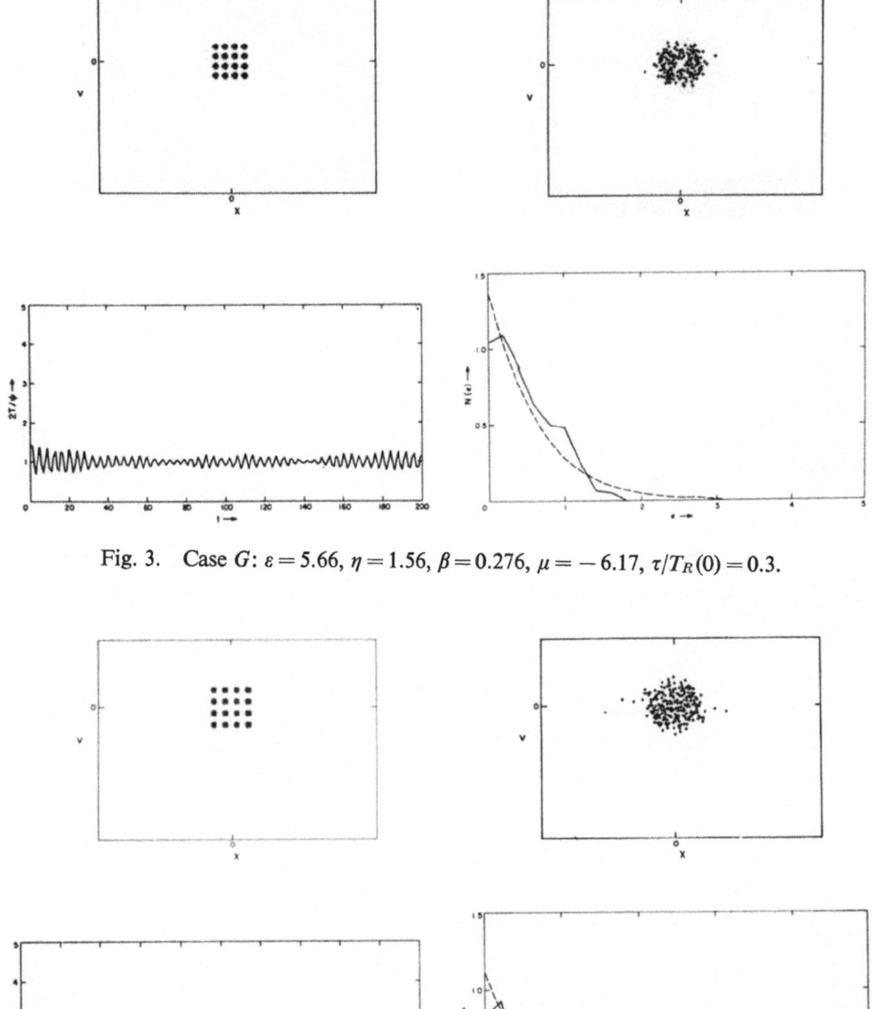

Fig. 3. Case G: $\varepsilon = 5.66$, $\eta = 1.56$, $\beta = 0.276$, $\mu = -6.17$, $\tau/T_R(0) = 0.3$.

Fig. 4. Case E: $\varepsilon = 5.35$, $\eta = 1.00$, $\beta = 0.293$, $\mu = -5.50$, $\tau/T_R(0) = 0.4$.

space volume, $f(\varepsilon)$, is plotted. The evaluation of the predicted distribution functions is described in the Appendices.

3. Conclusions

Figures 2 through 7 show that the agreement between theory and experiments becomes progressively worse as the relaxation becomes more violent. Large-amplitude

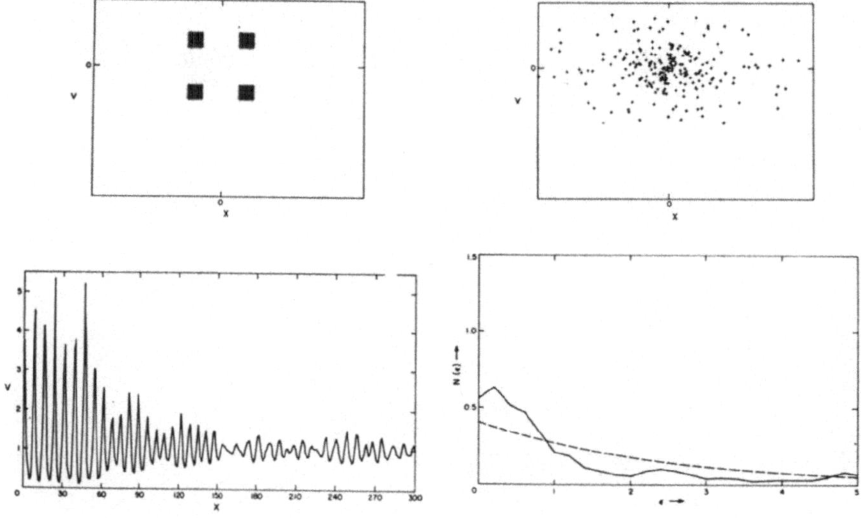

Fig. 5. Case I: $\varepsilon = 6.50$, $\eta = 0.250$, $\beta = 0.238$, $\mu = -8.12$, $\tau/T_R(0) = 1.3$.

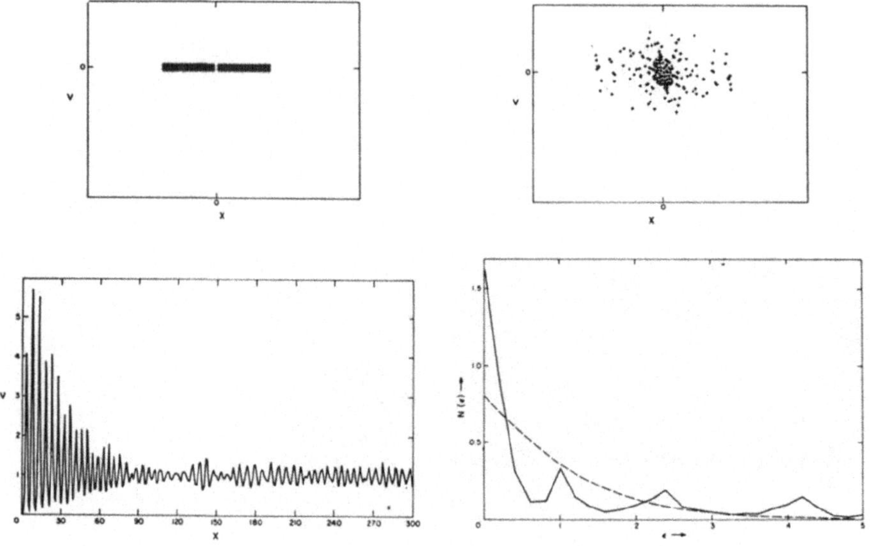

Fig. 6. Case N: $\varepsilon = 2.87$, $\eta = 0.246$, $\beta = 0.587$, $\mu = -0.754$, $\tau/T_R(0) = 1.4$.

oscillations result in higher energy halos and lower energy cores than the theory predicts. While the theory indicates that the entropy of the system would increase if the core expanded and the halos contracted, after the oscillations damp, the halo and core are almost completely decoupled.

In actual (three-dimensional) galaxies, most of the halo might escape or be torn off

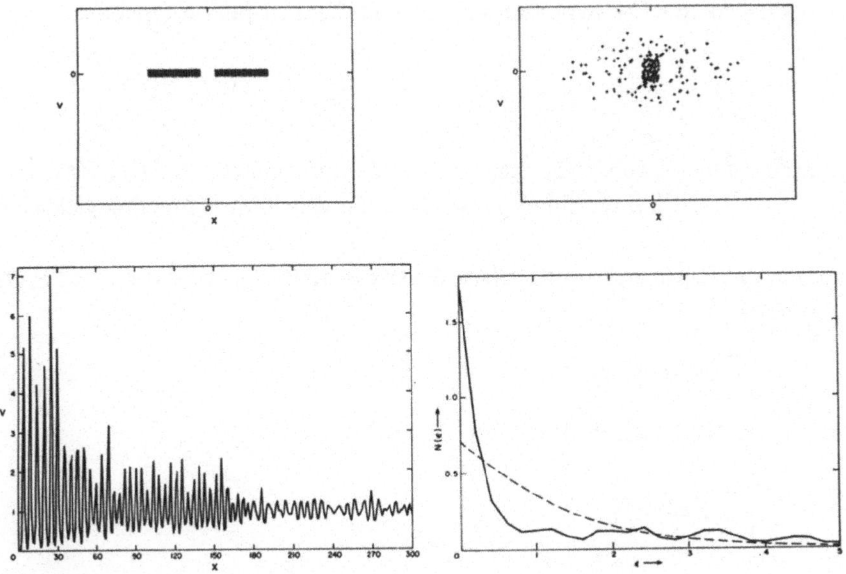

Fig. 7. Case P: $\varepsilon = 3.32$, $\eta = 0.246$, $\beta = 0.496$, $\mu = -1.49$, $\tau / T_R(0) = 1.3$.

during encounters with other galaxies. If that happens, a good approximation to the remaining galaxy might be the completely degenerate distribution function. We note that for the degenerate case,

$$\varrho \sim \int_0^{v_m} f v^2 \, \mathrm{d}v \sim v_m^3,$$

$$P \sim \int_0^{v_m} f v^4 \, \mathrm{d}v \sim v_m^5,$$

so $P \sim \varrho^{5/3}$. That is, the run of pressure and density is represented by a polytrope of index $n = \frac{3}{2}$ $(\gamma = \frac{5}{3})$. It would be useful to construct a series of models, including angular momentum, and compare them with observed elliptical galaxies.

If the initial phase density (at the time of star formation) was not uniform, such models are difficult to construct. It is perhaps worth noting that the initial maximum phase density η bounds the density ϱ. Because $\varrho \lesssim \eta v^3$, and with the use of the Virial Theorem, this implies that $\varrho < G^3 (M\eta)^2$, where M is the mass of the galaxy.

Appendix I

THE DETERMINATION OF β AND μ FROM E AND M

For a self-gravitating system in one dimension, Lynden-Bell's statistical mechanics are complete, in that the parameters β and μ in the distribution function

$$f(\varepsilon) = \frac{\eta}{e^{\beta(\varepsilon-\mu)}+1} \tag{I.1}$$

are determined by the conserved quantities M (the total mass) and E (the total energy). (This is not the case in three dimensions, where this distribution function yields infinite mass.)

In Equation (I.1), η is the initial phase density ($\eta\ \mathrm{d}x\ \mathrm{d}v \sim \mathrm{g\ cm}^{-2}$) and ε is the energy per unit mass,

$$\varepsilon = \tfrac{1}{2}v^2 + \psi(x) \qquad (\varepsilon \sim \mathrm{cm}^2\ \mathrm{s}^{-2}), \tag{I.2}$$

where ψ is determined from Poisson's equation

$$\frac{\mathrm{d}^2}{\mathrm{d}x^2}\psi = 4\pi G\varrho\,. \tag{I.3}$$

The density ϱ and the kinetic energy density $T=\frac{1}{2}\varrho\langle v^2\rangle$ are obtained by integrating f over velocity:

$$\varrho[\psi(x)] = \int_{-\infty}^{+\infty} f(\varepsilon)\,\mathrm{d}v = 2^{1/2}\int_0^\infty f(\varepsilon+\psi)\frac{\mathrm{d}\varepsilon}{\varepsilon^{1/2}}, \tag{I.4}$$

$$T[\psi(x)] = \int_{-\infty}^{+\infty} f(\varepsilon)\tfrac{1}{2}v^2\,\mathrm{d}v = 2^{1/2}\int_0^\infty f(\varepsilon+\psi)\,\varepsilon^{1/2}\,\mathrm{d}\varepsilon\,. \tag{I.5}$$

The quantities M and E result from integrating ϱ and T over x, but it is more convenient to use ψ as an independent variable. Let $a(\psi)=\mathrm{d}\psi/\mathrm{d}x$ as a function of ψ. Then,

$$\frac{\mathrm{d}^2\psi}{\mathrm{d}x^2} = \frac{\mathrm{d}}{\mathrm{d}\psi}\tfrac{1}{2}a^2 = 4\pi G\varrho\,,$$

so

$$a(\psi) = \left[8\pi G\int_0^\psi \varrho(\psi')\,\mathrm{d}\psi'\right]^{1/2}, \tag{I.6}$$

$$M = 2\int_0^\infty \varrho(\psi)\,a^{-1}(\psi)\,\mathrm{d}\psi \qquad (M \sim \mathrm{g\ cm}^{-2}), \tag{I.7}$$

$$E = 3\times 2\int_0^\infty T(\psi)\,a(\psi)^{-1}\,\mathrm{d}\psi \qquad (E \sim M\times \mathrm{cm}^2\ \mathrm{s}^{-2}), \tag{I.8}$$

where we have used the Virial Theorem $E=3T$.

As a first illustration of the determination of β and μ from M and E, we treat the extreme degenerate ('0 temperature') limit ($\beta \to \infty$). In this limit,

$$\begin{aligned} f &= \eta, \qquad \varepsilon \leqslant \mu \\ &= 0, \qquad \varepsilon > \mu. \end{aligned} \tag{I.9}$$

Equations (I.4) through (I.8) integrate to

$$\varrho(\psi) = 2^{1/2}\eta \int_0^{\mu-\psi} \frac{d\varepsilon}{\varepsilon^{1/2}} = 2^{3/2}\eta(\mu-\psi)^{1/2}, \tag{I.10}$$

$$T(\psi) = \frac{2^{3/2}}{3}\eta(\mu-\psi)^{3/2}, \tag{I.11}$$

$$a(\psi) = 4 \times 2^{1/4} \times \left(\frac{2\pi G\eta}{3}\right)^{1/2} [\mu^{3/2} - (\mu-\psi)^{3/2}], \tag{I.12}$$

$$M = \frac{4 \times 2^{1/4}}{3^{1/2}} \left(\frac{\eta}{2\pi G}\right)^{1/2} \mu^{3/4}, \tag{I.13}$$

and

$$E = \frac{2 \times 2^{1/4}}{3^{1/2}} \left(\frac{\eta}{2\pi G}\right)^{1/2} \beta\left(\tfrac{5}{3}, \tfrac{1}{2}\right) \mu^{7/4}, \tag{I.14}$$

where

$$\beta\left(\tfrac{5}{3}, \tfrac{1}{2}\right) = \frac{\Gamma\left(\frac{5}{3}\right) \times \Gamma\left(\frac{1}{2}\right)}{\Gamma\left[\left(\frac{5}{3}\right) + \left(\frac{1}{2}\right)\right]} = 1.478\,348\,320.$$

From Equations (I.13) and (I.14) we obtain

$$\varepsilon = \frac{E}{M} = \frac{3^{2/3}\beta\left(\frac{5}{3}, \frac{1}{2}\right)}{2^4} \left(\frac{2\pi G M^2}{\eta}\right)^{2/3}$$

$$\varepsilon_0 = 0.192\,193\,0265 \left(\frac{2\pi G M^2}{\eta}\right)^{2/3}. \tag{I.15}$$

As the '0-temperature' value of ε depends only on M and η, ε serves as a convenient unit of energy, and we refer to it henceforth as ε_0.

As a second illustration, consider the extreme nondegenerate limit ($\beta \to 0$, $\beta\mu \to -\infty$). In this limit,

$$f = \eta\, e^{\beta\mu} e^{-\beta\varepsilon}. \tag{I.16}$$

In this case also, Equations (I.4) through (I.8) are easily integrated to yield

$$\varrho(\psi) = \left(\frac{2\pi}{\beta}\right)^{1/2} \eta\, e^{\beta\mu} e^{-\beta\psi}, \tag{I.17}$$

$$T(\psi) = \frac{1}{2\beta}\left(\frac{2\pi}{\beta}\right)^{1/2} \eta\, e^{\beta\mu} e^{-\beta\psi}, \tag{I.18}$$

$$a(\psi) = 2\left(\frac{2\pi}{\beta^3}\right)^{1/4} (2\pi G\eta)^{1/2}\, e^{\beta\mu/2}\,(1 - e^{-\beta\psi})^{1/2}, \tag{I.19}$$

$$M = 2\left(\frac{2\pi}{\beta^3}\right)^{1/4} \left(\frac{\eta}{2\pi G}\right)^{1/2} e^{\beta\mu/2}, \tag{I.20}$$

and

$$E = \frac{3}{\beta}\left(\frac{2\pi}{\beta^3}\right)^{1/4} \left(\frac{\eta}{2\pi G}\right)^{1/2} e^{\beta\mu/2}. \tag{I.21}$$

From Equations (I.20) and (I.21), we obtain

$$\varepsilon = \frac{E}{M} = \frac{3}{2\beta} \tag{I.22}$$

and

$$\varepsilon = \frac{3}{2^{8/3}\pi^{1/3}}\left(\frac{2\pi GM^2}{\eta}\right)^{2/3} e^{-2\beta\mu/3} \tag{I.23}$$

or

$$\frac{\varepsilon}{\varepsilon_0} = \left(\frac{48}{\pi}\right)^{1/3} \frac{1}{\beta\left(\frac{5}{3}, \frac{1}{2}\right)} e^{-2\beta\mu/3} = 1.678\,496\,150\, e^{-2\beta\mu/3},$$

where

$$\mu = \varepsilon \ln\left(\frac{1.678\ldots}{\varepsilon/\varepsilon_0}\right). \tag{I.24}$$

Aside from these two limiting cases, the determination of β and μ is tedious but straightforward. The formulas are expressed compactly by making use of the Fermi-Dirac integrals

$$F_n(x) = \int_0^\infty \frac{\mathrm{d}t\, t^n}{e^{t-x} + 1}. \tag{I.25}$$

We note that

$$\frac{\mathrm{d}}{\mathrm{d}x} F_n(x) = nF_{n-1}(x).$$

In terms of the dimensionless variables

$$\xi = \beta\varepsilon, \qquad \zeta = \beta\psi, \qquad \nu = \beta\mu, \tag{I.26}$$

Equations (I.4) through (I.8) are written

$$\varrho(\zeta) = \left(\frac{2}{\beta}\right)^{1/2} \eta F_{-1/2}(\nu - \zeta), \tag{I.27}$$

$$T(\zeta) = \frac{1}{\beta}\left(\frac{2}{\beta}\right)^{1/2} \eta F_{1/2}(\nu - \zeta), \tag{I.28}$$

$$a(\zeta) = \left\{\left(\frac{2}{\beta}\right)^{3/2} 8\pi G\eta\left[F_{1/2}(\nu) - F_{1/2}(\nu-\zeta)\right]\right\}^{1/2} =$$

$$= \left\{\frac{2}{\beta} \times \frac{8\pi G\varrho(0)}{F_{-1/2}(\nu)}\left[F_{1/2}(\nu) - F_{1/2}(\nu-\zeta)\right]\right\}^{1/2}, \tag{I.29}$$

$$M = \tfrac{1}{2}\left(\frac{2}{\beta}\right)^{3/4}\left(\frac{\eta}{2\pi G}\right)^{1/2}\int_0^\infty \frac{F_{-1/2}(\nu-\zeta)\,\mathrm{d}\zeta}{\left[F_{1/2}(\nu) - F_{1/2}(\nu-\zeta)\right]^{1/2}} =$$

$$= 2\left(\frac{2}{\beta}\right)^{3/4}\left(\frac{\eta}{2\pi G}\right)^{1/2}\left[F_{1/2}(\nu)\right]^{1/2}, \tag{I.30}$$

and

$$E = \frac{3}{\beta}\left(\frac{2}{\beta}\right)^{3/4}\left(\frac{\eta}{2\pi G}\right)^{1/2} I(\nu), \tag{I.31}$$

where

$$I(\nu) = \tfrac{1}{2}\int_0^\infty \frac{F_{1/2}(\nu-\zeta)\,\mathrm{d}\zeta}{\left[F_{1/2}(\nu) - F_{1/2}(\nu-\zeta)\right]^{1/2}}.$$

From Equations (I.30) and (I.31) we obtain

$$\varepsilon = \frac{E}{M} = \frac{3}{2^{10/3}}\left(\frac{2\pi G M^2}{\eta}\right)^{2/3}\frac{I(\nu)}{\left[F_{1/2}(\nu)\right]^{7/6}}, \tag{I.32}$$

or

$$\frac{\varepsilon}{\varepsilon_0} = \frac{(12)^{1/3}}{\beta\left(\frac{5}{3},\frac{1}{2}\right)}\frac{I(\nu)}{\left[F_{1/2}(\nu)\right]^{7/6}} = 1.548\,639\,420\,\frac{I(\nu)}{\left[F_{1/2}(\nu)\right]^{7/6}}$$

and

$$\beta = \frac{3}{2\varepsilon}\frac{I(\nu)}{\left[F_{1/2}(\nu)\right]^{1/2}}. \tag{I.33}$$

Equation (I.32) determines ν (implicitly) since ε is known. With $\nu = \beta\mu$ known, Equation (I.33) determines β, and thus β and μ as a function of ε. The asymptotic values of $F_{1/2}(\nu)$ and $I(\nu)$ are as $\nu \to \infty$,

$$F_{1/2}(\nu) \to \tfrac{2}{3}\nu^{3/2}, \qquad I(\nu) \to \tfrac{1}{2}\left(\tfrac{2}{3}\right)^{3/2}\beta\left(\tfrac{5}{3},\tfrac{1}{2}\right)\nu^{7/4},$$

and as $\nu \to -\infty$,

$$F_{1/2}(\nu) \to \frac{\pi^{1/2}}{2}e^{\nu}, \qquad I(\nu) \to \left[F_{1/2}(\nu)\right]^{1/2},$$

from which the above-mentioned limiting cases are easily recovered.

Using the rational Chebyshev approximations of Cody and Thacher (1967), we have determined β and μ for $1.68 < \varepsilon < 14.00$. For $\varepsilon > 14.0$, the extreme nondegenerate formulas are sufficient. The results are presented in Table II and Figures 8 and 9.

TABLE II

Parameters of Lynden-Bell distribution function ε, β, μ in units of ε_0, $2\pi G = M = 1$

$e^{\beta\mu}$	ε	β	μ	$\ln\varepsilon$	$\ln\beta$
0.02	2.2994 +1	6.5537 −2	−5.9691 +1	3.1352 +0	−2.7251 +0
0.04	1.4619 +1	1.0355 −1	−3.1083 +1	2.6823 +0	−2.2676 +0
0.06	1.1258 +1	1.3508 −1	−2.0827 +1	2.4210 +0	−2.0018 +0
0.08	9.3763 +0	1.6290 −1	−1.5504 +1	2.2381 +0	−1.8145 +0
0.10	8.1513 +0	1.8820 −1	−1.2234 +1	2.0981 +0	−1.6702 +0
0.12	7.2808 +0	2.1160 −1	−1.0020 +1	1.9852 +0	−1.5530 +0
0.14	6.6257 +0	2.3349 −1	−8.4203 +0	1.8909 +0	−1.4545 +0
0.16	6.1122 +0	2.5415 −1	−7.2105 +0	1.8102 +0	−1.3698 +0
0.18	5.6973 +0	2.7376 −1	−6.2638 +0	1.7400 +0	−1.2955 +0
0.20	5.3541 +0	2.9246 −1	−5.5029 +0	1.6778 +0	−1.2294 +0
0.22	5.0648 +0	3.1037 −1	−4.8783 +0	1.6223 +0	−1.1699 +0
0.24	4.8172 +0	3.2758 −1	−4.3564 +0	1.5722 +0	−1.1159 +0
0.26	4.6025 +0	3.4416 −1	−3.9140 +0	1.5266 +0	−1.0666 +0
0.28	4.4144 +0	3.6017 −1	−3.5343 +0	1.4848 +0	−1.0211 +0
0.30	4.2479 +0	3.7565 −1	−3.2049 +0	1.4464 +0	−9.7907 −1
0.32	4.0995 +0	3.9066 −1	−2.9166 +0	1.4108 +0	−9.3990 −1
0.34	3.9662 +0	4.0522 −1	−2.6622 +0	1.3778 +0	−9.0330 −1
0.36	3.8457 +0	4.1938 −1	−2.4360 +0	1.3469 +0	−8.6897 −1
0.38	3.7363 +0	4.3315 −1	−2.2337 +0	1.3181 +0	−8.3665 −1
0.40	3.6363 +0	4.4657 −1	−2.0518 +0	1.2909 +0	−8.0614 −1
0.42	3.5447 +0	4.5965 −1	−1.8872 +0	1.2654 +0	−7.7727 −1
0.44	3.4602 +0	4.7242 −1	−1.7377 +0	1.2413 +0	−7.4986 −1
0.46	3.3822 +0	4.8490 −1	−1.6014 +0	1.2185 +0	−7.2380 −1
0.48	3.3098 +0	4.9709 −1	−1.4765 +0	1.1969 +0	−6.9896 −1
0.50	3.2425 +0	5.0902 −1	−1.3617 +0	1.1763 +0	−6.7525 −1
0.60	2.9653 +0	5.6513 −1	−9.0390 −1	1.0870 +0	−5.7069 −1
0.70	2.7585 +0	6.1623 −1	−5.7879 −1	1.0146 +0	−4.8412 −1
0.80	2.5976 +0	6.6326 −1	−3.3643 −1	9.5459 −1	−4.1058 −1
0.90	2.4684 +0	7.0688 −1	−1.4904 −1	9.0359 −1	−3.4689 −1
1.00	2.3622 +0	7.4759 −1	0.0000 +0	8.5961 −1	−2.9089 −1
1.10	2.2732 +0	7.8580 −1	1.2129 −1	8.2119 −1	−2.4105 −1
1.20	2.1973 +0	8.2181 −1	2.2185 −1	7.8726 −1	−1.9624 −1
1.30	2.1319 +0	8.5588 −1	3.0654 −1	7.5701 −1	−1.5562 −1
1.40	2.0747 +0	8.8822 −1	3.7881 −1	7.2984 −1	−1.1852 −1
1.50	2.0243 +0	9.1901 −1	4.4119 −1	7.0525 −1	−8.4449 −2
1.60	1.9795 +0	9.4840 −1	4.9557 −1	6.8288 −1	−5.2969 −2
1.70	1.9394 +0	9.7652 −1	5.4338 −1	6.6242 −1	−2.3753 −2
1.80	1.9033 +0	1.0034 +0	5.8574 −1	6.4361 −1	3.4756 −3
1.90	1.8705 +0	1.0293 +0	6.2353 −1	6.2625 −1	2.8948 −2
2.00	1.8407 +0	1.0542 +0	6.5745 −1	6.1016 −1	5.2858 −2
2.10	1.8134 +0	1.0782 +0	6.8807 −1	5.9520 −1	7.5370 −2
2.20	1.7882 +0	1.1014 +0	7.1583 −1	5.8125 −1	9.6623 −2
2.30	1.7650 +0	1.1238 +0	7.4113 −1	5.6820 −1	1.1673 −1
2.40	1.7436 +0	1.1454 +0	7.6428 −1	5.5596 −1	1.3582 −1
2.50	1.7236 +0	1.1664 +0	7.8553 −1	5.4445 −1	1.5396 −1
2.60	1.7050 +0	1.1867 +0	8.0512 −1	5.3361 −1	1.7124 −1
2.70	1.6877 +0	1.2065 +0	8.2323 −1	5.2337 −1	1.8774 −1

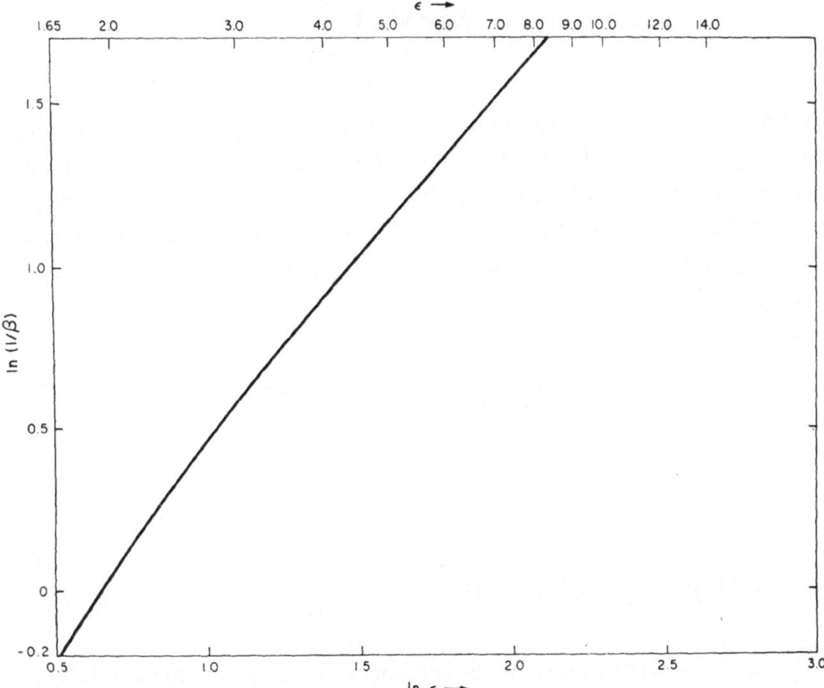

Fig. 8. Parameters of Lynden-Bell distribution function $\ln(1/\beta)$ vs $\ln(\varepsilon)$; β, ε in units of ε_0, $2\pi G = M = 1$.

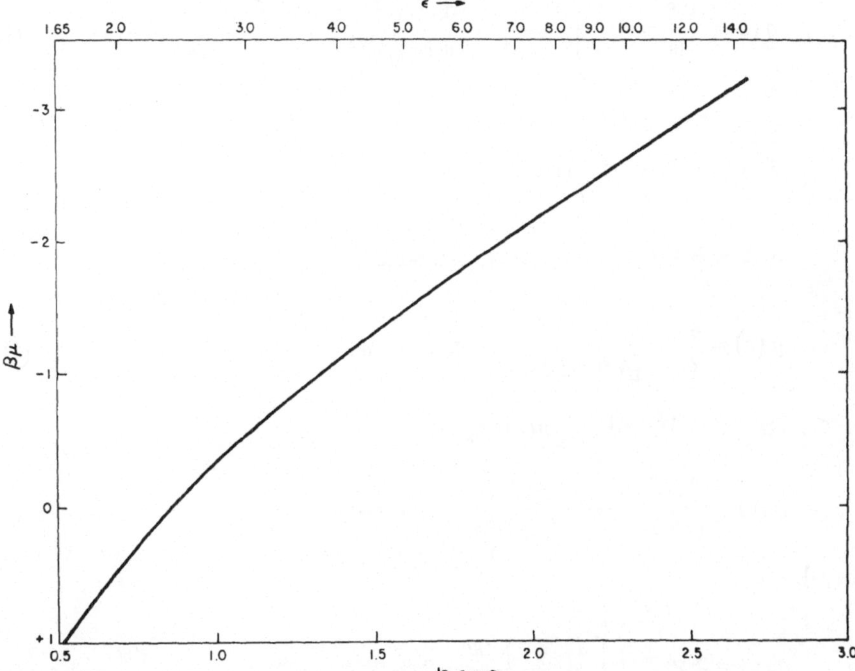

Fig. 9. Parameters of Lynden-Bell distribution function β, μ vs $\ln(\varepsilon)$; β, μ, ε in units of ε_0, $2\pi G = M = 1$.

Appendix II

$N(\varepsilon)$

In this Appendix, we calculate the number or mass of particles with energies in the interval ε to $\varepsilon + \mathrm{d}\varepsilon$. As $f[\varepsilon(x, y)]\,\mathrm{d}x\,\mathrm{d}v$ is the mass in $\mathrm{d}x\,\mathrm{d}v$, we must determine the area in phase space between the energy curves ε and $\varepsilon + \Delta\varepsilon$. (Both the potential and the kinetic energies increase monotonically, so the ε-curves are concentric.) If $\mathrm{d}A(\varepsilon)$ is the phase-space area between ε and $\varepsilon + \mathrm{d}\varepsilon$, then the required mass is $\mathrm{d}M(\varepsilon) = f(\varepsilon)\,\mathrm{d}A(\varepsilon)$.

The area interior to ε is

$$A(\varepsilon) = 4 \int_0^{\psi(x)=\varepsilon} \mathrm{d}x \int_0^{[2(\varepsilon-\psi)]^{1/2}} \mathrm{d}v = 4 \times 2^{1/2} \int_0^{\psi(x)=\varepsilon} \mathrm{d}x\,[\varepsilon - \psi(x)]^{1/2}. \tag{II.1}$$

Let

$$\mathrm{d}A(\varepsilon) = \frac{\mathrm{d}A(\varepsilon)}{\mathrm{d}\varepsilon}\,\mathrm{d}\varepsilon \equiv g(\varepsilon)\,\mathrm{d}\varepsilon.$$

Then $g(\varepsilon)$ is the 'statistical weight' of the energy state 'ε'. It turns out that $g(\varepsilon)$ is also the period of a particle with energy ε,

$$g(\varepsilon) = \frac{\mathrm{d}A}{\mathrm{d}\varepsilon} = 4 \int_0^{\psi(x)=\varepsilon} \frac{\mathrm{d}x}{\{2[\varepsilon - \psi(x)]\}^{1/2}} \quad 4 \int_0^{\psi(x)=\varepsilon} \frac{\mathrm{d}x}{v(x)} = \tau(\varepsilon). \tag{II.2}$$

As

$$\ddot{x} = -4\pi G \int_0^x \varrho(x')\,\mathrm{d}x',$$

$$\ddot{x} \cong -4\pi G\varrho(0)\,x = -\omega^2 x \quad \text{for small} \quad x,$$

so

$$g(\varepsilon) = \frac{2\pi}{\omega} = \frac{2\pi}{[4\pi G\varrho(0)]^{1/2}} \quad \text{as} \quad \varepsilon \to 0. \tag{II.3}$$

As $\varepsilon \to \infty$, $\ddot{x} \cong -2\pi GM$ and $\varepsilon \cong 2\pi GMX_m$, so

$$g(\varepsilon) \cong \frac{4(2X_m)^{1/2}}{(2\pi GM)^{1/2}} \cong \frac{4(2\varepsilon)^{1/2}}{2\pi GM} \quad \text{as} \quad \varepsilon \to \infty. \tag{II.4}$$

In general,

$$g(\varepsilon) = 2 \times 2^{1/2} \int_0^{\varepsilon} \frac{a^{-1}(\psi)\,\mathrm{d}\psi}{(\varepsilon - \psi)^{1/2}} = 2 \times 2^{1/2} \int_0^{\varepsilon} a^{-1}(\varepsilon - \psi) \times \frac{\mathrm{d}\psi}{\psi^{1/2}}. \tag{II.5}$$

For a Lynden-Bell distribution, in the notation of Appendix I,

$$g(\xi)=\frac{2^{1/2}}{[4\pi G\eta(2/\beta)^{1/2}]^{1/2}}\int_0^\xi\frac{d\zeta}{\zeta^{1/2}[F_{1/2}(\nu)-F_{1/2}(\nu-\xi+\zeta)]^{1/2}}=$$

$$=\frac{2^{1/2}[F_{-1/2}(\nu)]^{1/2}}{[4\pi G\varrho(0)]^{1/2}}\int_0^\xi\frac{d\zeta}{\zeta^{1/2}[F_{1/2}(\nu)-F_{1/2}(\nu-\xi+\zeta)]^{1/2}}. \quad \text{(II.6)}$$

We check that as $\xi\to 0$,

$$[F_{1/2}(\nu)-F_{1/2}(\nu-\xi+\zeta)]^{1/2}\to[\tfrac{1}{2}F_{-1/2}(\nu)\times(\xi-\zeta)]^{1/2}$$

and

$$g(\xi)\to\frac{2\pi}{[4\pi G\varrho(0)]^{1/2}},$$

while as $\xi\to\infty$,

$$[F_{1/2}(\nu)-F_{1/2}(\nu-\xi+\zeta)]^{1/2}\to[F_{1/2}(\nu)]^{1/2}$$

and

$$g(\xi)\to\frac{4(2\xi)^{1/2}}{2\pi GM}.$$

References

Cody, W. J. and Thacher, Jr., H. C.: 1967, *Math. Comput.* **21**, 30.
Cuperman, S., Goldstein, S., and Lecar, M.: 1969, *Monthly Notices Roy. Astron. Soc.* **146**, 161.
Hohl, F. and Campbell, J. W.: 1968, *Astron. J.* **73**, 611.
Lynden-Bell, D.: 1967, *Monthly Notices Roy. Astron. Soc.* **136**, 101.

A PHASE-SPACE BOUNDARY INTEGRATION OF THE VLASOV EQUATION FOR COLLISIONLESS ONE-DIMENSIONAL STELLAR SYSTEMS

S. CUPERMAN and A. HARTEN
Tel-Aviv University, Ramat-Aviv, Israel

and

M. LECAR
Smithsonian Astrophysical Observatory, Cambridge, Mass., U.S.A.

Abstract. The evolution of one dimensional (stratified) self-gravitating systems of stars with constant phase-space density ('water bag' model) is investigated by following the motion of the boundary curves defining the systems. The results are compared with those obtained by sheet-model computer experiments and good agreement is found. New aspects of the evolution, revealed by the present method, are discussed.

1. Introduction

The evolution of collisionless selfgravitating systems of stars is usually investigated analytically, by solving the self-consistent system of Vlasov-Poisson equations; or numerically, by carrying out computer experiments with 'superstars' obeying consistent Newton equations. The first method provides at most a linear solution. The second method represents, given the inevitably limited number N of 'superstars' used, only an approximation of the actual situation.

Recently an alternative method for the investigation of collective relaxations has been proposed by Roberts and Berk (1967) and applied to the one-dimensional plasma case (to study the 'two-stream instability'). The basis of this method is the following: Let us picture the distribution function $f(x, v, t)$ as the density of an incompressible 'phase fluid' moving in the two-dimensional (x, v) space. Then, by Liouville's theorem, it is sufficient to follow the motion of the boundary curve(s) in order to know the state of the system. Indeed, the theorem asserts that the density of each individual moving element must remain constant in time; that is, the boundary may strongly change with time, but inside the area enclosed by the (deformed) boundary, the density will be the same as it was initially. Consequently, this method of integration enables us to investigate a system possibly consisting of a very large number of stars (enclosed by the boundary curve) without having to treat them explicitly.

In this paper, we study the evolution of a one-dimensional, constant-phase space density selfgravitating system of stars by using a modified version of Roberts and Berk's method. The results are compared with those obtained from the integration of the same system by a sheet-model computer experiment.

New features in the evolution of the system are investigated. Included in this is the time evolution of the total length of the boundary curve, which only the present method

M. Lecar (ed.), Gravitational N-Body Problem, 276–289.

is capable of describing. In addition we find a relatively high degree of activity of the core at the well-advanced stages of evolution.

Finally, a system with two disconnected boundary curves is also integrated and the results compared with sheet-model calculations. This system represents a particular case of a two-density system to be investigated in the future (here, $f_1=\eta$ and $f_2=0$).

2. The Integration Method

As mentioned in the Introduction, the evolution of a collisionless one-dimensional self-gravitating system with constant phase-space density is completely defined by the evolution of its boundary curve. Therefore, the main effort has to be put into the most accurate determination of the continuously changing boundary curve.

A. BASIC EQUATIONS

The points on the boundary curve experience the collective forces due to the whole system. The corresponding accelerations are

$$a(x,t) = -2\pi G \int_{-\infty}^{x} \varrho(x,t)\,\mathrm{d}x + 2\pi G \int_{x}^{+\infty} \varrho(x,t)\,\mathrm{d}x, \tag{1}$$

where G is the gravitational constant and $\varrho(x,t)$ is the mass density defined as

$$\varrho(x,t) = \int_{-\infty}^{+\infty} f(x,v,t)\,\mathrm{d}v. \tag{2}$$

This is related to the total mass M by

$$M = \int_{-\infty}^{+\infty} \varrho(x,t)\,\mathrm{d}x. \tag{3}$$

We set $f=\text{constant}=\eta$ inside the domain defined by the boundary curve and $f=0$ outside it. Then, with the convenient normalizations $M=1$ and $2\pi G=1$ and denoting by A the area enclosed by the boundary curve ($A=M/\eta$), expressions (1)–(3) become:

$$a(x,t) = -\frac{1}{A}\int_{x_{\min}}^{x} \mathrm{d}x \int_{v_{\min}}^{v_{\max}} \mathrm{d}v + \frac{1}{A}\int_{x}^{x_{\max}} \mathrm{d}x \int_{v_{\min}}^{v_{\max}} \mathrm{d}v \tag{1'}$$

$$\varrho(x,t) = \frac{1}{A}\int_{v_{\min}}^{v_{\max}} \mathrm{d}v, \tag{2'}$$

and

$$M = 1 = \eta \int_{x_{\min}}^{x_{\max}} \mathrm{d}x \int_{v_{\min}}^{v_{\max}} \mathrm{d}v = \eta A. \tag{3'}$$

In expressions (1′)–(3′) we have also replaced the $\pm\infty$ limits of the integrations by the minimum and maximum values of x and v (which are, obviously, functions of time).

If the boundary curve is multivalued (it will necessarily become so, sooner or later), the integral over velocities in (1′)–(3′) actually represents a sum of integrals referring to phase-space regions full of matter. That is,

$$\int_{v_{min}}^{v_{max}} dv = \int_{v_{min}}^{v_1} dv + \int_{v_2}^{v_3} dv \ldots \int_{v_s}^{v_{max}}$$

where $v_1 - v_{min}$, $v_3 - v_2, \ldots$ $v_{max} - v_s$ $(v_{min} < v_1 < v_2 \ldots v_s < v_{max})$ represent the widths of possible filamentary-like regions of the system (where $f = \eta \neq 0$) at a given value of the x coordinate.

The equations of motion for the ith point on the boundary curve, defined by $R_i(t) = (x_i(t), v_i(t))$, are

$$x_i(t + \Delta t) = x_i(t) + v_i(t)\cdot\Delta t + \tfrac{1}{2}a_i[x_i(t)]\cdot(\Delta t)^2 + \\ + \tfrac{1}{6}\dot{a}_i[x_i(t)]\cdot(\Delta t)^3 + 0((\Delta t)^4) \tag{4}$$

and

$$v_i(t + \Delta t) = v_i(t) + a_i[x_i(t)]\cdot\Delta t + \tfrac{1}{2}\dot{a}_i[x_i(t)]\cdot(\Delta t)^2 + 0((\Delta t)^3), \tag{5}$$

with $a_i[x_i(t)]$ given by (1′) and $\dot{a}_i[x_i(t)]$ defined by

$$\dot{a}_i[x_i(t)] \equiv \frac{d}{dt} a_i[x_i(t)] = \frac{a_i[x_i(t)] - a_i[x_i(t - \Delta t)]}{\Delta t} + 0(\Delta t). \tag{6}$$

If an approximation to order $(\Delta t)^4$ is chosen for $x_i(t+\Delta t)$ given by the Taylor expansion (4), i.e. if terms $0((\Delta t)^4)$ are neglected in (4), one must correspondingly neglect terms of order $0((\Delta t)^3)$ in the expansion (5) and terms of order $0(\Delta t)$ in the expansion (6). Then, by (4), (5) and (6), to order $(\Delta t)^4$, the following consistent set of equations of motion is obtained:

$$x_i(t + \Delta t) = x_i(t) + v_i(t)\cdot\Delta t + \tfrac{1}{6}[4a_i(t) - a_i(t - \Delta t)]\,(\Delta t)^2 \tag{4′}$$

and

$$v_i(t + \Delta t) = v_i(t) + \tfrac{1}{2}[3a_i(t) - a_i(t - \Delta t)]\,\Delta t, \tag{5′}$$

where, for simplicity we have set $a_i(t) \equiv a_i[x_i(t)]$.

Since at $t=0$ the acceleration $a_i[x_i(-\Delta t)]$ is not known, we calculate $\dot{a}_i[x_i(t=0)]$ by the following iterative procedure:

Integrate Equations (4) and (5) with $\dot{a}_i(t=0)=0$ to find new positions $R_i^{(0)}(\Delta t)$, which enable the calculation of the accelerations $a_i^{(0)}(\Delta t)$. Then, using a forward expansion we obtain a first approximation to the $a_i(t=0)$:

$$\dot{a}_i^{(0)}(t = 0) = \frac{a_i^{(0)}(\Delta t) - a_i(0)}{\Delta t}. \tag{6′}$$

Next, integrate Equations (4) and (5) with $\dot{a}_i(t=0) = \dot{a}_i^{(0)}(t=0)$ given by (6′) to obtain new positions $R_i^{(1)}(\Delta t)$, which enable the calculation of the accelerations $a_i^{(1)}(\Delta t)$; again, by the same forward expansion we obtain a second approximation for $\dot{a}_i(t=0)$, namely

$$\dot{a}_i^{(1)}(t=0) = \frac{a_i^{(1)}(\Delta t) - a_i(0)}{\Delta t}. \tag{6''}$$

This iterative procedure converges very quickly.

The total kinetic and potential energies of the system are, respectively:

$$E_k(t) = \frac{1}{2A} \int_{x_{\min}}^{x_{\max}} \mathrm{d}x \int_{v_{\min}}^{v_{\max}} v^2 \mathrm{d}v \tag{7}$$

and

$$E_p(t) = - \int_{x_{\min}}^{x_{\max}} x \cdot \varrho(x, t)\, a(x, t)\, dx = - \frac{1}{A} \int_{x_{\min}}^{x_{\max}} x \cdot a(x, t) \int_{v_{\min}}^{v_{\max}} \mathrm{d}v. \tag{8}$$

B. PROCEDURE

Schematically, the method works as follows:

(1) The desired (x, v) configuration is defined through a boundary curve inside which the phase-space density $f = \eta =$ constant and outside which $f=0$. Sufficient points $R_i(t=0) = [x_i(t=0), v_i(t=0)]$ on the boundary curve are chosen in order to represent it accurately.

(2) The system is divided into n vertical strips by straight lines parallel to the v-axis and situated a distance Δx apart from each other, with $\Delta x = (x_{\max} - x_{\min})/n$. As we see below, the introduction of the vertical lines is very convenient for the calculation of the accelerations.

(3) The accelerations at the positions x_i (corresponding to the initial marking points R_i) are determined in the following way:

(i) One first finds the intersection points (ξ_l, η_j) of the boundary curve(s) with the vertical lines $\{\xi_l\}$ $(l = 1, 2 \ldots n)$, by following the curve(s) in a given direction. In this the boundary curve is generally represented by a parabola $v(x)$ passing through three successive points, though there are situations when a linear, rather than parabolic representation of the boundary curve is required. Then one arranges the intersection points of the boundary curve with the l-th vertical line, $\{\xi_l, \eta_j\}_{j=1,2\ldots J}$, in a decreasing order, so as to obtain the proper topological order.

(ii) Next, using (1′) and Simpson integrations one calculates the accelerations $a_l(t=0) \equiv a[\xi_l(t=0)]$ at the intersection positions $\xi_l(t=0)$. At this stage, the total kinetic and potential energies at $t=0$ are also calculated.

(iii) Finally, by linear interpolations between acceleration values $a(\xi_l)$ and $a(\xi_{l+1})$, corresponding to intersection positions ξ_l and ξ_{l+1} with $\xi_l \leqslant x_i \leqslant \xi_{l+1}$ (recall that $R_i = (x_i, v_i)$, $i = 1 \ldots K$ represent the initial marking points), one obtains the accelerations $a_i(t=0) \equiv a_i[x_i(t=0)]$. The use of linear interpolation for the calculation of the accelerations appears to be adequate. This is a consequence of the rather smooth dependence (almost two linear parts) of the acceleration as a function of the position, $a(x)$, as shown in Figure 5 of Goldstein *et al.*, 1969.

(4) The points $R_i = [x_i(t=0), v_i(t=0)]$ are developed through the equations of motion (4′) and (5′) to new values $R_i = [x_i(\Delta t), v_i(\Delta t)]$.

(5) The integration process is then repeated as described in steps (3) and (4).

(6) With time, the boundary curve stretches and the system spreads out. Consequently, more points are required to describe it accurately. On the other hand, because of the rapid change (with time) in the degree of curvature of various portions of the boundary curve, some regions may develop too large a density of points. This means unnecessary computer-time consumption, increased accumulation of round-off errors, and what is even more important from the point of view of the experiment, unnecessary storage in the limited computer memory at the expense of other regions in real need of supplementary points. Because of this, we have designed our computer program to systematically introduce necessary additional points or take out unnecessary ones according to local conditions. This is done by the use of a curvature-based criterion which systematically checks every possible set of three points and ensures that the difference between the parabolic and the triangle areas defined by these three points lies between fixed upper and lower limits. In this way the accuracy of the integration process is maintained at the desired level.

Special care is given to turning points of the boundary curve. (A turning point is a point at which, following the boundary, the x-projection changes direction.) In this case, a parabola $v(x)$ passing through three successive points $R_{i-1}(t)$, $R_i(t)$ and $R_{i+1}(t)$, with $R_i(t)$ the turning point, is not adequate to describe the curve. Instead, the program uses the two sets of points (R_{i-2}, R_{i-1}, R_i) and $(R_i, R_{i+1}$ and $R_{i+2})$ for this purpose.

3. Results for a One-Boundary-Curve System

The integration method described in Section 2 has been used to investigate the evolution of a selfgravitating system characterized by the following properties:

(1) The initial phase-space configuration is represented by a rectangular boundary, symmetric about both x and v axes. Inside the rectangle, the fine-grained phase-space density is constant and equal to η, while outside, it is zero (see Figure 1, $T=0$).

(2) The initial state is chosen to be far from equilibrium. Thus, the initial ratio of twice the total kinetic energy to the total potential energy is about 15, while its equilibrium (virial) value is one.

(3) The total energy of the system is about twice that corresponding to the minimum energy state.

The same configuration has been treated in the past by the sheet-model method (Goldstein *et al.*, 1969; Cuperman *et al.*, 1969). This enables one to make a detailed comparison between the results of our present method with those of the above.

Figure 1 shows the evolution in phase-space of this system, to be called system ‘C’. As seen, the present method enables a detailed and complete representation of the

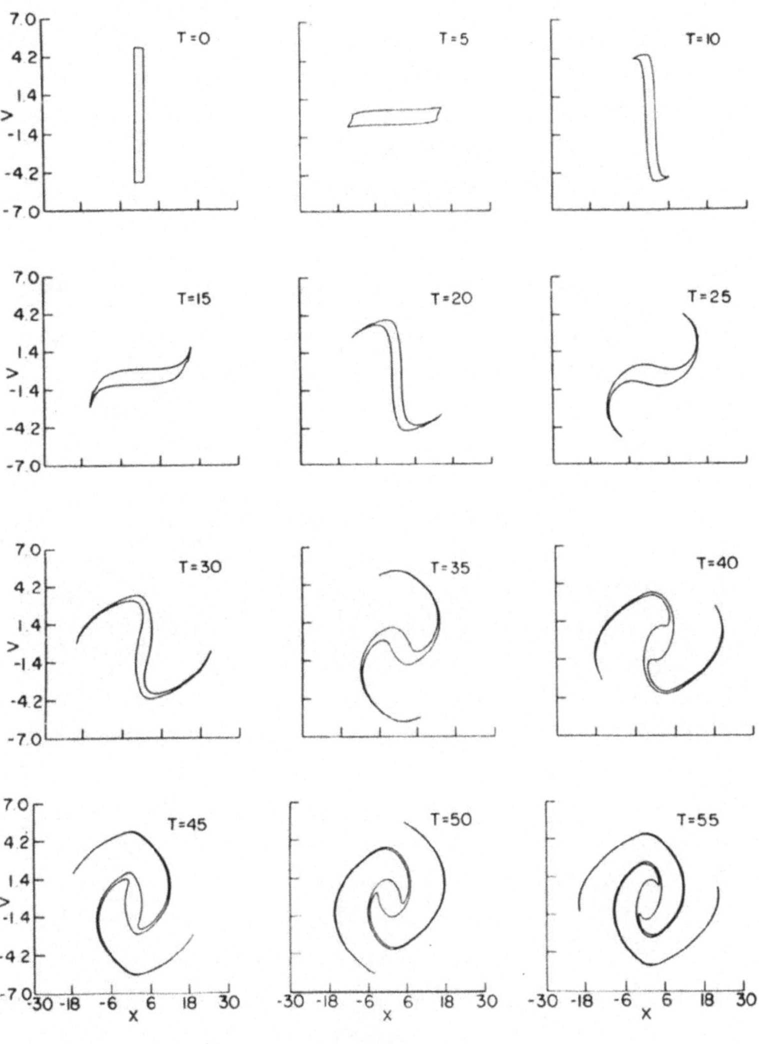

Fig. 1a.

Figs. 1a–d. The evolution in phase-space of system C. A collective period, τ_c (defined in the sense of the usual harmonic oscillator period), corresponds to about 22 time units.

system even in a relatively developed stage of evolution. The continuity of the boundary is perfect, and this is the main advantage of the method. It provides us with results corresponding to the ideal limiting case of large N (i.e., $N \to \infty$). These are the results with which those obtained by other methods have to be compared.

Figure 2 shows the evolution of the total kinetic energy of system C. For comparison, the corresponding results obtained from a sheet-model experiment ($N=320$) (Goldstein *et al.*, 1969) are also given.

From the results obtained by the present method and from comparison with the sheet-model results the following conclusions can be drawn:

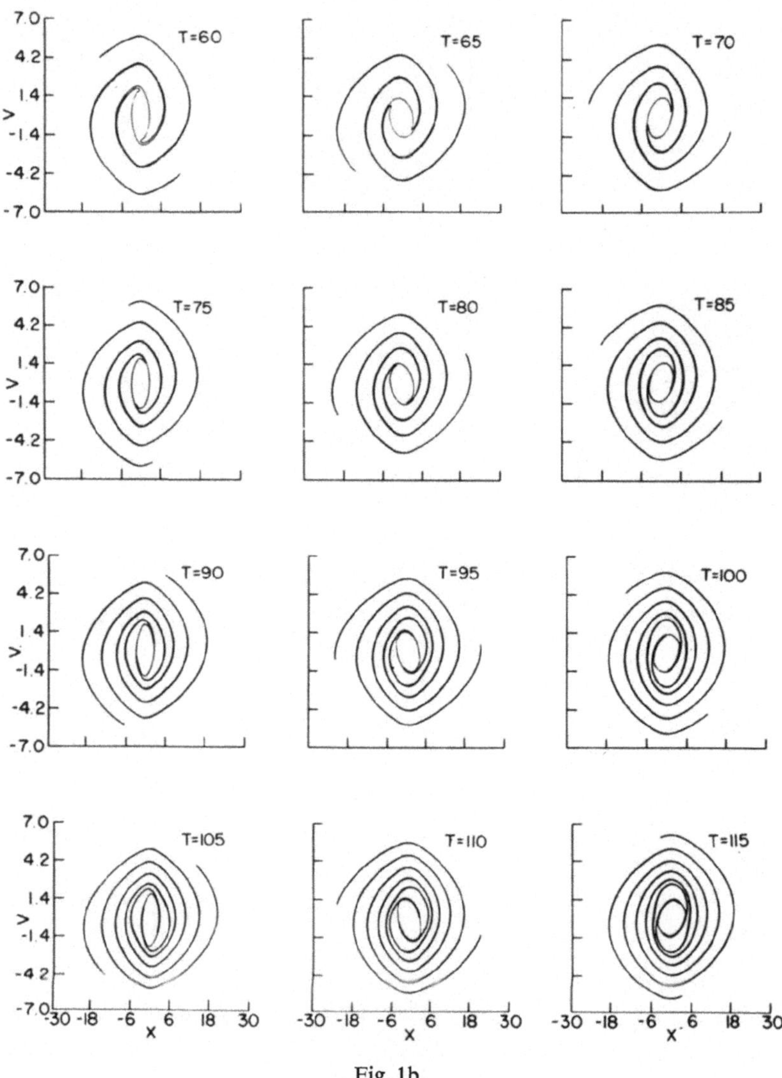

Fig. 1b.

(1) As far as the integration is carried out, there is a good agreement between the results obtained by the two methods. In fact, for about the first five collective periods the results are almost identical, as may be seen from Figure 2. Thereafter, some differences appear. These are a measure of the departure of the sheet-model results (using a rather limited number of sheets) from the correct values*. Consequently, for purposes not requiring too large an accuracy, a sheet-model experiment with only a few hundred superstars, having the advantage of being faster and cheaper, is sufficient.

* Actually doubling the number of sheets results in differences of the same type in the evolution of the total kinetic energy (see Goldstein *et al.*, 1969).

T=120
T=125
T=130
T=135
T=140
T=145
T=150
T=155
T=160
T=165
T=170
T=175
7.0
4.2
1.4
-1.4
-4.2
-7.0
V
-30 -18 -6 6 18 30
X

Fig. 1c.

T = 175
T = 180
7.0
3.5
0.0
-3.5
-7.0
V
-30 -15 0.0 15 30
X

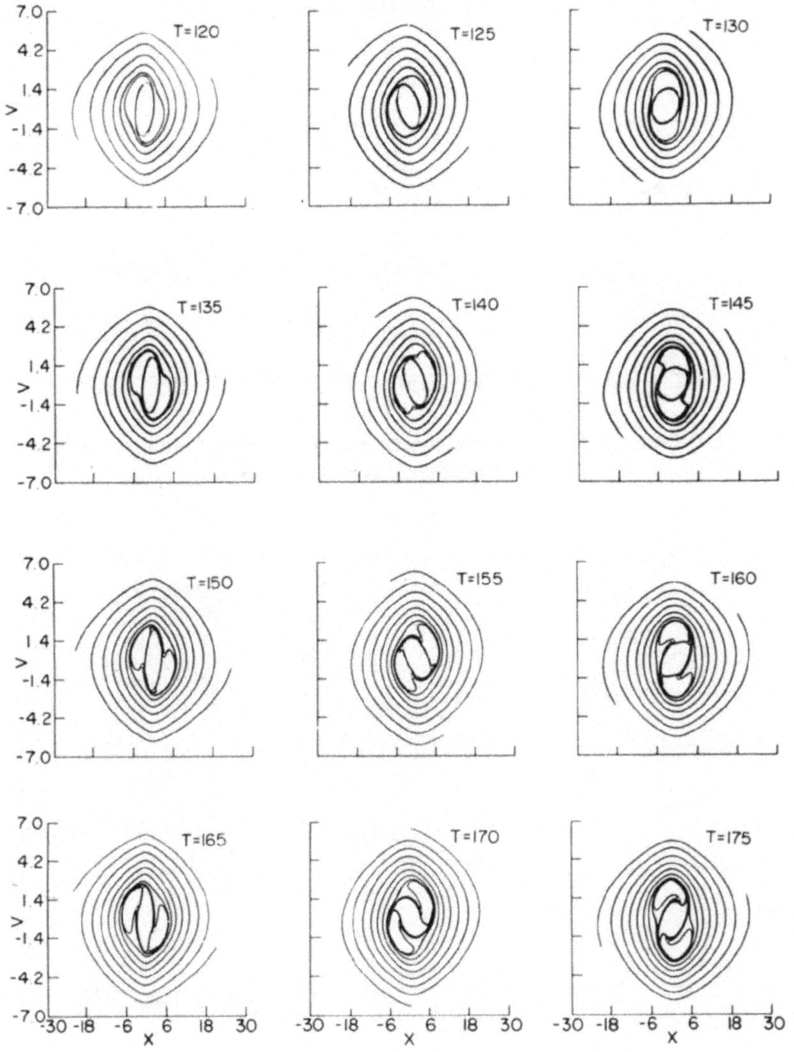

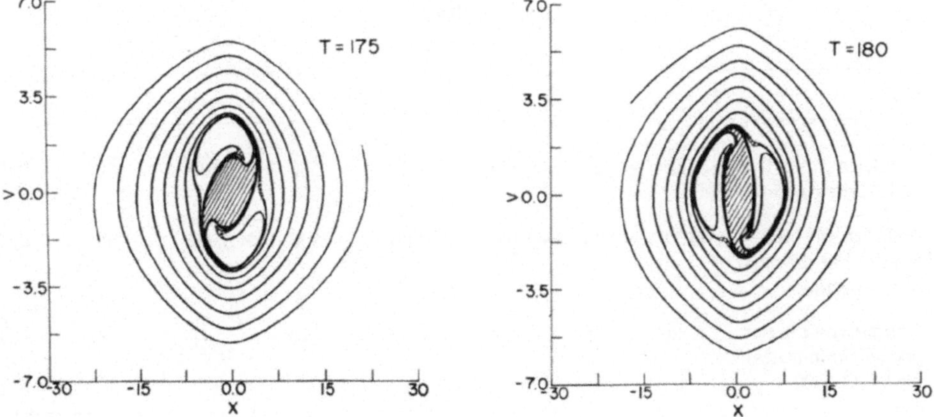

Fig. 1d.

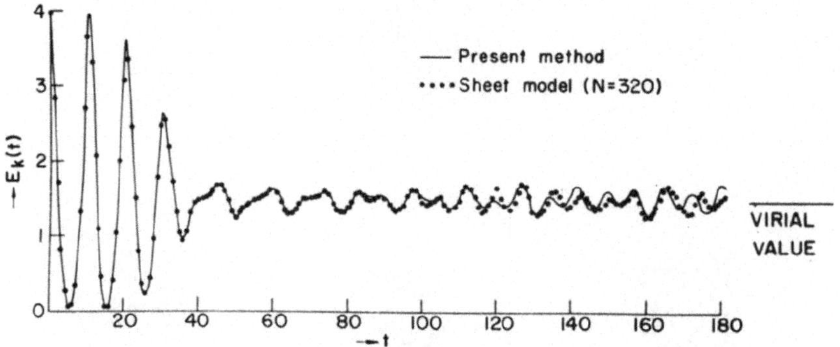

Fig. 2. The evolution of the total kinetic energy of system C and comparison with corresponding sheet-model results.

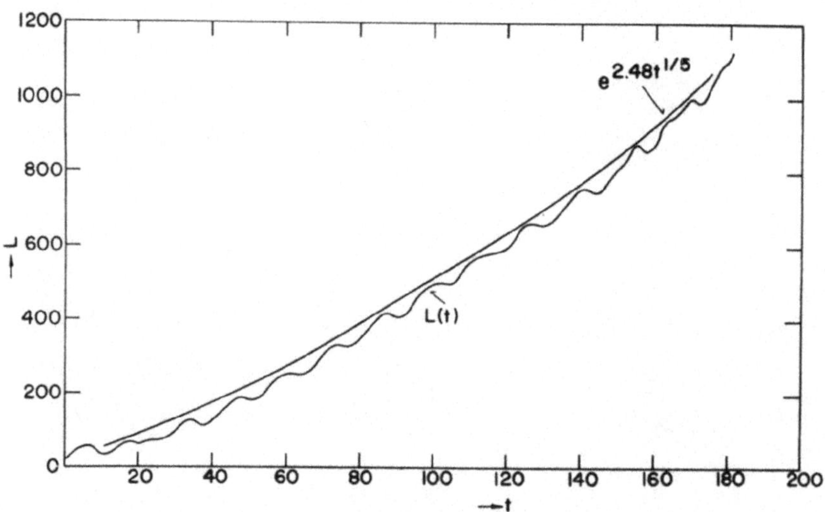

Fig. 3. The evolution of the total length of system C.

TABLE I

Case 'C'

t/τ_c	1	2	3	4	5	6	7	8
Total length at the end of the period	~ 80	~ 180	~ 280	~ 420	~ 560	~ 700	~ 860	~ 1050
Number of points required at the end of the period	~ 300	~ 800	~ 1500	~ 1900	~ 2500	~ 5500	~ 7500	~ 8500
Computer time per period (in minutes) on CDC 6400	4	5	7	9	13	18	27	35

(2) The present integration scheme reveals a rather interesting new feature: That even after about eight collective periods, the core of the system is still very active. Indeed, it exhibits a rather complicated and continuously changing structure and may therefore still produce star acceleration (deceleration). This is clearly seen in the snapshots at times 175 and 180 (Figure 1d).

(3) The availability at all times of a perfectly continuous boundary of the system permits the measurement and the representation as a function of time of the total length of the system. Though this quantity does not have, so far, a well-defined physi-

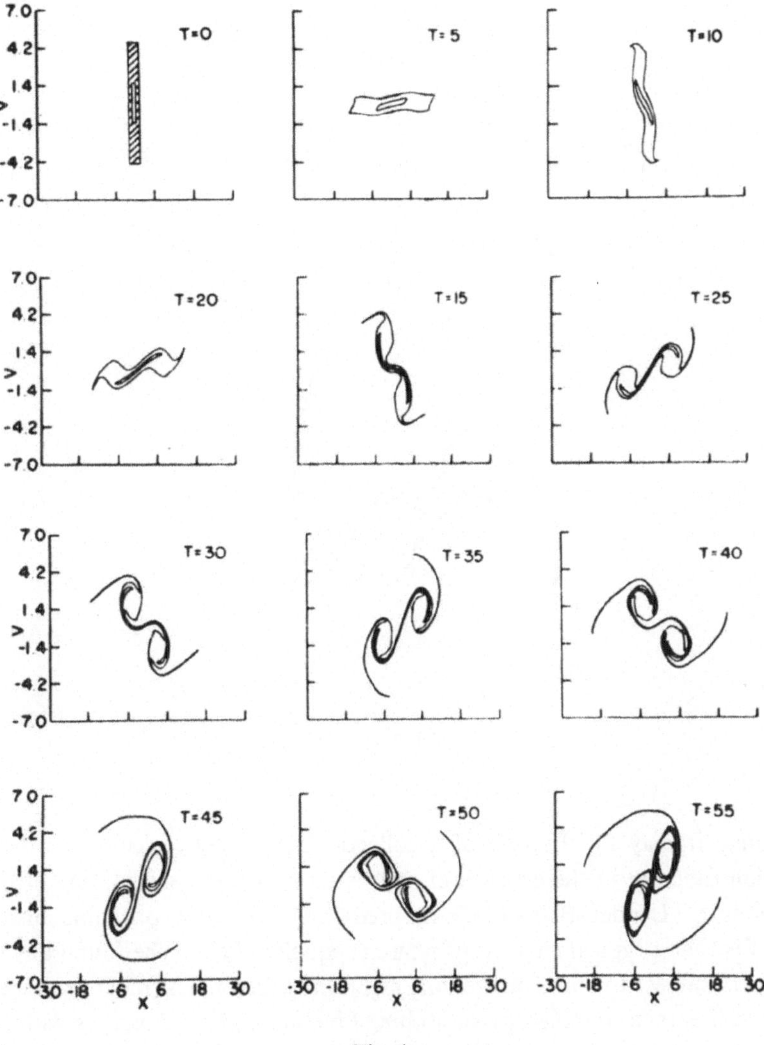

Fig. 4a.

Figs. 4a–c. The evolution in phase-space of system B. The collective period of system B is close to that of system C.

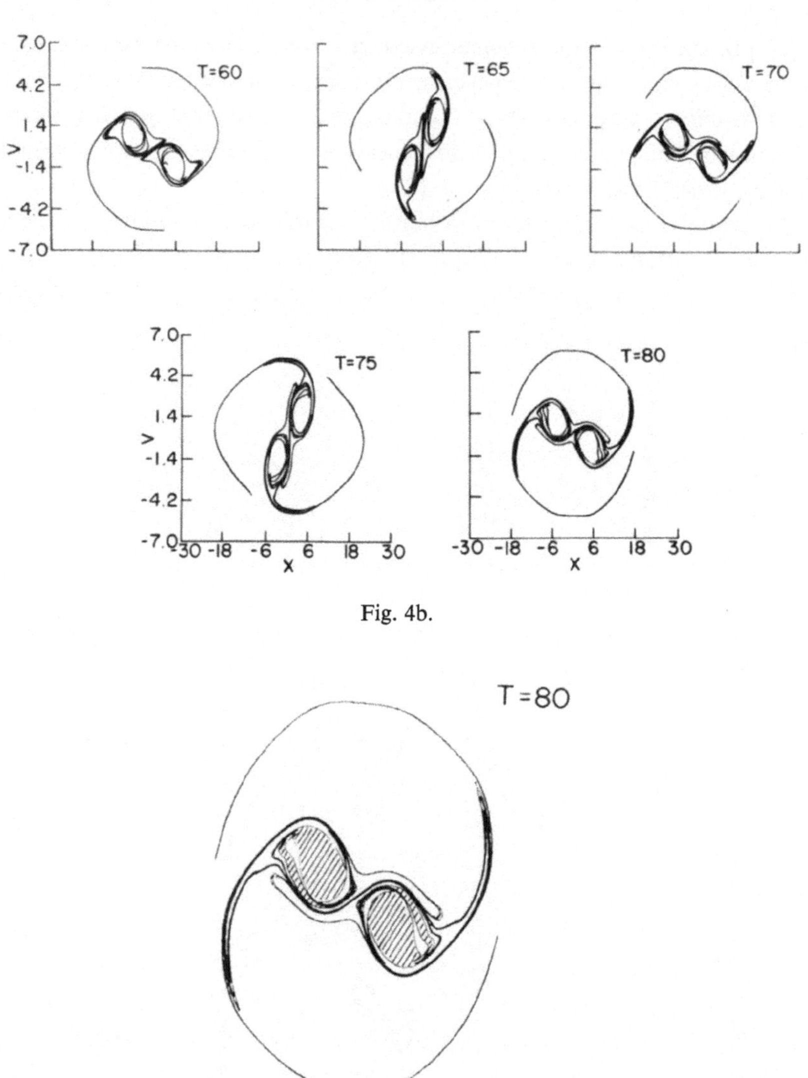

Fig. 4b.

Fig. 4c.

cal meaning, it may fill the role of a collective relaxation measure for the energy distribution function in the absence of any better criterion. Indeed, one of the basic assumptions of Lynden-Bell (1967) in predicting the most probable distribution function $\bar{f}$ is that a typical element of phase is equally likely to be found anywhere in phase-space (subject to some restrictions mentioned there). The most probable distribution $\bar{f}$ itself is then obtained by smoothing (averaging) f over regions with different phase-space densities. Thus, starting with a given boundary in phase-space inside which the density is uniform and equal to η (and zero outside), the system must increasingly stretch out becoming longer and thinner, wind up and spread out over the largest possible area in phase-space. It is in this sense that the total length of the system and its time evolution might be related to the evolution of the system toward the most probable state.

The time evolution of the total length for system C is given in Figure 3. As may be seen, the total length increases by a factor of 50 in about 8 collective periods, the rate of increase itself increasing slightly with time. Actually, the time dependence of the total length could be fitted by an expression of the type $\exp(2.48\, t^{1/5})$ (see Figure 3).

SOME TECHNICAL DETAILS

The program was first run on a CDC 3400 computer (about six collective periods) and then on a CDC 6400 computer. The computer-time consumption was found to be a rapidly increasing function of the number of points on the boundary curve required to keep the total error (as expressed in the accuracy of the total mass conservation) below 1%. A summary of relevant figures is given in Table I below.

4. Results for a Two-Boundary-Curves System

System C represents the case of a one-boundary configuration inside which the phase-space density is initially constant and continues to be so at any later time. Having a

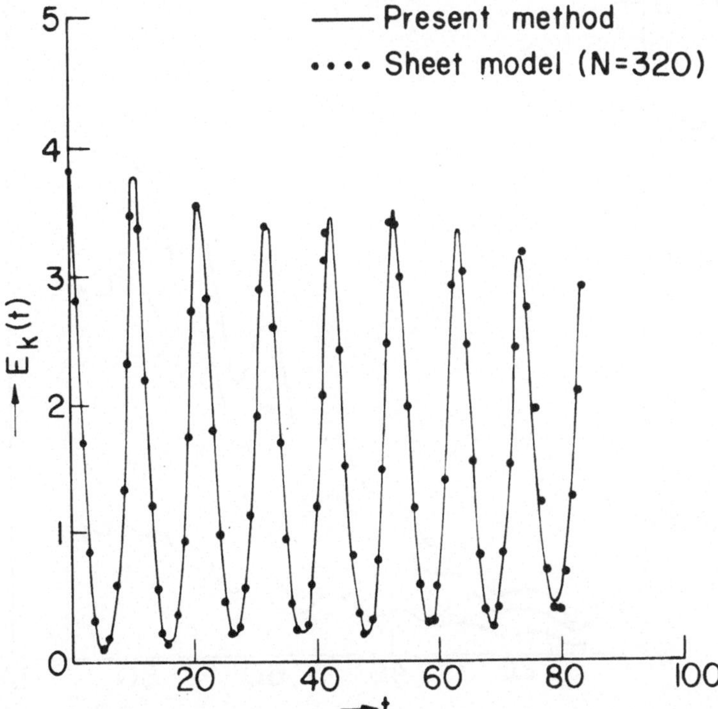

Fig. 5. The evolution of the total kinetic energy of system B and comparison with the corresponding sheet-model results.

single phase-space density is the simplest possible situation. The next step would be to have a two-boundary configuration enabling the consideration of two different phase-space densities. This problem will be treated in a subsequent paper (Cuperman *et al.*, 1970).

In this paper, still remaining within the frame-work of a *one*-phase-space density only, we also treat a two-boundary configuration consisting of two different regions, the interior one having zero density (see Figure 4, case B, at $T=0$). Since such a configuration can still be treated by a sheet-model experiment (and has been done so, in Goldstein *et al.*, 1969), we decided, for completeness (and for future reference) to treat it also by the present method.

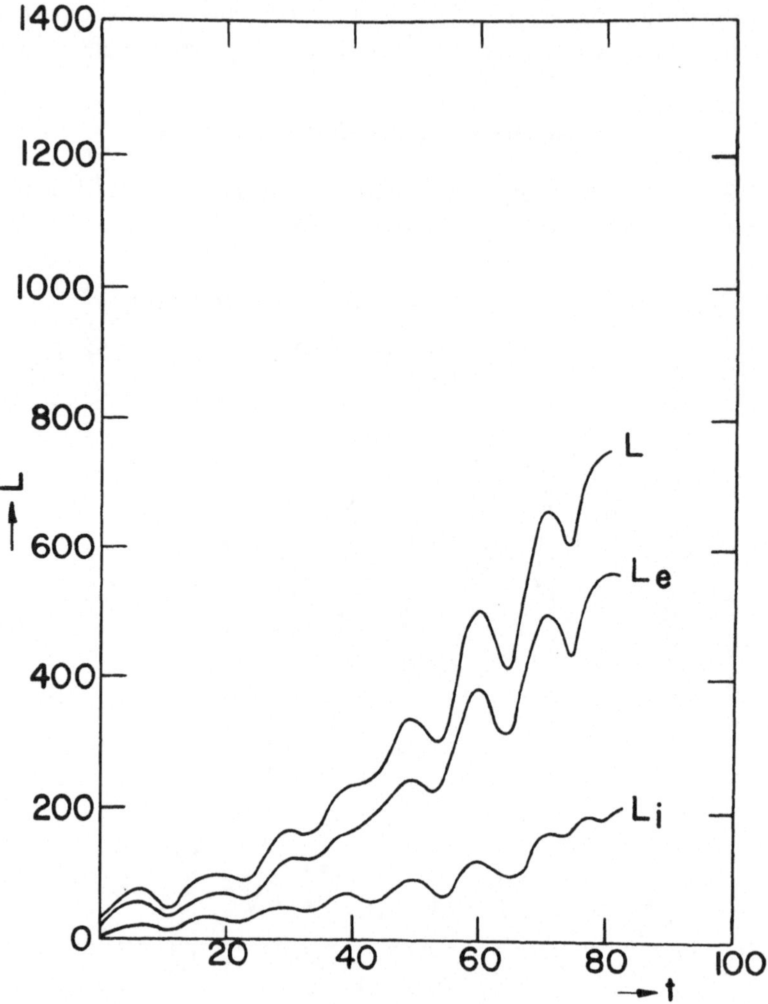

Fig. 6. The evolution of the total length of system B. L_i and L_e refer to the interior and exterior boundary curve, respectively.

The results of the direct integration of case B are given in Figure 4 (its evolution in the phase-space), Figure 5 (the evolution in time of its total kinetic energy and comparison with the sheet-model experiment) and Figure 6 (the evolution in time of its total length).

It was found that all conclusions obtained for case C also hold for case B. This gives us confidence that the direct-integration method for a multiple-boundary case, when applied to a multi-phase-space-density, will give reliable results.

Acknowledgements

Mr. P. Rosenau has also participated in the initial stages of this work. Helpful discussions with Mr. S. Goldstein are acknowledged.

This work was supported by grant SFC-8-7006 from the Smithsonian Institution, Washington, D.C.

References

Cuperman, S., Goldstein, S., and Lecar, M.: 1969, *Monthly Notices Roy. Astron. Soc.* **146**, 161.
Cuperman, S., Harten, A., and Lecar, M.: 1972, this volume, p. 290.
Goldstein, S., Cuperman, S., and Lecar, M.: 1969, *Monthly Notices Roy. Astron. Soc.* **143**, 209.
Lynden-Bell, D.: 1967, *Monthly Notices Roy. Astron. Soc.* **136**, 101.
Roberts, K. V. and Berk, H. L.: 1967, Symposium on Computer Simulation of Plasma and Many Body Problems, NASA SP-153, 91.

THE COLLECTIVE RELAXATION OF TWO-PHASE-SPACE-DENSITY COLLISIONLESS ONE-DIMENSIONAL SELFGRAVITATING SYSTEMS

S. CUPERMAN and A. HARTEN
Tel-Aviv University, Ramat-Aviv, Israel

and

M. LECAR
Smithsonian Astrophysical Observatory, Cambridge, Mass., U.S.A.

Abstract. The evolution of one-dimensional *two*-phase-space-density selfgravitating systems of stars is investigated by following the motion of the boundary curves of the systems in phase space.

A *qualitative* agreement with Lynden-Bell's theory predicting, for the most probable state, velocity dispersions inversely proportional to the phase space density of the component at the time of star formation, is found.

1. Introduction

The basic methods of investigation providing information on the evolution and especially on the most probable state of collisionless selfgravitating systems of stars are statistical-mechanical (i.e., Lynden-Bell's theory, 1967) or numerical (i.e., computer experiments or other integration schemes). While Lynden-Bell's theory applies to systems consisting of any number of phase-space densities, so far the numerical methods concentrated on the *one*-phase-space-density systems only. For this particular case, a partial agreement between Lynden-Bell's prediction and numerical results has been obtained (Hénon, 1964, 1967, 1968; Lecar, 1966, 1967; Lecar and Cohen, 1967, 1968; Hohl and Feix, 1967; Hohl and Campbell, 1968; Goldstein *et al.*, 1969; Cuperman *et al.*, 1969).

The *one*-phase-space-density case represents, however, a rather poor approximation to the actual systems of stars. In fact, the number of different phase-space densities should be very large and the corresponding most probable state should be a superposition of Maxwellian components whose velocity dispersions are inversely proportional to the phase space density of the component at star formation (Lynden-Bell, 1967).

In this paper one further step is taken: we here investigate numerically the evolution of one-dimensional *two*-phase-space-density selfgravitating systems of stars. The integration method – most suitable for the investigation of multidensity systems – consists in following the motion of the boundary curves enclosing the two regions, each of which is characterized by a different phase-space-density. (A detailed description of the method is given in Cuperman *et al.*, 1970; see also Roberts and Berk, 1967.)

A large variety of *initial* conditions (different phase-space configurations, phase-space densities, mean square velocities and different degrees of departure from the virial state) have been considered.

M. Lecar (ed.), Gravitational N-Body Problem, 290–310. *All Rights Reserved*

One particular relationship between the mean square velocites $\langle(v_{\rm I}^2\rangle$ and $\langle v_{\rm II}^2\rangle)$ predicted by Lynden-Bell as the most probable state for a two-phase-space-density ($\eta_{\rm I}$ and $\eta_{\rm II}$) system, namely $\langle v_{\rm I}^2\rangle/\langle v_{\rm II}^2\rangle=\eta_{\rm II}/\eta_{\rm I}$, is thoroughly investigated. A *qualitative* agreement between theory and numerical results is found.

2. Theory

The distribution function $f(\mathbf{r}, \mathbf{v}, t)$ of a collisionless selfgravitating system of stars, defined such that $f\mathrm{d}^3r\mathrm{d}^3v$ is the total mass of those stars in a volume d^3r about $\mathbf{r}$ flowing with velocities in the range d^3v about $\mathbf{v}$, has to be found by solving a consistent Vlasov-Poisson system of equations.

An alternative, statistical-mechanical approach to the problem is also possible. The Vlasov equation, $\mathrm{D}f/\mathrm{D}t=0$, states that the convective derivative of 'f' following the motion in phase space is zero, that is each element in phase space conserves its phase-density as it moves (Liouville's theorem). Thus, $m(f)\,\delta f$, the total mass of all those elements between f and $f+\delta f$, is conserved, and this provides an infinity of conserved quantities. Lynden-Bell (1967) approached the problem in statistical mechanics which allows for this conservation. He obtained the following solution for the coarse-grained phase-space density for a collisionless selfgravitating system:

$$\bar{f}=\sum_j \mu_j \frac{\exp\{-\beta_j(\varepsilon-\mu_j)\}}{1+\sum\limits_j \exp\{-\beta_j(\varepsilon-\mu_j)\}}. \tag{1}$$

Here, the μ_j's are the phase-space-densities of class 'j'; the $(1/\beta_j)$'s represent a measure of the random excitation above the minimum energy state (in which case $1/\beta_j=0$) and are related to the densities η_j's by the relation

$$\beta_j=\frac{\eta_j}{\bar{\eta}_j}\beta\equiv\frac{\eta_j}{(\sum\limits_j \eta_j M_j)/M}\beta \tag{2}$$

where $\bar{\eta}_j$ represents the average value of the η_j's and β is to be determined from the total energy condition. The other quantities appearing in (1) are: μ_j's, corresponding to the Fermi energies in the Fermi-Dirac statistics; ε, the energy per unit mass associated with a point x, v in phase-space; M_j, the mass of all regions characterized by η_j; and M, the total mass of the system. Essentially, expression (1) corresponds to a (fourth) statistics having combined features of classical statistics (the particles are distinguishable as in the Maxwell-Boltzmann statistics) and quantum statistics (the existence of an exclusion principle, as in Fermi-Dirac statistics). The exclusion principle now refers to elements of phase space and is a consequence of Lioville's theorem: no two elements of phase can overlap in phase space, for then the phase space density would be different in the region of overlap.

In the *extreme-non-degenerate* (E.N.D.) limit, when $\bar{f}\ll\eta_j$ for all j is expected, Lynden-Bell (1967) predicts for $\bar{f}$(1) a superposition of maxwellian components whose velocity dispersions $\langle v_j^2 \rangle (\equiv 1/\beta_j)$ are inversely proportional to the phase-space densities η_j of the component j at star formation:

$$\bar{f}_{\text{E.N.D.}} = \sum_j A_j \exp(-\beta_j \varepsilon) \tag{3}$$

$$\beta_{j,\text{E.N.D.}} = \frac{1}{\langle v_j^2 \rangle} \beta = \frac{\eta_j M}{\sum_j \eta_j M_j} \beta . \tag{4}$$

Therefore, in the case of violent collective relaxation of a *collisionless* selfgravitating system, a segregation by mass effect is expected. Obviously, this statistical-mechanical effect is different from the segregation by mass effect due to *collisional* equipartition of energy.

The simplest case in which this prediction could be checked is a one-dimensional two-phase-space-density system. In such a case (j=I, II), by (2) and (4) one has

$$\frac{\langle v_{\text{II}}^2 \rangle}{\langle v_{\text{I}}^2 \rangle} = \frac{\eta_{\text{I}}}{\eta_{\text{II}}} . \tag{5}$$

3. Integration and Results

To go beyond the one-phase-space-density case considered so far and, at the same time, to check expression (5), which is the simplest tractable prediction of Lynden-Bell's statistical mechanics of violent relaxation in stellar systems involving more than one phase-space-density, we investigated the evolution of one-dimensional *two*-phase-space-density selfgravitating systems.

The integration method used consisted in following the motion of the boundaries of each one of the two one-phase-space regions involved. As shown by Liouville's theorem, knowledge of the boundary curves completely defines the state of the system; therefore, integration of the equation of motion for all the staff of the system may be replaced by that for the boundary curve alone. This method, especially suitable for the treatment of systems consisting of more than one phase-space-density, has recently been applied to *one*-phase-space-density systems consisting of *one* and *two* boundary curves (Cuperman *et al.*, 1970). We refer the reader to that work for details.

Actually, three types of initial configurations have been considered:

A: Systems initially having a central 'light' phase-space-density surrounded by a 'heavier' phase-space-density;

B: Systems initially having a central 'heavy' phase space-density surrounded by a 'lighter' phase-space-density;

C: Systems initially having 'mixed' regions of heavy and light phase-space-densities.

For a better understanding of the problem, we investigated (for type A systems) the dependence of the evolution (and in particular of the sought-for effect) on the following two factors:

(i) the ratio of the two-phase-space-densities involved;

(ii) the initial value of the ratio of twice the total kinetic energy to the total potential energy.

We here present some of the most relevant results. For convenience, we refer to each one of Groups A–C separately, but next compare and summarize all the results.

GROUP A

Figure 1 (case 2 in Table I) represents a four-strip system consisting of *two* phase-space-densities η_I and η_{II}, with corresponding areas A_I and A_{II}, respectively. The ini-

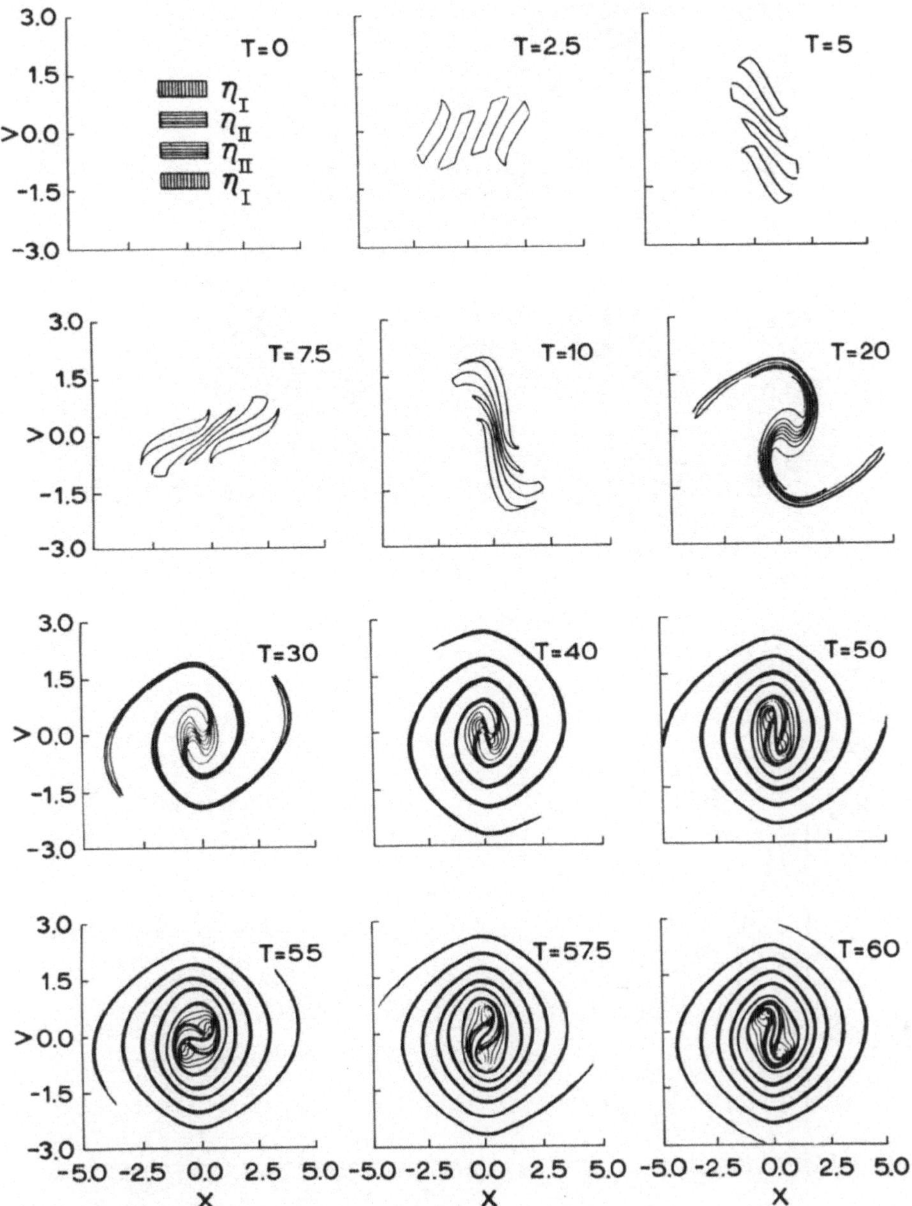

Fig. 1a. The evolution in phase space of the system '2'. A collective period, τ_c, corresponds to about 10 time units.

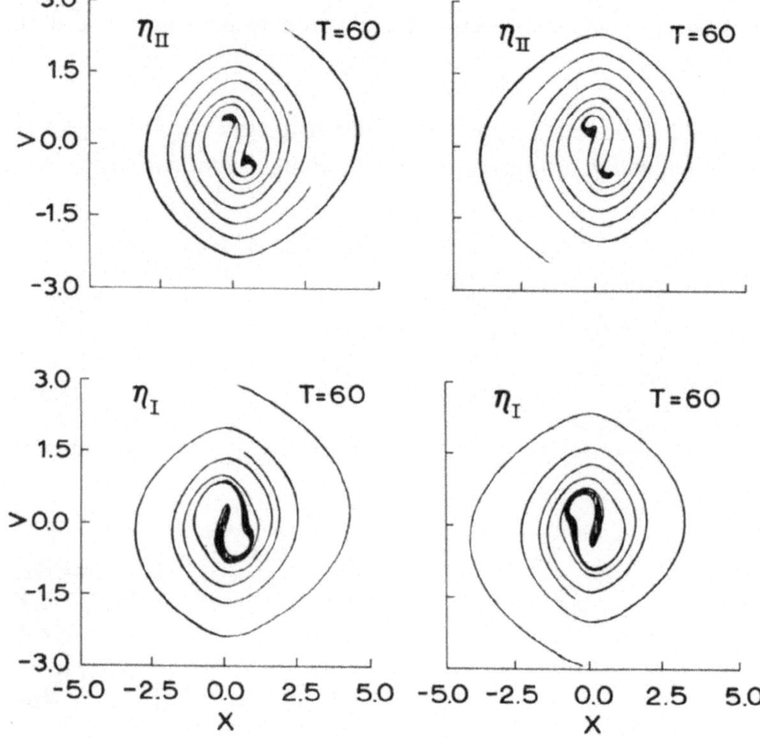

Fig. 1b. The states at time $T=60$ of the initial four strips of the system '2'.

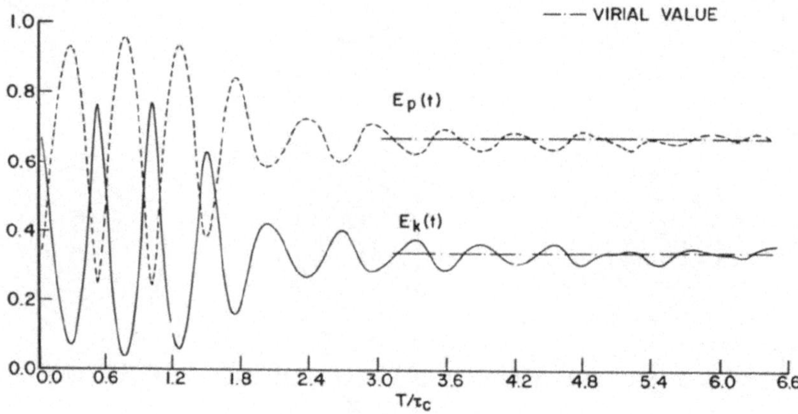

Fig. 1c. The evolution of the total kinetic and potential energies of the system '2', respectively.

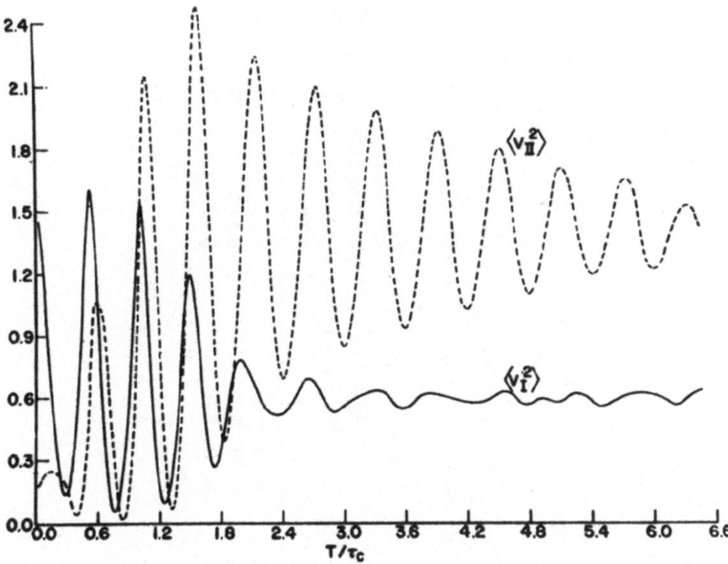

Fig. 1d. The evolution of the mean square velocities $\langle v_{\mathrm{I}}^2\rangle$ and $\langle v_{\mathrm{II}}^2\rangle$ for the system '2', respectively.

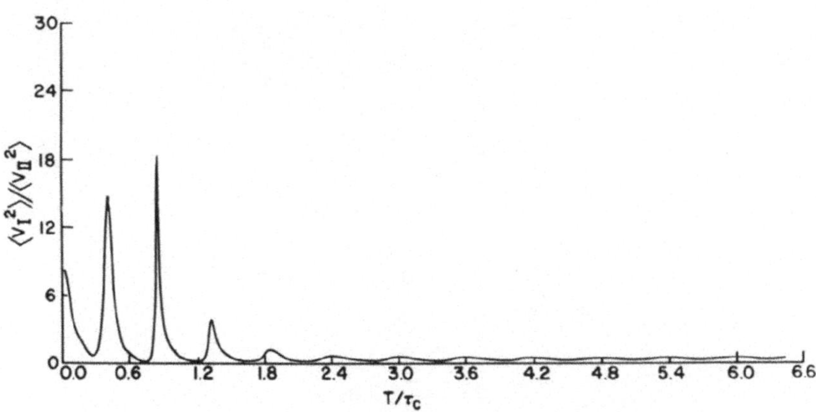

Fig. 1e. The evolution of the ratio $\langle v_{\mathrm{I}}^2\rangle/\langle v_{\mathrm{II}}^2\rangle$ for the system '2'.

tial value of the ratio of mean square velocities $\langle v_{\mathrm{I}}^2\rangle/\langle v_{\mathrm{II}}^2\rangle$ is taken to be rather large, 8.38. The initial value of $2E_k/E_p$ is 4, thus relatively far from equilibrium to provide violent collective relaxation. The system is integrated for several collective periods* and its evolution in phase space is given in Figures 1a and 1b, the evolution of its total kinetic energy and total potential energy in Figure 1c, the evolution of the mean square velocity of each separate phase in Figure 1d, and the evolution of the ratio

* A collective period, τ_c, is here defined in the sense of the usual harmonic oscillator period. With the convenient normalizations $M=1$ and $2\pi G=1$ ($M=$ the total mass, $G=$ the gravitational constant), $\tau_c \simeq 10$ time units in Case 2.

TABLE I

Summary of initial conditions and results for initial configurations of type A having different values of the ratio η_I/η_{II}

Case	Initial conditions								τ_c	Predicted $\langle v_I^2\rangle/\langle v_{II}^2\rangle$	Obtained average over the sixth period	L/L_0 at the end of the sixth period
	E_p	E_k	$2E_k/E_p$	$\langle v_I^2\rangle/\langle v_{II}^2\rangle$	η_I/η_{II}	A_I/A_{II}	v_0	η_I				
1	1/3	2/3	4	8.38	4	1	0.2097	1/2.0970	~ 10	0.25	0.54	21.1
2	1/3	2/3	4	8.38	10	1	0.2000	1/1.7600	~ 10	0.10	0.44	24.0
3	1/3	2/3	4	8.38	25	1	0.1949	1/1.6216	~ 10	0.04	0.40	25.5

$\langle v_{\rm I}^2\rangle/\langle v_{\rm II}^2\rangle$ in Figure 1e. As seen, a rather thorough mixing of the two phases (see Figure 1b) and a relative quick relaxation of the system toward the virial value (see Figure 1c) occur. The relaxation of the 'heavy' phase is *much* faster than that of the 'light' phase (see Figure 1d). The evolution of the ratio $\langle v_{\rm I}^2\rangle/\langle v_{\rm II}^2\rangle$ (Figure 1e) indicates that after only 2 collective periods a rather dramatic decrease (by a factor of 16) in the initial value occurs. The decrease slowly continues thereafter and the average value for the fifth or sixth collective period is 0.44 (the peak and lowest values are 0.54 and 0.34, respectively). Therefore, a *qualitative* agreement with Lyndon-Bell's prediction (expression 5) is obtained.

To investigate the dependence of the ratio of the asymptotic mean square velocities $\langle v_{\rm I}^2\rangle/\langle v_{\rm II}^2\rangle$ on the phase-densities ratio $\eta_{\rm I}/\eta_{\rm II}$, two other four-strip systems (case 1 and case 3 in Table I) have been investigated. Thus, cases 1, 2 and 3 have the *same* total energy E_k+E_p, *same* initial value of $2E_k/E_p$ and the *same* initial value of $\langle v_{\rm I}^2\rangle/\langle v_{\rm II}^2\rangle$. However, they constitute a sequence of decreasing values of the ratio $\eta_{\rm II}/\eta_{\rm I}$: 0.25, 0.1 and 0.04. Then, according to Lynden-Bell's theory, the asymptotic values of $\langle v_{\rm I}^2\rangle/\langle v_{\rm II}^2\rangle$ for cases 1, 2 and 3 should be 0.25, 0.1 and 0.04, respectively. Again, a qualitative agreement with the theory is obtained, as a sequence of slightly decreasing values is obtained for $\langle v_{\rm I}^2\rangle/\langle v_{\rm II}^2\rangle$ (see Table I).

Finally, we investigated the role of the degree of violence in the relaxation of the systems by considering a sequence of three similar configurations with different initial values of $2E_k/E_p$ (that is, with different initial departures from the virial value of unity; see cases 4, 5 and 2 in Table II). Thus, all three systems have the same density ratio $\eta_{\rm II}/\eta_{\rm I}$ (0.1), same initial mean square velocity ratio, $\langle v_{\rm I}^2\rangle/\langle v_{\rm II}^2\rangle$ (8.38), same initial potential energy, E_p (1/3); however, the sequence of the initial values for E_k is 1/6, 2/6 and

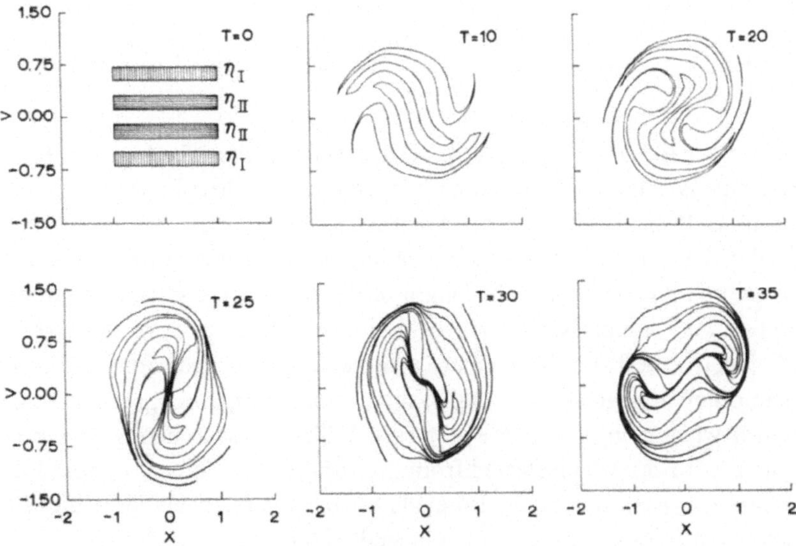

Fig. 2a. The evolution in phase space of the system '4'. A collective period, τ_c, corresponds to about 7 time units.

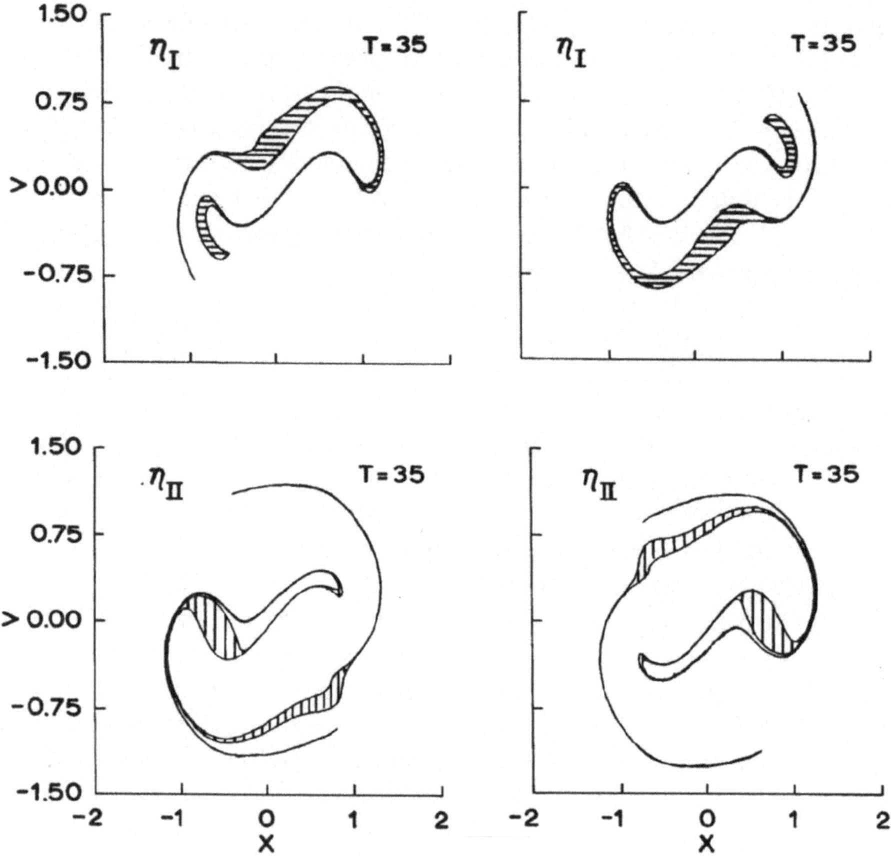

Fig. 2b. The states at time $T=35$ of the initial four strips of the system '4'.

4/6, resulting therefore in 1, 2 and 4, respectively as the initial values for $2E_k/E_p$. (Actually, this is achieved by building homothetic – in the v-coordinate – four-strip configurations. If the width of a strip, which is equal to the distance between two strips, is denoted by 2 v_0, then the values of v_0 leading to the desired initial configuration are given in column 8 of Table II. The length of the strips is the same in all three cases).

Inspection and comparison of the evolution in phase space of the three systems (cases 4, 5 and 2 in Table II) indicate a rather different behaviour in time and, in particular, different degrees of mixing of the two phases η_{I} and η_{II}, the mixing becoming stronger when going from case 4 to case 2. The weakest mixing occurs in case 4 (see Figures 2a to 2e), which started from the virial value. Here, as opposed to case 2, for instance, the weak mixing is also evident from the relatively large blobs of some phase-space-densities persisting up to the end of the run.

As seen from the results of Table II, the asymptotic value of $\langle v_{\text{I}}^2\rangle/\langle v_{\text{II}}^2\rangle$ is strongly dependent on the initial value of $2E_k/E_p$ approaching more closely the predicted value when $2E_k/E_p$ is larger, in (qualitative) agreement with Lynden-Bell's theory. Indeed, his statistical-mechanical prediction requires violent collective relaxation, that is, strongly changing potentials to favor maximum mixing, Moreover, relation (5) is strictly valid in the E.N.D. limit, when all the random excitation above the minimum energy state $1/\beta_j$ may be identified with the velocity dispersions $\langle v_j^2 \rangle$.

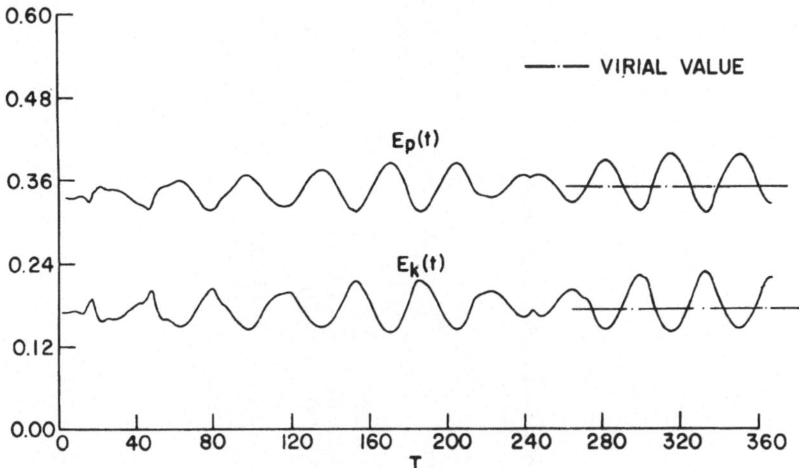

Fig. 2c. The evolution of the total kinetic and potential energies of the system '4', respectively.

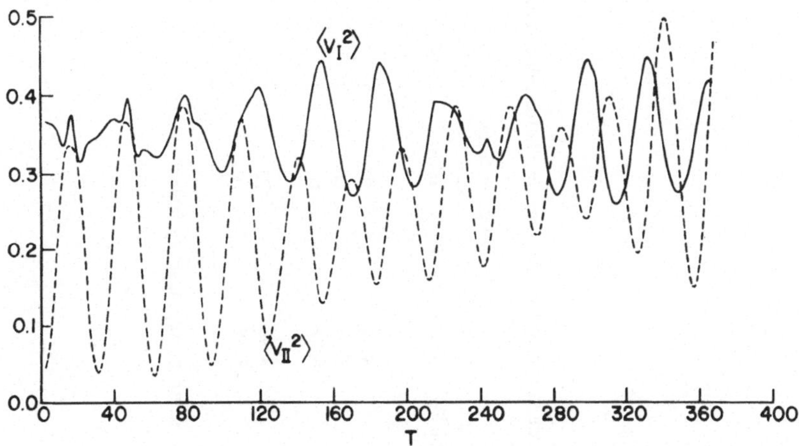

Fig. 2d. The evolution of the mean square velocities $\langle v_I^2 \rangle$ and $\langle v_{II}^2 \rangle$ for the system '4', respectively.

Since the agreement with the predicted relaxed value for $\langle v_I^2 \rangle / \langle v_{II}^2 \rangle$ improves when going to larger and larger initial values of $2E_k/E_p$, we considered initial configurations similar to those in the sequence 4, 2, 5 but with higher $2E_k/E_p$ values (cases 6 and 7). Unfortunately, when initially $2E_k/E_p \gtrsim 5$, the systems very soon become binary-like configurations and this significantly slows down the evolution of the system. Thus,

TABLE II

Summary of initial conditions and results for initial configurations of type A having different initial values of $2E_k/E_p$.

Case	Initial conditions								τ_c	Predicted $\langle v_I^2\rangle/\langle v_{II}^2\rangle$	Obtained average over the fifth* period	L/L_0 at the end of the fifth* period
	E_p	E_k	$2E_k/E_p$	$\langle v_I^2\rangle/\langle v_{II}^2\rangle$	η_I/η_{II}	A_I/A_{II}	v_0	η_I				
4	1/3	1/6	1	8.38	10	1	0.100	1/0.880	~7	0.1	1.32	4.9
5	1/3	1/3	2	8.38	10	1	0.141	1/1.240	~8.5	0.1	~0.93	11.1
2	1/3	2/3	4	8.38	10	1	0.200	1/1.760	~10	0.1	0.44	19.8
6	1/3	5/6	5	8.38	10	1	0.223	1/1.967	~11	0.1	0.53*	13.3*
7	1/3	1	6	8.38	10	1	0.244	1/2.147	~12	0.1	0.50*	14.3*

* For Cases 6 and 7, ‘fifth’ should be replaced by ‘third’.

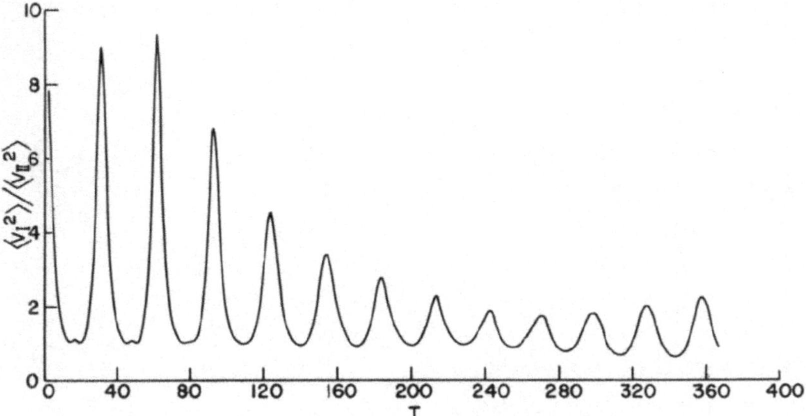

Fig. 2e. The evolution of the ratio $\langle v_{\mathrm{I}}^2 \rangle / \langle v_{\mathrm{II}}^2 \rangle$ for the system '4'.

because of technical reasons we could not continue the integration of this type of system very long. As an example, we present in Table II (case 7) and in Figure 3 the results pertaining to a binary-type system.

To summarize the results pertaining to initial configurations of type A:

(i) a general tendency of the systems to evolve toward a state with a value of the ratio of the mean square velocities predicted by Lynden-Bell's theory, that is, a *qualitative* agreement with the theory is observed;

(ii) the *quantitative* agreement with the theory improves with increasing initial value of $2E_k/E_p$;

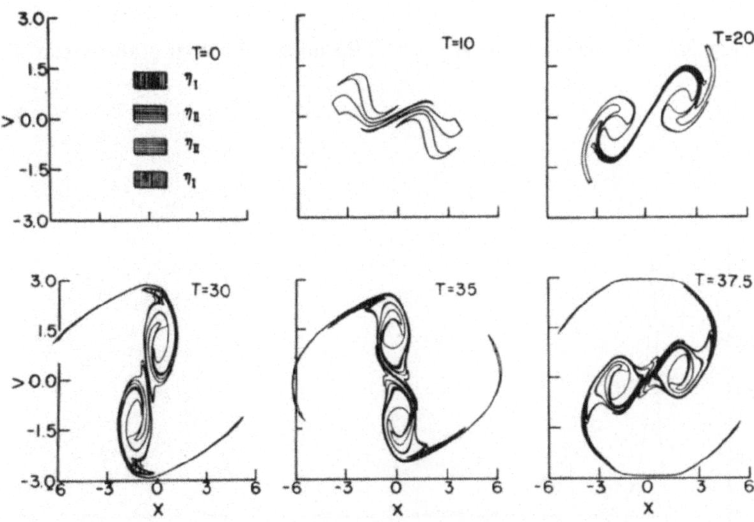

Fig. 3a. The evolution in phase space of the system '7'. A collective period, τ_c, corresponds to about 12 time units.

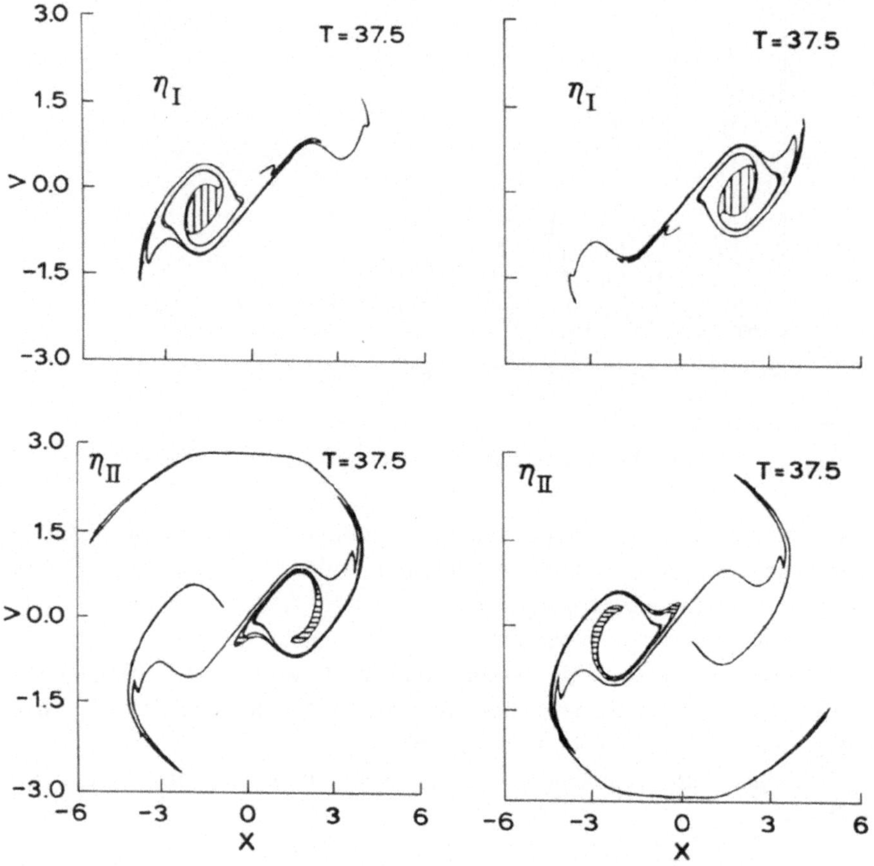

Fig. 3b. The states at time $T = 37.5$ of the initial four strips of the system '7'.

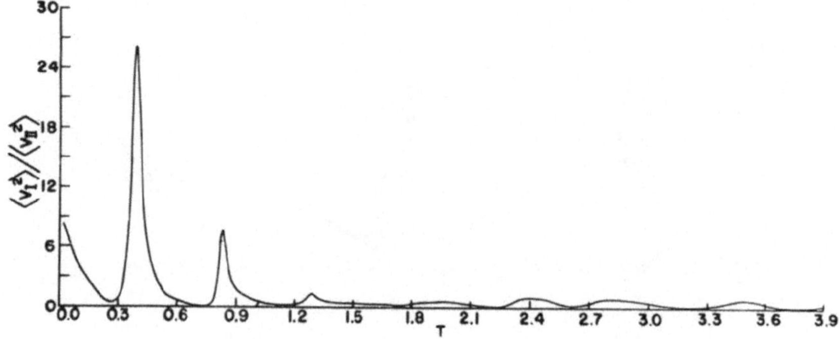

Fig. 3c. The evolution of the ratio $\langle v_I{}^2\rangle / \langle v_{II}{}^2\rangle$ for the system '7'.

TABLE III

Summary of initial conditions and results for a type '*B*' initial configuration.

Case	Initial Conditions							τ_c	Predicted $\langle v_I^2 \rangle / \langle v_{II}^2 \rangle$	Obtained average over the fifth period	L/L_0 at the end of the fifth period
	E_p	E_k	$2E_k/E_p$	$\langle v_I^2 \rangle / \langle v_{II}^2 \rangle$	η_I/η_{II}	A_I/A_{II}	η_I				
8	0.5	2/3	18	1	3	1.5	1/108	~ 26	1/3	~ 1/2	21.5

TABLE IV

Summary of initial conditions and results for a type '*C*' initial configuration.

Case	Initial Conditions							τ_c	Predicted $\langle v_I^2 \rangle / \langle v_{II}^2 \rangle$	Obtained average over the fifth period	L/L_0 at the end of the fifth period
	E_p	E_k	$2E_k/E_p$	$\langle v_I^2 \rangle / \langle v_{II}^2 \rangle$	η_I/η_{II}	A_I/A_{II}	η_I				
9	0.35	0.44	2.5	2.5	4	1/4	1/2.24	~ 9	0.25	~ 1.1	9.5

(iii) a *qualitative* dependence on the $\eta_{\rm I}/\eta_{\rm II}$ value as predicted by theory is also observed.

GROUP B

As opposed to group A, group B includes systems whose heavy phase is surrounded

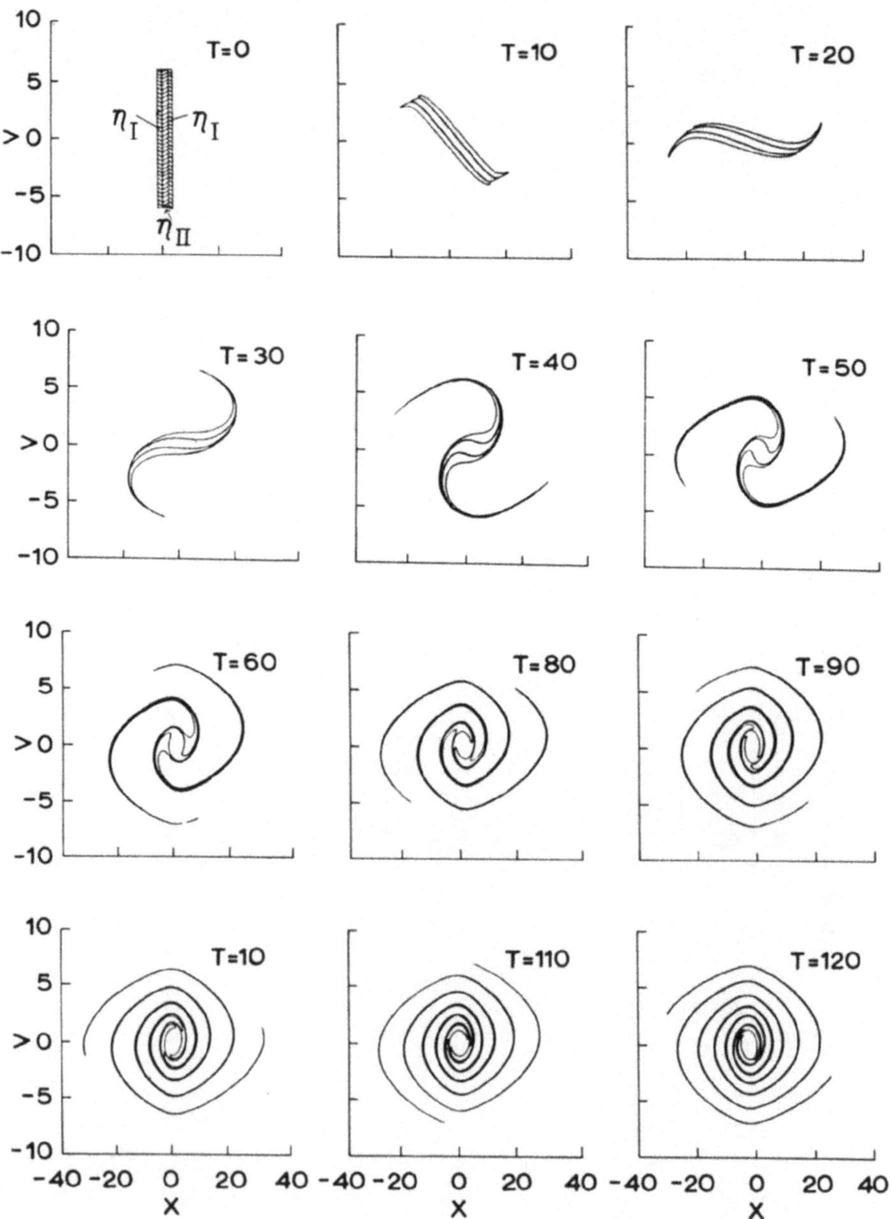

Fig. 4a. The evolution in phase space of the system '8'. A collective period, τ_c, corresponds to about 26 time units.

by a light phase. As an example, case 8 (Figure 4a–4e) represents a system with $\eta_{\rm I}/\eta_{\rm II}=1/3$ and $\{\langle v_{\rm I}^2\rangle/\langle v_{\rm II}^2\}_{T=0}=1$. The other characteristics are given in the first line of Table III. If Lynden-Bell's prediction is to be true, then an asymptotic value of the mean square velocity ratio of 0.33 is to be found. The system has been integrated for several collective periods, and the time evolution of its relevant characteristics are given in Figures 4a to 4e. After about five periods, the quantity $\langle v_{\rm I}^2\rangle/\langle v_{\rm II}^2\rangle$ oscillates about a value 0.5, which, though different from the predicted 0.33, again indicates a *qualitative* agreement with Lynden-Bell's prediction.

It might be that the better agreement with the theory in this case is due to the fact that this type of configuration starts with a state closer to the relaxed one than type A.

GROUP C

We here present the results for an initial configuration which is different from both those in Groups A and B. Thus, case 9 (see Figures 5a to 5e and Table IV) represents

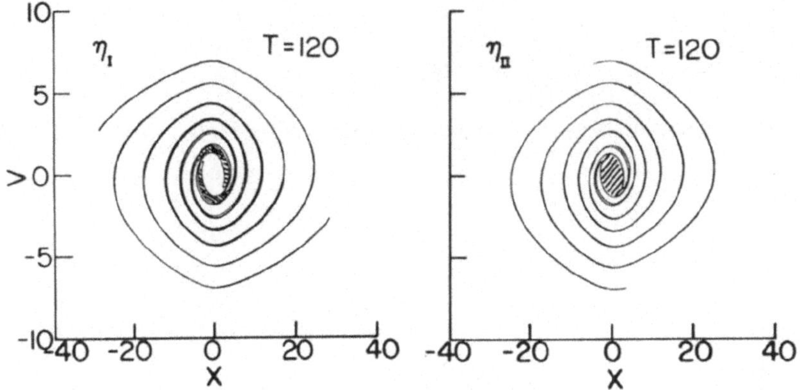

Fig. 4b. The states at time $T=120$ of the phase space regions having initially densities $\eta_{\rm I}$ and $\eta_{\rm II}$, respectively.

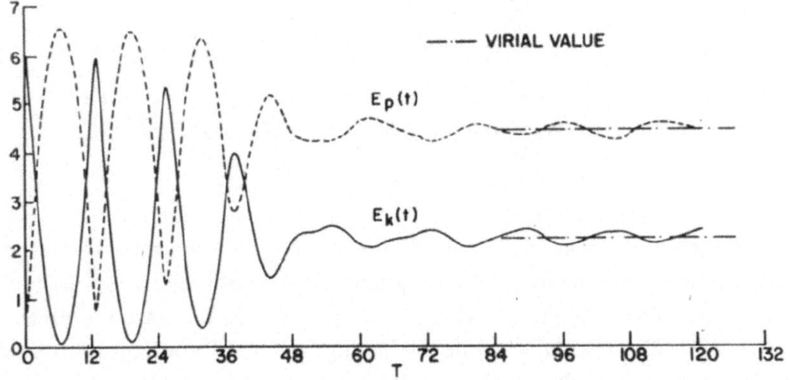

Fig. 4c. The evolution of the total kinetic and potential energies of the system '8', respectively.

a symmetric (in x, v) initial configuration consisting of seven heavy-density (η_{I}) squares surrounded by a lighter density (η_{II}); $\eta_{\mathrm{I}}/\eta_{\mathrm{II}}=4$. The initial value of $2E_k/E_p$ is about 2.5. A predicted value of 0.25 for the relaxed quantity $\langle v_{\mathrm{I}}^2\rangle/\langle v_{\mathrm{II}}^2\rangle$ is therefore expected. In fact, after about five collective periods, the system indicates (see Figures 5a and 5b)

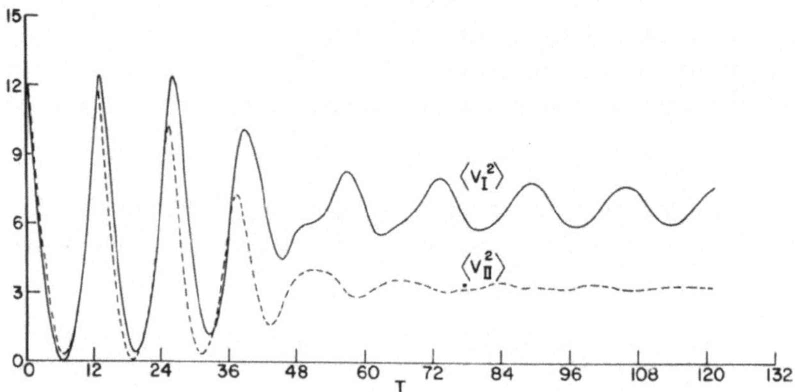

Fig. 4d. The evolution of the mean square velocities $\langle v_{\mathrm{I}}^2\rangle$ and $\langle v_{\mathrm{II}}^2\rangle$ for the system '8', respectively.

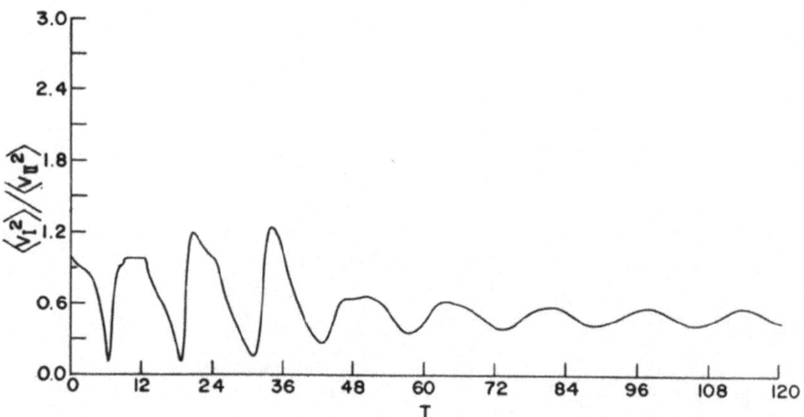

Fig. 4e. The evolution of the ratio $\langle v_{\mathrm{I}}^2\rangle/\langle v_{\mathrm{II}}^2\rangle$ for the system '8'.

that the major part of the areas belonging to η_{I} or to η_{II} did not mix. Thus, the system exhibits an oscillatory behaviour with weak tendency toward relaxation, as can be seen from Figures 5c to 5e. The initial value of $\langle v_{\mathrm{I}}^2\rangle/\langle v_{\mathrm{II}}^2\rangle$ (2.5) decreased to an average of about 1.10. Actually, case 9, in spite of having an initial configuration completely different from the relatively unrelaxed case 4 previously considered, exhibits rather similar features.

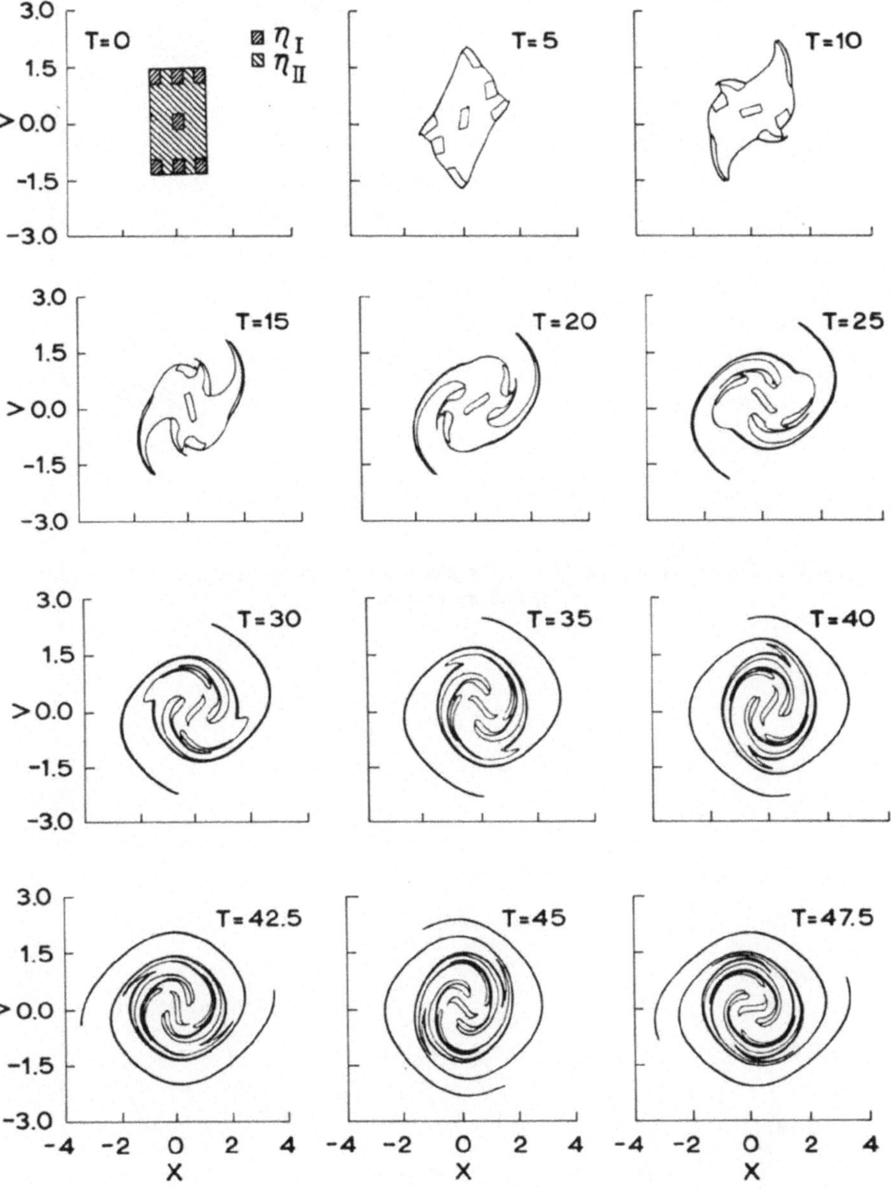

Fig. 5a. The evolution in phase space of the system '9'. A collective period corresponds to about 9 time units.

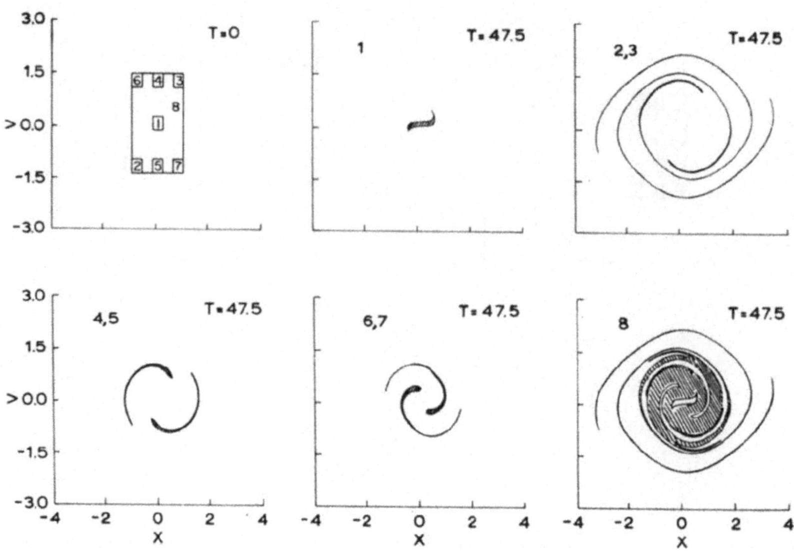

Fig. 5b. The states at time 47.5 of the phase space regions having initially densities η_{I} and η_{II} respectively.

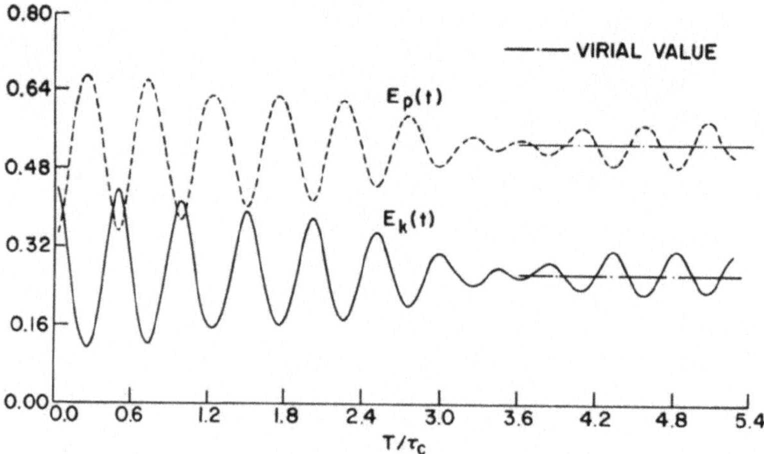

Fig. 5c. The evolution of the total kinetic and potential energies of the system '9', respectively.

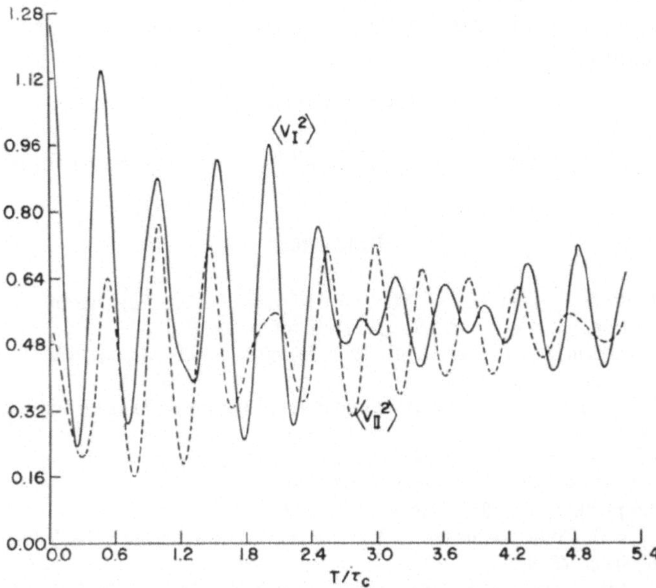

Fig. 5d. The evolution of the mean square velocities $\langle v_{\mathrm{I}}^2\rangle$ and $\langle v_{\mathrm{II}}^2\rangle$ for the system '9', respectively.

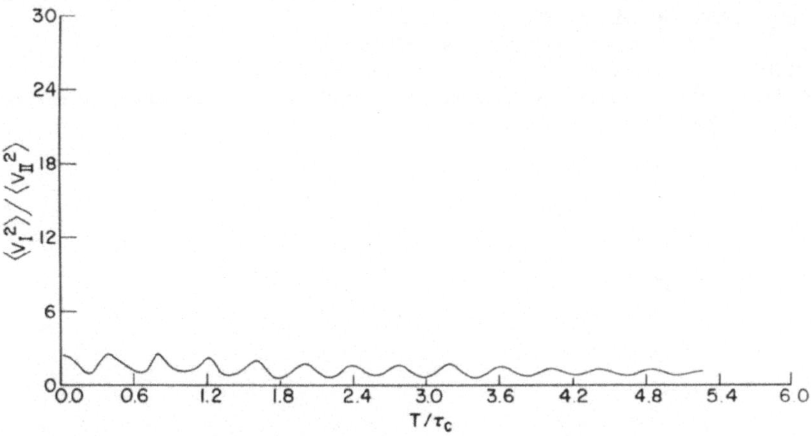

Fig. 5e. The evolution of the ratio $\langle v_{\mathrm{I}}^2\rangle/\langle v^2_{\mathrm{II}}\rangle$ for the system '9'.

4. Summary

The investigation of the evolution of one-dimensional *two*-phase-space density systems indicates a *qualitative* agreement with Lynden-Bell's statistical-mechanical prediction. This conclusion is valid for all types of initial configurations considered in this paper

The agreement is found to improve with increasing initial value of $2E_k/E_p$, i.e., with increasing degree of violent relaxation (leading to a corresponding degree of mixing between the two phases). A *qualitative* agreement concerning the dependence on the ratio $\eta_{\mathrm{I}}/\eta_{\mathrm{II}}$ is also found. Further investigations, especially for type B configurations, are required.

Acknowledgement

This work was supported by grant SFC-8-7006 from the Smithsonian Institution, Washington, D.C.

References

Cuperman, S., Goldstein, S., and Lecar, M.: 1969, *Monthly Notices Roy. Astron. Soc.* **143**, 209.
Cuperman, S., Harten, A., and Lecar, M.: 1972, this volume, p. 276.
Goldstein, S., Cuperman, S., and Lecar, M.: 1969, *Monthly Notices Roy. Astron. Soc.* **146**, 161.
Henon, M.: 1964, *Ann. Astrophys.* **27**, 83.
Henon, M.: 1967, Symposium on Computer Simulation of Plasma and Many-Body Problems, NASA SP-153, p. 349.
Hénon, M.: 1968, *Bull. Astron.*, Tome III, Fascicule I, in press.
Hohl, F. and Feix, M. R.: 1967, *Astrophys. J.* **147**, 1164.
Hohl, F. and Campbell, J. W.: 1968, *Astron. J.* **73**, 611.
Lecar, M.: 1966, in G. Contopoulos (ed.), 'The Theory of Orbits in the Solar System and in Stellar Systems', *IAU Symp.* **25**, 46.
Lecar, M.: 1967, Mémoires in $-8°$ de la Société Royale des Sciences de Liège, Cinquième Série, Tome XI, 227.
Lecar, M. and Cohen, L.: 1967, Symposium on Computer Simulation of Plasma and Many-Body Problems, NASA SP-153, p. 299.
Lecar, M. and Cohen, L.: 1967, Symposium on Computer Simulation of Plasma and Many-Body Problems, NASA SP-153, p. 309.
Lecar, M.: 1968, *Bull. Astron.*, Tome III, Fascicule I, in press.
Lynden-Bell, D.: 1967, *Monthly Notices Roy. Astron. Soc.* **136**, 101.
Roberts, K. V. and Berk, H. L.: Symposium on Computer Simulation of Plasma and Many Body Problems, NASA SP-153, 91 (1967).

NUMERICAL EXPERIMENTS WITH A ONE-DIMENSIONAL GRAVITATIONAL SYSTEM BY AN EULER-TYPE METHOD

G. JANIN

Summary. Evolutions of one-dimensional gravitational water-bag models are computed with an Euler-type method. More accurate than the Lagrange-type one, this method shows the separation of the system into a stable core of uniform phase density and a halo formed by thin filaments.

If our systems have a space extension greater than the Jeans length, the fragmentation into subsystems due to the Jeans instability appears only when the initial distribution of points is irregular.

A more detailed paper is published in *Astron. Astrophys.* **11** (1971), 188.

M. Lecar (ed.), Gravitational N-Body Problem, 311. *All Rights Reserved*

N-BODY PROBLEM AND GAS DYNAMICS IN ONE DIMENSION

F. NAHON

Observatoire de Paris

Abstract. The aim of this article is to study the behavior of selfgravitating, one dimensional, systems from the point of view of Gas Dynamics.

The gas associated with Liouville's equation behaves like a perfect gas of index 3. The entropy is linked with the central moment of order 3 of the frequency function.

The evolution of the water-bag model is described by a partial differential equation which generalizes that of Riemann for the propagation of waves.

The stability of the stationary solution for the water-bag model is investigated with the help of the Virial Theorem.

The paper has appeared in *Ann. Inst. Henri Poincaré* **14** (1971), 249–84.

M. Lecar (ed.), Gravitational N-Body Problem, 312.

PART III

NUMERICAL EXPERIMENTS AND ANALYTICAL TREATMENTS IN PLASMA PHYSICS

COMPUTER SIMULATION OF PLASMAS

JOHN M. DAWSON

Plasma Physics Laboratory, Princeton University, Princeton, New Jersey 08540, U.S.A.

During the last few years computer simulation of plasma has developed quite extensively, with many groups springing up in various places. Numerous approaches are reported in 'Methods of Computational Physics' (Vol. **9**, *Plasma Physics*, 1970). I shall not attempt to summarize all this work, but shall concentrate on the work that has been carried out at Princeton and at the Naval Research Laboratory, and work with which I have been associated. To a large extent this work has been devoted to electrostatic particle models. I believe that this has been the most successful model to date, and certainly most of the work has been devoted to it. There is also some work on the Fourier-Fourier transform method for solving the Vlasov equation which I should like to mention, and particularly I should like to show some comparison of the results with those from the particle method.

I should like to start with describing a number of techniques used in the simulation of plasmas, and then I will show a few examples of results.

With the electrostatic particle model we attempt to model a plasma by duplicating nature; i.e., by following the actual orbits of a set of particles interacting through electrostatic forces:

$$m_i\ddot{\mathbf{r}}_i = \sum_j \mathbf{F}_{ij}\,. \tag{1}$$

$$\mathbf{F}_{ij} = \frac{q_i q_j (\mathbf{r}_i - \mathbf{r}_j)}{|\mathbf{r}_i - \mathbf{r}_j|^3} \quad 3\,\text{D} \tag{2}$$

$$\mathbf{F}_{ij} = \frac{q_i q_j (\mathbf{r}_i - \mathbf{r}_j)}{|\mathbf{r}_i - \mathbf{r}_j|^2} \quad 2\,\text{D} \tag{3}$$

$$\mathbf{F}_{ij} = \frac{q_i q_j (\mathbf{r}_i - \mathbf{r}_j)}{|\mathbf{r}_i - \mathbf{r}_j|} \quad 1\,\text{D} \tag{4}$$

r_{ij}

We can do this in one, two, or three dimensions. Now, if we try to follow the motion of, say, N particles, then for each particle we must compute the force due to all other N particles and we must repeat this for each of the N particles. Thus, we have $N^2/2$ terms of the form shown to evaluate. This is completely impractical for more than a few hundred or perhaps a few thousand particles, except in one dimension, where the force is independent of distance, and depends only on the ordering of the particles i.e., depending on which particle precedes the other. Thus, if we tried to do the simplest, most straightforward thing, we could at most follow the motion of a few thousand particles, and even then the program would run very slowly.

M. Lecar (ed.), Gravitational N-Body Problem, 315–336.

In addition, such a program would suffer from an important physical effect. Because the model is composed of discrete particles, such a model exhibits collisional effects due to individual encounters between particles. Now the collision time in a plasma is given by the following formula:

$$\omega_{p_e}\tau = \begin{cases} n\lambda_D^3/\ln \Lambda & 3\,\mathrm{D} \\ n\lambda_D^2 & 2\,\mathrm{D} \\ n\lambda_D & 1\,\mathrm{D} \end{cases} \tag{5}$$

$$\omega_p^2 = \frac{4\pi n e^2}{m}, \qquad \lambda_D = \frac{v_T}{\omega_p}. \tag{6}$$

ω_p is the plasma frequency; it characterizes collective oscillations of the plasma and is a basic property of the plasma. If the model is to behave like a plasma at all, then $\omega_p\tau$ must be much greater than 1 – say, of the order of 100. Further, λ_D is the characteristic distance for the plasma. Over distances larger than this the plasma responds collectively, while for shorter distances the response is essentially that of a collection of individual particles. If we are to have a plasma at all, it must be many Debye lengths across (actually the size has to be large compared to $2\pi\lambda_D$) – say 50. Thus, to do a minimal calculation with point particles would require several hundreds of thousands of particles and would be impossible, or at any rate very expensive with todays computers.

What we should like to do is to reduce the collisional effects and at the same time speed up the calculation of the force. Fortunately, there is a technique which will accomplish both of these things simultaneously. This is what is known as the finite-size particle model. The collisional effects in a plasma are primarily due to encounters between particles at distances of less than a Debye length. Due to the singular nature of the force, close encounters lead to rapid variations of the force and to scattering of the particles. We should like to reduce the force between particles when they are close together, yet retain its Coulomb character when they are far apart. In this way we could retain the large scale collective motions or plasma oscillations while eliminating the short-range collisional effects. One way to do this is to use finite-size particles or charged clouds instead of particles. Such charged clouds interact via Coulomb forces when they are separated by large distances, but the force falls off to zero as they interpenetrate each other, as shown in Figure 1.

It is possible to carry out the statistical mechanics for systems of such particles and to compute how the finite size influences the collision rate. Such theories have been worked out by Okuda (1970), and Langdon and Birdsall (1970), and they show that substantial reductions in the collision frequency can be achieved by this means. For two-dimensional systems, using particles with radii of one Debye length, one can achieve roughly an order of magnitude reduction in collision rate. (I shall show this later). Particles of this size do not materially alter the behavior for wavelengths greater than 2π Debye lengths. An order of magnitude gain is equivalent to using an order of magnitude larger number of particles, or to increasing ones computing budget by an

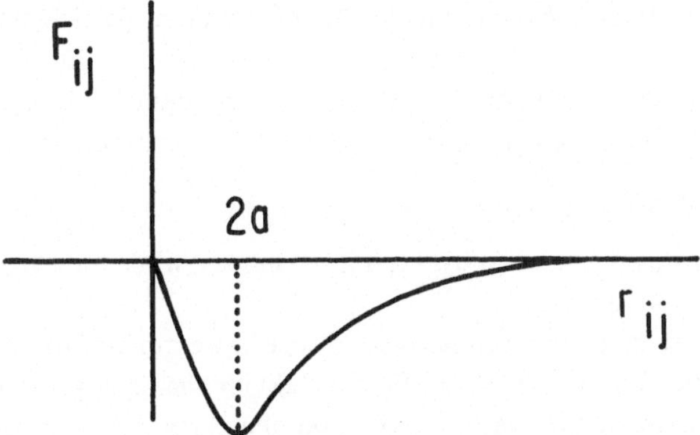

Fig. 1. Force between two finite-size charged particles.

order of magnitude. One can reduce the collision rate still further by using larger particles. However, this can start to influence the dispersion relation for long wavelength waves, and, in particular, it can alter their group velocity significantly, so one must proceed with some caution. For three-dimensional models the gains are even larger. This is fortunate, since three-dimensional problems are difficult and at present strain the state of the art.

If one uses finite-size particles, then one cannot resolve density fluctuations with wavelengths smaller than the size of the particle. This puts an upper limit on the wave numbers that it makes any sense to keep; this limit is given by

$$k_{\max} \approx 1/a\,. \tag{7}$$

To illustrate, consider a charged particle with a Gaussian distribution of charge

$$\varrho = \frac{\sigma}{(2\pi a^2)^{1/2}}\, e^{-(x-x_i)^2/(2a^2)}\,. \tag{8}$$

I have written this in one-dimensional form for simplicity σ is the charge on the particle, x_i is the position of its center, and a is its radius. The electric field due to this particle is obtained from Poisson's equation

$$\nabla \cdot E = 4\pi\varrho\,. \tag{9}$$

Fourier analyzing gives:

$$E(k) = \frac{i4\pi\,\varrho(k)\,\mathbf{k}}{k^2} = \frac{4\pi\mathbf{k}}{k^2}\, e^{ik\cdot x_i} e^{-k^2a^2/2}\,. \tag{10}$$

If ka is large, $e^{-k^2a^2/2}$ is very small and we can neglect such terms.

For an arbitrarily shaped particle we have

$$E(k) = -\frac{4\pi\mathbf{k}}{k^2}\, e^{ik\cdot x_i} S(k)\,,$$

where S is a form factor. $E(k)$ is equal to that for a point particle multiplied by the form factor.

There is also a longest wavelength that must be kept, namely, the size of the system. To solve for the electric field we need only keep a finite discrete set of k's:

$$k = 2\pi n/L, \qquad n = 1, \ldots, L/a. \tag{11}$$

For a two-dimensional case we must keep L^2/a^2 modes, while for a three-dimensional case we require L^3/a^3 modes.

Now, as we have many charged particles of size a, we can approximate their charge distribution by the following method. First, subdivide the region by a uniform spatial grid. Next, approximate the charge distribution of a particle by a charge plus a dipole (both made up of finite-size particles) at the nearest grid point. If the particle is larger than the grid spacing, then this is a very good approximation. One can, of course, add in quadrupole and higher order corrections. However, the computation time increases rapidly as one goes to higher approximations, and I doubt if it is worthwhile to go beyond the dipole order. If one stops at the first step and leaves out the dipole approximation, then we get what is called the nearest grid point approximation (Hockney, 1966) and the computation is very fast (about twice as fast as the dipole approximation). In many cases this is good enough.

Another technique, known as charge sharing (Birdsall and Fuss, 1968; Harlow, 1964; and Morse, 1970) has been used by a number of authors. If one expands the charge density to dipole order, then one can show that this method and the dipole are equivalent.

We are now in a position to rapidly calculate the electric field. Consider first the field due to the charges on the grid points. Treat these as if they are point particles located at the grid points. This constitutes a charge density due to charges placed on a uniform grid and one can solve for $E(k)$ by making use of fast Fourier transform methods (Gentlemen and Sande, 1966). One then obtains the $E(k)$ due to the extended particles by multiplying by the form factor $S(k)$. One can do a similar thing for the dipole contributions. One can then compute the field and its derivative on the grid points. One interpolates the field to find the force on a particle between grid points.

The time required to compute the fields for M grid points is proportional to

$$M \ln M. \tag{12}$$

The time required to put the charges on the grids, to interpolate the forces on the particles, and to move the particles, are proportional to N. If M is much smaller than N, then the force calculation is much quicker than for point interacting charges.

Generally M will be much smaller than the number of particles because the spatial modes must only give an adequate representation of the charge density distribution whereas the particles must represent the full distribution function in phase space, i.e., both positions and velocities.

Using the above technique we have written codes which at the dipole level require about 10 μsec per particle per time step in two dimensions. If one drops the dipole corrections and retains only the nearest grid point approximation the speed is roughly doubled. For many problems this last approximation has proven to be adequate.

This is the basic method used. There are a couple of important modifications which I should also like to mention. The first of these is the so-called quiet or ordered start. This technique has been most extensively developed by Jack Byers and Grewal, (1970). Basically, we can describe this technique as follows. Let me restrict myself to one dimension for simplicity. Consider the phase plane for such a model (Figure 2). Divide

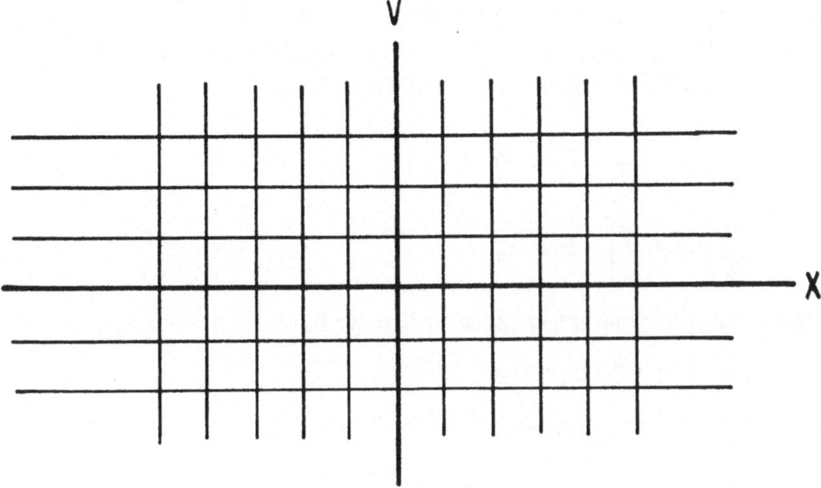

Fig. 2. Grid in phase space.

this phase plane into a number of cells. The initial state of the plasma is given by its distribution function f. We wish to approximate this by inserting the appropriate number of particles in each cell. One method for doing this has been to use random numbers that have the desired distribution. However, if we do this the fluctuation or the deviation from the desired f, is quite large compared to those that would be found in a real plasma. The fluctuations in f in any cell will be of the order of the square root of the number of particles to be expected in that cell. Since we will at most have a few hundred particles in one cell we will have fluctuation of the order of 10% or more which for many problems is unacceptable.

We do not have to start with such a random distribution of particles. We can instead put precisely the number of particle desired in each cell and thus avoid these fluctuations. Starting the system out in this way gives a quiet start.

This method works all right for regions where there are many particles per cell. However, it is clear that in regions where there are few particles per cell, or even less than one per cell, this method will also fail. This problem is serious only if there are regions in the phase space which are important to the problem but in which f is small. One example of such a situation is the instability associated with a low-density beam propagating through the plasma. For example, if we had a beam whose density was

10^{-3} of the main plasma, then even if we used 10^5 particles the beam would be represented by only 100 particles, which would not be very good. A way around this deficiency is to use a number of different types of particles with different charges and masses but all with the same charge-to-mass ratio. One can show that if one does this, then at least to the Vlasov approximation the system behaves the same. We have investigated this and the technique seems to work quite well. This more or less summarizes the techniques used in the particle model.

I should also like to mention one other technique for solving the Vlasov equation. This is the Fourier-Fourier transform method. It was originated by George Knorr (1963) and has been developed to a high degree during the last year by Denavit (1969) at the Naval Research Laboratory. One starts with the Vlasov equation for f and E

$$\frac{\partial f}{\partial t} + v\frac{\partial f}{\partial x} - \frac{eE}{m}\frac{\partial f}{\partial v} = 0 \tag{13}$$

$$\frac{\partial E}{\partial x} = -4\pi e \int f \mathrm{d}v - n_0 . \tag{14}$$

First, transform the equation in space so that we have

$$f(x, v, t) = \sum_n f_n(v, t)\, e^{inkox}, \tag{15}$$

$$E(x, t) = \sum_n E_n(t)\, e^{inkox}, \tag{16}$$

$$\frac{\partial f_n}{\partial t}(v, t) + ink_0 v f_n + \frac{e}{m}\sum_l E_l \frac{\partial}{\partial v} f_{n-l}(v, t) = 0, \tag{17}$$

$$ik_0 n E_n = -4\pi e \int f_n \,\mathrm{d}v . \tag{18}$$

Next, Fourier-analyze in velocity.

$$f_n(v, t) = \frac{1}{2\pi}\int_{-\infty}^{\infty} F_n(y, t)\, e^{ivy}\,\mathrm{d}y ; \tag{19}$$

$$F_n(y, t) = \int_{-\infty}^{\infty} f_n(v, t)\, e^{-ivy}\,\mathrm{d}v . \tag{20}$$

Substituting into the Fourier transformed Vlasov equation gives

$$\frac{\partial F_n}{\partial t}(y, t) + nk_0\frac{\partial F_n(y, t)}{\partial y} + iy\sum E_l(t)\, F_{n-l}(y, t) = 0 . \tag{21}$$

The first two terms are the total derivative of F_n along the characteristic lines $y = y_0 + {}$ $+ nk_0 t$. This equation is solved by integrating along the characteristics. Fast-Fourier transform methods are used for evaluating the convolution (Denavit and Kruer, 1970).

Other transform methods have been used. Fourier-Hermite transforms (Armstrong, 1967; Sadowski, 1967; and Grant and Feix, 1967) have been a popular method. However, I believe the Fourier-Fourier method has been the most successful method and the other methods have some fundamental difficulties for following the motion of the system for long times.

Now I should like to present a number of examples of things we have done with our plasma codes.

1. Reversibility

One test that can be applied is that of reversibility. Such a test was made of one of the codes by running out and reversing a violent two-stream instability. We use 10^4 particles in a 1-D periodic system 256 Debye lengths long. The result is shown in Figure 3. The top of the figure shows phases pace at $\omega_{pe}t = 5$. We set the two Maxwellian beams, separated in velocity by 6 thermal velocities. The instability exponentiates many times, until the wave energy is -5% of the total energy. Even so, the code con-

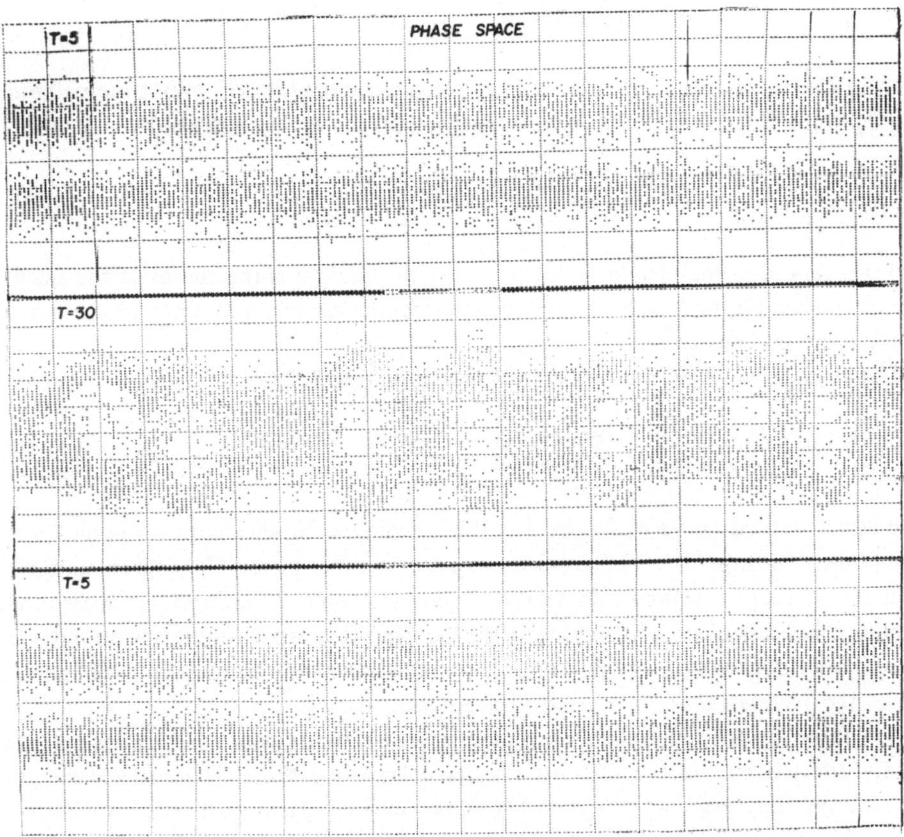

Fig. 3. Phase space at three different times.

served energy to within 0.03%. The center of the figure shows phase space at $\omega_p t = 35$. The instability has saturated; vortices have formed and indeed are even beginning to coalesce. At this time we reverse the velocities of the particles. Finally, the bottom of the figure shows phase space after the system has traced itself back in time to $\omega_{pe} t = 5$. The top and bottom picture are identical in detail (reversing the velocities interchanges the top and bottom beams). This test gives a good check on the numerical stability of the code as well as checking the accuracy of the programs.

2. Effects of Finite-Size Particles

A. INVESTIGATIONS OF FLUCTUATIONS ABOUT THERMAL EQUILIBRIUM

One of the first problems we investigated with the finite-size particle model was that of fluctuations about thermal equilibrium to see if they behaved as expected. Figure 4 shows a plot of the amplitude of the rms electric field fluctuations vs mode number for charge clouds with $a = 2\lambda_D$ (λ_D is the Debye length), and with $k_{max}\lambda_D$ equal to 2. The solid curve is the theoretically predicted curve for Gaussian charge clouds. This curve is predicted from the formula

$$P(E_k)\,\mathrm{d}E_k \propto e^{-\psi_k(E_k)/KT}\,\mathrm{d}E_k, \tag{22}$$

where $P(E_k)$ is the probability of finding the electric field in dE_k about the value E_k, and ψ_k is the work required to create the fluctuations E_k; ψ_k is given by

$$\psi_k = \frac{E_k^2 L}{16\pi}\left(1 + k^2\lambda_D^2 e^{k^2a^2}\right). \tag{23}$$

(L is the length of the system.) The first term on the right is the energy in the electric field, the second term is that required to compress the gas of cloud centres isothermally to the required density.

The average value of E^2L obtained from Equations (22) and (23) is

$$\left\langle\frac{E_k^2 L}{8\pi}\right\rangle = \frac{KT}{2\left\{1 + k^2\lambda_D^2 e^{k^2a^2}\right\}}. \tag{24}$$

The upper dashed curve in Figure 4 is that predicted for sheets, i.e., a equals zero. The points are those obtained from the numerical experiment. They agree quite well with the theoretically predicted values. There are some deviations for small mode numbers, but this is most likely due to the fact that the initial conditions do not start these modes out with energy KT, and they take a long time to relax to their thermal value (the averages used here are time averages).

As can be seen from Figure 4 the theoretical fluctuations at long wavelength are hardly affected by the use of finite-size particles while those at short wavelengths are strongly suppressed as expected. We have run other cases with different values of a and always find similar agreement. The suppression of short wavelength fluctuations reduces collisional effects.

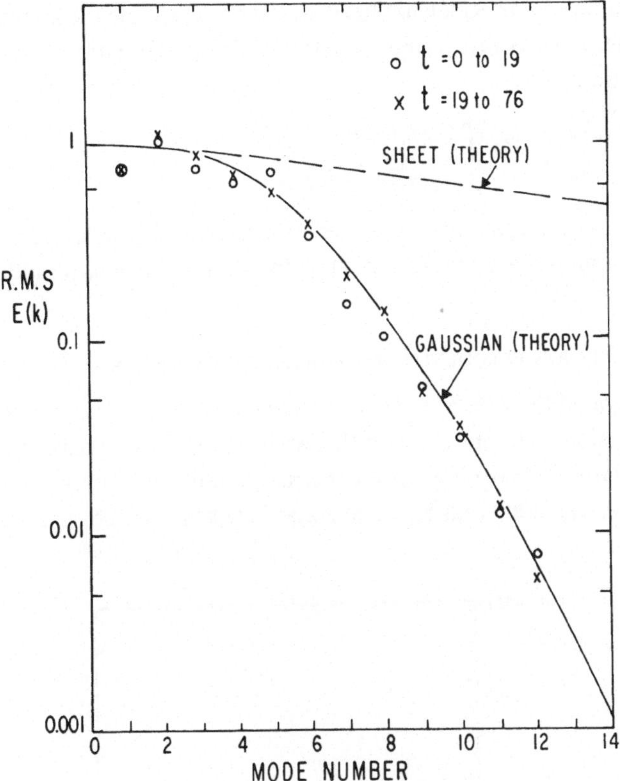

Fig. 4. rms E field vs mode number.

Another interesting thing we can do is derive the dispersion relation for the finite-size particle model. The collisionless Boltzmann equation for these particles is given by

$$\frac{\partial f}{\partial t} + v\frac{\partial f}{\partial x} + \frac{F}{m}\frac{\partial f}{\partial v} = 0, \tag{25}$$

where $f(x, v)$ is the distribution function of cloud centers and velocities. The force on the particles obtained is by integrating $E(\xi)\varrho(\xi, x)$ over ξ, x is the position of the center of the charge cloud,

$$F(x) = 4\pi\sigma^2 i \int \frac{\mathrm{d}k e^{-ikx-k^2a^2}}{(2\pi)^{1/2}k} \int f(k, v)\,\mathrm{d}v. \tag{26}$$

This is the same expression one obtains for point particles except for the factor $e^{-k^2a^2}$. If we linearize Equation (25) and Fourier analyze Equations (25) and (26) in space and time, we obtain

$$f(k, \omega) = \frac{iF(k, \omega)\,\partial f_0/\partial v}{m(\omega + kv - i\varepsilon)} \tag{27}$$

$$ikF = 4\pi\sigma^2 e^{-k^2a^2} \int f(k, v)\,\mathrm{d}v. \tag{28}$$

Here ε is a small damping which has been added to determine the direction of integration around the poles. Substituting Equation (27) into Equation (28) we obtain the dispersion relation

$$1 = \frac{4\pi\sigma^2}{mk} e^{-k^2a^2} \int \frac{\partial f_0/\partial v \, \mathrm{d}v}{(\omega + kv - i\varepsilon)}. \tag{29}$$

This is the same as the usual dispersion relation except for the factor $e^{-k^2a^2}$. Thus the long-wavelength modes are unaffected while the short-wavelength modes are strongly modified.

B. EFFECTIVE COLLISION FREQUENCY FOR FINITE-SIZE PARTICLES

We have investigated the effective collision frequency in a plasma of finite-size charged rods (two-dimensional charges). To do this we have measured the collisional damping of plasma oscillations. Working with a thermal equilibrium plasma we correlate the modes of the electric field; that is, we average $E_k(t)\, E_k(t+\tau)$ over t. The correlation

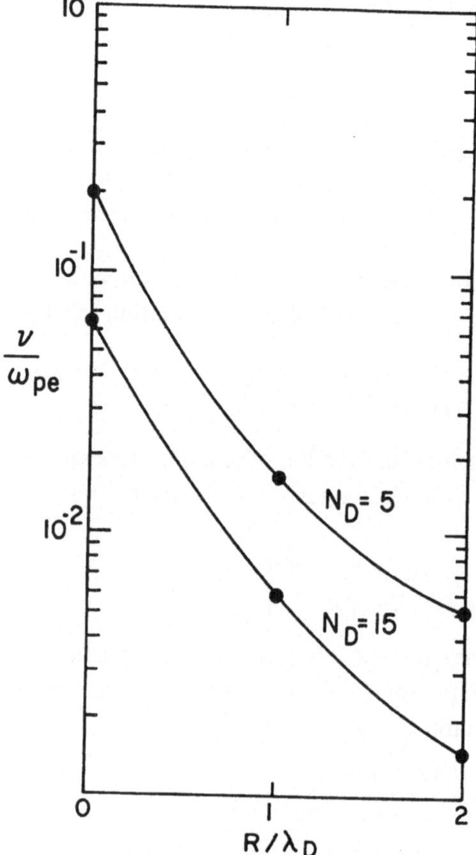

Fig. 5. Collision frequency vs particle size.

function is a slowly decaying oscillation; the frequency of the oscillation yields the frequency of the plasma mode and its decay gives the collisional damping.

For these runs we used a two-species plasma in a system $32\lambda_D$ by $32\lambda_D$ (where λ_D is the Debye length). We varied both the particle size and the total number of particles used. The results for a number of runs are shown in Figure 5, where we have plotted collision frequency versus the ratio of the particle size to its Debye length. The upper curve is for the number of particles per Debye circle equal to 5, and the lower curve for that number equal to 15. The $R=0$ points are taken from a empirical formula deduced by Hockney (1969) from a large number of runs with essentially zero-size particles. The reduction of collision frequency with increasing particle size is evident. A particle size of λ_D leads to a reduction of about a factor of 10. The magnitude of this reduction agrees reasonably with a recent theory by Okuda (1969).

3. Quiet Start and Echoes

A quiet start simply refers to beginning the calculation with very regular initial con-

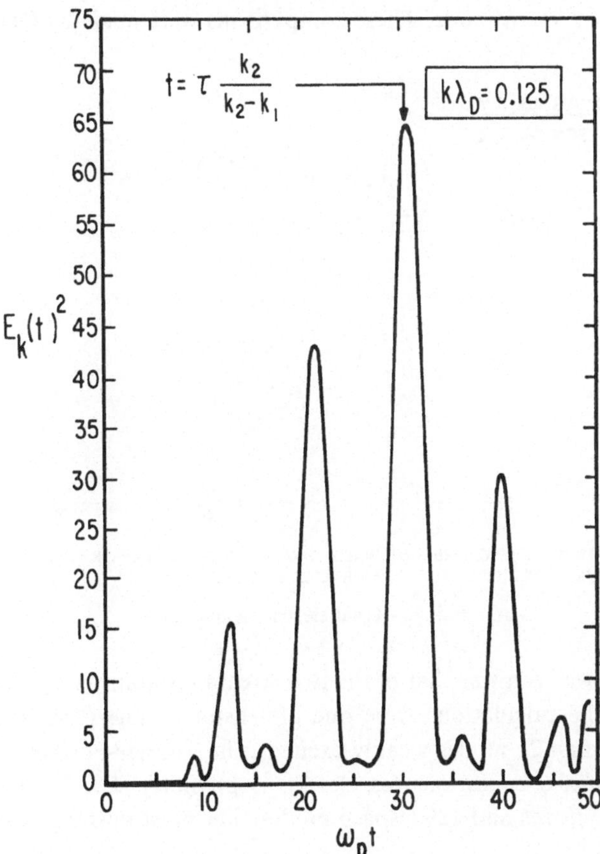

Fig. 6. The wave energy of mode $k\lambda_0 = 0.125$ vs $\omega_{pe}t$.

ditions. Basically, one uses no random numbers in the initial conditions. This leads to a marked suppression of the noise level in the plasma. Such techniques have proven useful in obtaining realistic noise levels without the use of a large number of particles (Birdsall, 1969; Mason, 1969; and Byers and Grewal, 1970).

We have observed a temporal echo using a quiet start (as is well-known, echoes are very sensitive to noise level). The result is shown in Figure 6. We use a one-dimensional system with 2×10^4 particles in a periodic system 256 λ_D (where λ_D is the Debye length). Initially the particles are uniformly spaced and given a Maxwellian velocity distribution by uniform loading (*Handbook of Mathematical Functions*, 1964). At $\omega_p t = 1$ the electric field mode, $k\lambda_D = 0.5$, is excited. At $\omega_p t = 7$ the mode $k\lambda_D = 0.625$ is excited. We then expect an 'echo'; that is, we expect to see mode $k\lambda_D = 0.125$ rise at $\omega_p t = 30$ (a second-order echo). This is just what is shown in the figure.

4. Comparison of Results from a Particle Code and a Fourier-Fourier Vlasov Code

A comparison has been made between results obtained from a particle code and from a Fourier-Fourier Vlasov code. These comparisons were made by Denavit and Kruer (1970).

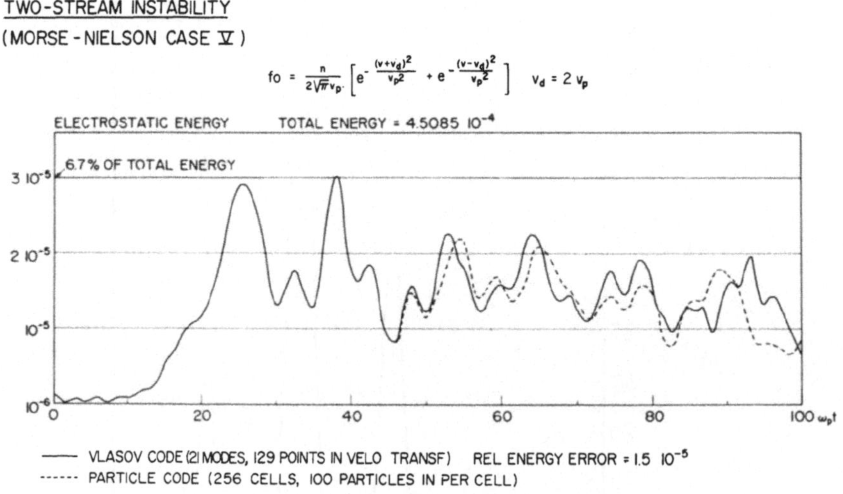

Fig. 7. Total electric field vs time.

The case that was run was that of the two-stream instability with two equally dense warm beams. The calculations were one dimensional. The particle code was quite started with the first 21 modes weakly excited. The Fourier-Fourier code was started with the same initial conditions (as closely as possible). The Fourier-Fourier code used 21 x space modes and 129 v space modes. The most unstable x space mode was mode 3.

Figure 7 shows a plot of the total electric field energy versus time for the two codes.

This energy grows up and saturates in both calculations. Up to time $\omega_p t = 50$ the results are virtually identical. After this time there is some change. However, the features even here are quite similar and probably represent some differences in the phases of the various waves in the two calculations. Such phase shifts are almost impossible to avoid, they simply arise because slight shifts in the orbits of the particles will ultimately results in their getting out of phase. When the initial conditions for the Fourier-Fourier code were altered slightly by an amount comparable to the difference for the two codes (since the two codes were different starting them in exactly the same initial condition was difficult) it lead to comparable differences in results at the latter times. The good agreement obtained here gives one confidence that these two codes are correct and that their results are meaningful.

5. Simulation of Experiments

SIDEBAND INSTABILITY

In computer experiments on one-dimensional sheet plasmas we have investigated the production of sidebands on a large-amplitude electrostatic wave. We drive a large wave into an initially Maxwellian plasma and follow the evolution of the wave spectrum. The energy in the lower sidebands (wave numbers less than that of the large-amplitude wave) exponentiates over nearly two orders of magnitude. The upper sidebands grow less (growth about an order of magnitude). These numerical results are consistent, both with a recent experiment (Wharton *et al.*, 1968) and with recent theories (Kruer *et al.*, 1969; and Goldman, 1970), predicting a sideband instability.

In a recent paper, Wharton *et al.* (1968) described the excitation of large-amplitude plasma waves by means of an electrostatic probe immersed in a warm plasma. The experiment showed the growth of sidebands to the frequency of the large-amplitude wave. The frequency separation of these sidebands was found to be proportional to the square root of the large wave amplitude. This strongly suggests that trapped particles bouncing in the potential trough at a frequency $\omega_B = [(eEk)/m]^{1/2}$ (E is the wave field, k the wave number) play an important role in the generation of these sidebands.

To explain these sidebands, we have proposed a model for a new instability resulting from particles trapped in a large-amplitude electrostatic wave (Kruer *et al.*, 1969). These particles move along with the wave, oscillating back and forth in the wave troughs. A number of physical arguments suggest that an instability is possible. For a simple model we mocked up the trapped particles as a bunched beam of harmonic oscillators streaming through a warm plasma. This model predicts the growth of sidebands on the large wave.

To further investigate the sideband production we have performed numerical experiments on an externally driven one-dimensional sheet plasma (Dawson, 1962). Beginning with a plasma having a Maxwellian distribution, we drive a single mode of the electric field in the plasma to a large amplitude, and then observe the evolution of the wave spectrum. The results agree reasonably with the real experiments and the simple model. First we will describe the experiments and then present the results in detail.

We start with a Maxwellian distribution, which is the initial distribution for the Wharton *et al.* (1968) experiment. We drove the plasma in time with an external (not self-consistent) force at a given frequency and wave number. Thus during some prescribed interval of time each particle experiences an external acceleration given by $a_i = \varepsilon \sin(kx_i - \omega t)$, where x_i is the particle position and a_i the acceleration. This driver corresponds in some degree to perturbations caused by a series of grids inserted in the plasma.

For these experiments we used a 5000-sheet system with 20 sheets per Debye length. We drove mode 10 (ten wavelengths in the system) at a frequency of $\omega = 1.09\ \omega_p$ (the Bohm-Gross value), resulting in a wave with a phase velocity of $4v_T$. The driving started at $t = 1\ \omega^{-1}$ and ended at $t = 32\ \omega^{-1}$. Three values were used for the amplitude of the driver, resulting in large-amplitude waves of 3200 kT, 2400 kT, and 1250 kT. For each driving amplitude we averaged two separate runs, with differing random numbers used for the initial conditions.

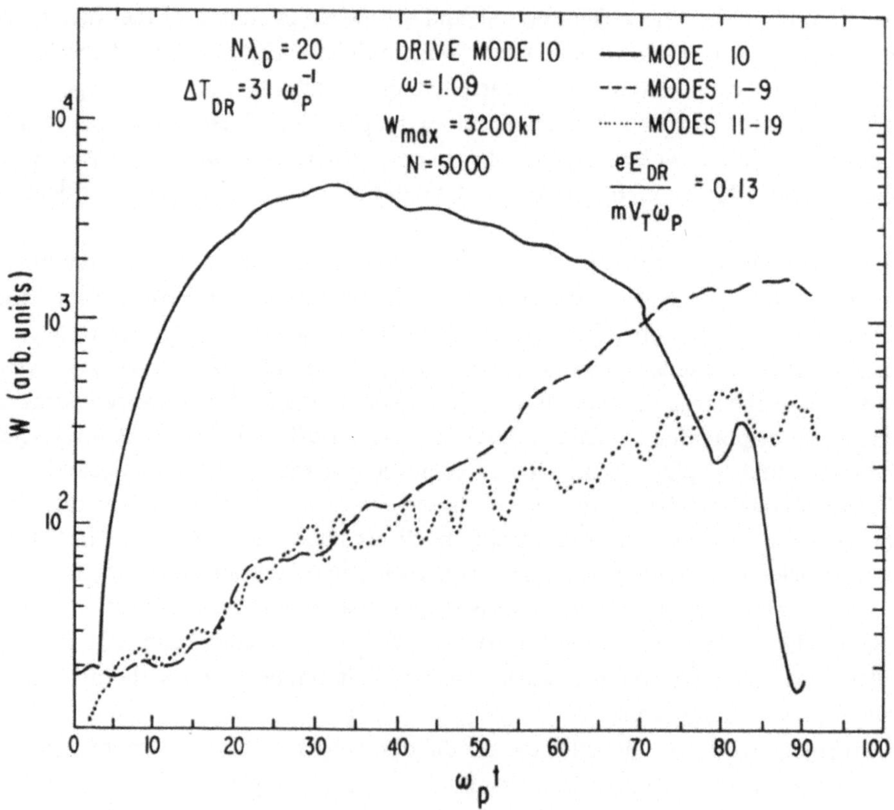

Fig. 8. The energy in the large wave and the upper and lower sidebands vs $\omega_{pe}t$ $(eE_{DR}/m\omega\, v_T = 0.13,\ \omega_p \Delta T_{DR} = 31)$.

The evolution of the wave energy in time is shown in Figure 8 for the case of the largest driver. The solid curve shows the energy in mode 10 (the large-amplitude wave), the dashed curve the total energy in modes 1 to 9, and the dotted curve the energy in modes 11 to 19. Mode 10 steadily rises to a maximum until the driving is stopped; then it steadily decays as the energy in the adjacent modes exponentiates. Note that the energy in modes 1 to 9 exponentiates over roughly two orders of magnitude. The higher k modes also grow (about a factor of ten), but the phase velocity of these modes falls in a region of negative slope of the distribution function, and hence must compete with Landau damping. As we expect, the unstable modes saturate when they become comparable in magnitude to the decaying large wave. At that time the original wave shows a catastrophic decay. Phase-space plots show that the orderly swirls of trapped particles are then broken up by the instability.

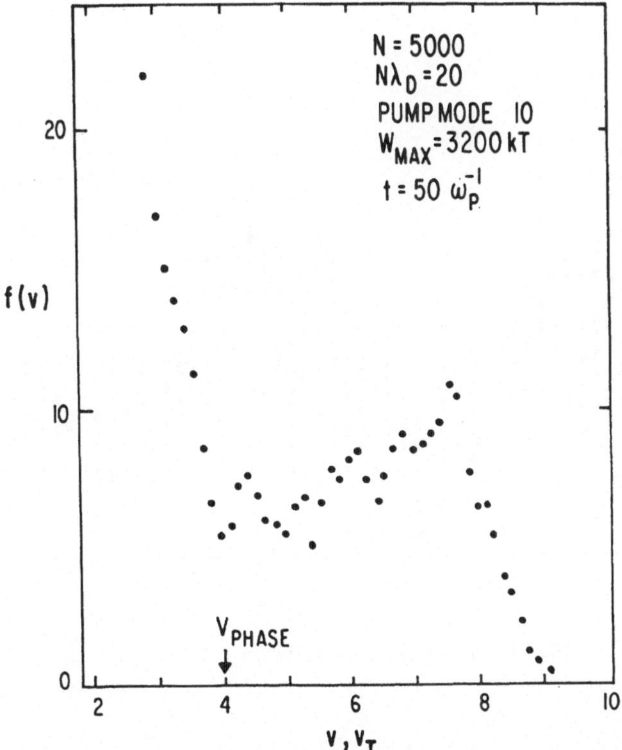

Fig. 9. The tail of the distribution function ($eE_{\mathrm{DR}}/M\omega_p v_T = 0.13$, $\omega_{pe}t = 50$).

Figure 9 shows a typical plot of the distribution function (originally Maxwellian) averaged over a plasma oscillation at about $t = 50\ \omega^{-1}$. The large wave has pulled out a long tail with an apparent bump at its upper end. A typical phase-space plot, Figure 10, shows that this long tail has a very interesting structure, corresponding to the trapped particle oscillations in phase space. The spatial modulation due to the large-amplitude wave is clearly shown. We see that the bump in the distribution function (which is a spatial average of the phase-space plot) is due to the fact that there are more particles near the maximum velocity, because of the hollow nature of the vortex. It is clear that this bump is not an ordinary one but has a strong spatial dependence. Furthermore, the strong acceleration of the particles cannot be ignored.

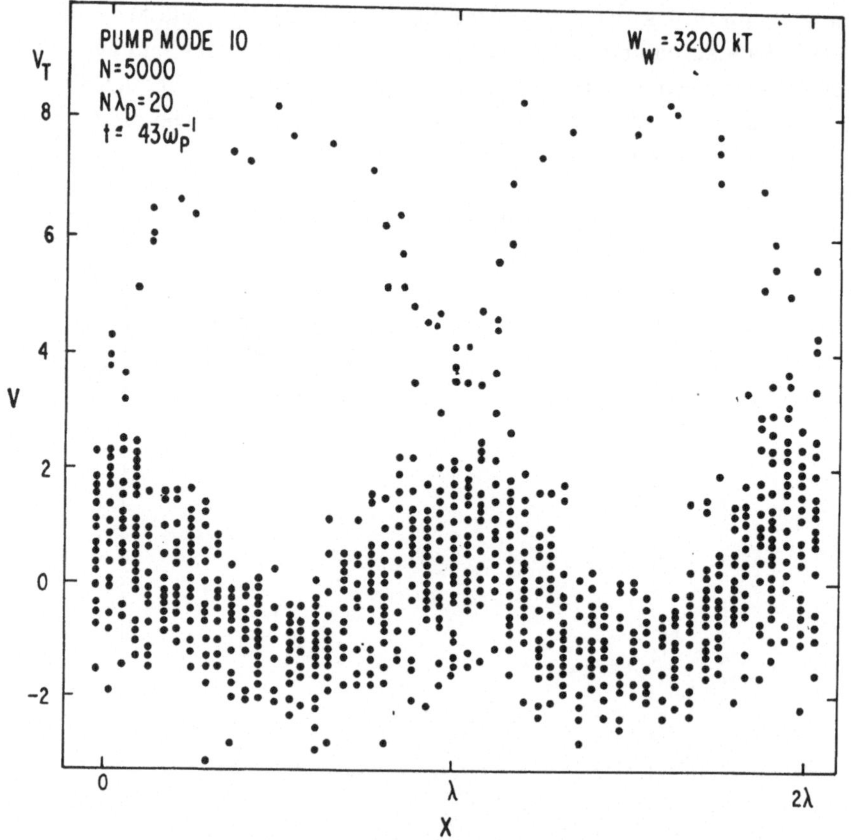

Fig. 10. Phase space ($eE_{\mathrm{DR}}/m\omega_p v_\tau = 0.13$, $\omega_{pe}t = 43$).

Hence, a treatment such as that of the simple model presented by Kruer *et al.* (1969), is more appropriate than the simple two-stream description. Note that the swirls of trapped particles are fairly circular in phase space, but the excursions of the particles are rather large (large amplitude of oscillation in the wave through). Thus we can expect only semiquantitative agreement with our small excursion theory.

The magnitude of the growth rate of the lower sidebands is roughly $\lambda = 0.03\ \omega_p$ (our simple model predicts $\gamma \sim 0.06\ \omega_p$). The large excursions mean that the bounce frequency is less, and so the smaller growth rates are reasonable. The frequencies of the unstable waves are shifted away from that of the large-amplitude wave. For example, the frequency of mode 7 is $\simeq 0.9\ \omega_p$, while that of the large-amplitude wave is $-1.09\ \omega_p$. The magnitude of the frequency shift is roughly 50% ω_B.

The results for the other two driving amplitudes are shown in Figures 11 and 12. The large wave is weaker in these two cases, corresponding to a maximum wave energy

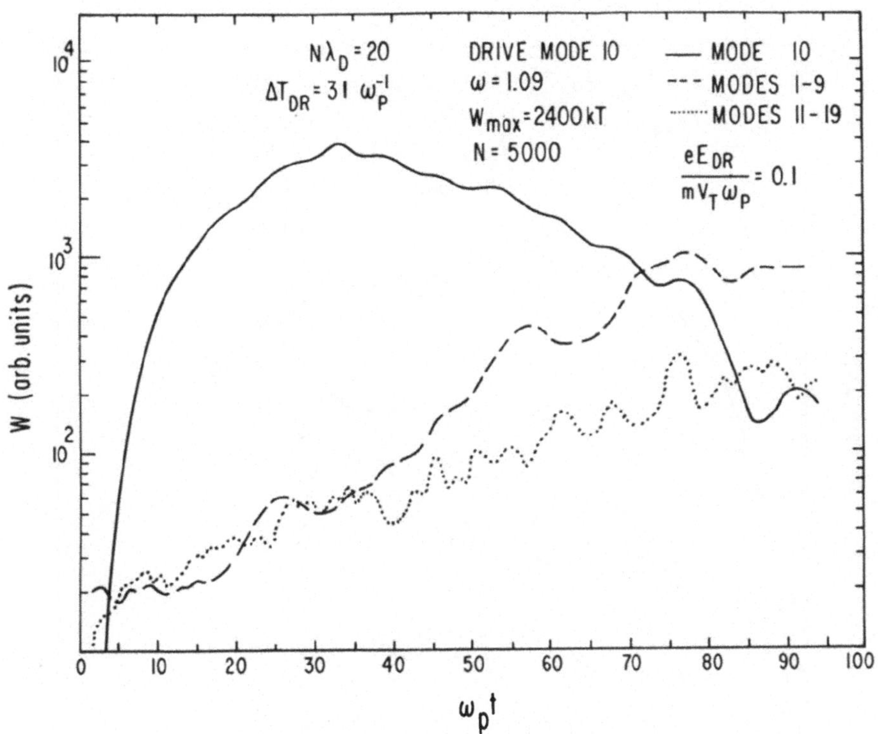

Fig. 11. The energy in the large wave and the upper and lower sidebands vs $\omega_{pe}t\,(eE_{DR}/m\omega_p v_\tau = 0.10,\ \omega_p \Delta T_{DR} = 31)$.

of 2400 kT and 1250 kT, respectively. Hence there are fewer trapped particles, and the statistical fluctuations are larger than in Figure 8. The growth rates and frequency separations are less in these weaker-amplitude cases, and scale roughly as $\sqrt{E}$. It should be noted that such a scaling is tentative, since the variation in E in these experiments is less than a factor of 2. It is interesting that for the 1250 kT case the large-amplitude wave does not experience catastrophic decay at the time that the lower sidebands acquire comparable energy.

As shown in Figures 8–11, the energy of the large-amplitude wave shows beats. However, these beats are not simply at the bounce frequency, as are those observed in the Wharton *et al.* (1968) experiment. Rather, there is mixed in a faster oscillation at roughly the plasma frequency. This faster oscillation is more characteristic of the beats found in other calculations for a distribution function with a fast particle tail (Shanny *et al.*, 1968).

Beats at the bounce frequency are due to the reversal – at roughly the bounce time – of the net imbalance of particles, taking energy from the wave and returning it. A trapped particle has an oscillation frequency in the large wave trough which depends on its energy. The trapped particles with large amplitude and smaller oscillation frequencies get out of step rather soon with the bulk of the trapped particles. They

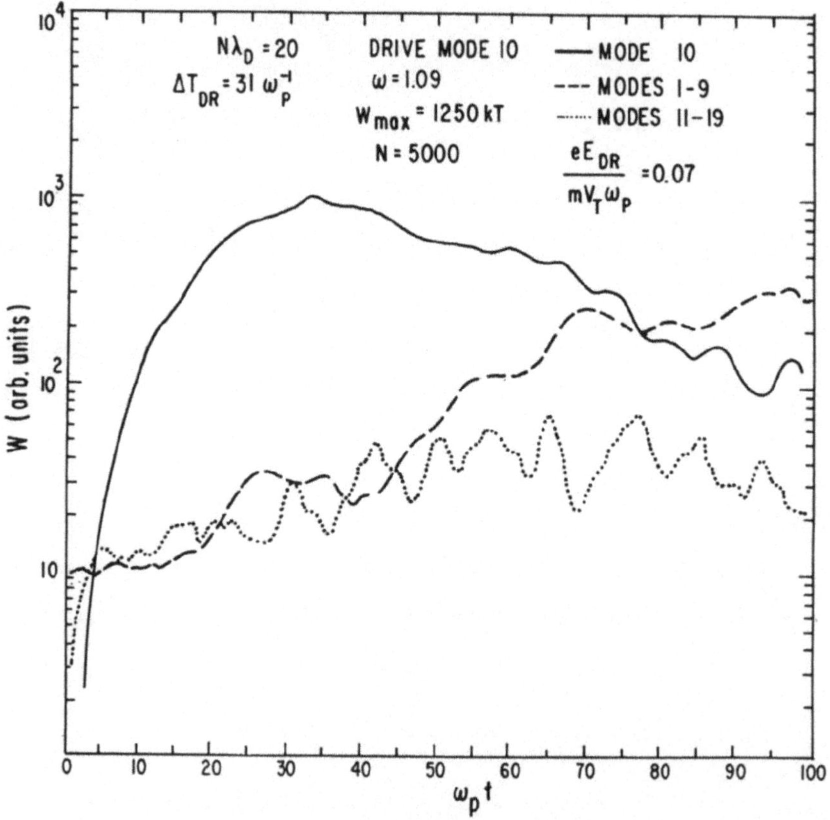

Fig. 12. The energy in the large wave and the upper and lower sidebands vs $\omega_{pe}t\,(eE_{\mathrm{DR}}/m\omega_p v_\tau = 0.07,\ \omega_p \Delta T_{\mathrm{DR}} = 31)$.

may be taking energy from the wave as other trapped particles are returning energy; hence, the beats in the large wave energy should show up most clearly for the first bounce time or two.

For these experiments we drove the large wave into the plasma rather slowly. During the driving time the trapped particles oscillated in the wave trough almost twice. The phase-space plots show that many of the trapped particles are out of step by this time. If we drive the large wave into the plasma more quickly [this, in fact, corresponds better to the Wharton *et al.* experiment (1968)], we should see the beats at the bounce frequency more clearly.

To test this hypothesis we used a 5000-particle system with 40 sheets per Debye length. We drove mode 5, which again resulted in a large wave with phase velocity $4v_T$. However, we drove the plasma three times harder for a time one-third as long. The results are shown in Figure 13. Indeed, beating of the large wave energy at roughly the bounce frequency clearly shows. When the driving is stopped, the large wave first loses one-third of its energy, and then recovers this energy in the course of a bounce

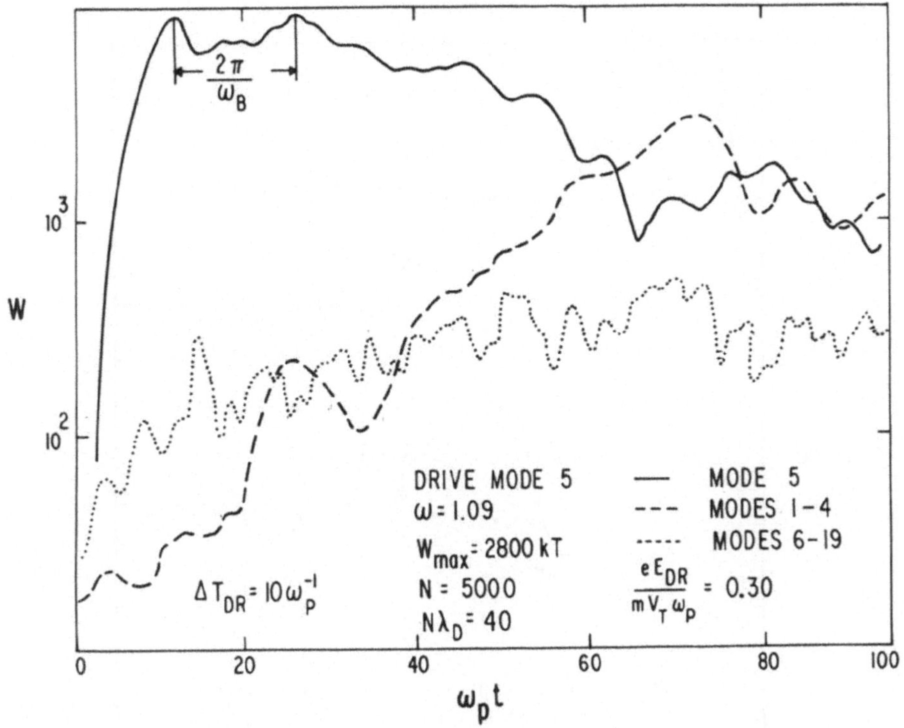

Fig. 13. The energy in the large wave and the upper and lower sidebands vs $\omega_{pe}t\,(eE_{\mathrm{DR}}/m\omega_p v_\tau = 0.3,\ \omega_p \Delta T_{\mathrm{DR}} = 10)$.

time. The next beat can be seen, although it is not so well defined. We again see the higher frequency interference. Thus, with this more rapid pumping we see the modulation of the driven wave at the bounce frequency that is characteristic of the Wharton *et al.* (1968) experiment.

Finally, we have carried out some experiments using many particles on a faster machine. A finite-size particle model (Dawson, 1962) is used instead of the sheet model A sample run is shown in Figures 14 and 15. Here we use 40000 particles in a periodic system 256 λ_D across ($n\lambda_D \simeq 150$). A large wave with energy 10800 kT and phase velocity 4 thermal velocities is driven into the plasma in 6 ω_p^{-1}. Figure 14 shows the wave energy as a function of time. The behavior – of both the main wave and side bands is essentially the same as that given by the sheet model. The beats in the large wave energy show up more clearly. Figure 15 shows a portion of phase space. Now the trapped particle swirls are quite well populated, and therefore discrete particle effects should be small.

In conclusion, we have investigated with numerical experiments the production of sidebands on the large-amplitude wave. The numerical simulation gives results consistent with both the Wharton *et al.* experiment (1968) and with recent theories predicting a sideband instability.

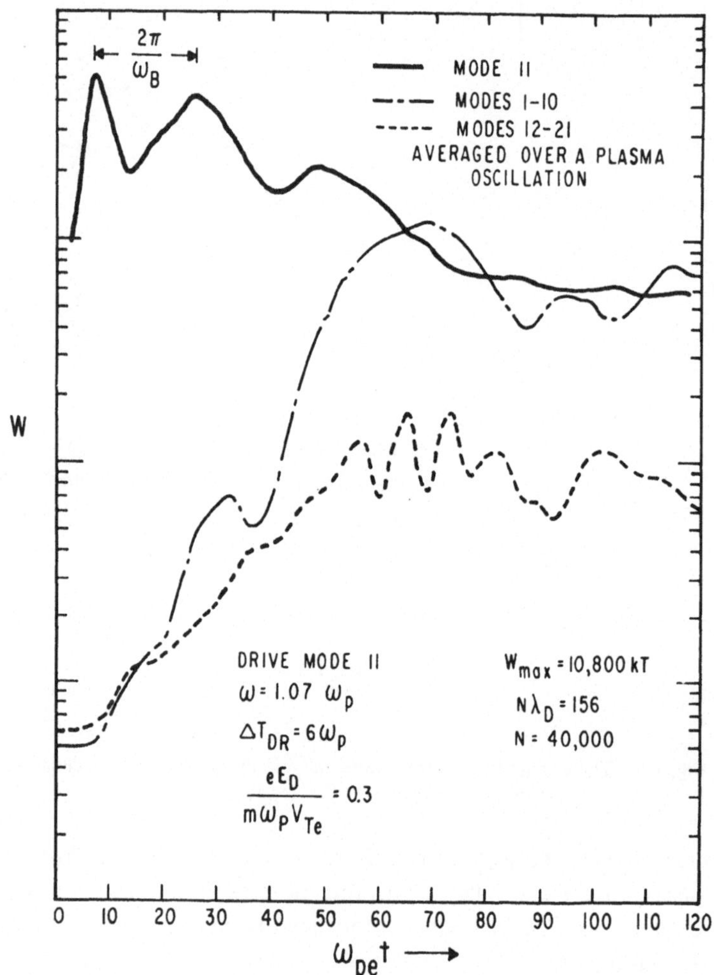

Fig. 14. The energy in the large wave and the upper and lower sidebands vs $\omega_{pe}t\,(eE_{\mathrm{DR}}/m\omega_p v_T = 0.3,\ \omega_p \Delta T_{\mathrm{DR}} = 6$, finite-size particle model).

Acknowledgements

The author wishes to acknowledge the fact that much of the work reported here was performed by Dr. W. Kruer and this article owes much to his efforts. The author is

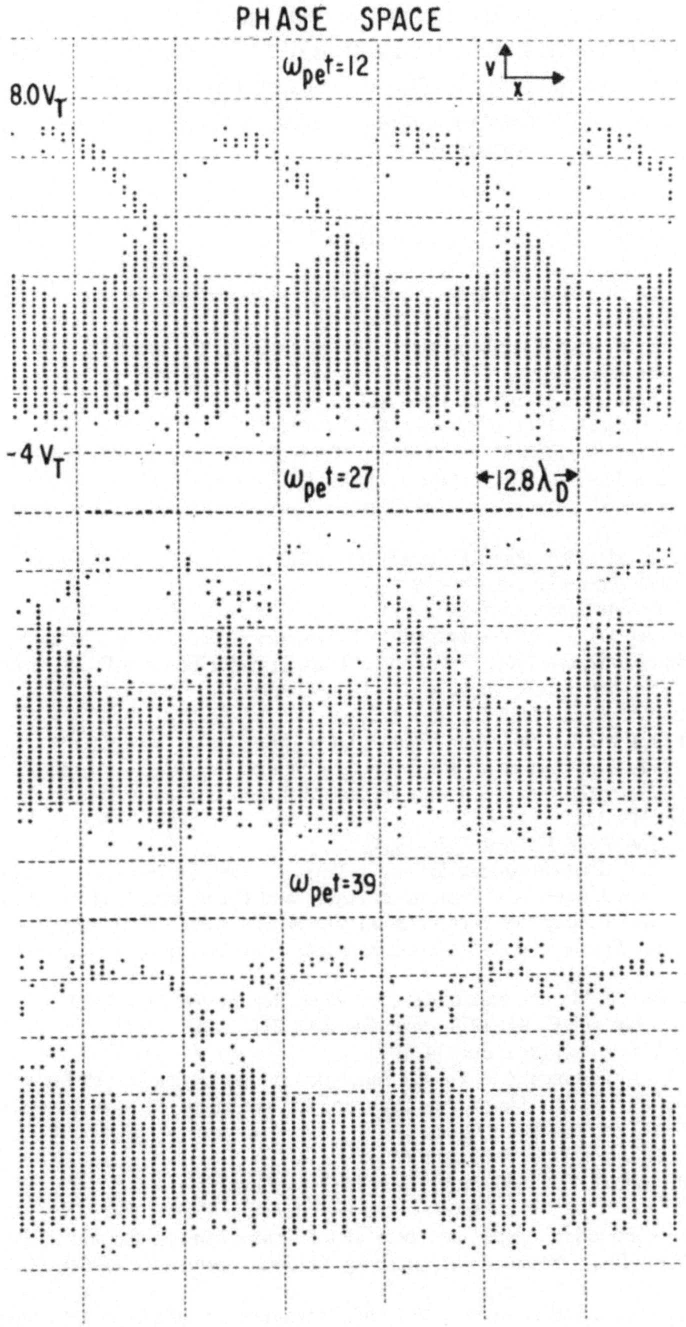

Fig. 15. Phase space at three different times (finite-size particle model).

also indebted to Dr. J. Denavit who allowed reporting on the comparison of this Fourier-Fourier method with the author's particle method.

This work was performed under the auspices of the U.S. Atomic Energy Commission, Contract AT (30-1)-1238; also Naval Research Laboratory, Contract N 00014-67-A-0151-0021. Use was made of computer facilities supported in part by National Science Foundation Grant NSF-GP 579.

References

Abramowitz, M. and Stegun, I. A.: 1964, *Handbook of Mathematical Functions*, National Bureau of Standards, Washington, D.C. p. 952.

Alder, B., Fernbach, S., and Rotenberg, M.: 1970, 'Methods in Computational Physics', *Plasma Physics*, Vol. **9**, Academic Press, N.Y.

Armstrong, T. P.: 1967, *Phys. Fluids* **10**, 1269–1280.

Birdsall, C. K. and Fuss, D.: 1968, *J. Comput. Phys.* **3**, 494–511.

Birdsall, C. K., Byers, J. A., Fuss, D., and Grewal, M.: 1969, 'Unstable Plasma Waves Propagating Perpendicular to a Magnetic Field: Small Amplitude Growth and Nonlinear Saturation from Computer Experiments', paper presented at Sherwood Theoretical Meeting, Gatlinburg, Tennessee, April 24–25, 1969.

Byers, J. and Grewal, M.: 1970, *Phys. Fluids* **13**, 1819–1830.

Dawson, J. M.: 1962, *Phys. Fluids* **5**, 445-459.

Denavit, J.: 1969, private communications.

Denavit, J. and Kruer, W. O.: 1971, submitted to *The Physics of Fluids* for publication.

Gentleman, W. and Sande, G.: 1966, 'Fast Fourier Transforms for Fun and Profit', *Proc. Fall Joint Computer Conf.* Vol. **29**, p. 563, Mcmillan Co. Ltd., London.

Goldman, M. V.: 1970, *Phys. Fluids* **13**, 1281–1289.

Grant, F. C. and Feix, M. R.: 1967, *Phys. Fluids* **10**, 696–702.

Harlow, F. H.: 1964, 'The Particle-*n*-Cell Computing Method for Fluid Dynamics', *Methods in Computational Physics* (B. Alder, S. Fernbach, and M. Rotenberg, eds.), Academic Press, New York, Vol. **3** pp. 319–343.

Hockney, R. W.: 1966, *Phys. Fluids* **9**, 1826–1835.

Hockney, R. W.: 1969, 'Particle Models in Plasma Physics', *Proc. of Conference on Computational Physics*, Culham, Vol. **I**, paper II. C (Institute of Physics and The Physical Society, CLM-CP 1969).

Knorr, G.: 1963, 'Zur Lösung der Nicht-linearen Vlasov Gleichung', *Rept. MPI/PA-34/63.* Max Planck Institut für Physik und Astrophysik, München (1963A); *also Z. Naturforsch.* **18a** (1963) pp. 1304–1315.

Kruer, W. L., Dawson, J. M., and Sudan, R.: 1969, *Phys. Rev. Letters* **23**, 838–841.

Langdon, A. B. and Birdsall, C. K.: 1970, *Phys. Fluids* **13**, 2115–2122.

Mason, R. J.: 1969, *Bull. Am. Phys. Soc.* **14**, 1043.

Morse, R. L.: 1970, 'Multidimensional Plasma Simulation by the Particle-in-Cell Method', *Methods in Computational Physics* (B. Alder, S. Fernbach, and M. Rotenberg, eds.), Academic Press, New York, Vol. **9** (1970) pp. 213–239.

Okuda, H.: 1969, *Bull. Am. Phys. Soc.* **14**, 1065.

Okuda, H. and Birdsall, C. K.: 1970, *Phys. Fluids* **13**, 2123–2134.

Sadowski, W. L.: 1967, 'One Some Aspects of the Eigenfunction Expansion of the Solution of the Nonlinear Vlasov Equation', paper presented at the Symposium on Computer Simulation of Plasmas and Many-Body Problems (Williamsburg, Virginia, April 19–21, (1967), NASA SP-153), pp. 149–150.

Shanny, R., Kruer, W. L., and Dawson, J. M., 1968, 'Investigations of Electrostatic Oscillations of a Non-Maxwellian Plasma', Proceedings of the Topical Conference on Numerical Simulation of Plasma, Sept. 18–20, Los Alamos, New Mexico (Los Alamos Scientific Laboratory LA-3990) paper A-2.

Wharton, C. B., Malmberg, J. H., and O'Neil, T. M.: 1968, *Phys. Fluids* **11**, 1761–1763.

ENHANCEMENT OF RELAXATION PROCESSES BY COLLECTIVE EFFECTS

RUSSELL M. KULSRUD

Plasma Physics Laboratory, Princeton University, Princeton, New Jersey 08540, U.S.A.

1. Introduction

It is well-known that in a plasma the rate of relaxation toward equilibrium can be enhanced by collective processes. Since the gravitational force is analogous to the Coulomb force, the question naturally arises whether a similar enhancement may occur in stellar dynamics. That such an enhancement may well occur can be inferred, for example, from the observation of well-developed velocity distributions in stellar systems such as elliptical galaxies, in which ordinary two-body collision processes are so slow that they could not possibly lead to any appreciable relaxation, even during the age of the universe. Further, rapid relaxation of stellar clusters in galactic nuclei can lead to a compact cluster of stars in which violent processes may occur, such as those observed in Seyfert galaxies and quasars (Spitzer and Saslaw, 1966). Thus, it is of considerable interest to enquire what processes can possibly lead to such a rapid collapse.

In this paper I wish to review, in a general fashion, the various collective processes in plasma physics, with an emphasis on their possible relevance to stellar dynamics. One can say that relaxation can occur in roughly two different ways: First, by ordinary collision between two particles, generally referred to as two-body collisions. These are well-known and are described by the ordinary Fokker-Planck equation (into which some appropriate cutoff is introduced for the divergent Coulomb cross section (Rosenbluth *et al.*, 1957). Second, relaxation may occur by collective collisions between one particle and many others.

Two-body collisions are essentially random collisions that produce small effects that accumulate slowly. Each collision can lead in two directions, increasing or decreasing the energy of one of the particles, so the accumulated effect is a random walk of each particle in momentum space that gradually takes the complete system toward thermal equilibrium.

Collective processes are completely analogous to two-body collisions, except that one particle collides with many which are collected together by some coherent process such as a wave. Again the process is random, but usually much stronger and leads to a rapid random walk of the particles that rapidly takes the complete system toward thermal equilibrium.

As a trivial but familiar example, consider a plasma which consists of fully stripped ions of charge $ze \gtrsim R$. Then the scattering angle θ is given by

$$\Delta\theta \sim (ze^2/b\varepsilon) \qquad (1)$$

M. Lecar (ed.), Gravitational N-Body Problem, 337–346.

where b is the impact parameter and ε the energy. The accumulated scattering, after N such collisions, is given by

$$\Sigma(\Delta\theta)^2 = Nz^2(e^2/b\varepsilon)^2. \tag{2}$$

This is to be compared with the value for Nz collisions in a plasma with protons, $Nz(e^2/b\varepsilon)^2$. By 'collecting' the protons into nuclei (i.e., into bunches of z protons) we have enhanced the scattering processes by z.

This is a rather extreme example, but as it turns out it represents the actual situation quite well. In a turbulent plasma the protons or electrons are collected together into bunches by waves. Here z represents the excess number of protons over the mean number in a volume of a wavelength cubed. The only differences are that the impact parameter can never be taken much smaller than a wavelength and the collected bunch of charge moves with the group velocity of the wave, rather than with the thermal velocity of the particle.

In this paper I should like to:

(a) Give some examples of enhanced relaxation from plasma physics.

(b) Describe the results which quasilinear theory predicts for the relaxation rates.

(c) Give a crude model for scattering, based on the above example.

(d) Discuss the possible extension of these results to the gravitational case.

2. Examples of Rapid Relaxation

As a first example let us consider the problem of confinement in a mirror machine; that is, in an axial magnetic $B_0(z)\,\hat{z}$ field which is a minimum at the center and increases in both directions in z (Rosenbluth *et al.*, 1957). In this case all particles moving perpendicular to B_0 are trapped, while those moving along z are untrapped and leave. As a consequence, the distribution of particles in momentum space is as sketched in Figure 1, in which the shaded region indicates the region of trapping. Thus, the density of particles outside this region is very low, and the length of confinement time is just the time for ions in the shaded region to move into the unshaded region, either by ordinary two-body or by collective collisions (i.e., it is just the ordinary relaxation time). Now in such an equilibrium there is a strong tendency for ion oscillations with wave fronts parallel to the z axis to be unstable, with a growth rate $\gamma \sim \omega_{p_i} = (4\pi\, ne^2/M)^{1/2}$, where n is the particle density of ions and M is the ion mass. Since the scattering rate is proportional to the square of the amplitude of these waves, and the waves are increasing in amplitude at an exponential rate $\exp \gamma t$, it is clear (on a moment's reflection) that the effective collision frequency ν_{eff} must also be

$$\nu_{\text{eff}} \sim \gamma \sim \omega_{p_i}. \tag{3}$$

This is larger than the ordinary collision rate ν by roughly $n\mathrm{D}^3$, where $\mathrm{D} = (KT/4\pi ne^2)^{1/2}$ is the Debye length. This factor can be of order 10^4–10^6, so a very large enhancement of the relaxation rate can occur (see, for example, Sagdeev and Galeev, 1969).

As a supplement to this first example, we may consider the particles of the Van

Allen belt trapped in the earth's dipole magnetic field. Again the particles are trapped by the mirror effect and the situation is very similar to that shown in Figure 1. In this case particles are lost by collective interactions, the lost particles producing the precipitation of particles into the ionosphere that causes various auroral phenomena. In this case, however, the more significant waves causing enhancement are probably the whistler waves propagating along the lines (Kennel and Petschek, 1966). The sit-

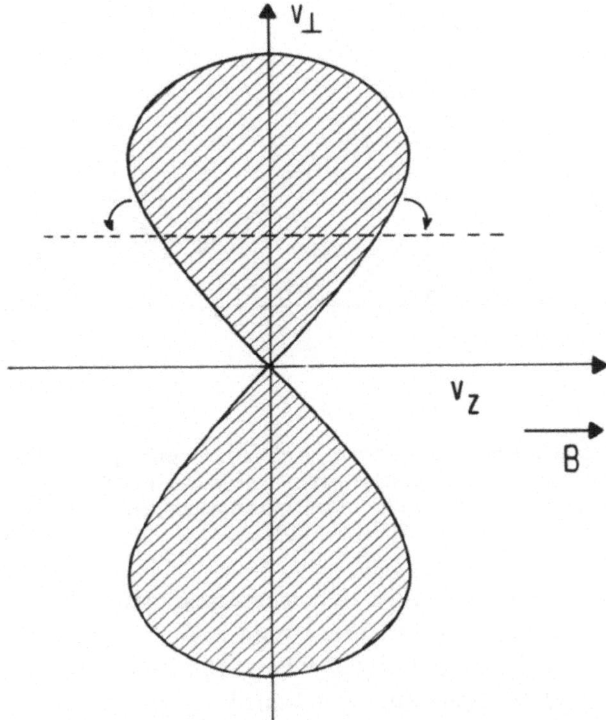

Fig. 1. The loss cone distribution in a magnetic mirror. Particles in the shaded region are trapped by the magnetic mirrors; particles in other regions leave freely. Relaxation toward equilibrium causes particles to enter the untrapped region and be lost. This is enhanced by the unstable ion oscillation propagating perpendicular to **B**, whose phase front is indicated by the dotted line. Particles on this line resonate with the wave and are scattered rapidly; those near the edge of the trapped region are thus lost.

uation is different in one essential way from the mirror machine example. In the first case one had an equilibrium established which lasted only for a finite time. In the case of the Van Allen belts, one obviously has a steady state. Particles are being continuously supplied to the trapped distribution in a way not yet completely understood. In this case the waves must be maintained at just such a level as will maintain this steady state by enhanced scattering; that is, at just such a rate as to remove particles at the rate at which they are introduced. Thus, the waves cannot be allowed to exponentiate in time, and therefore the equilibrium must adjust to a marginal state at which the waves are neither growing nor damping. Such a state can be reached if (1) the necessary enhanced diffusion is not comparable with the growth rate, and (2) the situation is stable in the sense that any tendency to enhance the collision rate leads to a modified equilibrium in which the waves are damped, while any tendency to depress the wave amplitude leads to an equilibrium in which the waves grow. (See, for more detail, Sagdeev and Galeev, 1969.)

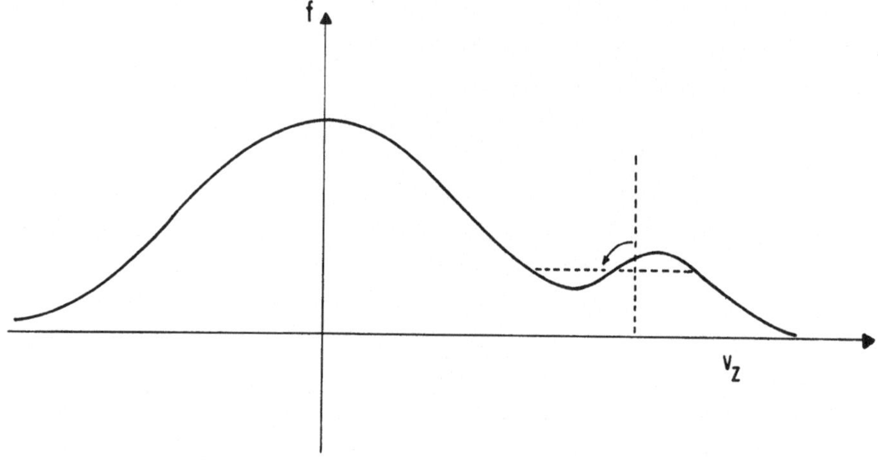

Fig. 2. The two-stream instability; the distribution of particles with v_z, the velocity in the direction of the beam as a function of v_z. The unstable wave whose phase velocity is indicated by the dotted line leads to rapid collisions which make the distribution function tend to the dashed line for which the wave is stable.

As a second example let us consider an unmagnetized plasma in which only electrostatic modes are present. Let us introduce a beam of electrons into this plasma with a velocity exceeding the thermal velocity of either. (This is the familiar situation of the two-beam instability.) The equilibrium is indicated in Figure 2, where the distribution of particles with velocity v_z is plotted. Without collective effects, the distribution would relax in a collision time or so toward a Maxwellian distribution. However, a plasma oscillation with phase velocity approximately that of the beam (indicated by a dotted line in Figure 2) is unstable, with growth rate $\gamma \sim \omega_{pe} n_B/n$, where $\omega_{pe} = (4\pi ne^2/m)^{1/2}$ where m is the electron mass and n_B/n is the ratio of the beam density to the plasma density. Thus, again one expects an enhancement of the particle collision rate by wave particle scattering. In this case the enhanced particle scattering will not lead to a Maxwellian distribution, but rather to the flattened distribution indicated in Figure 2 by a dotted curve (Drummond and Pines, 1962). This distribution is such that the wave is no longer unstable. Thus, as before, one expects to relax to this distribution in a time γ^{-1}, which is a factor $n_B D^3$ shorter than ordinary relaxation by two-body collisions.

Again this situation can occur in two ways: a time-dependent way in which a beam is actually injected into a plasma, as in the case of a collisionless shock (Krall and Book, 1969), or a steady state situation in which a large electric field is applied to a plasma, such as occurs in turbulent heating of a plasma (Hamberger *et al.*, 1967). (In the latter case the actual situation is one in which the electrons are displaced relative to the ions, so that the two beams are electrons and ions rather than both electrons).

These examples are only two of the many possible cases in which enhancement of relaxation can occur in a plasma. However, in some sense they are typical, and the following comments about them are relevant:

(1) These instabilities are in some sense weak. Strong instabilities, such as the Raleigh-Taylor instability, tend to destroy the equilibrium. These weak instabilities work on a smaller scale, leading to a rapid relaxation toward a situation in which the equilibrium is no longer unstable.

(2) These instabilities can occur in a time-dependent situation, in which case they usually lead to a relaxation at a rate comparable with the growth rate of the instability, or in a steady state, in which case they lead to a marginally stable equilibrium with a relaxation rate in balance with whatever outside forces tend to disturb it.

(3) The processes of growth of the waves and relaxation of the equilibrium are intimately related. They can generally be interpreted as a maser in which the unstable equilibrium corresponds to an overpopulation of the emitting states for the wave. The induced emission process is then the scattering process, as well as the process that makes the wave grow.

(4) The wave scattering is very much a collective process in which the particles are colliding with bunches of particles. (This will appear below.)

3. Quasilinear Theory

The quantitative theory for a description of the above processes is the quasilinear theory (Drummond and Pines, 1962). It has roughly two halves: (1) Given a distribution, how does one determine the amplitude of the waves, (2) Given a wave spectrum, how does one determine the evolution of the equilibrium?

The first half of the theory, problem (1), can be answered rather simply, even for inhomogeneous equilibrium, if the scale of the waves is sufficiently small and the periods sufficiently short. One introduces the 'number' of waves $N_{\mathbf{K}}=\varepsilon_{\mathbf{K}}/\omega_{\mathbf{K}}$ which would be proportional to the number of phonons where $\varepsilon_{\mathbf{K}}$ is the energy density of waves in **K** space and $\omega_{\mathbf{K}}$ is their frequency. $N_{\mathbf{K}}$ tends to be conserved and satisfies a continuity equation in **K**, **x** space.

$$\frac{\partial N_{\mathbf{K}}}{\partial t}+\nabla\left(\frac{\partial\omega_{\mathbf{K}}}{\partial\mathbf{K}}N_{\mathbf{K}}\right)-\frac{\partial}{\partial\mathbf{K}}\left(\frac{\partial\omega_{\mathbf{K}}}{\partial\mathbf{x}}N_{\mathbf{K}}\right)=2\gamma_{\mathbf{K}}N_{\mathbf{K}}, \tag{4}$$

where $\gamma_{\mathbf{K}}$ is the growth rate and $\omega_{\mathbf{K}}$ and $\gamma_{\mathbf{K}}$ depend on F, the equilibrium distribution function. For more details see Dewar (1970) or Whitham (1965).

The second half of the problem – the evolution of F – is given by a diffusion-type equation. For homogeneous systems it is usually written in the form

$$\frac{\partial F}{\partial t}=\frac{\partial}{\partial\mathbf{v}}\cdot\left(\mathrm{D}(v)\cdot\frac{\partial F}{\partial v}\right), \tag{5}$$

where

$$D \sim \left(\frac{\Delta v \Delta v}{2t} \right) \tag{6}$$

is a tensor which represents the diffusion rate in velocity space. It is the main task of quasilinear theory to express D in terms of the wave spectrum $\varepsilon_{\mathbf{K}}$. This has been done in a number of cases (in particular, for the examples discussed above). For an unmagnetized electrostatic plasma, in which only the electrons are considered,

$$D = \pi (e/m)^2 \int \hat{\mathbf{K}}\hat{\mathbf{K}} \, |E_{\mathbf{K}}|^2 \, \delta(K \cdot v - \omega_{\mathbf{K}}) \, d^3K \sim \nu_{\text{eff}} v^2 . \tag{7}$$

Here $E_{\mathbf{K}}$ is the Fourier transform of the electric field for plasma waves, while the δ function represents the fact that the waves must be in resonance with the wave, and $\hat{\mathbf{K}}$ is a unit vector.

To get a rough order of magnitude of the size of ν_{eff}, let us consider a spectrum of waves (such as indicated in Figure 3), in **K** space. The shaded region of width ΔK is filled with waves with total energy $\mathscr{E}$. In this case it is easy to show that for an electron with v perpendicular to this slab,

$$\nu_{\text{eff}} \sim \pi \omega_p \frac{2\mathscr{E}}{(3/2nKT)} \frac{K}{\Delta K} \tag{8}$$

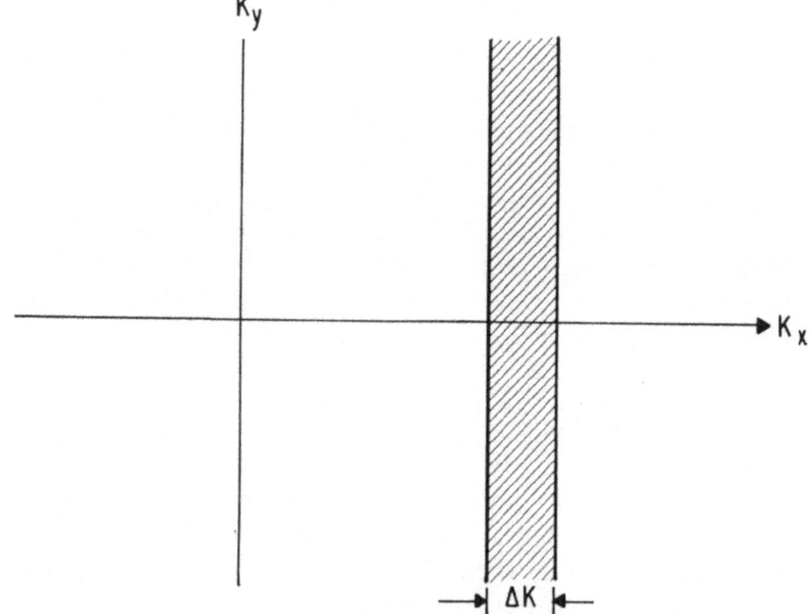

Fig. 3. The distribution of wave energy in K space. The shaded region of width ΔK indicates the region of **K** space where E_K is nonzero for the example considered in Equation (8).

($\frac{3}{2}nKT$ is the thermal energy in the plasma). The diffusion in velocity space is partly in angle and partly in energy. From Equation (8) it can be seen that when the wave energy $\mathscr{E} \sim \Delta K/K$ times the thermal energy, the scattering rate is of order ω_p.

The effective rate of scattering can also be expressed in another way: From Equation (8) it is seen that the rate of wave scattering is proportional to the energy density of waves in **K** space. This energy density is $KT/8\pi^3$ for a thermal plasma. If we introduce a fictitious wave temperature T_w such that the energy density is $KT_w/8\pi^3$, then Equation (8) can be written as

$$\nu_{\text{eff}} = \frac{1}{8\pi^2} \frac{\omega_p}{n\mathrm{D}^3} \frac{T_w}{T} \frac{v_t}{v} \sim \frac{\nu}{10} \frac{T_w}{T} \frac{v_t}{v}, \tag{9}$$

where ν is the ordinary collision rate for thermal particles, $v_t = (KT/m)^{1/2}$ is the thermal velocity, and v is the velocity of the particle. Equation (9) has been derived, from kinetic considerations, by Rosenbluth and Rostoker (1960). We see that if $T_w = T$, the collective collision rate is somewhat smaller than ordinary collisions but falls off more slowly at large velocities than two-body collisions. The enhancement in relaxation is simply $T_w/10\,T$ for $v \sim v_t$.

To the order in the amplitude of the waves to which quasilinear theory is valid, no interaction between the waves themselves occurs but the second-order effect of the interaction between waves and particles is included. As one considers the situation in which the amplitude is larger, it is expected that even this interaction is not well-represented by the theory. Since the theory basically assumes the orbits of the particles to be perturbed by small amounts from straight-line orbits, one can guess that a limit on the theory will occur when the amplitude of the waves is large enough to trap the particles. Therefore, if the bounce period of a particle trapped in a wave is shorter than the time a particle spends in a wave packet, one could reasonably expect the theory to be inaccurate. The situation is illustrated in Figure 4, where a wave packet is shown in the case that the group velocity v_g is less than the phase velocity v_ϕ. Thus,

Fig. 4. Motion of a particle in a wave packet. The wave packet at two different times, for the case $0 < v_g < v_\phi$. The dot indicates the position of the particle.

a resonant particle that moves with velocity $v = v_\phi$ gradually works itself to the right in the wave packet, and finally leaves in a time τ given by

$$\tau = \frac{1}{\omega} \frac{k}{\Delta k} \frac{v_\phi}{|v_\phi - v_g|}. \tag{10}$$

If the time to execute a bounce oscillation $\tau_B = [k(e\phi/m)^{1/2}]^{-1}$ is longer than τ, then we expect the approximation of straight-line motion to be good and the quasilinear theory to be valid. This reduces to the sufficient condition

$$\Delta k/k > (k\mathrm{D})^{1/2} (\mathscr{E}/nT)^{1/4} \tag{11}$$

for validity of the quasilinear theory. On the other hand, $\tau_B \gtrsim \tau_{\text{life}}$, where the wave has a life (or correlation time) τ_{life}, is also a sufficient condition.

Finally, it is possible that the particle resonates with no one wave but that it can resonate with two waves simultaneously; e.g., if the condition

$$v \sim \frac{\omega_1 + \omega_2}{K_1 + K_2} \tag{12}$$

is satisfied. Under these circumstances the particle tends to scatter energy from one of the waves to the other, suffering a recoil that enhances its own scattering rate. This is to be contrasted with ordinary scattering, which is basically an absorption and (as in atomic physics) is a much stronger process. The scattering process is usually referred to as non-linear Landau damping (Kadomtsev, 1965; Drummond, 1965).

4. A Crude Picture of Collective Processes

So far we have quoted results which are derived on the basis of a rather sophisticated treatment of the Vlasov equation. But these processes are usually referred to as collective processes, and it seems worthwhile to give a simple, crude derivation of formula (7) based directly on this collective idea.

In order to do this we continue the picture of the Introduction, in which waves are regarded as gigantic nuclei of charge z, where now

$$z = \delta n b^3 \tag{13}$$

and b is the size of the wave packet. For simplicity we restrict ourselves to wave packets with a single wave, so that $\Delta k \sim k$. For a single collision with a wave packet we have, for the scattering angle $\Delta\theta$,

$$mv\Delta\theta \approx \frac{ze^2}{b^2} \frac{b}{|v - v_g|} = \frac{ze^2}{b\,|v - v_g|}. \tag{14}$$

The first factor represents the mean force, that for a spherical charge is the same as at the edge of the charge. The second factor is the time of interaction, since the relative velocity is $v - v_g$ (where group velocity v_g is the velocity of the wave packet). The effective cross section for such interactions is

$$\sigma = \pi b^2 (\Delta\theta)^2, \tag{15}$$

since it takes $(\Delta\theta)^{-2}$ such collisions to give a 90° scattering. Finally,

$$\nu_{\text{eff}} = N\,|v - v_g|\,\langle\sigma\rangle, \tag{16}$$

where N is the number density of wave packets. Since the Coulomb energy of a single wave packet is $(ze)^2/b$, we have, for the energy density of the waves.

$$\mathscr{E}_{\text{waves}} = N(ze)^2/b\,. \tag{17}$$

Combining Equations (13)–(17), we find

$$\mathrm{D} = v_{\text{eff}} v^2 \sim \pi (e/m)^2 \frac{\mathscr{E} b}{|v - v_g|}\,. \tag{18}$$

This is in approximate agreement with Equation (7) if we carry out the K integral and set $E^2/8\pi \sim \mathscr{E}/k \sim b\mathscr{E}$.

5. The Gravitational Case

In plasma physics one usually deals with systems many orders of magnitude larger than the scale of the waves contributing to collective processes, namely D. This makes the analysis comparatively simple, since one can work with Fourier modes. Unfortunately this is not the case for gravitation where the size of the system is about the same as the wavelength scale. This makes it necessary to treat the waves as eigenmodes in the system. For example, for a spherical cluster one can generally specify the angular dependence of the waves, but a differential equation must be solved to determine their radial dependence.

Nevertheless, with this proviso it seems possible to carry over most of the formal apparatus of the quasilinear theory. We do not attempt to do this in detail in this picture, but merely sketch how this may be done by analogy with the plasma theory. Let us assume that the gravitational potential can be expanded in normal modes

$$\phi_n \sim \hat{\phi}_n(r) \exp i(m\theta - \omega t), \tag{19}$$

where θ represents, in a rather loose way, the angular dependence. The equilibrium solution for the distribution function is generally represented by constants of the motion $c_1 \ldots c_n$, rather than x and v. Now, after a particle executes an orbit and returns to its original radial position and velocity, it will have suffered a change in its constants of the motion c by an amount Δc in a time Δt. Therefore, the mean value of $\dot{c}$ is

$$\bar{\dot{c}} = \Delta c/\Delta t = \eta_n a_n \exp i(m\langle\theta\rangle - \omega t), \tag{20}$$

where $\langle\theta\rangle$ is a mean value of θ during the orbit; η_n is the result of this calculation for a normal mode of unit amplitude, $a_n = 1$, and a_n is the amplitude of the normal mode. These changes are to be calculated for all normal modes and c's, but $\langle\dot{c}\rangle$ is to be compared with the equation

$$\dot{v} = (e/m)\, E_K \exp i(Kx - \omega t)\,. \tag{21}$$

The analogies to be drawn are

$$v \to c\,, \quad e/m \to \eta\,, \quad E_K \to a\,, \quad \mathrm{K}\cdot v \to \langle\dot{\theta}\rangle = \Delta\theta/\Delta t\,.$$

Thus, one would expect the equation for enhanced evolution of the Vlasov equation to read

$$\frac{\partial F}{\partial t} = \pi \frac{\partial}{\partial c_i} \left\{ \sum_m \eta_m^i \eta_m^j |a_m|^2 \, \delta(\omega - m \langle \dot{\theta} \rangle) \frac{\partial f}{\partial c_j} \right\}. \tag{22}$$

Of course these calculations must be carried out in a more precise way.

Acknowledgement

Work in preparation for this paper was supported by Air Force Office of Scientific Research Contract AF-F44620-70-C-0033.

References

Dewar, R. L.: 1970, *Phys. Fluids* **13**, 270.

Drummond, W. E.: 1965, 'Quasi-Linear Theory of Plasma Turbulence', *Plasma Physics* (International Atomic Energy Agency, Vienna, 1965) pp. 527–541.

Drummond, W. E. and Pines, D.: 1962, *Nucl. Fusion*, Suppl. 3, 1049–1057.

Hamberger, S. M., Malein, A., Adlam, J. A., and Friedman, M.: 1967, *Phys. Rev. Letters* **19**, 350–352.

Kadomtsev, B. B.: 1965, 'Plasma Turbulence General Topics', *Plasma Physics* (International Atomic Energy Agency, Vienna, 1965) pp. 543–554.

Kennel C. F. and Petschek, H. E.: 1966, *Geophys. Res.* **71**, 1–28.

Krall, N. A. and Book, D. L.: 1969, *Phys. Fluids* **12**, 347–355.

Rosenbluth, M. N., Mac Donald, W. M., and Judd, D. L.: 1957, *Phys. Rev.* **107**, 1–6.

Rosenbluth, M. N. and Post, R. F.: 1965, *Phys. Fluids* **8**, 547–550.

Rosenbluth, M. N. and Rostoker, N.: 1960, *Phys. Fluids* **3**, 1–14.

Sagdeev, R. Z. and Galeev, A. A.: 1969, *Nonlinear Plasma Theory*, W. A. Benjamin, Inc. New York.

Spitzer, Jr., L. and W. Saslaw: 1966, *Astrophys. J.* **143**, 400–419.

Whitham, G. B.: 1965, *Fluids Mech. J.* **22**, 273–283.

STABILITY PROPERTIES FOR ENCOUNTERLESS SELF-GRAVITATIONAL STELLAR GAS AND PLASMA

M. R. FEIX, J. P. DOREMUS, and G. BAUMANN
Groupe Physique Théorique et Plasma, Université de Nancy, 54, Nancy, France

Abstract. The stability properties of the collisionless plasma and encounterless stellar gas described by the Vlasov equations are studied. The introduction of the 'multiple Water Bag' model allows, for one-dimensional plane geometry, a treatment of the general case and removes some of the difficulties connected with the formulation of the energy variation. From this last result it can be deduced that both plasma and stellar systems steady state described by a monotonically decreasing distribution $F(\varepsilon)$ are stable. The demonstration is extended to the spherically symmetric case for self-gravitating gas. Next the constraint of a monotonically decreasing $F(\varepsilon)$ is relaxed and it is supposed that the instability appears through the point $\omega=0$. This is known to be true for some type of plasma instabilities (two streams) but is a simple working hypothesis in the gravitational case. For this marginal mode the N 'bags' equations degenerate into a single wave equation and the stability of the system is given by the sign of the eigenvalues of a Schroedinger type operator. A simple physical picture is obtained for the plasma case where the quantitity $\int(\mathrm{d}F/\mathrm{d}\varepsilon)\,\mathrm{d}V$ (the square of the local maximum wavenumber of instability) is introduced. A virtual variation of this quantity indicates if the initial steady state was stable or unstable.

1. Introduction

The stability properties of the time-independent solutions of an encounterless stellar gas moving in its own self-consistent gravitational field are practically unknown. It is consequently interesting to obtain solutions in some of the simplest possible geometry (one dimensional and plane or spherical symmetry). For these two last cases the steady state distribution function is $F(\varepsilon)$, where ε is the total energy, the only invariant considered in this problem among the six possible.

The plasma physicists face, of course, the same problem. They have the advantage that a large class of equilibria – namely the homogeneous in space steady state – can be completely studied by the formalism of the dielectric constant. Then general stability criteria as the Penrose criteria can be obtained (Penrose, 1960).

From this homogeneous plasma analysis perturbation methods allow the treatment of small inhomogeneities. Unfortunately such a homogeneous situation has no equivalence in the gravitational case.

The treatment of a strongly inhomogeneous case (and we must point out that all gravitational equilibria are of this form) implies, to get the eigenfrequencies, the solution of an integral equation involving a very complicated kernel and very often numerical techniques must be used to obtain partial results.

In recent years some progress has been made through the introduction of new concepts where additional information is used to simplify the problem.

A first step in this direction was introduced in Hohl and Feix, 1967, 1968; Antonov,

M. Lecar (ed.), Gravitational N-Body Problem, 347–364.

1961; and Hohl, 1969. The idea is to look at all possible values of the perturbation δF and to compute the energy variation δW. Of course, some constraints on δF must be taken into account corresponding to the fact that the system must change according to the Boltzmann–Vlasov equation in its own self-gravitating field. The idea is that by introducing some of the aspects of the solution of δF, δW is shown to be positive and consequently all possible evolutions from F to another state are ruled out, implying stability.

In this type of study an always used property is the conservation of the phase space density along the trajectories of the particles. Such conservation is very conveniently expressed in the so called 'Water Bag' model. In such a model we draw in phase space the lines f=constant and suppose that between two neighboring lines the distribution function is a constant. By doing so we approximate the function $F(\varepsilon)$ by succession of steps. If we let the number of steps go to infinity we can approximate any function as closely as we like, and we will consider usually the model of N bags as the general case. Although the case of 1 or 2 bags introduces a severe limitation of the possible initial situation they are sometimes worth a special study because of the simplicity of the calculations. We will come back to this point later.

Lynden Bell (1969) and Lynden Bell and Sanitt (1969) have also obtained an expression of δW for the continuum case. We will connect their calculation to the results of the Water Bag model. In the above-mentioned calculation of δW, in addition to the property of conservation of F along the phase space trajectories, the self-consistency of δE (the field variation) and δF must be taken into account to compute δW. We will see that these two properties are sufficient to establish the stability properties of some steady states both in the plasma and in the gravitational case.

For some situations, these two properties are nevertheless insufficient to establish the stability known to exist. Two concepts may be introduced: the concept of the dynamic of a marginal stability mode and its self-consistency. We will see that these two new properties will allow us, in the plasma case, to compute the stability property for a two stream inhomogeneous situation. A marginal stability mode is a mode growing as $\exp(\gamma t)$ with $\gamma \to 0$. For some situations it is known that the first unstable mode is of that type while for other (beam plasma) the first unstable mode to appear is a overstable mode, i.e. a mode varying as $\exp(\gamma t + i\omega t)$ with $\gamma \to 0$ and ω finite.

For the plasma case a physical picture for the appearance of the instability can be obtained: we compute the marginal instability mode and try to fit it in the plasma. If to be able to fit it, we must at every point lower the stability property of the plasma, we can conclude that the considered plasma was stable and vice versa. The sign of the eigenvalue of an operator will tell us what should be done. In the gravitational case a marginal stability mode is always found in the system*; the physical picture is not so clear and more work is needed on this question. It must be pointed out that the class of steady state solutions for the plasma is much richer than for the gravitational case due to the presence of the ions which provide a fixed but otherwise quite arbitrary inhomogeneous background.

* Only for the plane case.

Finally a last point to be examined is a connection between the eigenfrequencies ω^2 and a quantity which is somewhat related to δW. We will see that the sign of ω^2 is indeed connected to the sign of δW.

Consequently we will consider in this paper the following points.

Introduction of the Water Bag Calculation of δW and connection to the continuum case.

Stability property of a decreasing function of the energy in the plasma and gravitational case. For plane geometry we show that the limiting case must be studied through the multiple Water-Bag concept.

Introduction of the marginal stability mode. Computation of the stability property for the plasma case through a virtual modification of the property of the medium.

2. Water Bag Model and Computation of δW

Figure 1 shows the 'Water Bag' concept. A certain number of curves delineates regions where f=constant. Coming from the center we find curves 1, 2, 3 etc. With the notation of Figure 1 it can be seen that we are really dealing with N 'bags', each bag with a density A_j extending from a value $V_j^-(x)$ to a value V_j^+. We notice that some of the A_j can be negative (corresponding to a 'bag' of holes) if in some region $f(V)$ is decreasing. Let us now compute explicitly δW. Both for plasma and gravitational cases we will use rationalised units and we will have $e=m=1$ and $4\pi G=1$. At equilibrium the contours V_j^+ and V_j^- are symmetrical $V_j^+=-V_j^-=V_j$ and the distribution F is

$$F=\sum_j A_j\{\gamma[V+V_j(x)]-\gamma[V-V_j(x)]\}, \tag{1}$$

where γ is the unit step function, $\gamma(x)=0$, if $x<0$ and $\gamma(x)=1$ if $x>0$.

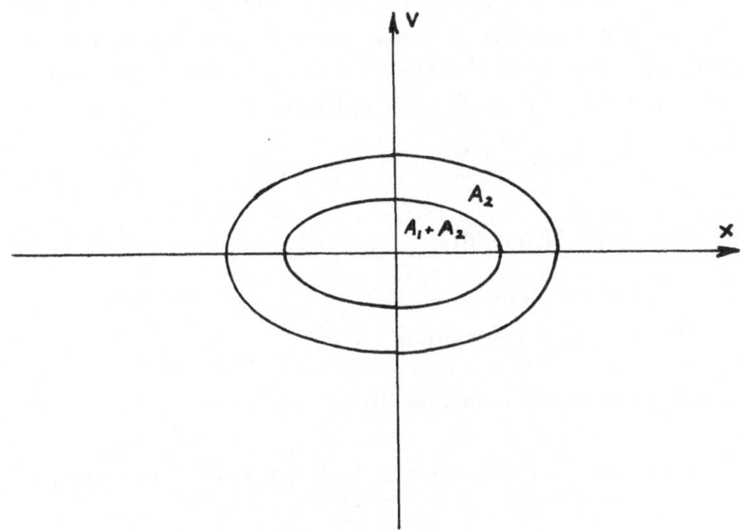

Fig. 1. Illustration of the Water Bag model.

The perturbed contours are

$$\begin{aligned} V_j^+ &= V_j + \delta V_j^+ , \\ V_j^- &= - V_j + \delta V_j^- . \end{aligned} \tag{2}$$

The kinetic energy density $\mathrm{d}K/\mathrm{d}x$ is

$$\frac{\mathrm{d}K}{\mathrm{d}x} = \tfrac{1}{2} \int f V^2 \,\mathrm{d}V = \sum_j \frac{A_j}{2} \int_{V_j^-}^{V_j^+} V^2 \,\mathrm{d}V = \sum_j \frac{A_j}{6} \left[(V_j^+)^3 - (V_j^-)^3 \right]. \tag{3}$$

Taking into account (2) and keeping terms up to the second order in $\delta V_j^{\pm}$ and integrating on x we find the variation in the kinetic energy.

$$\delta K = \sum_j \frac{A_j}{2} \int \left\{ (\delta V_j^+ - \delta V_j^-) V_j^2 + V_j \left[(\delta V_j^+)^2 + (\delta V_j^-)^2 \right] \right\} \mathrm{d}x . \tag{4}$$

The potential energy density is

$$\mathrm{d}P/\mathrm{d}x = \pm \tfrac{1}{2} E^2 , \tag{5}$$

where we take the upper $(+)$ sign for the plasma case and the lower $(-)$ sign for the gravitational case. From now we will use the convention that the upper and lower sign refer respectively to the plasma and the gravitational cases. The variation δP of the potential energy is

$$\delta P = \pm \int E \,\delta E \,\mathrm{d}x \pm \tfrac{1}{2} \int (\delta E)^2 \,\mathrm{d}x ; \tag{6}$$

up to now we have taken into account the conservation of F in the phase space. The advantage of the W.B. model is that such a conservation is automatically fulfilled. Next we use the fact that the field variation δE and the density variation $\delta \varrho$ must be self-consistent, i.e. must fulfil the Poisson equation.

$$(\partial/\partial x)\, \delta E = \pm\, \delta \varrho = \pm \sum_j A_j (\delta V_j^+ - \delta V_j^-) . \tag{7}$$

We integrate by parts $\int E \,\delta E \,\mathrm{d}x$ with

$$\begin{aligned} E \,\mathrm{d}x &= \mathrm{d}\alpha \Rightarrow \alpha = - \Phi \\ \delta E &= \beta \Rightarrow \mathrm{d}\beta = (\partial/\partial x)\, \delta E = \pm\, \delta \varrho . \end{aligned}$$

At the two extremities $\delta E = 0$. Consequently

$$\int E \,\delta E \,\mathrm{d}x = \pm \int \Phi \,\delta \varrho \,\mathrm{d}x = \pm \sum A_j \int \Phi \,(\delta V_j^+ - \delta V_j^-) \,\mathrm{d}x . \tag{8}$$

Now we compute $\delta W = \delta K + \delta P$. We first combine the first terms of the right hand side of (4) and (6) taking (8) into account

$$\sum_j \frac{A_j}{2} \int V_j^2 (\delta V_j^+ - \delta V_j^-) \, \mathrm{d}x + \sum_j A_j \int \Phi (\delta V_j^+ - \delta V_j^-) \, \mathrm{d}x = \\ = \sum_j A_j \left(\frac{V_j^2}{2} + \Phi \right) (\delta V_j^+ - \delta V_j^-) \, \mathrm{d}x . \qquad (9)$$

But $V_j^2/2 + \Phi$ is the energy W_j characterising the trajectory and is constant along this trajectory. It can be taken out of the $\int$ sign. Moreover taking into account the fact that the number of particles in each bag is conserved we have

$$\int (\delta V_j^+ - \delta V_j^-) \, \mathrm{d}x = 0 ,$$

and (9) equal to zero. The first order term in the total energy variation cancels and we are left with the expression

$$2\delta W = \sum_j A_j \int V_j [(\delta V_j^+)^2 + (\delta V_j^-)^2] \, \mathrm{d}x \pm \int (\delta E)^2 \, \mathrm{d}x , \qquad (10)$$

an expression already obtained by Hohl in (4). It is now obvious that if $F(\varepsilon)$ is a decreasing function of ε all the A_j are positive and (10) is positive for the plasma case. So no motion with $\delta W = 0$ (imposed by the Vlasov equation) is possible and such a plasma is stable.

The gravitational case is not so obvious because of the $-$sign in front of $\int (\delta E)^2 \, \mathrm{d}x$. We will come back to that point later.

3. Connection to the Continuum Case

Now we want to make contact with a calculation of Lynden Bell (1969) in the continuum case. For completeness we quickly give his calculation.

If we consider the distribution function we know that in the Vlasov equation f is conserved along the trajectory. If J is a function of f we may write

$$\int J(f) \, \mathrm{d}x \, \mathrm{d}V = \text{Constant} .$$

Let us consider the evolution from the steady state F to the new state $F + \delta F$. We have

$$\int J(F) \, \mathrm{d}x \, \mathrm{d}V = \int J(F + \delta F) \, \mathrm{d}x \, \mathrm{d}V , \qquad (11)$$

we take into account the δF expansion of $J(F + \delta F)$ up to the second order in δF. We obtain:

$$J(F + \delta F) = J(F) + \delta F J'(F) + \tfrac{1}{2} (\delta F)^2 J''(F) . \qquad (12)$$

Combining (11) and (12) we get

$$\int J'(F)\,\delta F\,\mathrm{d}x\,\mathrm{d}V + \tfrac{1}{2}\int J''(F)\,(\delta F)^2\,\mathrm{d}x\,\mathrm{d}V = 0\,. \tag{13}$$

We specialise to the case $J'(F)=\varepsilon$ and $J''(F)=\mathrm{d}\varepsilon/\mathrm{d}F=1/(\mathrm{d}F/\mathrm{d}\varepsilon)$ (13) is now written

$$\int \varepsilon F\,\mathrm{d}x\,\mathrm{d}V = \tfrac{1}{2}\int (\delta F)^2/-(\mathrm{d}F/\mathrm{d}\varepsilon)\,\mathrm{d}x\,\mathrm{d}V\,. \tag{14}$$

As for the W.B. case we compute

$$\delta K = \tfrac{1}{2}\int V^2\,\delta F\,\mathrm{d}x\,\mathrm{d}V\,, \tag{15}$$

$$\delta P = \pm\int E\,\delta E\,\mathrm{d}x \pm \tfrac{1}{2}\int (\delta E)^2\,\mathrm{d}x\,. \tag{16}$$

We have already shown {Formula (8)}

$$\int E\,\delta E\,\mathrm{d}x = \pm\int \Phi\,\delta\varrho\,\mathrm{d}x = \pm\int \phi\,\delta F\,\mathrm{d}x\,\mathrm{d}V\,. \tag{17}$$

Combining (15), (16) and (17), we obtain

$$\delta W = \int ((V^2/2)+\Phi)\,\delta F\,\mathrm{d}x\,\mathrm{d}V \pm \tfrac{1}{2}\int (\delta E)^2\,\mathrm{d}x\,. \tag{18}$$

Taking (14) into account we recover the Lynden Bell formula

$$2\,\delta W = \int (\delta F)^2/(-\,\mathrm{d}F/\mathrm{d}\varepsilon)\,\mathrm{d}x\,\mathrm{d}V \pm \int (\delta E)^2\,\mathrm{d}x\,. \tag{19}$$

If we consider the ways (13) and (19) have been obtained we notice the similarity in the two derivations. Let us show that indeed for the Water Bag case (19) reduces to (13). We will use freely the Dirac δ function although strict demonstration involves the use of distribution theory.

From

$$F = \sum_j A_j\left[\gamma(V - V_j^-) - \gamma(V - V_j^+)\right],$$

we deduce that variations δV_j^- and δV_j^+ imply

$$\delta F = \sum_j A_j\left[-\,\delta(V - V_j^-)\,\delta V_j^- + \delta(V - V_j^+)\,\delta V_j^+\right]. \tag{20}$$

We square (20) and take into account the fact that the δ functions have a very narrow range and consequently $\delta(V-V_j)\,\delta(V-V_k)=0$ if $j\neq k$. Moreover for the steady state $V_j^- = -V_j^+ = -V_j$,

$$(\delta F)^2 = \sum_j A_j^2\left[\delta^2(V+V_j)\,(\delta V_j^-)^2 + \delta^2(V-V_j)\,(\delta V_j^+)^2\right]. \tag{21}$$

Now we compute $\mathrm{d}F/\mathrm{d}\varepsilon = [(\mathrm{d}F/\mathrm{d}V)/(\mathrm{d}\varepsilon/\mathrm{d}V)] = (1/V)(\mathrm{d}F/\mathrm{d}V)$

$$\frac{\mathrm{d}F}{\mathrm{d}\varepsilon} = \sum_j A_j [\delta(V + V_j) - \delta(V - V_j)] \frac{1}{V} =$$

$$= \sum_j \frac{(-A_j)}{V_j} [\delta(V + V_j) + \delta(V - V_j)]. \tag{22}$$

Again to compute $(\delta F)^2/(\mathrm{d}F/\mathrm{d}\varepsilon)$ we take into account the fact that the δ functions have a very narrow range.

$$(\delta F)^2/(-\mathrm{d}F/\mathrm{d}\varepsilon) = \sum A_j V_j [\delta(V + V_j)(\delta V_j^-)^2 + \delta(V - V_j)(\delta V_j^+)^2]. \tag{23}$$

Integrating (23) on V we see that for the W.B. case (19) degenerates into (10).

This last calculation is not as academic as it looks. To get (19) it must be supposed that $-\mathrm{d}F/\mathrm{d}\varepsilon$ is always a positive quantity. Indeed if ε is a multiple-valued function of F the change of variable in (13) is impossible. On the other hand we see that the W.B. equation allows negative A_j without any difficulty and we will use this equation to study two streams instability in plasma.

4. Multiple Water Bag Equations

The linearised equations for the two contours of the jth bag are

$$\left.\begin{aligned} (\partial/\partial t)\,\delta V_j^+ + (\partial/\partial x)(V_j\,\delta V_j^+) = \delta E\,, \\ (\partial/\partial t)\,\delta V_j^- - (\partial/\partial x)(V_j\,\delta V_j^-) = \delta E\,. \end{aligned}\right\} \tag{24}$$

(24) must be completed with the Poisson equation

$$(\partial/\partial x)\,\delta E = \pm\,\delta\varrho\,. \tag{25}$$

We write δE as the sum of δE_j each δE_j, corresponding to the field created by the jth bag. Consequently

$$(\partial/\partial x)\,\delta E_j = \pm\,A_j(\delta V_j^+ - \delta V_j^-). \tag{26}$$

We suppose a time variation in $\exp i\omega t$ and consequently $\partial/\partial t = i\omega$. Combining (24), (25) and (26), we get the following equation for the jth bag.

$$\omega^2\,\delta E_j = \pm\,\omega_j^2\,\delta E - V_j(\mathrm{d}/\mathrm{d}x)[V_j(\mathrm{d}/\mathrm{d}x)\,\delta E_j] \tag{27}$$

where ω_j^2 is the local plasma (or Jean's) frequency of the jth bag with $\omega_j^2 = 2A_jV_j$. The coupling comes from the fact that $\delta E = \sum_j \delta E_j$. Let us establish a connection between the sign of ω^2 and a quantity closely connected to δW. First we multiply (27) by $\delta E_j/\omega_j^2$. We obtain

$$\omega^2((\delta E_j)^2/\omega_j^2) = \pm\,\delta E\,\delta E_j - (\delta E_j/2A_j)(\mathrm{d}/\mathrm{d}x)[V_j(\mathrm{d}/\mathrm{d}x)\,\delta E_j]. \tag{28}$$

We multiply (28) by dx and integrate on x. The last expression in the right hand side of 28) is integrated by parts taking into account the fact that at the two extremities of each bag $\delta E_j = 0$. We get

$$\omega^2 \int ((\delta E_j)^2/\omega_j^2)\,\mathrm{d}x = \pm \int \delta E\,\delta E_j\,\mathrm{d}x + \int (V_j/2A_j)\,((\mathrm{d}/\mathrm{d}x)\,\delta E_j)^2\,\mathrm{d}x. \tag{29}$$

We introduce a squared wavenumber k_j^2 connected to the jth bag by the relation

$$k_j^2 = 2A_j/V_j. \tag{30}$$

Now we sum the N Equations (29) corresponding to each bag and we get

$$\omega^2 \sum_j \int \frac{(\delta E_j)^2}{\omega_j^2}\,\mathrm{d}x = \pm \int (\delta E)^2\,\mathrm{d}x + \sum_j \int \frac{1}{k_j^2}\left(\frac{\mathrm{d}}{\mathrm{d}x}\,\delta E_j\right)^2 \mathrm{d}x = Q. \tag{31}$$

If an unstable mode can increase as $\exp\gamma t$ (with γ real) then all the δE_j are real and ω^2 is real and negative. Consequently the sign of ω^2 and the sign of Q are the same and if the plasma is unstable Q is negative. If Q is positive the plasma is stable. Let us show that Q is closely connected to the quantity $2\,\delta W$ as given by (10).

We can write

$$(\delta V_j^+)^2 + (\delta V_j^-)^2 = \tfrac{1}{2}(\delta V_j^+ + \delta V_j^-)^2 + \tfrac{1}{2}(\delta V_j^+ - \delta V_j^-)^2,$$

and consequently we write instead of (10) the inequality

$$2\,\delta W \geqslant \sum_j \int \frac{A_j V_j}{2}\,(\delta V_j^+ - \delta V_j^-)^2\,\mathrm{d}x \pm \int (\delta E)^2\,\mathrm{d}x. \tag{32}$$

Taking into account (26) and (30), (32) can be written

$$2\,\delta W \geqslant \int \left\{\sum_j \frac{1}{k_j^2}\left(\frac{\mathrm{d}}{\mathrm{d}x}\,\delta E_j\right)^2\right\} \mathrm{d}x \pm \int (\delta E)^2\,\mathrm{d}x = Q. \tag{33}$$

5. Stability in the Gravitational Case for a Decreasing $F(\varepsilon)$

If $F(\varepsilon)$ is always decreasing all the A_j and consequently all the k_j^2 are positive. We have to study the sign of Q. The term $\sum_j \int k_j^{-2}((\partial/\partial x)\,\delta E_j)^2\,\mathrm{d}x$ is always positive and $-\int(\delta E)^2\,\mathrm{d}x$ always negative. We use the Schwarz inequality

$$\left(\sum_j \frac{\xi_j^2}{k_j^2}\right)\left(\sum_j k_j^2\right) \geqslant \left(\sum \xi_j\right)^2 \tag{34}$$

and write

$$Q \geqslant \frac{\left(\frac{\partial}{\partial x}\sum_j \delta E_j\right)^2}{\sum_j k_j^2}\,\mathrm{d}x - \int (\delta E)^2\,\mathrm{d}x. \tag{35}$$

Introducing $\sum_j k_j^2 = k^2(x)$ we get

$$Q \geqslant \int \left\{ \frac{\left(\frac{\partial}{\partial x} \delta E\right)^2}{k^2} - (\delta E)^2 \right\} \mathrm{d}x. \tag{36}$$

It is interesting to look at the expression of $k^2(x)$ in the continuum case. We introduce the quantity

$$\xi^2(x) = -\int \frac{\mathrm{d}F}{\mathrm{d}\varepsilon} \mathrm{d}V = -\int \frac{\partial F}{\partial V} \frac{1}{V} \mathrm{d}V. \tag{37}$$

In the W.B. limit (37) becomes, taking into account (1):

$$\xi^2(x) = -\sum_j A_j \int \frac{\delta(V + V_j) - \delta(V - V_j)}{V} \mathrm{d}V = \sum_j \frac{2A_j}{V_j} = k^2(x). \tag{38}$$

We notice that the steady state solution F satisfies

$$V(\partial F/\partial x) + E(\partial F/\partial V) = 0. \tag{39}$$

We divide (39) by V then multiply by $\mathrm{d}V$ and integrate

$$\frac{\partial \varrho}{\partial x} + \left[\int \frac{1}{V} \frac{\partial F}{\partial V} \mathrm{d}V\right] E = 0. \tag{40}$$

We use the Poisson law $\partial E/\partial x = -\varrho$ and the definition of $\xi^2 = k^2$ to obtain from (40) the equation for the steady state gravitational field

$$(\mathrm{d}^2 E/\mathrm{d}x^2) + k^2(x) E = 0. \tag{41}$$

Now let us go back to the question of the sign of Q. We have to minimize an expression of the form

$$Q = \int \mathscr{L}(\delta E, (\partial/\partial x) \delta E, x) \mathrm{d}x.$$

The Euler-Lagrange equation (Whittaker and Watson, 1966) gives the equation that δE must fulfil in order to get an extremum

$$\frac{\mathrm{d}}{\mathrm{d}x} \frac{\partial \mathscr{L}}{((\partial/\partial u) \delta E)} - \frac{\partial \mathscr{L}}{\partial(\delta E)} = 0. \tag{42}$$

Introducing $\mathscr{L}$ as given by (36) in (42) we get the equation

$$\frac{\partial}{\partial x} \left[\frac{(\partial/\partial x) \delta E}{k^2(x)}\right] + \delta E = 0. \tag{43}$$

Moreover due to the positiveness of the coefficient of $((\partial/\partial x) \delta E)^2$ this extremum is a minimum. Now we show that the solution of (43) is connected to the solution of (41). More precisely, α being an arbitrary constant

$$\delta E = \alpha \varrho = -\alpha(\mathrm{d}E/\mathrm{d}x) \tag{44}$$

is a solution of (43) if E is solution of (42). From (44) and (41) we get

$$(\partial/\partial x)\,\delta E = -\,\alpha\,(\mathrm{d}^2E/\mathrm{d}x^2) = \alpha k^2(x)\,E$$

and

$$\frac{\partial}{\partial x}\left[\frac{(\partial/\partial x)\,\delta E}{k^2(x)}\right] = +\,\alpha\,\frac{\mathrm{d}E}{\mathrm{d}x},$$

and (44) is consequently a solution of (43). Moreover, the boundaries conditions $\delta E(a)=\delta E(-a)=0$ and $\varrho(a)=\varrho(-a)=0$ are the same.

To compute Q we introduce the solution (44) in the right hand side of (36)

$$Q \geqslant \int\left\{\frac{((\partial/\partial x)\,\delta E)^2}{k^2} - (\delta E)^2\right\}\mathrm{d}x = \int\left\{\frac{\alpha^2}{k^2}\left(\frac{\mathrm{d}^2E}{\mathrm{d}x^2}\right)^2 - \alpha^2\left(\frac{\mathrm{d}E}{\mathrm{d}x}\right)^2\right\}\mathrm{d}x. \quad (45)$$

Taking (41) into account (45) is written

$$Q \geqslant \alpha^2\int\left[-\,E\,(\mathrm{d}^2E/\mathrm{d}x^2) - (\mathrm{d}E/\mathrm{d}x)^2\right]\mathrm{d}x. \quad (46)$$

But integrating by parts $E(\mathrm{d}^2E/\mathrm{d}x^2)$ and remembering that at the two extremities $\mathrm{d}E/\mathrm{d}x=0$, we obtain

$$-\int E\,(\mathrm{d}^2E/\mathrm{d}x^2)\,\mathrm{d}x = \int(\mathrm{d}E/\mathrm{d}x)^2\,\mathrm{d}x.$$

As a consequence, if $\delta E=\alpha\varrho$, then the right hand side of (36) is zero. It remains to be shown that this minimum value (zero) is an absolute minimum. To show it, (Kulsrud and Mark, 1970), we try an expression of the form:

$$\delta E = \alpha(x)\,\varrho \quad (47)$$

where $\alpha(x)$ is now a function of x. We notice that (47) implies that α is a continuous function of x since ϱ vanishes only at the two extremities where δE also vanishes. Taking (47) into account we have

$$\int\left[k^{-2}((\partial/\partial x)\,\delta E)^2 - (\delta E)^2\right]\mathrm{d}x = \int\left[(\mathrm{d}\alpha/\mathrm{d}x)^2\,(\varrho^2/k^2) + \right.$$
$$\left. + \,2\alpha\,(\mathrm{d}\alpha/\mathrm{d}x)\,(\varrho/k^2)\,(\mathrm{d}\varrho/\mathrm{d}x) + (\alpha^2/k^2)\,(\mathrm{d}\varrho/\mathrm{d}x)^2 - \alpha^2\varrho^2\right]\mathrm{d}x. \quad (48)$$

We use the steady state relations derived from Poisson's law and (41)

$$(\mathrm{d}\varrho/\mathrm{d}x) = k^2E - \alpha^2\varrho^2 = \alpha^2\varrho\,(\mathrm{d}E/\mathrm{d}x)$$

to transform the last three terms of the right hand side integral of (48).

We get

$$2\alpha\,(\mathrm{d}\alpha/\mathrm{d}x)\,(\varrho/k^2)\,(\mathrm{d}\varrho/\mathrm{d}x) + (\alpha^2/k^2)\,(\mathrm{d}\varrho/\mathrm{d}x\;^2 - \alpha^2\varrho^2$$
$$= \alpha^2\,(\mathrm{d}/\mathrm{d}x)\,(E\varrho) + E\varrho\,(\mathrm{d}\alpha^2/\mathrm{d}x) = (\mathrm{d}/\mathrm{d}x)\,[\alpha^2\varrho E].$$

In the integration the contribution of these three last terms is indeed zero since ϱ

vanishes at the two extremities. We obtain

$$\int [k^{-2}((\partial/\partial x)\,\delta E)^2 - (\delta E)^2]\,\mathrm{d}x = \int k^{-2}\varrho^2\,(\mathrm{d}\alpha/\mathrm{d}x)^2\,\mathrm{d}x,$$

A quantity obviously positive except if α is a constant where we recover the solution given by (44). The gravitational system is consequently stable. We notice that the fundamental point in the demonstration was that, $\varrho(x)$ being always positive, $\alpha(x)$ is a continuous arbitrary function and the integrations are permitted.

We now look for a demonstration of the stability property in a spherically system. The first idea is to consider, in analogy with the plane case, the equation:

$$\mathscr{L}(r) = \int (-(\mathrm{d}F/\mathrm{d}\varepsilon))\,[\delta F/(-(\mathrm{d}F/\mathrm{d}\varepsilon)) + \delta\Phi]^2\,\mathrm{d}V$$

the energy variation δW can be written:

$$\delta W/2\pi = \int_0^a \mathscr{L}(r)\,r^2\,\mathrm{d}r + I,$$

with I given by:

$$I = \int_0^a [-2\,\delta\varrho\,\delta\phi - k^2(\delta\phi)^2 + \delta\varrho\,\delta\phi]\,r^2\,\mathrm{d}r \tag{49}$$

with

$$k^2(r) = -\int (\mathrm{d}F/\mathrm{d}\varepsilon)\,\mathrm{d}V.$$

Now we integrate by parts $-\int_0^a \delta\varrho\,\delta\phi r^2\,\mathrm{d}r$ using the Poisson equation

$$(1/r^2)(\mathrm{d}/\mathrm{d}r)[r^2(\mathrm{d}\,\delta\phi/\mathrm{d}r)] = \delta\varrho.$$

(49) can now be written

$$I = \int_0^a [((\mathrm{d}\,\delta\phi/\mathrm{d}r))^2 - k^2(r)(\delta\phi)^2]\,r^2\,\mathrm{d}r. \tag{50}$$

To prove stability it is sufficient to show that I given by (50) is positive.

The Euler-Lagrange equation corresponding to the extremization of (50) is

$$(1/r^2)(\mathrm{d}/\mathrm{d}r)[r^2(\mathrm{d}\,\delta\phi/\mathrm{d}r)] + k^2(r)\,\delta\phi = 0. \tag{51}$$

The solution of (51) is a perturbation δF given by

$$\delta F = (\mathrm{d}F/\mathrm{d}\varepsilon)\,\delta\Phi. \tag{52}$$

For such a perturbation taking the definition of $k^2(r)$ into account

$$\delta\varrho = -\,k^2(r)\,\delta\Phi\,,$$

and the Poisson equation $\Delta\,\delta\phi = \delta\varrho$ is nothing else but Equation (51) (Δ being the Laplacian operator for a spherically symmetric system). $\delta\phi$ must fulfil.

$$\Delta\,\delta\phi + k^2(r)\,\delta\phi = 0\,. \tag{53}$$

For $\delta\phi$ satisfying Equation (52), $I=0$. We notice that Equation (53) is the spherical equivalent of Equation (43). But the analogy stops here. A solution of Equation (43) was known because the steady state electric field fulfills a similar equation: Equation (41). Then we could give a proof of positiveness of Q by introducing

$$\delta E = \alpha(x)\cdot\varrho(x)\,. \tag{54}$$

Unfortunately, the steady state field for a spherical geometry does not obey Equation (53) but

$$\Delta E + \left(k^2(r) - 2/r^2\right) E = 0\,. \tag{55}$$

The unfortunate term: $(2/r^2)\,E$ destroys the equivalence and now we should study the solution of Equation (53). This is a difficult problem and while progress could be obtained by selecting a model (i.e. a well defined function $k^2(r)$) it is obvious that nothing can be said in the general case.

A way out of this difficulty has been given by Doremus *et al.* (1971) using the multiple Water-Bag concept. The contours delineating the bags were built in such a way that for a given mass the total energy is a minimum, then it was proved that these contours fulfill the steady state Vlasov equation. It was finally proved that there is no marginal mode corresponding to $\delta W=0$.

6. Comment on the Marginal Mode

We have seen that one mode (corresponding to a δE fulfilling Equation (44) in the plane case) can exist with a growth rate $\gamma=0$, and for the spherical case such a mode does not exist.

We show that this mode is a trivial mode corresponding to the fact that the equations of motion are Galilean invariant and that if we add to each particle a very slight average velocity the system will drift. We are going to give the demonstration for the one-dimensional plane geometry case. Again we will suppose that F is a monotonically decreasing function of the energy ε. We introduce:

$$\mathscr{L}(x) = \int \left(-\,(\mathrm{d}F/\mathrm{d}\varepsilon)\right) \left[\frac{\delta F}{\left(-\,(\mathrm{d}F/\mathrm{d}\varepsilon)\right)} + \delta\phi\right]^2 \mathrm{d}V\,. \tag{56}$$

(19) can now be written

$$2\,\delta W = \int \mathscr{L}(x)\,\mathrm{d}x + I\,. \tag{57}$$

With I given by

$$I = -2\int \delta\varrho\,\delta\Phi\,\mathrm{d}x - \int \left[(\delta E)^2 + k^2(x)\,(\delta\Phi)^2\right]\mathrm{d}x\,. \tag{58}$$

To get (58) we introduce the definition of $k^2(x)$ given by (37) and (38). Now we use the Poisson equation $\partial^2(\delta\Phi)/\partial x^2 = \delta\varrho$ to integrate by parts the first integral in the rhs of (58).

Moreover we use the boundaries conditions $\delta E(\pm a) = -\delta(\delta\phi)/\partial x = 0$. We obtain

$$I = \int \left[(\delta E)^2 - k^2(x)\,(\delta\phi)^2\right]\mathrm{d}x\,. \tag{59}$$

(59) is the equivalent of Equation (49), when considering the spherical case $\mathscr{L}(x)$ being obviously positive (since $-\mathrm{d}F/d\varepsilon > 0$), we will prove stability if $I > 0$. We minimize (59) using the Euler-Lagrange method, $\delta\phi$ must be a solution of

$$(\partial^2/\partial x^2)\,\delta\phi + k^2(x)\,\delta\phi = 0 \tag{60}$$

with the boundaries conditions $\partial(\delta\phi)/\partial x|_{x=\pm a=0}$.

But we already encountered Equation (60) for the steady state of the field {Equation (41)} with the same boundaries conditions. (The density ϱ must indeed be zero at the two extremities.) For a $\delta\phi$ fulfilling (60) it is easily shown that $I=0$. Moreover for the mode defined by

$$\delta F = (\mathrm{d}F/\mathrm{d}\varepsilon)\,\delta\phi \tag{61}$$

$\delta W = 0$. For all the other modes $\delta W > 0$ and the conservation of energy implied by the Vlasov equation excludes all these modes. We must now show that this always present marginal mode will not destroy the stability property for the system with $\mathrm{d}F/\mathrm{d}\varepsilon < 0$. In fact as we already announced this mode is a trivial solution which corresponds to a space displacement of the whole system, a motion clearly allowed by the Galilean invariance of the Vlasov equation, which obviously has not to be considered as an instability. Indeed for a displacement Δ

$$\delta F = -(\partial F/\partial x)\,\Delta = -(\mathrm{d}F/\mathrm{d}\varepsilon)\,(\partial\varepsilon/\partial x)\,\Delta = (\mathrm{d}F/\mathrm{d}\varepsilon)\,E\Delta \tag{62}$$

taking into account $\varepsilon = [\tfrac{1}{2}]\,V^2 + \phi$ and $E = -\mathrm{d}\Phi/\mathrm{d}x$, we have also

$$\delta\Phi = -(\mathrm{d}\phi/\mathrm{d}x)\,\Delta = E\Delta\,. \tag{63}$$

Combining (62) and (63) we see that for a displacement Δ, δF satisfies (61). Moreover, (63) shows that $\delta\phi$ for this mode is proportional to E and fulfils the same equation and boundaries conditions as the marginal mode defined by (60). This shows the identity of this last mode with a trivial displacement mode.*

* The authors are indebted to Dr Howard Bloomberg for pointing out the identity of these two modes.

In the plasma problem such a mode will not be found in the case of an inhomogeneous ion background since the equation for the perturbed potential $\delta\phi$ will not be identical with the equation of the steady state electric field. Indeed we cannot displace the electron population through such a fixed ion background without creating an electric field (unless of course N_i=constant).

7. Stability for a Two Streams Inhomogeneous Plasma

In the plasma case the ions provide a fixed, inhomogeneous, globally neutralizing background and in this calculation we suppose that all trajectories extend from $x=-\infty$ to $x=+\infty$ (which means that we suppose no trapped particles). Moreover we consider a velocity distribution function $F(V)$ of the type indicated on Figure 2 while the trajectories in phase space are shown in Figure 3.

A useful staring point is provided by the well know property of an infinite homogeneous plasma. Then k^2 as given by (37) and (38) is independent of x and its sign indicates stability ($k^2>0$ for a stable plasma, $k^2<0$ for an unstable plasma).

But k^2 has also another interesting meaning. If we plot the growth rate γ as a function of the wavenumber K we get the result indicated on Figure 4. $(-k^2)^{\frac{1}{2}}$ is the maximum wavenumber of instability. Consequently $K_m^2=-k^2$ gives also some indication about the length necessary for the instability to develop. Indeed if L is such that $K_m L\gg 1$ we will suspect the system to have properties close to those of the infinite system while for $K_m L\ll 1$ we know that very likely no instability will develop.

Obviously no information can be obtained about the sign of Q. Since k^2 is negative in some region the right hand side of (36) can be negative. Consequently we must proceed to a finer analysis. We will suppose that the system goes unstable through the point $\omega^2=0$. For such a mode (called the marginal stability mode) we have the N differential equations of second order corresponding to the N bags. This system is given by (27) where we equal the left hand side to zero.

$$\omega_j^2\,\delta E - V_j(\mathrm{d}/\mathrm{d}x)\,\{V_j(\mathrm{d}\,\delta E_j/\mathrm{d}x)\} = 0\,. \tag{64}$$

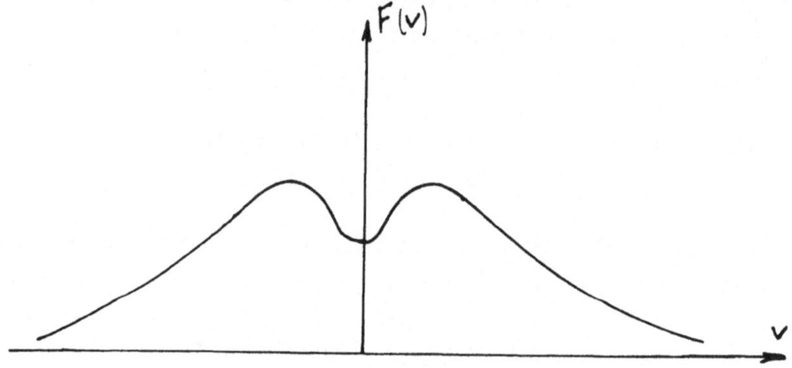

Fig. 2. Two streams instability: velocity distribution.

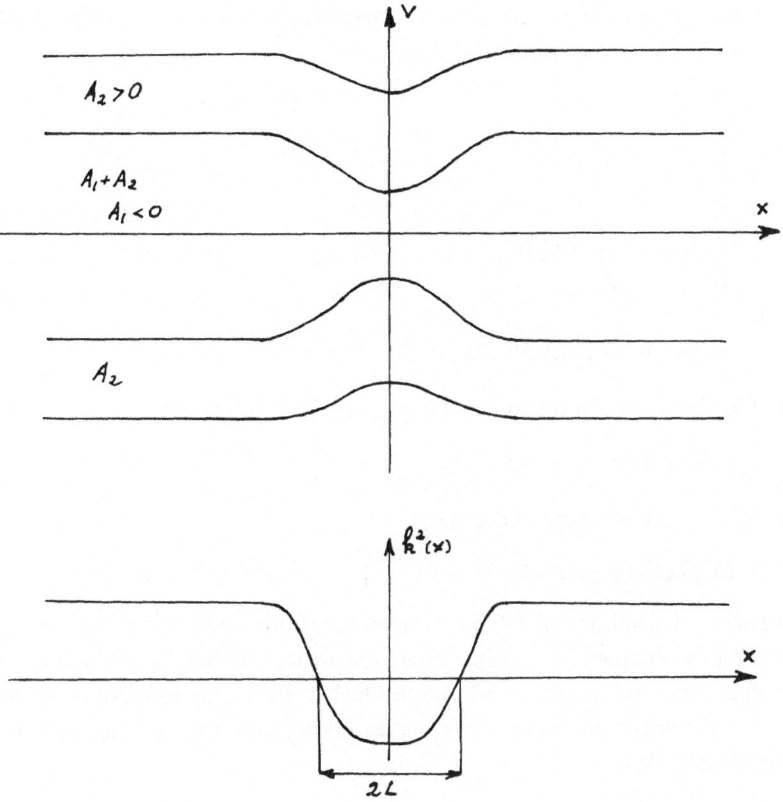

Fig. 3. Two streams inhomogeneous plasma: phase space representation and variation of $k^2(x)$ with x.

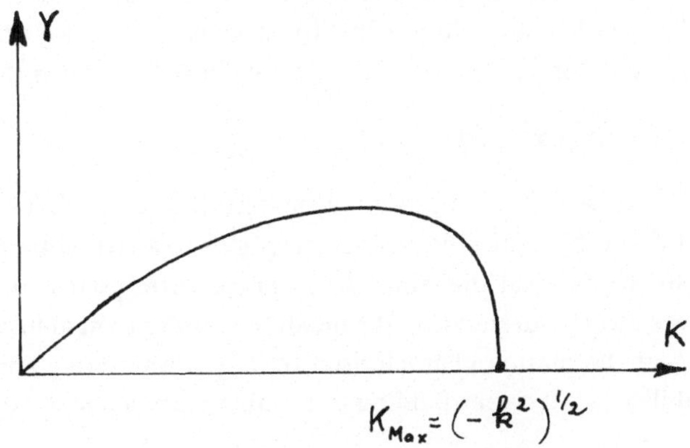

Fig. 4. Growth rate γ function of wavenumber K.

We show that for this mode the system of N Equations (64) degenerates into one second order equation. To show that we multiply (64) by $\mathrm{d}x/V_j$ and integrate on x from $-\infty$ to x. Moreover we introduce

$$\begin{cases} \delta\phi = -\int\limits_{-\infty}^{x} \delta E \, \mathrm{d}x\,, \\ \delta\phi_j = -\int\limits_{-\infty}^{x} \delta E_j \, \mathrm{d}x\,. \end{cases} \tag{65}$$

We obtain

$$-2A_j \, \delta\phi - V_j (\mathrm{d}\, \delta E_j/\mathrm{d}x) = 0\,. \tag{66}$$

Dividing (58) by V_j and introducing k_j^2 as given by (30) we get

$$(\partial^2/\partial x^2)\, \delta\Phi_j - k_j^2 \, \delta\phi = 0\,. \tag{67}$$

We sum (67) for all the bags and obtain

$$(\partial^2/\partial x^2)\, \delta\phi - k^2(x)\, \delta\phi = 0\,, \tag{68}$$

which we will solve subject to the boundaries conditions $\delta\phi(\pm\infty)=0$. This equation has been already obtained by Santini (1969) and we notice that if we look in the plasma case for the equation of the steady state electric field (corresponding to formulas (39)–(41) for the gravitational case) we obtain taking into account the space variation of the ion density $N_i(x)$

$$(\partial^2 E/\partial x^2) - k^2(x)\, E + (\mathrm{d}N_i/\mathrm{d}x) = 0\,. \tag{69}$$

We see that the equation for $\delta\phi$ {Equation (68)} is not equivalent to the equation of the steady state {Equation (69)}. This is due to the presence of the ions which brings an extra degree of freedom in the problem.

Now to determine the stability property of the system let us consider a given profile of $k^2(x)$. Usually (68) has no solution but for a given profile we can compute the eigenvalue of the operator $\{k^2(x) - \partial^2/\partial x^2\}$, i.e. we look for λ such that

$$[k^2(x) - (\mathrm{d}^2/\mathrm{d}x^2)]\, \delta\phi = \lambda\, \delta\phi\,. \tag{70}$$

We see that (70) corresponds to a marginal mode fitting a profile characterised not by $k^2(x)$ but $k^2(x) - \lambda$. If λ is positive this corresponds to a virtual medium less stable in all points than the original one since k^2 negative corresponds to an unstability. Consequently if we have to 'destabilize' the medium in order to fit at least one marginal mode it means that the plasma characterised by $k^2(x)$ was stable and the sign of λ indicates the stability ($\lambda > 0$ corresponding to a stable plasma and $\lambda < 0$ to an unstable one).

In the gravitational case things are a little bit more complicated. We have, always present, the trivial displacement mode studied in the preceding paragraph and which must be considered as a marginal stability mode. Equation (68) is now changed to

$$(\partial^2/\partial x^2)\,\delta\phi + k^2(x)\,\delta\phi = 0. \tag{71}$$

(71) should be really changed into an equation for $\delta E = -(\partial\delta\phi/\partial x)$ with the boundaries conditions $\delta E(\pm a)=0$. We already noticed that Equation (41) is just that equation with E playing the role of $\delta\phi$ and $\varrho = -(\partial/\partial x)\,E$ the role of δE.

δE as ϱ should banish at the two boundaries. Finally let us show that for this mode all the inequalities obtained in paragraph 5 turn into equalities. First of all we have for each bag

$$J_j \pm (\partial E_j/\partial t) = 0,$$

$\omega \to 0$ and consequently $J_j = 0$.

Taking into account

$$J_j = (A_j/2)\left[(V_j + \delta V_j^+)^2 - (-V_j + \delta V_j^-)^2\right] = A_j V_j (\delta V_j^+ + \delta V_j^-) = 0$$

and we should take the equality sign in (33) with $Q = 2\delta W$.

Finally the Schwarz inequality (34) will turn also into an equality if ξ_j/k_j is proportional to k_j which means

$$(\partial/\partial x)\,\delta E_j \quad \text{proportional to} \quad k_j^2 = (2A_j/V_j)$$

but from Equation (66) of the marginal mode we saw that:

$$(1/k_j^2)\,(\partial\,\delta E_j/\partial x) = -\,\delta\phi. \tag{72}$$

This last result precise the proof of stability given in 5.

Indeed a trivial displacement marginal mode $\omega=0$ is possible and for this mode $\delta W=0$. But for all the other modes $\delta W>0$ and such motions are not allowed indicating that the system is stable.

8. Conclusion

In this paper we have presented two kinds of results. For function F decreasing with the energy, the energy variation δW can be easily shown to be positive taking into account the phase conservation of the Liouville equation and the self-consistency of the system. For plane systems (see also Kulsrud and Mark, 1970) the Schwarz inequality allows a simple treatment. For spherical systems (studied in Doremus *et al.*, 1971) we must be more cautious and avoid the too strong Schwarz inequality and the proof proceeds in a different way.

For function F not always decreasing with energy we get an easily understandable picture for the plasma case. The starting point is the well-known fact that for some situations the instability will come through the point $\omega=0$. For such a mode the dynamics are easily computed and the N bags equations degenerate into one second order differential equation. The following physical picture is obtained. The plasma is locally defined by a coefficient $k^2(x)$ and we try to fit the unstable mode with the largest wavenumber. This corresponds to possible 'propagation' of the instability in the region where k^2 is negative and absorption in the region where k^2 is positive and the problem of the possible existence of instability is identical to fitting an electromagnetic mode in a propagating region surrounded by two absorbing media.

Finally the sign of the virtual modification of the 'propagation of instability' constant of the medium gives information about its stability property and the mathematical problem turns out to be the computation of the eigenvalues of a Schroedinger operator. Similar results must be looked for in the gravitational case although we lost the physical picture built from our knowledge in the plasma case of the homogeneous problem.

References

Antonov, V. A.: 1961, *Sov. Astron.-AJ* **4**, 859.
Doremus, J. P., Feix, M. R., and Baumann, G.: 1971, *Phys. Rev. Letters* **26**, 725.
Hohl, F.: 1969, *Phys. Fluids* **12**, 230.
Hohl, F. and Feix, M. R.: 1967, *Astrophys. J.* **147**, 1164.
Hohl, F. and Feix, M. R.: 1968, *Astrophys. J.* **151**, 783.
Kulsrud, R. M. and Mark, J. W. K.: 1970, *Astrophys. J.* **160**, 471.
Lynden Bell, D.: 1969, *Monthly Notices Roy. Astron. Soc.* **144**, 189.
Lynden Bell, D. and Sanitt, N.: 1969, *Monthly Notices Roy. Astron. Soc.* **143**, 167.
Penrose, O.: 1960, *Phys. Fluids* **3**, 258.
Santini, F.: 1969, *Plasma Phys.* **12**, 230.
Whittaker, E. T. and Watson, G. N.: 1966, *A Course of Modern Analysis*, Cambridge University Press.

PART IV

SUMMARY OF THE COLLOQUIUM

SUMMARY

In summarizing this third IAU Colloquium on the Gravitational N-Body Problem, I will confine my remarks to the numerical experiments. The analytical treatments are – I think – more straightforward and speak for themselves.

Because the numerical experiments are often as difficult to interpret as the observations of the astronomical objects themselves, I would like to add to the justification for this difficult effort. First of all, numerical simulation can delineate the relevant physics. As we control the physics we put into the experiment, we know what is responsible for the phenomena we get out. Even qualitative results can tell us whether Newtonian Gravitation is adequate, or whether magnetic fields, radiation, dissipation, mass loss, etc. are also important. And while astronomical observations of the projected densities can often be made consistent with almost any theory, the computer results provide detailed information about the distribution function. Finally, and perhaps most important, in the computer we can watch the systems evolve.

There are two time scales in the Gravitation N-Body Problem. They are:

'The crossing time' $= Tc = R/V$;
'The relaxation time' $= Tr = \text{constant} \times N/\ln N \times Tc$.

These times, for representative astronomical systems, are presented in Table I.

TABLE I

Crossing times and relaxation times for representative astronomical systems

Description	N	$\varrho(M_\odot/\text{pc}^3)$	Tc(yr)	Tr(yr)	Age(yr)
Galactic cluster	10^2–10^3	1	10^7	10^7–6×10^7	10^7–10^8
Cluster of galaxies	10^2–10^3	10^{-4}	10^9	10^9–6×10^9	10^{10}
Globular cluster	10^5–10^7	10^2	10^6	4×10^8–2×10^{10}	10^{10}
Galaxy	10^{10}–10^{12}	10^{-1}	3×10^7	6×10^{14}–4×10^{16}	10^{10}

Specifically, I have set $Tc = 10^7$ yrs$/\sqrt{\varrho}$ with ϱ in Solar Masses per cubic parsec, and $Tr/Tc = N/50 \log_{10} N$. One could argue for a factor of five in both constants, but the ϱ and N dependence are correct. In any case, as the evolution of a collisionally dominated system proceeds, ϱ increases and N decreases, so both Tc and Tr get shorter; the evolution 'runs away' (on a time scale of 10^7–10^9 yrs).

M. Lecar (ed.), Gravitational N-Body Problem, 367–370.

Collisional relaxation is certainly important for Galactic and Globular Clusters and certainly not important for Galaxies; it's not clear whether Clusters of Galaxies are partially relaxed or not. I will discuss the numerical simulation of these systems in order of increasing N.

A. Galactic Clusters

We can now simulate a system containing a few hundred stars by direct integration. The computing time is approximately $(N/100)^2$ 7094-hr per crossing time. Wielen was able to match the observed disruption times to within a factor of three, but this required considerable extrapolation of his numerical experiments. He established clearly the importance of the tidal field of the Galaxy. However, in judging the effectiveness of encounters with gas clouds, Wielen's results differed from those of Bouvier and Janin.

As this problem is within reach of direct numerical simulation, one hopes that more experiments will be performed with realistic mass spectrums and tidal fields. Perhaps for the nearby clusters, more quantitative comparisons using density profiles and velocity dispersions can be made.

B. Clusters of Galaxies

In spite of their importance for cosmology, there were no simulations of clusters of galaxies presented at the colloquium. I wonder if numerical experiments might indicate the observational consequences of various forms of the 'dark matter'. For example, I would suspect that dwarf ellipticals would lead to relatively high central density, while a uniform density background would tend to a flat density profile with perhaps even a hollowed out center.

C. Globular Clusters

These objects are our favorites because they look so tantalizingly similar; they look 'relaxed'. I even strongly suspect that they are globular because they are relaxed. Two new methods to simulate globular clusters were presented at the colloquium; the Monte Carlo Method by Hénon and the Fluid Dynamical method by Larson. Both methods can be scaled to simulate any (large) number of stars. As Hénon predicted, the end state of globular clusters is one of infinite central density. Larson's models, for various initial conditions, reached this end in 30–150 initial relaxation times. In time, both Hénon's and Larson's models forgot their initial conditions, but in Larson's models, they still remembered at present day central densities of $10^3\ M_\odot/\mathrm{pc}^3$. Both methods give the velocity distribution in the core tending towards isotropy, while the velocity distribution in the halo develops a progressive radial bias. Hénon also showed that with a mass spectrum, the evolution is away from global equi-partition. The various masses do not occupy the same volume; instead the heavier masses tend to concentrate towards the center and develop higher *velocities* than the lighter masses.

Both of these methods are very fast, can simulate large N, are statistically stable, and can follow the evolution for many relaxation times. Neither methods treats close encounters accurately, nor binaries at all, which brings us to Aarseth's direct integration of 250 and 500 bodies. The binary is still there and continues to soak up 50–90% of the total binding energy of the cluster. For comparison, two stars separated by the mean distance between stars ($N^{-1/3} R$) have on the average $N^{-5/3}$ (3×10^{-5} for $N=500$) of the total binding energy. Many arguments have been given to indicate that binaries should become less important with large N. For example, a typical argument to show that the number of binaries should not increase with N, goes as follows. The average phase space volume per star is $r^3v^3=N(Gmr_0)^{3/2}$, with $r_0=N^{-1/3} R$ and $v^2=GNm/R$. The phase space volume per binary star is $(Gma)^{3/2}$ where 'a' is the separation and $v^2=Gm/a$. As 'a'$\leqslant r_0$, the chance of two stars sharing the binary volume goes as $1/N^2$; since there are N^2 pairs, the chance of finding a binary is independent of N. This argument is probably not far off as the experiments do confirm that the number of binaries formed does not increase with N. But one is enough! And as this phenomenon has persisted for N going from 25 to 500, I don't think we're free to dismiss it. The binary serves both as an energy sink (which a high central concentration could mimic in Hénon's or Larson's models) and as a source of energetic encounters (which is difficult to duplicate in their models). It would be usefull if Aarseth, Hénon and Larson compared their time scales by calculating the same initial model to the time when 25% of the mass escaped, or when the radius containing half the mass increased by some fixed factor.

D. Galaxies

An impressive effort was made by Miller, Prendergast and Quirk (MPQ) and Hohl to simulate disk galaxies and search for spiral patterns. Both methods use 50–100000 test particles and calculate the mean field on a 64×64 to 256×256 grid. Both models have verified the existence of 'density waves' (versus material arms). But both models are typically 'too hot'; the velocity dispersions being 1.5 to 2 times the velocity dispersions to stabilize axi-symmetric disturbances (according to Toomre). Scaled to our galaxy, the models have velocity dispersions of 75 km/s vs the 35–40 km/s 'observed'.

At the colloquium, the discussions tended to fault the numerical procedures in the experiments; e.g., a random component in the force. But Hohl did in fact achieve a static, self-consistent model which did not further heat up. In addition, the heating seemed to be associated with the formation of a central bar; when the bar settled down, the heating stopped. I am inclined to take the experiments at face value and to suggest that Toomre's velocity dispersion is just not high enough to stabilize against non-axi-symmetric disturbances, namely the bar. However, only a limited class of initial distribution functions have so far been studied, and that class should be considerably widened before we are forced to accept that higher velocity dispersions are necessary for stability.

Concluding Remarks

'Experimental Stellar Dynamics', to use Frank Hohl's phrase, has now survived it's first decade. Like most vigorous experimental programs, there has been a tendency to 'skim the cream'. This tendency not only results from the experimenters impatience to get on to new finds, but is encouraged by the expense and limited availability of computing time. I hope that now that the field has matured and to some extent proven itself, more time will be available for systematics, and for critical comparisons of the results of the various methods of numerical simulation.

I have become more convinced that numerical experiments provide the most reliable guide to further extensions of the analytical theory. I am thus encouraged by the continued vitality of this field, and look forward to new stimulation from computer simulation.

M. LECAR

PART V

APPENDIX: METHODS OF COMPUTER SIMULATION OF THE GRAVITATIONAL N-BODY PROBLEM

DIRECT INTEGRATION METHODS OF THE N-BODY PROBLEM

S. J. AARSETH
Institute of Theoretical Astronomy, Cambridge, England

Abstract. A fourth-order polynomial method for the integration of N-body systems is described in detail together with the computational algorithm. Most particles are treated efficiently by an individual time-step scheme but the calculation of close encounters and persistent binary orbits is rather time-consuming and is best performed by special techniques. A discussion is given of the Kustaanheimo-Stiefel regularization procedure which is used to integrate dominant two-body encounters as well as close binaries. Suitable decision-making parameters are introduced and a simple method is developed for regularizing an arbitrary number of simultaneous two-body encounters.

1. Introduction

Numerical integrations of the gravitational N-body problem may in principle be performed by the simple method of advancing all equations of motion stepwise in time using constant forces. Any desired accuracy may be obtained by an appropriate choice of the time interval as long as rounding errors remain small. In view of the lengthy force calculation for large particle numbers it is desirable to employ high-order schemes which allow the choice of greater intervals. More powerful methods have also been developed for dealing with special configurations which would otherwise lead to serious loss of accuracy, or at best require very time-consuming calculations.

The present paper describes in some detail an ordinary fourth-order polynomial method as well as the Kustaanheimo-Stiefel regularization procedure for studying close two-body encounters. Although the former is very efficient in general circumstances, it is desirable to include special treatments of critical cases. Alternative regularization formulations are discussed elsewhere in this volume (Szebehely and Bettis, 1972; Heggie, 1972). A classical perturbation treatment of close binaries is also available, but this method is more suitable for small perturbations (Aarseth, 1970). Several types of ordinary integration schemes have been used for direct N-body calculations but high-order polynomial methods appear to be the most efficient tried so far (Lecar, 1968).

2. Individual Time-Step Method

The motivation for introducing an individual time-step method stems from the desire to solve the equations of motion to the same relative accuracy in the absence of rounding errors. The resulting saving of force summations speeds up the calculations by a large factor, while the additional requirement of co-ordinate prediction only represents 20–30% of the total computing time depending on the order used.

M. Lecar (ed.), Gravitational N-Body Problem, 373–387.

In the present formulation which follows an earlier derivation (Aarseth, 1968) we make explicit use of the first four terms of a fitting polynomial, but an additional correction term is included. Practical tests indicate that there is no significant gain in efficiency when going to higher orders, but an equivalent general derivation is available (Wielen, 1967).

We begin by writing the extrapolating force polynomial for an arbitrary body as an expansion about the reference time t_0, with the interval $t_r = t - t_0$,

$$\mathbf{F} = \mathbf{F}_0 + \mathbf{B}' t_r + \mathbf{C}' t_r^2 + \mathbf{D}' t_r^3 + \mathbf{E}' t_r^4 . \tag{1}$$

The coefficients $\mathbf{B}'$, $\mathbf{C}'$, $\mathbf{D}'$ are obtained by fitting the polynomial at three previous times and can be expressed in terms of higher divided differences at the time $t = t_0$ weighted by the corresponding intervals, while $\mathbf{F}_0$ represents the force per unit mass. Let the three preceding time-steps be denoted by Δt_1, Δt_2, Δt_3 in sequential order such that Δt_3 is the most recent interval. The force expansion (1) is then considered valid over the time interval $-(\Delta t_1 + \Delta t_2 + \Delta t_3) \leqslant t_r \leqslant \Delta t_4$, but the coefficient $\mathbf{E}'$ is not known until the end of the fourth step Δt_4 when its contribution is added. This procedure may be referred to as a semi-iteration since the main part of the improvement is achieved without recalculating the force which is based on the predicted position. In this way almost one extra order of integration is included at very little additional effort and only one force calculation is needed for each interval.

It is more convenient for computational purposes to write Equation (1) in the form

$$\begin{aligned} \mathbf{F} = \mathbf{F}_0 + \mathbf{B} t_r + \mathbf{C}(\Delta t_3 + t_r)\, t_r + \mathbf{D}(\Delta t_2 + \Delta t_3 + t_r)(\Delta t_3 + t_r)\, t_r \\ + \mathbf{E}(\Delta t_1 + \Delta t_2 + \Delta t_3 + t_r)(\Delta t_2 + \Delta t_3 + t_r)(\Delta t_3 + t_r)\, t_r . \end{aligned} \tag{2}$$

Explicit expressions for the coefficients $\mathbf{B}$, $\mathbf{C}$, $\mathbf{D}$ may then be obtained in terms of the divided backwards differences $\hat{\mathbf{F}}_0$, $\hat{\hat{\mathbf{F}}}_0$, $\hat{\hat{\hat{\mathbf{F}}}}_0$ defined by

$$\hat{\mathbf{F}}_0 = \frac{\mathbf{F}_0 - \mathbf{F}_{-3}}{\Delta t_3}, \quad \hat{\hat{\mathbf{F}}}_0 = \frac{\hat{\mathbf{F}}_0 - \hat{\mathbf{F}}_{-3}}{\Delta t_3}, \quad \hat{\hat{\hat{\mathbf{F}}}}_0 = \frac{\hat{\hat{\mathbf{F}}}_0 - \hat{\hat{\mathbf{F}}}_{-3}}{\Delta t_3}, \tag{3}$$

where the force at time $t_r = -\Delta t_3$ is denoted by $\mathbf{F}_{-3}$. The adopted expressions take the final form

$$\mathbf{B} = \hat{\mathbf{F}}_0, \quad \mathbf{C} = \frac{\Delta t_3}{\Delta t_2 + \Delta t_3} \hat{\hat{\mathbf{F}}}_0, \tag{4}$$

$$\begin{aligned} \mathbf{D} = {} & \frac{\Delta t_2 \Delta t_3}{(\Delta t_1 + \Delta t_2 + \Delta t_3)(\Delta t_1 + \Delta t_2)} \hat{\hat{\hat{\mathbf{F}}}}_0 - \\ & - \frac{\Delta t_2^2 - \Delta t_1 \Delta t_3}{(\Delta t_1 + \Delta t_2 + \Delta t_3)(\Delta t_1 + \Delta t_2)(\Delta t_2 + \Delta t_3)} \hat{\hat{\mathbf{F}}}_0 , \end{aligned}$$

$$\mathbf{E} = \frac{\dfrac{\Delta t_4}{\Delta t_3 + \Delta t_4}\hat{\hat{\mathbf{F}}}_4 - \dfrac{\Delta t_3}{\Delta t_2 + \Delta t_3}\hat{\hat{\mathbf{F}}}_0}{(\Delta t_1 + \Delta t_2 + \Delta t_3 + \Delta t_4)(\Delta t_2 + \Delta t_3 + \Delta t_4)} - \frac{\mathbf{D}}{(\Delta t_1 + \Delta t_2 + \Delta t_3 + \Delta t_4)},$$

where the coefficient $\mathbf{E}$ contains the divided difference $\hat{\hat{\mathbf{F}}}_4$ equivalent to $\hat{\hat{\mathbf{F}}}_0$ evaluated at $t_r = \Delta t_4$.

3. Computational Algorithm

The integrations may be started by first calculating the Taylor series derivatives from explicit differentiation of the equation of motion for each particle i,

$$\ddot{\mathbf{r}}_i = -\sum_{\substack{j=1 \\ j \neq i}}^{N} m_j \frac{\mathbf{r}_i - \mathbf{r}_j}{|\mathbf{r}_i - \mathbf{r}_j|^3}, \tag{5}$$

where $\mathbf{F}_0 = \ddot{\mathbf{r}}_i$ by the notation above. Denoting Taylor series derivatives by dots and writing $\mathbf{r}_{ij} = \mathbf{r}_i - \mathbf{r}_j$, $r_{ij} = |\mathbf{r}_{ij}|$, we obtain the relations (Gonzalez and Lecar, 1968)

$$\dot{\mathbf{F}}_0 = -\sum_j m_j \left\{ \frac{\dot{\mathbf{r}}_{ij}}{r_{ij}^3} - \frac{3\mathbf{r}_{ij}(\mathbf{r}_{ij}\cdot\dot{\mathbf{r}}_{ij})}{r_{ij}^5} \right\},$$

$$\ddot{\mathbf{F}}_0 = -\sum_j m_j \left\{ \frac{\ddot{\mathbf{r}}_{ij}}{r_{ij}^3} - \frac{6\dot{\mathbf{r}}_{ij}(\mathbf{r}_{ij}\cdot\dot{\mathbf{r}}_{ij})}{r_{ij}^5} + \frac{3\mathbf{r}_{ij}}{r_{ij}^5}\left[\frac{5(\mathbf{r}_{ij}\cdot\dot{\mathbf{r}}_{ij})^2}{r_{ij}^2} - \dot{\mathbf{r}}_{ij}\cdot\dot{\mathbf{r}}_{ij} - \mathbf{r}_{ij}\cdot\ddot{\mathbf{r}}_{ij} \right] \right\}, \tag{6}$$

$$\dddot{\mathbf{F}}_0 = -\sum_j m_j \left\{ \frac{\dddot{\mathbf{r}}_{ij}}{r_{ij}^3} - \frac{9\ddot{\mathbf{r}}_{ij}(\mathbf{r}_{ij}\cdot\dot{\mathbf{r}}_{ij})}{r_{ij}^5} - \frac{\dot{\mathbf{r}}_{ij}}{r_{ij}^5}\left[9\dot{\mathbf{r}}_{ij}\cdot\dot{\mathbf{r}}_{ij} + 9\mathbf{r}_{ij}\cdot\ddot{\mathbf{r}}_{ij} - \frac{45(\mathbf{r}_{ij}\cdot\dot{\mathbf{r}}_{ij})^2}{r_{ij}^2} \right] - \right.$$
$$\left. - \frac{\mathbf{r}_{ij}}{r_{ij}^5}\left[3\mathbf{r}_{ij}\cdot\dddot{\mathbf{r}}_{ij} + 9\dot{\mathbf{r}}_{ij}\cdot\ddot{\mathbf{r}}_{ij} - \frac{45(\mathbf{r}_{ij}\cdot\dot{\mathbf{r}}_{ij})(\mathbf{r}_{ij}\cdot\ddot{\mathbf{r}}_{ij})}{r_{ij}^2} - \frac{45(\mathbf{r}_{ij}\cdot\dot{\mathbf{r}}_{ij})(\dot{\mathbf{r}}_{ij}\cdot\dot{\mathbf{r}}_{ij})}{r_{ij}^2} + \frac{105(\mathbf{r}_{ij}\cdot\dot{\mathbf{r}}_{ij})^3}{r_{ij}^4} \right] \right\}.$$

Thus the second and third derivatives are readily determined once all the current forces and the corresponding first derivatives have been calculated.

Initial time-steps must now be allocated to each particle and a simpler form of the general criterion is used for this purpose. Usually it is sufficient to adopt

$$\Delta t_4^3 = \eta \frac{F_0}{\frac{1}{6}\dddot{F}_0}, \tag{7}$$

with $F_0 = |\mathbf{F}_0|$, etc. The parameter η specifies the permissible relative change of force during the new step as contributed by the last known term and hence controls the convergence of the Taylor series expansion. The definition (7) is independent of mass and has the desired property of preserving the relative accuracy of each orbit during close encounters by reducing the integration interval. Although the integration proper starts at the time $t=0$, it is necessary to initialize the previous steps in order to make use of the general formulation; hence we put

$$\Delta t_k = \Delta t_4 \quad (k=1,2,3), \tag{8}$$

where again the index i has been suppressed. A consistent conversion to polynomial derivatives is then readily obtained from the general relations

$$\begin{aligned}
\hat{\mathbf{F}}_0 &= \dot{\mathbf{F}}_0 - \tfrac{1}{2}\ddot{\mathbf{F}}_0 \Delta t_3 + \tfrac{1}{6}\dddot{\mathbf{F}}_0 \Delta t_3^2 , \\
\hat{\hat{\mathbf{F}}}_0 &= \tfrac{1}{2}\ddot{\mathbf{F}}_0 (\Delta t_2 + \Delta t_3)/\Delta t_3 - \tfrac{1}{6}\dddot{\mathbf{F}}_0 (\Delta t_2 + 2\Delta t_3)(\Delta t_2 + \Delta t_3)/\Delta t_3 , \\
\hat{\hat{\hat{\mathbf{F}}}}_0 &= \hat{\hat{\mathbf{F}}}_0 \frac{\Delta t_2^2 - \Delta t_1 \Delta t_3}{(\Delta t_2 + \Delta t_3)\, \Delta t_2\, \Delta t_3} + \tfrac{1}{6}\dddot{\mathbf{F}}_0 \frac{(\Delta t_1 + \Delta t_2 + \Delta t_3)(\Delta t_1 + \Delta t_2)}{\Delta t_2\, \Delta t_3} .
\end{aligned} \tag{9}$$

In order to proceed with the individual time-step scheme, each particle must be assigned the time t_i of the most recent force computation. The remaining part of the interval is introduced as an auxiliary quantity

$$\Delta \tilde{t}_i = t_i + \Delta t_i - t , \tag{10}$$

where the current step is denoted by Δt_i rather than the previous definition Δt_4.

The integration is continued by finding the index α with the smallest value of $\Delta \tilde{t}_i$,

$$\Delta \tilde{t}_\alpha = \min_i (\Delta \tilde{t}_i), \tag{11}$$

which determines the next particle to be considered. Advancing the time t to $t + \Delta \tilde{t}_\alpha$, all quantities $\Delta \tilde{t}_i$ are subtracted by $\Delta \tilde{t}_\alpha$ in order to be consistent with the definition (10). The body $i=\alpha$ now requires a new force determination but first all co-ordinates are predicted by low-order extrapolation. Increments to the positions are written as

$$\Delta \mathbf{r}_i = \left[\left(\tfrac{1}{6}\hat{\mathbf{F}}_0 \Delta t_i' + \tfrac{1}{2}\mathbf{F}_0 \right) \Delta t_i' + \dot{\mathbf{r}}_0 \right] \Delta t_i' , \tag{12}$$

where $\Delta t_i' = t - t_i$ is the time interval since the previous force computation. If Δt_α is small, the calculation of dominant force terms may be improved by including one extra order in the co-ordinate prediction for any other particles with small steps.

A more accurate position for the body $i=\alpha$ is obtained by integrating twice the increments from the coefficients C and D defined by Equation (4), after which the new force is calculated by the summation (5). The whole contribution from the fifth term of Equation (2) is now added as an improvement and the new velocity is obtained in a consistent manner by integrating the force polynomial once. At this stage it is convenient to initialize the individual reference time; i.e., $t_\alpha = t$, and update the quantities Δt_1, Δt_2, Δt_3 for the particle considered. Finally, the next time-step is predicted from the relative criterion

$$\Delta t_\alpha^3 = \eta \left[\frac{F_0 + \dot{F}_0 \Delta t_3}{\tfrac{1}{6}\dddot{F}_0 + \tfrac{1}{24} F_0^{(\mathrm{IV})} \Delta t_3} \right], \tag{13}$$

where two terms have been added to the expression (7) in order to ensure proper convergence in exceptional cases. The extra terms are usually small and only the third-order Taylor series derivative is converted from the polynomial expression (9). No upper limit is used but the new value is not allowed to increase by more than a factor of 1.4 in order to safeguard the numerical stability. The treatment of the body α is completed by initializing the auxiliary interval, hence $\Delta\tilde{t}_\alpha = \Delta t_\alpha$. A new cycle is entered at Equation (11) which determines the next particle to be considered and the calculations continue as discussed above.

The integration procedure for a high-order scheme using individual time-steps is simple since all decision-making is controlled by the auxiliary variable $\Delta\tilde{t}_i$. It may be noted that the velocity is available only at the end of each interval and proper care must be exercised when evaluating sensitive quantities at a different time, as for instance when calculating new integrals of motion. The co-ordinate prediction (12) makes it necessary to preserve all positions at the beginning of each dynamical step. It may also be remarked that the extrapolation of co-ordinates within an interval are not directly useful to the integration since the information is lost at the next cycle, but this device forms an essential part of the scheme which permits the simultaneous use of widely different time-steps.

A minimum of 28 N storage locations is required by the present method which explicitly includes third-order force differences. It is desirable to make use of extended precision when the computer word contains less than about 10 decimals or if extreme accuracy is intended. In the case of the I.B.M. 360/44 additional precision is used for all co-ordinates and velocities at the beginning of a time-step; $\mathbf{r}_0$, $\dot{\mathbf{r}}_0$, as well as the times t_i and current positions $\mathbf{r}_i$. This requires an additional 10 N storage locations but a 32000 word direct access store would still allow about 600 particles to be studied unless further variables or special treatments are included. The additional time requirement is less than 30% for the same number of integration steps, but the gain in accuracy is considerable when using seven figure precision. Fortunately, more powerful methods are available for treating critical encounters.

The actual choice of the time-step parameter η can only be determined from integration tests, but a value near 2×10^{-4} leads to satisfactory solutions in the absence of extremely close encounters.* A corresponding computing requirement of about one hour per mean crossing time may then be achieved for $N = 100$ when the basic addition time is 3.8 μs; similar times for other particle numbers scale as N^2. Alternatively, the time per individual step is given approximately as $1.0 \times 10^{-3}\, N$ s for $N \gg 2$. Nearly all the computing time is accounted for by Equations (5) and (12) together with Equation (11) and the subtraction of the quantity $\Delta\tilde{t}_\alpha$ from all $\Delta\tilde{t}_i$.

The ordinary polynomial method may be used separately or in combination with the regularization treatment discussed below. It is convenient to control the overall integration accuracy by the integrals of motion, in particular the relative error of

* Relative energy errors for a binary with eccentricity 0.92 are then $\Delta E/E = 2 \times 10^{-5}$ and 7×10^{-6} per revolution, using the two alternative orders of prediction with 234 steps for each component.

total energy is sensitive to the adopted time-step parameter. An additional and more detailed check of the numerical solutions is provided by the time reversibility of the equations of motion. The ability to deal with close encounters without special treatments may be tested by studying eccentric binary orbits over many revolutions. Accurate solutions are available for a critical three-body case studied by regularization techniques (Szebehely and Peters, 1967); a substantial part of the evolution may also be reproduced reasonably well by the ordinary method.

4. Two-Body Regularization

The Levi-Civita regularization of the plane two-body problem has only recently been generalized to three dimensions where it can be used to study close encounters in stellar dynamics (Kustaanheimo and Stiefel, 1965). It has already been demonstrated that this elegant formulation is extremely efficient in dealing with a variety of three-body configurations (Peters, 1968). In particular, the improvement over ordinary methods becomes apparent in critical two-body encounters since the new equations of motion are non-singular. It is therefore natural to investigate whether this regularization procedure is also effective for the integration of close encounters in larger systems. An extension to large particle numbers requires a suitable formulation of the basic equations of motion as well as the ability to deal with an arbitrary number of two-body encounters at the same time. In addition, suitable regularization criteria must be developed in order to combine efficiently the special treatment with the direct method described above.

In the following we make full use of the Hamiltonian formulation of Peters, preserving some of the notation. The four-dimensional equations of motion for the transformed relative co-ordinates $\mathbf{u}$ and momenta $\mathbf{v}$ are given by

$$\mathbf{u}' = \frac{1}{4\mu}\,\mathbf{v}, \tag{14}$$

$$\mathbf{v}' = 2(E - V)\,\mathbf{u} + 2\mu R \mathscr{L}^T (\mathbf{F}_k - \mathbf{F}_l), \tag{15}$$

where primes denote differentiation with respect to the fictitious time τ and μ is the reduced mass of the two particles being considered, subsequently to be identified by the individual masses m_k and m_l. The total energy of the system is denoted by E while the perturbing function of the relative motion is given by

$$V = \tfrac{1}{2}(m_k + m_l)\,\dot{\mathbf{Q}}^2 + \tfrac{1}{2}\sum_{\substack{i=1\\ i\neq k,l}}^{N} m_i \dot{\mathbf{r}}_i^2 - \tfrac{1}{2}\sum_{i=1}^{N}\sum_{\substack{j=1\\ j\neq i,k,l}}^{N} m_i m_j/(|\mathbf{r}_i - \mathbf{r}_j|), \tag{16}$$

which contains the kinetic energy of the centre of mass motion and excludes the contribution $m_k m_l$ from the double summation. The second term of Equation (15) contains the relative Newtonian perturbation

$$\mathbf{F}_k - \mathbf{F}_l = -\sum_{\substack{j=1\\ j\neq k,l}}^{N} m_j \left[\frac{\mathbf{r}_k - \mathbf{r}_j}{|\mathbf{r}_k - \mathbf{r}_j|^3} - \frac{\mathbf{r}_l - \mathbf{r}_j}{|\mathbf{r}_l - \mathbf{r}_j|^3}\right], \tag{17}$$

and $\mathscr{L}^T$ is the transpose of the generalized Levi-Civita matrix

$$\mathscr{L} = \begin{bmatrix} u_1 & -u_2 & -u_3 & u_4 \\ u_2 & u_1 & -u_4 & -u_3 \\ u_3 & u_4 & u_1 & u_2 \end{bmatrix}. \tag{18}$$

Finally, the particle separation R is obtained from the transformed co-ordinates by the relation

$$R = \sum_{j=1}^{4} u_j^2, \tag{19}$$

subsequently written in the form $\mathbf{u}\cdot\mathbf{u}$.

It is readily recognized that an application of Equation (15) to systems with large particle numbers would be prohibitive because of the double summation term in the expression for the perturbing function. Instead we combine the explicit expressions for E and V which simplify to

$$E - V = \frac{1}{2\mu}\mathbf{P}\cdot\mathbf{P} - \frac{m_k m_l}{R}, \tag{20}$$

where $\mathbf{P}$ is the momentum of the relative motion. This relation also follows directly from the regularized Hamiltonian which should be zero (cf. Peters). Thus Equation (20) introduces the binding energy per unit mass of the two-body motion,

$$h = (E - V)/\mu. \tag{21}$$

In order to evaluate Equation (20) when working with regularized quantities we make use of the transformation property

$$\mathbf{P}\cdot\mathbf{P} = \frac{1}{4R}\mathbf{v}\cdot\mathbf{v}. \tag{22}$$

Substituting $\mathbf{u}'$ for $\mathbf{v}$ from Equation (14) and dividing by μ finally gives the desired expression

$$h = [2\,\mathbf{u}'\cdot\mathbf{u}' - (m_k + m_l)]\,\frac{1}{R}. \tag{23}$$

The relative binding energy per unit mass is obtained at the expense of introducing a division by the particle separation and this procedure does not allow cases to be studied where $R \to 0$. The probability of near collisions, however, is extremely low in simulated clusters and the expression (23) may therefore be used in most calculations of practical interest, instead of the more accurate form (21).

The eight first-order equations of motion (14) and (15) may be combined into four equations of second order for the regularized co-ordinates, giving

$$\mathbf{u}'' = \tfrac{1}{2}h\,\mathbf{u} + \tfrac{1}{2}R\mathscr{L}^T(\mathbf{F}_k - \mathbf{F}_l). \qquad (24$$

The fictitious time is related to the ordinary integration time t by the regularizing transformation

$$\mathrm{d}t = R\,\mathrm{d}\tau\,. \tag{25}$$

In the present derivation each close pair is treated independently and all fictitious intervals $\Delta\tau$ are converted to ordinary time in order to provide a common frame of reference. This approach differs significantly from that of Peters and allows the introduction of multiple regularizations without affecting the integration of other particles. Such a scheme has many advantages and only requires accurate treatments of the conversion of fictitious time to ordinary time and vice versa.

Actual calculations show that the binding energy formulation leads to a numerical instability in special cases of rapidly varying separations. This difficulty arises because the predicted value of the relative binding energy appears in the equation of motion (24). An analysis of the corresponding unperturbed expression shows that the relative error in binding energy continues to grow if the parameter

$$\varepsilon \simeq \frac{u_0 u_0' \Delta\tau}{u_p^2} \tag{26}$$

exceeds unity. In this first-order derivation which uses the scalar approximation the quantities u_0 and u_0' represent the transformed co-ordinate and velocity at the beginning of the interval $\Delta\tau$, while u_p is the predicted co-ordinate. It is usually sufficient to ensure that the predicted integration step does not violate the adopted stability condition $\varepsilon \lesssim 0.5$, but growing oscillations of the binding energy still appear at extremely small separations. This undesirable behaviour is finally eliminated by using the previously calculated binding energy, rather than the predicted value which is known to one order less accuracy. This procedure is now dynamically consistent for very small perturbations and is only required on rare occasions.

An alternative formulation as yet untried makes it possible to retain the advantage of the binding energy description while avoiding the stability considerations entirely. Thus it has been suggested that an additional equation should be introduced for the binding energy itself (Stiefel, 1967). A convenient expression is obtained for h' by differentiating Equation (23), making use of the equation of motion (24) and the relation $R' = 2\,\mathbf{u}\cdot\mathbf{u}'$, finally giving

$$h' = 2\mathbf{u}' \cdot \mathscr{L}^T (\mathbf{F}_k - \mathbf{F}_l)\,. \tag{27}$$

This equation is completely regular and the right-hand side contains the perturbation term required by the equation of motion (24). Again the new binding energy must be obtained by prediction but the perturbation contribution remains well behaved for small separations as in the classical expression $\dot{h} = \dot{\mathbf{R}} \cdot (\mathbf{F}_k - \mathbf{F}_l)$. In addition, it may be noted that the procedure of integrating the binding energy separately gains one order of accuracy compared to Equation (23) which involves the regularized velocity. The subsequent discussion therefore assumes the integration treatment based on Equation (27), where the initial value is given by Equation (23).

5. Transformations

Assuming that the pair m_k, m_l, has been selected for special treatment, we introduce the relative co-ordinates and momenta*

$$\mathbf{R} = \mathbf{r}_k - \mathbf{r}_l\,, \tag{28}$$

$$\mathbf{P} = \mu (\dot{\mathbf{r}}_k - \dot{\mathbf{r}}_l)\,. \tag{29}$$

The transformation to regularized co-ordinates takes two forms depending on the sign of the first component of the separation vector (X, Y, Z). Thus for $X>0$ the initial components of the four-vector $\mathbf{u}$ are given by

$$\begin{aligned} u_1 &= [\tfrac{1}{2}(R + X)]^{1/2}\,, \\ u_2 &= Y/2u_1\,, \\ u_3 &= Z/2u_1\,, \\ u_4 &= 0\,, \end{aligned} \tag{30}$$

while for $X<0$ the proper choice is

$$\begin{aligned} u_2 &= [\tfrac{1}{2}(R - X)]^{1/2}\,, \\ u_1 &= Y/2u_2\,, \\ u_3 &= 0\,, \\ u_4 &= Z/2u_2\,. \end{aligned} \tag{31}$$

In both cases the inverse relations are

$$\begin{aligned} X &= u_1^2 - u_2^2 - u_3^3 + u_4^2\,, \\ Y &= 2(u_1 u_2 - u_3 u_4)\,, \\ Z &= 2(u_1 u_3 + u_2 u_4)\,. \end{aligned} \tag{32}$$

It can readily be seen that Equation (32) implies the relation (19).

The regularized momentum vector $\mathbf{v}$ is transformed according to

$$\mathbf{v} = 2\mathscr{L}^T \mathbf{P}\,, \tag{33}$$

where the transpose matrix is given by

$$\mathscr{L}^T = \begin{bmatrix} u_1 & u_2 & u_3 \\ -u_2 & u_1 & u_4 \\ -u_3 & -u_4 & u_1 \\ u_4 & -u_3 & u_2 \end{bmatrix}. \tag{34}$$

The relation (33) implies the non-holonomic condition

$$u_4 v_1 - u_3 v_2 + u_2 v_3 - u_1 v_4 = 0\,. \tag{35}$$

* The definition of relative co-ordinates and momenta used by Peters should be reversed in order to be consistent with the final equation of motion. A correct derivation is given in the original Ph.D. thesis, Yale University, 1968.

The integration accuracy can be checked using Equations (27) or (35). In practice it is more convenient to work with $\mathbf{u}$ and $\mathbf{u}'$ where the latter is defined by Equation (14). The equation of motion (24) then plays the role of Equation (5) and we can make use of the polynomial method discussed above for the integration.

The complete solution of the regularized motion is obtained by introducing the centre of mass co-ordinates

$$\mathbf{Q} = \frac{m_k \mathbf{r}_k + m_l \mathbf{r}_l}{m_k + m_l}, \tag{36}$$

with the corresponding equation of motion

$$\ddot{\mathbf{Q}} = \frac{m_k \mathbf{F}_k + m_l \mathbf{F}_l}{m_k + m_l}. \tag{37}$$

It may be noted that only the two perturbations enter in the centre of mass acceleration since the dominant terms cancel analytically. The proposed two-body regularization procedure requires in all 16 equations as compared to 12 equations for the ordinary method. Even so there is no loss of efficiency since the perturbation calculation is by far the most time-consuming for large particle numbers. The longer intervals permitted by the regularized solution therefore represent a net gain over standard integration schemes.

The original quantities may be obtained at any time from the transformations

$$\begin{aligned}
\mathbf{R} &= \mathscr{L}\mathbf{u}, \\
R &= \mathbf{u}\cdot\mathbf{u}, \\
\mathbf{P} &= 2\mu\mathscr{L}\mathbf{u}'/R, \\
\mathbf{r}_k &= \mathbf{Q} + \mu\mathbf{R}/m_k, \\
\mathbf{r}_l &= \mathbf{Q} - \mu\mathbf{R}/m_l, \\
\dot{\mathbf{r}}_k &= \dot{\mathbf{Q}} + \mathbf{P}/m_k, \\
\dot{\mathbf{r}}_l &= \dot{\mathbf{Q}} - \mathbf{P}/m_l,
\end{aligned} \tag{38}$$

with the reduced mass

$$\mu = \frac{m_k m_l}{m_k + m_l}. \tag{39}$$

The present integration procedure requires an accurate determination of the ordinary time corresponding to a fictitious time-step. Several methods may be used for the conversion of regularized time; here we make use of the Taylor series expansion

$$\Delta t = t_0' \,\Delta\tau + \tfrac{1}{2} t_0'' \,\Delta\tau^2 + \tfrac{1}{6} t_0''' \,\Delta\tau^3 + \tfrac{1}{24} t_0^{(\mathrm{IV})} \,\Delta\tau^4 + \tfrac{1}{120} t_0^{(\mathrm{V})} \,\Delta\tau^5. \tag{40}$$

The desired coefficients evaluated at the beginning of the interval $\Delta\tau$ are readily obtained by successive differentiations of the second Equation (38) with respect to the fictitious time, using the definition (25). The first three terms are given by

$$\begin{aligned} t_0' &= \mathbf{u}\cdot\mathbf{u}, \\ t_0'' &= 2\mathbf{u}'\cdot\mathbf{u}, \\ t_0''' &= 2\mathbf{u}''\cdot\mathbf{u} + 2\mathbf{u}'\cdot\mathbf{u}'. \end{aligned} \tag{41}$$

All derivatives of $\mathbf{u}$ required by Equation (40) are known from the high-order integration scheme, hence the conversion to ordinary time is very efficient. It may be noted that the regularized polynomial derivatives should be converted to actual Taylor series derivatives at the time τ_0 by the equivalent procedure of Equation (9). Numerical tests show that the adopted expansion converges rapidly since all the terms are well behaved.

An inverse relation is also required for extrapolation within an interval $\Delta\tau$ in order to transform the regularized co-ordinates at times not coinciding with the end-points. In this case less accuracy is needed and we adopt the expansion

$$\Delta\tilde{\tau} = \dot{\tau}_0\,\Delta t + \tfrac{1}{2}\ddot{\tau}_0\,\Delta t^2 + \tfrac{1}{6}\dddot{\tau}_0\,\Delta t^3, \tag{42}$$

using the definition (25),

$$\dot{\tau}_0 = 1/R. \tag{43}$$

The two higher derivatives can be expressed in terms of the quantities (41) as

$$\begin{aligned} \ddot{\tau}_0 &= -\frac{t_0''}{R^3}, \\ \dddot{\tau}_0 &= \left[\frac{3t_0''^2}{R} - t_0'''\right]\frac{1}{R^4}. \end{aligned} \tag{44}$$

The division by small values of R may be permitted since it does not effect the integration of the relative motion. Thus the procedure (42) is only required for the purpose of calculating $\ddot{\mathbf{Q}}$ or $\ddot{\mathbf{r}}_i$; the former may just as well be evaluated at the nearest end-point, while it is dynamically consistent to use the centre of mass approximation for very small separations when computing the force contribution to other particles.

The regularized components are also integrated by the fourth-order polynomial method since the acceleration calculations are still time-consuming. The formulation described by Equations (6)–(10) may be used to obtain a consistent starting procedure for the centre of mass motion. Coefficients representing polynomial derivatives are first determined for each component, excluding the dominant contribution, and the desired expressions are combined in the manner of Equation (37). The centre of mass co-ordinates and velocities are initialized and the integration proceeds as in the ordinary method, except that the co-ordinate transformation (38) must be performed in order to calculate the acceleration (37).

It is not possible to make use of the explicit starting scheme for the relative motion, however. Instead a fitting procedure is employed for obtaining Taylor series derivatives which are then converted in the usual manner. Denoting the regularized acceleration at time τ_0 by $\mathbf{G}_0$ instead of $\mathbf{u}''$, we write an expansion in terms of $\tau_r = \tau - \tau_0$ as

$$\mathbf{G} = \mathbf{G}_0 + \mathbf{G}_0'\tau_r + \tfrac{1}{2}\mathbf{G}_0''\tau_r^2 + \tfrac{1}{6}\mathbf{G}_0'''\tau_r^3 . \tag{45}$$

Successive accelerations $\mathbf{G}_j$, $j=1, 2, 3$ are determined from Equation (24) at three equal intervals $\delta\tau$ by advancing all co-ordinates appropriately, including the centre of mass. The fitting intervals $\delta\tau$ are chosen such that $3\delta\tau = 0.5\Delta\tau_0$, where $\Delta\tau_0$ is the initial time-step to be used by the high-order integration. The resulting coefficients are given by

$$\begin{aligned}
\mathbf{G}_0' &= [-\tfrac{11}{6}\mathbf{G}_0 + 3\mathbf{G}_1 - \tfrac{3}{2}\mathbf{G}_2 + \tfrac{1}{3}G_3]\frac{1}{\delta\tau}, \\
\mathbf{G}_0'' &= [\mathbf{G}_0 - \tfrac{5}{2}\mathbf{G}_1 + 2\mathbf{G}_2 - \tfrac{1}{2}\mathbf{G}_3]\frac{1}{\delta\tau^2}, \\
\mathbf{G}_0''' &= [-\tfrac{1}{6}\mathbf{G}_0 + \tfrac{1}{2}\mathbf{G}_1 - \tfrac{1}{2}\mathbf{G}_2 + \tfrac{1}{6}\mathbf{G}_3]\frac{1}{\delta\tau^3} .
\end{aligned} \tag{46}$$

Finally, the conversion to polynomial derivatives is performed in analogy with the procedure of Equations (8) and (9) and the regularized integration may be continued by the usual method. Starting coefficients for the right-hand side of Equation (27) are determined in a similar manner, writing Equation (45) as an expansion for the function $H \equiv h'$. Subsequent values of the binding energy are then obtained by integrating the corresponding polynomial once.

6. Regularization Parameters

Close encounters between ordinary particles lead to a shortening of the corresponding time-steps which is essentially independent of the relative binding energy for small separations. It is therefore natural to specify a critical interval $\Delta t_{\min}$ and a corresponding separation $R_{\min}$ to be used as an indication of suitable cases for special treatment. Once a particle m_k satisfies the condition $\Delta t_k < \Delta t_{\min}$ and the time-step is decreasing, a search is made for the body m_l which contributes the greatest force component, at the same time noting all other particles j inside a somewhat larger separation, say, $3R_{\min}$. The pair m_k, m_l is accepted for regularization, provided that $R < R_{\min}$ and the relative force is dominant. The latter requirement implies that

$$\frac{m_k + m_l}{|\mathbf{r}_k - \mathbf{r}_l|^2} > \frac{m_l + m_j}{|\mathbf{r}_l - \mathbf{r}_j|^2} \tag{47}$$

for all $j, \neq k, l$. The condition (47) allows for the possibility of body m_k being close to another regularized pair. It is also prudent to include approaching particles only in view of the additional calculations for the starting procedure.

The special treatment may be terminated in several ways, depending on the circumstances. It is convenient to make use of the invariant two-body perturbation defined by

$$\gamma = \frac{|\mathbf{F}_k - \mathbf{F}_l| \, R^2}{m_k + m_l}. \tag{48}$$

The latter quantity is particularly useful for deciding when to end the regularization of close binaries. Thus it is natural to replace the components m_k, m_l by, say, the pair m_l, m_j when the latter particle gives rise to the perturbation $\gamma \gtrsim 1$ since the original binary motion is then no longer dominant.

Regularization of hyperbolic cases or wide binaries is normally terminated by the combined criterion

$$R > R_0, \quad \gamma > \gamma_{\max}, \tag{49}$$

where R_0 denotes the initial separation. The second condition (49) makes it possible to continue the special treatment outside the initial separation; this feature is particularly useful for hyperbolic orbits with $R_0 \ll R_{\min}$, or when dealing with eccentric binaries where the apocentre distance exceeds the critical value $R_{\min}$. The regularization criteria discussed above are completely general; consistent values may be chosen from trial integrations or invariant definitions of close encounters.

The integration interval for regularized binary motion may be determined by reference to the orbital period. Bearing in mind the form of the equation of motion and the time transformation (25), we adopt an expression of constant time-step modified by perturbations,

$$\Delta\tau = \frac{2\pi}{\mathcal{N}} \left[\frac{1}{2|h|}\right]^{1/2} \frac{1}{(1 + 1000\,\gamma)^{1/3}}, \tag{50}$$

where the parameter $\mathcal{N}$ denotes the number of integration steps during one unperturbed revolution. The correction term is usually small but allows for a significant reduction of step-size in the presence of strong perturbations which lead to rapid variations of the binding energy.

Equation (50) is equally suitable for treating hyperbolic motion. The precaution is taken of replacing the unperturbed part of the predicted time-step by a constant β in all cases which would otherwise give $\Delta\tau > \beta$ in the absence of perturbations. An additional safety measure is included by using half the predicted value of the initial step $\Delta\tau_0$; subsequent intervals are allowed to increase by a factor of 1.2. The adopted time-step definition is not unique and alternative expressions may therefore be tried. It may be noted that the period of the regularized equation of motion (24) corresponds to twice the Keplerian value. This fundamental property of the Kustaanheimo-Stiefel transformation demonstrates clearly the effectiveness of the method.

Direct calculations of close binary orbits are also time-consuming when using the regularization description. In such cases the number of binary revolutions per crossing time may be very large, while the corresponding perturbation is often sufficiently small to be neglected (Aarseth, 1970). The simplification of unperturbed motion may also be introduced here since the Keplerian period is known in ordinary time units. Instead we reduce the number of interacting bodies to include the nearest neighbours which contribute the main part of the fluctuating force field. Thus the effect of distant particles tends to cancel when the relative motion is integrated over one complete revolution. Furthermore, it is only by replacing the variables $\mathbf{r}_k$, $\mathbf{r}_l$ by the set $\mathbf{Q}$, $\mathbf{R}$ that this technique can be used advantageously since the total centre of mass acceleration is usually required much less frequently.

The perturbation is included from all particles i which satisfy the condition

$$\frac{m_i}{|\mathbf{Q} - \mathbf{r}_i|^3} \gtrsim \frac{1}{\kappa^3 R^3}, \tag{51}$$

where κR represents the limiting separation for bodies of mean mass unity. Conversely, the force contribution to ordinary particles may be calculated by the centre of mass approximation when the distance to regularized pairs exceeds κR. It may be noted that the adopted expression is consistent with the tidal limit approximation of the perturbation (48) and corresponds to $\gamma \simeq \kappa^{-3}$, neglecting the mass dependence. The list of neighbours is updated at every apocentre passage as determined from the change in sign of the radial velocity R'.

Numerical values of the regularization parameters have not been discussed above since the choice is to some extent arbitary and depends on the desired integration accuracy. One example with eccentricity 0.7 shows that $\mathcal{N}=50$ is sufficient to keep the relative binding energy error per revolution below 1×10^{-6}, while $\mathcal{N}=63$ improves the accuracy to 3×10^{-7}, when all calculations are performed in extended precision. The adopted parameters for the first large N-body integration using regularization are given for completeness; i.e., $\Delta t_{\min}=1 \times 10^{-5}$, $R_{\min}=0.01$, $\gamma_{\max}=0.01$, $\mathcal{N}=50$, $\beta=0.01$, $\kappa=100$ for one case $N=500$, employing the energy scaling $E=-\frac{1}{4}N^2$ with mass units $\sum m_i=N$. The fast perturbation calculation is only used for close binaries satisfying the condition $h<-N$, but this procedure would be equally consistent for all regularizations. Finally, we remark that the additional transformations required when treating other particles inside the distance κR partly offsets the advantages of regularizing large separations, hence the conservative choice of the first three parameters.

7. Special Considerations

It is essential to organize tables of variables in a systematic way in order to facilitate the simultaneous treatment of ordinary particles and regularized pairs. For convenience we distinguish between global quantities g_i such as $\mathbf{r}_i$, $\dot{\mathbf{r}}_i$, etc. and regularized variables ρ_j denoting $\mathbf{u}$, $\mathbf{u}'$. The sequential arrays $\{g_i\}$ are modified to include the regularized components first. The last member of a global particle array is then g_N as usual but the centre of mass corresponding to the first regularization is added as g_{N+1}. The extension to an arbitrary number of regularized pairs follows quite naturally in the present treatment. Thus an alteration to the existing situation is performed by moving all relevant quantities up or down in the tables and deleting or adding the corresponding centre of mass.

Consider a general situation with n separate close pairs. The particle arrays $\{g_i\}$ where $i \leqslant 2n$ then represent the transformed components with corresponding relative parameters $\{\rho_j\}$, $j \leqslant n$. Subsequent locations $2n+1, \ldots, N$ are assigned to ordinary particles, followed by the centre of mass arrays $\{g_{N+j}\}$ with $j \leqslant n$. It is therefore quite simple to distinguish between the different procedures required by the three cases $\alpha \leqslant 2n$, $2n < \alpha \leqslant N$, $\alpha > N$, where α is determined by Equation (11). In addition, co-ordinate predictions and force calculations are more efficient when similar quantities are stored sequentially.

Regularization treatments are terminated by the transformations (38) after which the component co-ordinates and velocities are restored to the original locations. At the same time any quantities g_{N+j} and ρ_j introduced more recently are updated consistently. Finally, the starting procedure described by Equations (5)–(10) is applied to each component and the integration proceeds normally. A minimum of reorganization is achieved by arranging all arrays $\{g_i\}$ in terms of decreasing mass initially since heavy bodies are most frequently involved in close encounters. It is then numerically advantageous to perform the force summation (5) in reverse order.

In conclusion, it may be emphasized that a considerable programming effort is required in order to make efficient use of the methods described above. At the same time the introduction of two-body regularization represents a significant improvement of technique which permits more critical configurations to be studied. Some efficiency is lost, however, when integrating multiple close encounters if there are no dominant pairs which may be selected for regularization. An alternative treatment is available for such cases (Heggie, 1972) but the introduction of a third procedure has not yet been attempted. Further programming details are available upon request*.

Acknowledgements

The author has benefited from discussing problems of regularization with Dr D. G. Bettis, M. C. Froeschlè, Mr D. C. Heggie and Professor V. G. Szebehely.

References

Aarseth, S. J.: 1968, *Bull. Astron.* **3**, 105.
Aarseth, S. J.: 1970, *Astron. Astrophys.* **9**, 64.
Bettis, D. G. and Szebehely, V. G.: 1972, this volume, p. 388.
Gonzalez, C. C. and Lecar, M.: 1968, *Bull. Astron.* **3**, 209.
Heggie, D. C.: 1972, this volume, p. 148.
Lecar, M.: 1968, *Bull. Astron.* **3**, 91.
Kustaanheimo, P. and Stiefel, E.: 1965, *Math.* **218**, 204.
Peters, C. F.: 1968, *Bull. Astron.* **3**, 167.
Stiefel, E.: 1967, NASA Report CR-769.
Szebehely, V. G. and Peters, C. F.: 1967, *Astron. J.* **72**, 876.
Wielen, R.: 1967, *Veröff. Astron. Rechen-Inst. Heidelberg*, No. 19.

* The treatment based on Equation (27) has now been adopted. Energy errors for a binary with eccentricity 0.92 then improve from $\Delta E/E = -5 \times 10^{-6}$ to -8×10^{-8} per revolution, if $\mathcal{N} = 50$. An iterative solution of Equation (40) may be used in plase of Equation (42).

TREATMENT OF CLOSE APPROACHES IN THE NUMERICAL INTEGRATION OF THE GRAVITATIONAL PROBLEM OF *N* BODIES

D. G. BETTIS AND V. SZEBEHELY
The University of Texas at Austin

Abstract. One of the main difficulties encountered in the numerical integration of the gravitational n-body problem is associated with close approaches. The singularities of the differential equations of motion result in losses of accuracy and in considerable increase in computer time when any of the distances between the participating bodies decreases below a certain value. This value is larger than the distance when tidal effects become important, consequently, numerical problems are encountered *before* the physical picture is changed. Elimination of these singularities by transformations is known as the process of regularization. This paper discusses such transformations and describes in considerable detail the numerical approaches to more accurate and faster integration. The basic ideas of smoothing and regularization are explained and applications are given.

1. Introduction

The equations of motion for the gravitational n-body problem are

$$\frac{d^2 r_i}{dt^2} = G \sum_{\substack{i=1 \\ 1 \neq j}}^{n} m_j \frac{r_j - r_i}{R_{ij}^3}, \tag{1}$$

where r_i is the position vector and m_i is the mass of the ith body. When

$$R_{ij} = |r_i - r_j|,$$

becomes small (during a close encounter of the ith and jth bodies) the above differential equation becomes singular. The purpose of this paper is the formulation of an algorithm such that these equations are regular, as $R_{ij} \to 0$, thus avoiding any loss of significant digits.

First the regularizing transformation for a single pair of bodies is developed and then this result is extended to n pair of bodies.

2. Basic Principles and a Simple Example of Regularization

This section is dedicated to the uninitiated to the mysteries of close approaches and to the astronomer who encounters numerical (or analytical) difficulties when integrating the differential equations of dynamical systems.

The following few didactic lines are neither 'practical' nor pleasing to the purist. The simplest possible two-body encounter shall be investigated and an attempt will be made to convert the reader to the ever-increasing number of believers in the method of regularization. The spotlight will be on the essential aspects and neither

M. Lecar (ed.), Gravitational N-Body Problem, 388–405.

precision-mathematics nor astronomical applications will be allowed to deviate the attention from the simple and straight-forward matters at hand. The practical aspects bring complications which will be handled in the later sections and the mathematical sophistications are well established and available in the literature.

First, an over-simplified model is described. Second, the problem to be solved is outlined. Third, the roads available to handle the problem are outlined. Fourth, one of the possible avenues is selected and the problem is rendered trivial.

In stellar dynamics and in celestial mechanics collision is a physical, rather than a mathematical problem. The centers of two bodies do not occupy the same place at the same time, at least not while still central Newtonian gravitational forces dominate. Higher order gravitational harmonics present in the description of the gravitational field of a planet, for instance, become important when a satellite is in close approach. And in stellar dynamics, for instance, the tidal forces acting between the members of close binaries alter the central Newtonian gravitational forces. *The principal question is whether numerical difficulties are encountered as two bodies approach each other before the law of force changes or alters.* The answer to this question is in the affirmative both in the field of celestial mechanics *and* in stellar dynamics. In other words, the tidal effects are still negligible when the close approach destroys the accuracy and efficiency of the numerical integration if conventional formulations are used.

In spite of the fact that collisions do not occur in stellar dynamics in the mathematical sense, the over-simplified model used here is still a collision-model in order to emphasize the computational difficulties and in order to present a simple problem. If the techniques proposed can handle the numerical problems encountered because of a collision, they certainly will be able to treat close approaches. Consequently, the model has two-point masses in a straight line approaching each other. The assumption of point masses instead of finite bodies is, once again, of no importance and allows concentration on the real issues. As the two bodies move, they will approach their (fixed) center of mass. If x is the distance between the two bodies, the equation describing the dynamics of the problem, with properly selected units, $[G(m_1+m_2)=1]$, becomes

$$(\mathrm{d}^2x/\mathrm{d}t^2) = -(1/x^2). \tag{2}$$

As x decreases and approaches zero, the bodies approach each other (and the center of mass of the system). The integral of energy becomes

$$\tfrac{1}{2}(\mathrm{d}x/\mathrm{d}t)^2 = (1/x) + C \tag{3}$$

where C is the constant of energy and it is evaluated from the initial conditions. For simplicity's sake, the motion shall be started with zero velocity when the bodies are apart at a distance x_0. Consequently, $C = -(1/x_0)$.

Both of the above equations show singular behavior as $x \to 0$. The velocity as well as the acceleration approach infinity. Standard numerical integration techniques fail close to the singularity since the accurate evaluation of the right side of the equations is not feasible.

The elimination of such singular behavior is called *regularization*. The regularized equation has no singularity. It is not obvious that the above equations may be regularized, in fact, regularization of a differential equation is not possible in general. When the elimination of singularities is not feasible, often a smoothing of the singular behavior is introduced. Such a smoothing does not regularize but it may ease the numerical problems encountered during integration. The purposes of regularization and smoothing as applied to the differential equations of stellar dynamics are to enhance the process of numerical integration. Consequently, certain regularization techniques which are successful from an analytical point of view may not be satisfactory at all from the point of view of the numerical analyst.

Regularizations and smoothings are accomplished by transformations of the variables occurring in the differential equations of motion. The original regularization of the problem of two bodies as proposed by Sundman (1913) used transformations of the independent (time) variable only. It is recognized today that such transformations while rendering the equations of motion regular, do not offer the best formulation for numerical work. Transformations involving both the dependent and the independent variables are utilized to put the pertinent equations of motion in their optimum form for numerical integration.

Returning to Equations (2) and (3), first a transformation of the time will be introduced to show its effect in the simplest case. The new independent variable is introduced through a differential relation, following Sundman:

$$\mathrm{d}\tau = \mathrm{d}t/x\,. \tag{4}$$

Note that the new 'time' τ depends on the original time t as well as on the dependent variable x. Equation (4) is suggestive of a technique for changing the step-size or time-step during numerical integration. As x becomes smaller, the time-step of integration, Δt, decreases also in order to accommodate the large changes occuring on the right hand side of the equations to be integrated. In the method of using variable time-steps, no new variable is introduced, but, if one were to appear, it would abide by Equation (4). As x and Δt decrease during integration, their ratio may be smooth, or even constant. Consequently, if the variable τ were used in the differential equations of motion instead of the variable t, the steps $(\Delta\tau)$ used may be more even. The essential difference between integration with variable time steps and regularization is that in the latter technique the differential equations are rewritten in terms of the new independent variable. After the regularizing transformations the new equations are integrated. The method of integrating with variable time steps does not change the equations to be integrated; it does not eliminate the singularity; in fact, it does not even smooth the right side of the equations to be integrated. Such a technique must necessarily fail when a singularity is encountered, since as the bodies approach collision, the number of integration steps tends to increase beyond limit. Every step involves truncation and round-off errors, consequently, there is always a lower limit on the distance between the bodies below which no integration step can penetrate, still furnishing meaningful results.

Note that Equation (4) itself may introduce a (new) singularity. If the integral of the right side is not a 'proper' one, in other words, if it is not convergent, then the evaluation of the new independent variable τ may not be accomplished. Leaving the resolution of this problem to a later state, the new 'time' variable will now be introduced into Equations (2) and (3). For this purpose the new 'velocity' $\mathrm{d}x/\mathrm{d}\tau$ is computed as follows:

$$x' = (\mathrm{d}x/\mathrm{d}\tau) = (\mathrm{d}x/\mathrm{d}t)\,(\mathrm{d}t/\mathrm{d}\tau) = \dot{x}x\,. \tag{5}$$

The new velocity is the product of the distance between the bodies (x) and the original velocity. Since, when $x \to 0$, $\dot{x} \to \infty$ as shown by Equation (2), their product x may be a smoothly behaving quantity. Indeed, if Equation (5) is substituted into Equation (3), we have

$$(x')^2 = 2x + 2Cx^2\,, \tag{6}$$

as the new, regular equation of energy. As $x \to 0$, the new velocity approaches zero.

If an exponent of x other than 1 were introduced in Equation (4), the equation for the new velocity would have been different. The selection of the exponent of x in Equation (4) is one of the interesting numerical problems encountered in the numerical integration of the gravitational problem of n bodies and it will be discussed later.

The transformation of Equation (2) is more important than that of Equation (3) since that is the equation of motion to be integrated. In the special and over-simplified example treated in this section, the integral of energy, Equation (3), describes the problem completely and its solution will furnish the solution of Equation (2) also. This special situation is the consequence of treating a motion in one dimension.

The second derivative is evaluated as follows:

$$x'' = (\mathrm{d}x'/\mathrm{d}t)\,(\mathrm{d}t/\mathrm{d}\tau) = x(\mathrm{d}/\mathrm{d}t)\,(\dot{x}x)\,,$$

or

$$x'' = \ddot{x}x^2 + (\dot{x})^2 x\,,$$

from which

$$\ddot{x} = [x'' - (x')^2\,x^{-1}]\,x^{-2}\,. \tag{7}$$

Combining Equations (2) and (7), we obtain

$$x'' - ((x')^2/x) + 1 = 0\,. \tag{8}$$

Analytically speaking, Equation (8) is regular since

$$(x')^2/x = 2 + 2Cx \tag{9}$$

according to Equation (6). Nevertheless, if Equation (6) is not used, because an integral of energy is not available, numerical integration of Equation (8) will present difficulties. In the two and three-dimensional cases the elimination of terms corresponding to $(x')^2/x$ in Equation (8) may present serious difficulties. This is one of the main reasons for combining the transformation of the independent variable with a transformation of the dependent variables in more complicated situations.

Substituting now Equation (9) into Equation (8), we obtain the regularized differential equation of motion:

$$x'' - 2Cx - 1 = 0. \tag{10}$$

First, observe that this equation represents a harmonic oscillator if $C<0$, that is, if the motion is 'elliptic'. By this we mean that if the bodies would be slightly distrubed from their straight line collision orbits, they would describe ellipses and the collision would become a close approach. Note that using the initial conditions postulated before, $C=-1/x_0$. Consequently, C *is* negative since x_0 is the initial distance between the bodies.

The solution of Equation (10) with negative energy and satisfying the postulated initial conditions is

$$x = (x_0/2)\left(1 + \cos\sqrt{(2/x_0)}\,\tau\right). \tag{11}$$

The tacitly assumed initial condition for the new time variable is that $\tau=0$ when $t=0$.

The dependence of the new velocity on the new time may be obtained either by integrating Equation (6) or by differentiating Equation (11):

$$x' = -\sqrt{(x_0/2)}\sin\sqrt{(2/x_0)}\,\tau. \tag{12}$$

The relation between the new and original 'time' variables is obtained by integrating Equation (4):

$$t = \int_0^\tau x\,\mathrm{d}\tau = (x_0/2)\left[\tau + \sqrt{(x_0/2)}\sin\sqrt{(2/x_0)}\,\tau\right]. \tag{13}$$

At this point the observant reader will have discovered that τ is playing the role of the eccentric anomaly and Equation (13) is essentially Kepler's equation. The fact that the eccentric anomaly (and, in fact, also the true anomaly) are regularizing variables is interesting but probably the most significant result is that Equation (10) replaces Equation (2). In other words, the original differential equation of motion containing a $1/x^2$ type singularity is replaced by a linear differential equation without singularity.

In conclusion, we recall that the above example did not intend to be either practical or precise but was arrived at presenting regularization divested from all possible complications and from mathematical niceties. The essential features are the existence of a singularity in the original equation of motion, the transformation of the independent variable and the resulting new regular equation of motion.

3. Jacobian Transformation of the Coordinates

Assume that bodies k and l are the closest of the n bodies. Defining the vectors $\mathbf{Q}$ and $\mathbf{R}$ by

$$\left.\begin{aligned} \mathbf{Q} &= \frac{m_k\mathbf{r}_k + m_l\mathbf{r}_l}{m_k + m_l}, \\ &\text{and} \\ \mathbf{R} &= \mathbf{r}_k - \mathbf{r}_l, \end{aligned}\right\} \tag{14}$$

there results two differential equations relating the kth and the lth bodies. This process is discussed in detail by Szebehely (1968) and by Peters (1968) where it is shown that the equations of motion in this (Jacobian) coordinate system become

$$\ddot{\mathbf{Q}} = \frac{1}{m_k + m_l} \sum_{\substack{i=1 \\ i\neq k \\ i \neq l}}^{n} m_i \left[m_k \frac{\mathbf{r}_i - \mathbf{r}_k}{R_{ik}^3} + m_l \frac{\mathbf{r}_i - \mathbf{r}_l}{R_{il}^3} \right] \tag{15}$$

and

$$\ddot{\mathbf{R}} = -(m_k + m_l) \frac{\boldsymbol{r}_k - \boldsymbol{r}_l}{R_{kl}^3} + \sum_{\substack{i=1 \\ i\neq k \\ i \neq l}}^{n} m_i \left[\frac{\boldsymbol{r}_i - \boldsymbol{r}_k}{R_{ik}^3} - \frac{\boldsymbol{r}_i - \boldsymbol{r}_l}{R_{il}^3} \right], \tag{16}$$

or

$$\ddot{\mathbf{R}} = -(m_k + m_l)(\mathbf{R}/R^3) + \mathbf{F}, \tag{17}$$

where

$$\mathbf{F} = \sum_{\substack{i=1 \\ i\neq k \\ i \neq l}}^{n} m_i \left[\frac{\mathbf{r}_i - \mathbf{r}_k}{R_{ik}^3} - \frac{\mathbf{r}_i - \boldsymbol{r}_l}{R_{il}^3} \right].$$

Equation (15) will not suffer from a near singularity since it does not contain R_{kl}. Thus it may be considered as an auxiliary equation and solved with the other equations of motion. In order to determine $\mathbf{r}_k$ and $\mathbf{r}_l$ from Equations (14), $\mathbf{R}$ has to be found. This will be accomplished by a regularizing transformation. If we have m distinct pair of bodies that are near each other, then m transformations according to Equations (14) may be performed to yield a set of equations of the form (15) and (16). This gives us the capability of handling m close approaches.

4. Smoothing Transformations

First we will consider transformations of only the independent variable:

$$dt = g\,d\tau,$$

where g is some function and τ is the new independent variable. If

$$g = R^{\alpha}, \tag{18}$$

where α is an unspecified constant, then

$$(\mathrm{d}/\mathrm{d}t) = (1/R^{\alpha})\,(\mathrm{d}/\mathrm{d}\tau)$$

and

$$(\mathrm{d}^2/\mathrm{d}t^2) = (1/R^{2\alpha})\,(\mathrm{d}^2/\mathrm{d}\tau^2) - (\alpha/R^{1+2\alpha})\,(\mathrm{d}R/\mathrm{d}\tau)\,(\mathrm{d}/\mathrm{d}\tau).$$

Thus, Equation (17) becomes

$$\mathbf{R}'' - \frac{\alpha}{R}\mathbf{R}'R' + \frac{(m_k + m_l)}{R^{3-2\alpha}}\mathbf{R} = R^{2\alpha}\mathbf{F}, \tag{19}$$

where the prime denotes differentiation with respect to the new independent variable τ.

Expressed as a first order system, Equation (19) becomes

$$\left.\begin{aligned} \mathbf{R}' &= \mathbf{S}R^{\alpha}, \\ \mathbf{S}' &= \left[-\frac{(m_k + m_l)}{R^3}\mathbf{R} + \mathbf{F}\right]R^{\alpha}. \end{aligned}\right\} \tag{20}$$

No matter what value of α is selected, there still remains a singularity in Equation (19). However, even though the singularity has not been removed, its severity has been reduced, i.e., the term R^3 appears in the denominator of Equation (17), while in Equation (19), with $\alpha = \frac{3}{2}$, there is only a factor of R as the divisor. A transformation of this type that only reduces the effect of the singularity will be referred to as a smoothing transformation.

For unperturbed Keplerian motion, the choice $\alpha = 1$ corresponds to the use of the eccentric anomaly as the independent variable, and $\alpha = 2$ is equivalent to using the true anomaly. For perturbed Keplerian motion numerical experiments indicate that the choice $\alpha = \frac{3}{2}$ is better than $\alpha = 1$ or $\alpha = 2$. In fact, numerical investigations of orbits of artificial satellites show that the true anomaly has advantages near perigee, and that near apogee the eccentric anomaly is more efficient. Therefore, the use of $\alpha = \frac{3}{2}$ may be considered a compromise for the sallite problem. Experiments with n-body problems described by Szebehely and Bettis (1970) corroborate these findings and demonstrate that the selection $\alpha = \frac{3}{2}$ is the most advantageous choice for the smoothing transformation given by Equation (18).

Other choices for the smoothing function g have been suggested by various investigators. Szebehely (1967) discusses the use of the inversion of the velocity vector as the function g. Heggie (1970) has recently investigated the use of the potential energy and of the kinetic energy for g. The classical choice by Sundman (1912) is $\alpha = 1$ which regularizes the problem of two bodies but leaves much to be desired regarding its applicability to numerical work.

5. Regularizing Transformations

A. MOTION IN TWO DIMENSIONS

In order to remove the singularity in Equation (17) that occurs when R becomes small as well as to offer a method useful for numerical work, Levi-Civita (1903) introduced a coordinate transformation in addition to a transformation of the independent variable for two-dimensional motion. Kustaanheimo and Stiefel (1965) generalized Levi-Civita's transformation to the case when motion takes place in three dimensions.

Levi-Civita proposed Sundman's transformation for the independent variable:

$$dt = R d\tau . \tag{21}$$

This gives Equation (19) with $\alpha = 1$ as expected:

$$\mathbf{R}'' - \frac{R'}{R}\mathbf{R}' + \frac{(m_k + m_l)}{R}\mathbf{R} = R^2 \mathbf{F} . \tag{22}$$

In addition, Levi-Civita introduced the coordinate transformation

$$\begin{aligned} R_1 &= u_1^2 - u_2^2 , \\ R_2 &= 2u_1 u_2 , \end{aligned} \tag{23}$$

where R_1 and R_2 are the components of $\mathbf{R}$, and u_1 and u_2 are the new dependent variables. The transformation given by Equations (23) may be written as

$$\mathbf{R} = \mathscr{L}(\mathbf{u})\, \mathbf{u} , \tag{24}$$

where

$$\mathscr{L}(\mathbf{u}) = \begin{Bmatrix} u_1 & -u_2 \\ u_2 & u_1 \end{Bmatrix}$$

and $\mathbf{u}(u_1, u_2)$.

Note that Levi-Civita's transformation may be written as

$$z = w^2 ,$$

where $z = R_1 + iR_2$ and $w = u_1 + iu_2$ are complex vectors. In the following, with one exception, the matrix formulation is used because of its potential for generalization to higher dimensions. While the complex notation renders an opportunity to be generalized to four-dimensions by quarternions, it was felt that a straight forward, real matrix notation might be more widely understood without special preparation.

As the idea of the transformation of the independent as well as the dependent variables is introduced, a remark regarding the relation between these might be appropriate. Equation (22), the new form of the equation of motion, is the result of transforming only the independent variable. This equation may be considerably simplified *or* complicated by transforming also the dependent variables. Besides regularization, a possible simplification and stabilization of the equations is desirable. In general, the two transformations, using complex notation, may be written as

$$\mathrm{d}t = g(z)\,\mathrm{d}\tau$$

and

$$z = f(w).$$

The proper selection of the function $f(w)$ which introduces the new dependent variable is crucial, since the form of the function $g(z)$ is well established. The choice of f, once g is given, is arbitrary in principle. In order to obtain a simple form of the transformed equations of motion, it may be shown (Szebehely, 1966) that the relation

$$g = a\left|\frac{\mathrm{d}f}{\mathrm{d}w}\right|^2$$

must be satisfied. Here a is an arbitrary real constant. It should be emphasized that satisfying this relation is not pertinent to the regularization of the equation of motion but it has considerable practical importance.

The afore-mentioned transformation of the time is

$$g = R = \sqrt{R_1^2 + R_2^2} = |z|$$

and Levi-Civita's transformation is $z = w^2$. Consequently,

$$\left|\frac{\mathrm{d}f}{\mathrm{d}w}\right| = 2|w|$$

and the relation to be satisfied between f and g is

$$g = 4a|w|^2,$$

or

$$g = 4a|z|.$$

Comparing this with the function $g = R = \sqrt{R_1^2 + R_2^2}$ used for the time-transformation as given above or by Equation (21), we have $a = \frac{1}{4}$.

Now, we return to Equation (22) and introduce the new dependent variables given by Equations (23). First, we compute the derivative of R. Differentiating Equation (24) with respect to the new independent variable we obtain

$$\mathbf{R}' = 2\mathscr{L}(\mathbf{u})\,\mathbf{u}'. \tag{25}$$

From Equations (23) follows the important relation

$$R = u_1^2 + u_2^2 = (R_1^2 + R_2^2)^{1/2},$$

i.e., in the **u**-space the expression for the relative distance R does not require the calculation of a square root. Furthermore, angles at the origin of the **R**-space are doubled in the new **u**-space. Consequently, if in the **R**-space one body makes one revolution about the other, then in the **u**-space, this body will make only one-half of a revolution. The absence of the computation of the square root, and the halving of the angles at the origin result in considerable computational advantage when the Levi-Civita transformation, or its generalization the Kustaanheimo-Stiefel transformation is used.

Levi-Civita's matrix $\mathscr{L}(\mathbf{u})$ has the following properties (Stiefel and Scheifele, 1970):

$$\mathscr{L}^T(\mathbf{u})\,\mathscr{L}(\mathbf{u}) = RI\,, \tag{26a}$$

$$\mathscr{L}'(\mathbf{u}) = \mathscr{L}(\mathbf{u}')\,, \tag{26b}$$

$$\mathscr{L}(\mathbf{u})\,\mathbf{v} = \mathscr{L}(\mathbf{v})\,\mathbf{u}\,, \tag{26c}$$

and

$$(\mathbf{u}\cdot\mathbf{u})\,\mathscr{L}(\mathbf{v})\,\mathbf{v} - 2(\mathbf{u}\cdot\mathbf{v})\,\mathscr{L}(\mathbf{u})\,\mathbf{v} + (\mathbf{v}\cdot\mathbf{v})\,\mathscr{L}(\mathbf{u})\,\mathbf{u} = 0 \tag{26d}$$

where I is the unit matrix, **u** and **v** are arbitrary vectors and the scalar product is defined by the notation $(\mathbf{u}\cdot\mathbf{u})$. We proceed now to express Equation (22) in terms of the new dependent variable **u**. From Equation (25) it follows that

$$\mathbf{R}'' = 2\mathscr{L}(\mathbf{u})\,\mathbf{u}'' + 2\mathscr{L}'(\mathbf{u})\,\mathbf{u}',$$

or

$$\mathbf{R}'' = 2\mathscr{L}(\mathbf{u})\,\mathbf{u}'' + 2\mathscr{L}(\mathbf{u}')\,\mathbf{u}', \quad \text{since} \quad \mathscr{L}'(\mathbf{u}) = \mathscr{L}(\mathbf{u}')\,.$$

Using this expression for $\mathbf{R}''$ and Equations (24) and (25), Equation (22) becomes

$$2(\mathbf{u}\cdot\mathbf{u})\,\mathscr{L}(\mathbf{u})\,\mathbf{u}'' + 2(\mathbf{u}\cdot\mathbf{u})\,\mathscr{L}(\mathbf{u}')\,\mathbf{u}' - 4(\mathbf{u}\cdot\mathbf{u}')\,\mathscr{L}(\mathbf{u})\,\mathbf{u}' + \\ + (m_k + m_l)\,\mathscr{L}(\mathbf{u})\,\mathbf{u} = (\mathbf{u}\cdot\mathbf{u})^3\mathbf{F}\,, \tag{27}$$

where R' was eliminated by the relation

$$R' = 2(\mathbf{u}\cdot\mathbf{u}')\,.$$

With the aid of the Equation (26d), Equation (27) may be expressed as

$$2(\mathbf{u}\cdot\mathbf{u})\,\mathscr{L}(\mathbf{u})\,\mathbf{u}'' - 2(\mathbf{u}'\cdot\mathbf{u}')\,\mathscr{L}(\mathbf{u})\,\mathbf{u} + (m_k + m_l)\,\mathscr{L}(\mathbf{u})\,\mathbf{u} = (\mathbf{u}\cdot\mathbf{u})^3\mathbf{F}\,.$$

Multiplying this expression by $\mathscr{L}^{-1}(\mathbf{u})$ and by using Equation (26a), we obtain

$$\mathbf{u}'' + \frac{(m_k + m_l) - 2(\mathbf{u}'\cdot\mathbf{u}')}{2(\mathbf{u}\cdot\mathbf{u})}\,\mathbf{u} = \frac{(\mathbf{u}\cdot\mathbf{u})}{2}\,\mathscr{L}^T(\mathbf{u})\,\mathbf{F}\,. \tag{28}$$

Observe that the coefficient of **u** is one-half of the negative of the two-body binding energy per unit mass h, since, in the **R**-space,

$$h = \frac{(\dot{\mathbf{R}}\cdot\dot{\mathbf{R}})}{2} - \frac{(m_k + m_l)}{R} < 0 \tag{29a}$$

and in the **v** space

$$h = \frac{2(\mathbf{u}'\cdot\mathbf{u}') - (m_k + m_l)}{(\mathbf{u}\cdot\mathbf{u})}. \tag{29b}$$

The transformed equation for **R** may now be written as

$$\mathbf{u}'' - \frac{h}{2}\mathbf{u} = \frac{(\mathbf{u}\cdot\mathbf{u})}{2}\mathscr{L}^T(\mathbf{u})\,\mathbf{F}. \tag{30}$$

Equation (29b) contains a singularity. Consequently, if it is used to compute the binding energy, h, then Equation (30) will have a singularity. A regular differential equation may be obtained for the binding energy. If Equation (17) is multiplied by $\dot{\mathbf{R}}$ we have

$$(\dot{\mathbf{R}}\cdot\ddot{\mathbf{R}}) = -\frac{(m_k + m_l)}{R^3}(\dot{\mathbf{R}}\cdot\mathbf{R}) + (\dot{\mathbf{R}}\cdot\mathbf{F}). \tag{31}$$

Since

$$(\dot{\mathbf{R}}\cdot\mathbf{R}) = \dot{R}R,$$

Equation (31) becomes

$$\dot{\mathbf{R}}\cdot\ddot{\mathbf{R}} + (m_k + m_l)(\dot{R}/R^2) = \dot{\mathbf{R}}\cdot\mathbf{F},$$

or

$$\frac{\mathrm{d}}{\mathrm{d}t}\left[\frac{\dot{\mathbf{R}}\cdot\dot{\mathbf{R}}}{2} - \frac{(m_k + m_l)}{R}\right] = \dot{\mathbf{R}}\cdot\mathbf{F}.$$

The expression in the brackets above is the two-body binding energy per unit mass, therefore

$$(\mathrm{d}h/\mathrm{d}t) = \dot{\mathbf{R}}\cdot\mathbf{F},$$

or,

$$h' = \mathbf{R}'\cdot\mathbf{F}.$$

In the **u**-space this becomes

$$h' = 2(\mathscr{L}(\mathbf{u})\,\mathbf{u}'\cdot\mathbf{F})$$

or,

$$h' = 2(\mathbf{u}'\cdot\mathscr{L}^T(\mathbf{u})\,\mathbf{F}). \tag{32}$$

Equations (21), (30), and (32) form a system of regular differential equations which may be solved for **u** and for the time, t. Then **R** may be obtained by using Equation (24).

B. MOTION IN THREE DIMENSIONS

In order to extend Levi-Civita's transformation to three dimensions, Kustaanheimo and Stiefel (1965) introduced two four-dimensional vectors **R** and **u** which are connected by

$$\mathbf{R} = \mathscr{L}(\mathbf{u})\,\mathbf{u},$$

where

$$\mathscr{L}(\mathbf{u}) = \begin{pmatrix} u_1 & -u_2 & -u_3 & u_4 \\ u_2 & u_1 & -u_4 & -u_3 \\ u_3 & u_4 & u_1 & u_2 \\ u_4 & -u_3 & u_2 & -u_1 \end{pmatrix}$$

and where

$$\mathbf{R} = \begin{pmatrix} R_1 \\ R_2 \\ R_3 \\ R_4 \end{pmatrix}, \quad u = \begin{pmatrix} u_1 \\ u_2 \\ u_3 \\ u_4 \end{pmatrix}.$$

The components of R are

$$\left.\begin{aligned} R_1 &= u_1^2 - u_2^2 - u_3^2 + u_4^2, \\ R_2 &= 2(u_1u_2 - u_3u_4), \\ R_3 &= 2(u_1u_3 + u_2u_4), \\ R_4 &= 0, \end{aligned}\right\} \tag{33}$$

and its magnitude is

$$R = \sqrt{\mathbf{R}\cdot\mathbf{R}} = u_1^2 + u_2^2 + u_3^2 + u_4^2. \tag{34}$$

Hurwitz (1933) has shown that the generalization of Levi-Civita's transformation to three dimensions is not possible, but that the transformation may be extended to four dimensions. Indeed, by using the four-by-four matrix $\mathscr{L}(\mathbf{u})$, Equations (30) and (32) remain valid.

Since one of the four components of the vector **u** is arbitrary, the question arises as to how to select the initial values of the components of **u** where **R** is given. Adding the first of Equations (33) to Equation (34) we have

$$u_1^2 + u_4^2 = \tfrac{1}{2}(R_1 + R). \tag{35a}$$

Keeping in mind the arbitrariness of one of the components of u, we will select u_1 and u_4 such that this relation is satisfied. For simplicity we will select either u_1 or u_4 identically equal to zero or u_1 equal to u_4. From the second and third equations of Equations (33) there results

$$u_2 = \frac{R_2u_1 + R_3u_4}{R_1 + R}, \quad u_3 = \frac{R_3u_1 - R_2u_4}{R_1 + R}. \tag{35b}$$

When $R_1 \geqslant 0$, Equations (35a) and (35b) are used to determine the initial values of the components of u. In order to avoid the loss of significant digits when $R_1 < 0$, it is advantageous to use the set of relations

$$u_2^2 + u_3^2 = \tfrac{1}{2}(R - R_1), \tag{36a}$$

$$u_1 = \frac{R_2 u_2 + R_3 u_3}{R - R_1}, \quad u_4 = \frac{R_3 u_2 - R_2 u_3}{R - R_1}, \tag{36b}$$

choosing either u_2 or u_3 arbitrarily.

Once the components of u are determined, then u' is determined from Equations (25) and (26a):

$$\mathbf{u}' = (1/2R)\,\mathscr{L}^T(\mathbf{u})\,\mathbf{R}',$$

or

$$\mathbf{u}' = \tfrac{1}{2}\mathscr{L}^T(\mathbf{u})\,\dot{\mathbf{R}}. \tag{37}$$

To obtain the time, t, it is necessary to solve the differential equation

$$t' = R, \tag{38}$$

where R is computed from Equation (34). It is assumed at this point that the differential equation of motion, Equation (30) has been solved, so $\mathbf{u}(\tau)$ is known. An alternative, which may be used either as a check on the accuracy of the computation of R or to obtain R for using it in Equation (38), is to solve a differential equation for R. By differentiating Equation (34) twice with respect to τ, there results

$$R'' = 2(\mathbf{u}''\cdot\mathbf{u}) + 2(\mathbf{u}'\cdot\mathbf{u}').$$

By using Equations (29b) and (30) this becomes

$$R'' - 2hR = (m_k + m_l) + R\left[\mathbf{u}\cdot\mathscr{L}^T(\mathbf{u})\,\mathbf{F}\right]. \tag{39}$$

If the perturbation forces $\mathbf{F}$ are zero, Equations (30) and (39) become the differential equations of a harmonic oscillator which are stable in the Liapunov sense. It is well known that the original equation of motion, Equation (17), for $\mathbf{R}$ is unstable in the Liapunov sense. Since $\mathbf{F}$ is not zero in our n-body problem, strict Liapunov stability no longer exists, but the transformed equations represent a much more stable system of differential equations than the original.

6. Summary of the K–S Formulation

This section lists the computational steps to be performed when the K–S method of three dimensional regularization is used. It is assumed that the components of $\mathbf{R}$ and $\dot{\mathbf{R}}$ are given.

First compute R from

$$R = \sqrt{R_1^2 + R_2^2 + R_3^2}.$$

If $R_1 \geqslant 0$, let $u_4 = 0$, and compute the remaining components of u as:

$$u_1 = \sqrt{\tfrac{1}{2}(R_1 + R)},$$
$$u_2 = (R_2/2u_1), \quad u_3 = (R_3/2u_1).$$

If $R_1 < 0$, let $u_3 = 0$ and use the relations

$$u_2 = \sqrt{\tfrac{1}{2}(R - R_1)}$$
$$u_1 = (R_2/2u_2), \quad u_4 = (R_3/2u_2).$$

Then to compute u' from Equation (37) use

$$u_1' = \tfrac{1}{2}(u_1\dot{R}_1 + u_2\dot{R}_2 + u_3\dot{R}_3),$$
$$u_2' = \tfrac{1}{2}(-u_2\dot{R}_1 + u_1\dot{R}_2 + u_4\dot{R}_3),$$
$$u_3' = \tfrac{1}{2}(-u_3\dot{R}_1 - u_4\dot{R}_2 + u_1\dot{R}_3),$$
$$u_4' = \tfrac{1}{2}(u_4\dot{R}_1 - u_3\dot{R}_2 + u_2\dot{R}_3).$$

The initial value of the binding energy as given by Equation (29a) is

$$h = \frac{\dot{\mathbf{R}}\cdot\dot{\mathbf{R}}}{2} - \frac{(m_k + m_l)}{R}.$$

Using the initial values obtained above for $\mathbf{u}$, $\mathbf{u}'$ and h, solve the system of differential equations:

$$u_1'' - \frac{h}{2}u_1 = \frac{R}{2}(\mathscr{L}^T\mathbf{F})_1,$$
$$u_2'' - \frac{h}{2}u_2 = \frac{R}{2}(\mathscr{L}^T\mathbf{F})_2,$$
$$u_3'' - \frac{h}{2}u_3 = \frac{R}{2}(\mathscr{L}^T\mathbf{F})_3,$$
$$u_4'' - \frac{h}{2}u_4 = \frac{R}{2}(\mathscr{L}^T\mathbf{F})_4,$$
$$R'' - 2hR = (m_k + m_l) + R\sum_{i=1}^{4}(\mathscr{L}^T\mathbf{F})_i u_i,$$
$$h' = 2\sum_{i=1}^{4}(\mathscr{L}^T\mathbf{F})_i u_i'$$
$$t' = R.$$

Here

$$(\mathscr{L}^T\mathbf{F})_1 = u_1F_1 + u_2F_2 + u_3F_3,$$
$$(\mathscr{L}^T\mathbf{F})_2 = -u_2F_1 + u_1F_2 + u_4F_3,$$
$$(\mathscr{L}^T\mathbf{F})_3 = -u_3F_1 - u_4F_2 + u_1F_3,$$
$$(\mathscr{L}^T\mathbf{F})_4 = u_4F_1 - u_3F_2 + u_2F_3.$$

The formulae necessary for the transformation from the **u**-space to the **R**-space are

$$R_1 = u_1^2 - u_2^2 - u_3^2 + u_4^2,$$
$$R_2 = 2(u_1u_2 - u_3u_4),$$
$$R_3 = 2(u_1u_3 + u_2u_4),$$

and

$$\dot{R}_1 = \frac{2}{R}(u_1u_1' - u_2u_2' - u_3u_3' + u_4u_4'),$$

$$\dot{R}_2 = \frac{2}{R}(u_2u_1' + u_1u_2' - u_4u_3' - u_3u_4'),$$

$$\dot{R}_3 = \frac{2}{R}(u_3u_1' + u_4u_2' + u_1u_3' + u_2u_4').$$

During the solution of the system of differential equations, the following checks may be employed:

(i) the distance

$$R = \sum_{i=1}^{4} u_i^2,$$

(ii) the binding energy

$$h = \frac{2\sum_{i=1}^{4} u_i'^2 - (m_k + m_l)}{R}.$$

7. The Use of Several Independent Variables

Whenever transformations of the independent variable are performed there will be at least two 'time' variables which have to be related. For example, assume that bodies 1 and 2 on one hand and bodies 3 and 4 on the other hand, form two pairs so that these two pairs are *not* close to each other. Also assume that the other bodies 4, 5, ..., n are far from these pairs. Denoting the relative distance of bodies 1 and 2 by R_{12} and that of bodies 3 and 4 by R_{34}, we have for the new independent variables

$$\mathrm{d}t = R_{12}\mathrm{d}\tau_{12} = R_{34}\mathrm{d}\tau_{34}.$$

By selecting the independent variable with the smallest factor, for example τ_{12} we may relate the other times to this τ_{12} by

$$\mathrm{d}t = R_{12}\mathrm{d}\tau_{12} \quad \text{and} \quad \mathrm{d}\tau_{45} = (R_{12}/R_{34})\,\mathrm{d}\tau_{12}.$$

This idea may be extended to any distant pair of bodies. By writing the equations of motion as a system of first order differential equations, the correlation of the different times may be accomplished by multiplying all of the equations by the suitable factor of R_{ij}. That is, if the original equations of motion,

$$(\mathrm{d}^2\mathbf{r}_i/\mathrm{d}t^2) = f_i(\mathbf{r}_1, \ldots, \mathbf{r}_n)$$

are written as

$$\begin{aligned}
&(\mathrm{d}\mathbf{r}_1/\mathrm{d}\tau_{12}) = \mathbf{V}_1, \quad (\mathrm{d}\mathbf{V}_1/\mathrm{d}\tau_{12}) = \mathbf{f}_1,\\
&(\mathrm{d}\mathbf{r}_2/\mathrm{d}\tau_{12}) = \mathbf{V}_2, \quad (\mathrm{d}\mathbf{V}_2/\mathrm{d}\tau_{12}) = \mathbf{f}_2,\\
&(\mathrm{d}\mathbf{r}_3/\mathrm{d}\tau_{34}) = \mathbf{V}_3, \quad (\mathrm{d}\mathbf{V}_3/\mathrm{d}\tau_{34}) = \mathbf{f}_3,\\
&(\mathrm{d}\mathbf{r}_4/\mathrm{d}\tau_{34}) = \mathbf{V}_4, \quad (\mathrm{d}\mathbf{V}_4/\mathrm{d}\tau_{34}) = \mathbf{f}_4,\\
&\quad(\mathrm{d}\mathbf{r}_j/\mathrm{d}t) = \mathbf{V}_j, \quad (\mathrm{d}\mathbf{V}_j/\mathrm{d}t) = \mathbf{f}_j, \quad j = 5, 6, \ldots, n,
\end{aligned}$$

then by multiplying all but the first four of these equations by the appropriate factors of R_{ij}, we obtain the system

$$\begin{aligned}
&(\mathrm{d}\mathbf{r}_1/\mathrm{d}\tau_{12}) = \mathbf{V}_1, \quad (\mathrm{d}\mathbf{V}_1/\mathrm{d}\tau_{12}) = \mathbf{f}_1,\\
&(\mathrm{d}\mathbf{r}_2/\mathrm{d}\tau_{12}) = \mathbf{V}_2, \quad (\mathrm{d}\mathbf{V}_2/\mathrm{d}\tau_{12}) = \mathbf{f}_2,\\
&(\mathrm{d}\mathbf{r}_3/\mathrm{d}\tau_{12}) = \mathbf{V}_2\,(R_{12}/R_{34}), \quad (\mathrm{d}\mathbf{V}_3/\mathrm{d}\tau_{12}) = \mathbf{f}_3\,(R_{12}/R_{34}),\\
&(\mathrm{d}\mathbf{r}_4/\mathrm{d}\tau_{12}) = \mathbf{V}_4\,(R_{12}/R_{34}), \quad (\mathrm{d}\mathbf{V}_4/\mathrm{d}\tau_{12}) = \mathbf{f}_4\,(R_{12}/R_{34}),\\
&(\mathrm{d}\mathbf{r}_j/\mathrm{d}\tau_{12}) = \mathbf{V}_j R_{12}, \quad (\mathrm{d}\mathbf{V}_j/\mathrm{d}\tau_{12}) = \mathbf{f}_j R_{12}.
\end{aligned}$$

Thus the single independent variable τ_{12} may be used during the numerical integration, even though several different time variables were introduced.

8. Numerical Solution of the Differential Equations

Traditionally in celestial mechanics, finite difference methods have been used for the numerical solution of differential equations, such as the Störmer-Cowell or the Adams-Bashforth-Moulton methods. More recently methods based on power series gained popularity, such as straight forward Taylor series or recurrent power series. The choise of the particular method is dependent upon the speed and the memory capacity of the electronic computer being used. In general, the limitations of the computer become more critical as the number of bodies increases.

A special problem emerges when numerical integration is performed with m pair of bodies regularized by the K–S transformation. Everytime the configuration (i.e., the participating bodies) changes, the K–S transformation must be re-initialized. This is of no consequence to the self-starting power series methods, but it is difficult for the multi-step methods which require knowledge of the solution of the differential equations for several previous values of the independent variable. However, if the configuration is rather stable, the K–S transformation does not need to be re-initialized, and a starting routine does not need to be called.

In order to reach a compromise between computational accuracy and computer time during the close encounter of m pair of bodies, some criterion must be established as to when to regularize. In general, a close pair should be regularized by the K–S transformation only when there is reason to suspect that there is, or will be, a loss of significant digits in the calculation. While establishing such a criterion one must remember the most apparent advantages of the K–S transformation, (i) regular differential equations for a pair of bodies, (ii) no computation of a square root in order to determine, R and (iii) angles at the origin are doubled. These advantages should be weighed against the following disadvantages: (i) the differential equations are not regular if there are more than two bodies in the encounter, (ii) there are additional differential equations to be solved, and (iii) time consuming transformations to the Jacobian coordinates and to the K–S coordinates must be executed and vice-versa.

9. Numerical Methods

For an explanation of the use of the finite difference methods we refer the reader to any or all of the following articles: Henrici (1962), Schubart and Stumpff (1966), Aarseth (1970), Bettis (1970), and Stiefel and Bettis (1969).

It has been found that if the number of bodies is of the order of one-hundred to two-hundred, high order Runge-Kutta methods are very efficient when the configuration of the m pair of regularized bodies is rapidly changing. In particular, a Runge-Kutta method by Fehlberg (1968) has proved very efficient with a CDC 6600 Computer for accurate work as shown by Szebehely and Bettis (1970).

For the sake of the convenience of the reader, Fehlberg's algorithm is outlined. For a system of differential equations of the first order,

$$(\mathrm{d}y/\mathrm{d}t) = f(t, y)$$

the solution for the value $t = t_0 + s$, Fehlberg gives

$$f_0 = f(t_0, y_0),$$

$$f_\kappa = f\left(t_0 + \alpha_\kappa s, \quad y_0 + s \sum_{\lambda=0}^{\kappa-1} \beta_{\kappa\lambda} f_\lambda\right), \quad \kappa = 1, 2, 3, \ldots, 12,$$

$$y = y_0 + s \sum_{k=0}^{12} C_\kappa f_\kappa + 0(s^9),$$

where s is the step-size.

Even though this method requires thirteen evaluations of the function $f(t, y)$ for each step, the precise estimate of the step sizes of Fehlberg's method compensates for this disadvantage. The truncation error for this method is

$$\tfrac{41}{840}(f_0 + f_{10} + f_{11} + f_{12})\, s.$$

From this relation, the step-size s may be estimated. The reader is referred to the original article by Fehlberg (1968) for further details and for the coefficients α_x, $\beta_{x\lambda}$, and C_x.

Acknowledgements

Partial support of a Contract No. N00014-67-A-0126-0007 from the Office of Naval Research is acknowledged. Fruitful discussions are acknowledged with great pleasure conducted between Dr S. J. Aarseth, Mr D. Heggie, and the authors.

References

Aarseth, S.: 1972, this volume, p. 29.
Bettis, D. G.: 1970, *Numer. Math.* **14**, 421–434.
Fehlberg, E.: 1968, NASA TR R-287, also 1969, *Computing* **4,** 93.
Heggie, D.: 1972, this volume, p. 148.
Henrici, P.: 1962, *Discrete Variable Methods in Ordinary Differential Equations*, John Wiley and Sons.
Hurwitz, A.: 1933, *Math. Werke*, Vol. **2**, Birkhäuser, Basel.
Kustaanheimo, P. and Stiefel, E.: 1965, *J. Math.* **218**, 204.
Levi-Civita, T.: 1903, *Ann. Math.* **9**, 1.
Peters, C. F.: 1968, *Bull. Astron.* **3**, 167.
Schubart, J. and Stumpff, P.: 1966, *Veröff. Astron. Rechen-Inst. Heidelberg*, No. 18, Karlsruhe.
Stiefel, E. and Bettis, D. G.: 1969, *Numer. Math.* **13**, 154–75.
Stiefel, E. and Scheifele, G.: 1970, *Linear and Regular Celestial Mechanics*, Springer-Verlag.
Sundman, K. F.: 1912, *Acta Math.* **35**, 105.
Szebehely, V.: 1967, *Theory of Orbits*, Academic Press, New York.
Szebehely, V.: 1968, *Bull. Astron.* **3**, 91.
Szebehely, V. and Bettis, D. G.: 1972, this volume, p. 136.

THE MONTE CARLO METHOD

M. HÉNON
Observatoire de Nice

Abstract. We give here a detailed technical description of a Monte Carlo scheme for the dynamical evolution of spherical stellar systems. The philosophy of the method, as well as a few illustrative results, are given elsewhere (Hénon, 1972, hereafter called I).

1. Superstars

Each actual computation will be made with a set of n objects, with n having some definite value. On the other hand, we would like the results to be applicable to a system with an arbitrary number of stars N. We shall therefore make the convention that each object in the computation is a *superstar*, representing K actual stars, with: $K=N/n$. If the various units are suitably chosen (see below), K does not appear in the numerical computations, and is thus left as a freely adjustable parameter.

Let r be the distance of a star to the centre, v_r its radial velocity, v_t its transverse velocity, and m its mass. A superstar will be characterized at any given time by the four quantities, r, v_r, v_t, m, and will be understood as a collection of K stars having these coordinates. Thus the mass of the superstar is: Km. The two angular coordinates defining the position of each star on the sphere with radius r are not specified; the stars are assumed to be randomly distributed on the sphere. Similarly, the angular coordinate defining the orientation of the transverse velocity is assumed to be randomly distributed.

In this way, the introduction of the superstars has another important advantage: it makes the system exactly spherically symmetrical.

2. Initial Conditions

The computation is started by selecting the four quantities r, v_r, v_t, m for each of the n superstars at time $t=0$. This is done by prescribing an initial distribution function $f(\mathbf{r}, \mathbf{v}, m, 0)$, and then selecting the coordinates randomly in accordance with this distribution.

The choice of the quasi-random number generator requires some care since a large quantity of these numbers is used in the course of the computation (typically about 10^6 for one run with $n=1000$). We have used, to our entire satisfaction, one of the generators studied by Coveyou and Macpherson (1967):

$$n_{i+1} = \lambda n_i + 1 \pmod{2^{35}}, \tag{1}$$

with

$$\lambda = 273\,673\,163\,155 \text{ (octal)}. \tag{2}$$

M. Lecar (ed.), Gravitational N-Body Problem, 406–422.

The n_i are integers. Quasi-random numbers with a uniform distribution between 0 and 1 are obtained by: $X_i = 2^{-35}\, n_i$.

One restriction on the form of the initial distribution function is that it must be a steady state; the present scheme is not equipped to handle the initial phase of rapid collective motions, since it proceeds by time steps much larger than the crossing time t_c, and we assume that these motions have already died out at time $t=0$. In the case of spherical symmetry, this means that the distribution function has the form:

$$f = \Psi(E, A, m), \tag{3}$$

where E and A are the energy per unit mass and the angular momentum per unit mass:

$$\begin{aligned} E &= U(r) + \tfrac{1}{2}(v_r^2 + v_t^2), \\ A &= r v_t ; \end{aligned} \tag{4}$$

$U(r)$ is the gravitational potential.

The function Ψ in (3) can be arbitrary. In practice, however, the computation of the initial conditions will be easier if Ψ has a simple analytical form; in particular it is preferable to choose a form for which the corresponding potential $U(r)$ can be obtained explicitly.

3. Potential

We need to know the mean, smoothed-out potential, in order to compute the unperturbed motion of the stars. Smoothing in the transverse directions is immediately achieved by forgetting the discrete nature of each superstar and assimilating it to a spherical shell of radius r, of mass Km. Smoothing in the radial direction would also be required in principle. This, however, raises some practical difficulties. One of them is the precise definition of the smoothed potential in that case. Another is that the kinetic energy of a star, computed from its total energy by (4a), may turn out to be negative in some cases if the actual potential is replaced by the smoothed-out potential. On the other hand, experience indicates that radial smoothing is not really necessary (see Figure 4). The reason is probably that a system of concentric shells is already smoothed to a considerable degree; in particular, since it has spherical symmetry, the two integrals (4) of the motion of a star are exactly conserved.

Therefore we take the mean potential as being that of the system of n concentric shells. To compute it, we sort first the n shells in order of increasing radius. Let r_k and Km_k be the radius and mass of the shell of rank k from the centre. We also define: $r_0 = 0$ and $r_{n+1} = \infty$. Then, in the interval $r_k \leqslant r \leqslant r_{k+1}$, the potential is:

$$U(r) = KG\left(-\frac{1}{r}\sum_{i+1}^{k} m_i - \sum_{i=k+1}^{n} \frac{m_i}{r_i}\right). \tag{5}$$

Thus U, plotted against $1/r$, is a succession of straight segments, as schematically shown by Figure 1 for the case $n=3$. In consequence, it is sufficient to store the values

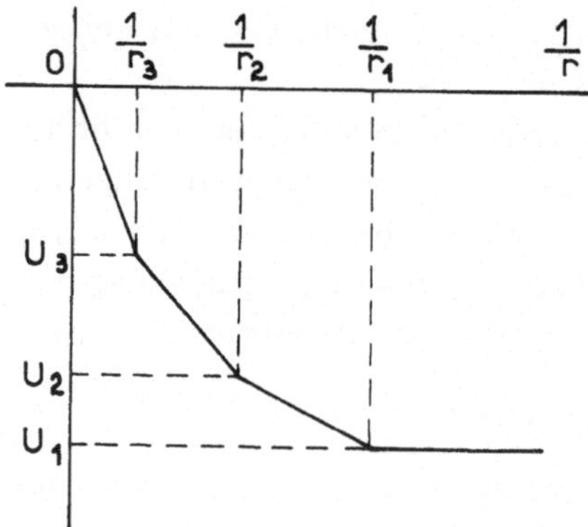

Fig. 1. Gravitational potential U as a function of $1/r$, for a system of three shells.

$U_k = U(r_k)$ of the potential on the shells in order to be able to compute quickly the potential at any point. These values are computed recursively by:

$$U_{n+1} = 0,$$

$$M_n = M = K \sum_{i=1}^{n} m_i,$$

$$\left.\begin{aligned} U_k &= U_{k+1} - GM_k\left(\frac{1}{r_k} - \frac{1}{r_{k+1}}\right), \\ M_{k-1} &= M_k - Km_k. \end{aligned}\right\} \quad (k = n, \ldots, 1) \tag{6}$$

M_k is the mass inside of and including the shell of rank k. M is the total mass of the system. The potential at any point is then given by:

$$U(r) = U_k + \frac{1/r_k - 1/r}{1/r_k - 1/r_{k+1}} (U_{k+1} - U_k), \quad \text{for} \quad r_k \leqslant r \leqslant r_{k+1}. \tag{7}$$

4. Time Step

We let now the system evolve during a time Δt, which must be small compared to the relaxation time. The crossing time is given as a function of the total mass M and the total energy $\mathscr{E}$ of the system by:

$$t_c = C_2 G M^{5/2} |\mathscr{E}|^{-3/2}, \tag{8}$$

where C_2 is a numerical constant (van Albada, 1968), and the relaxation time is related to the crossing time by:

$$t_r/t_c = C_1 N/\ln N, \tag{9}$$

where C_1 is another numerical constant (Chandrasekhar, 1942). Therefore we take for the time step:

$$\Delta t = b \frac{N}{\ln N} G M^{5/2} |\mathscr{E}|^{-3/2}, \tag{10}$$

where b is a small constant. Experience shows that $b = 0.005$ gives reasonable results.

Δt as given by (10) is not constant but decreases markedly with time because N and M decrease as a result of the escape of stars.

5. Relaxation

We assume that the mean potential does not change during Δt. Then the unperturbed motion of each star would be a rosette motion, characterized by constant values of E and A. In fact, however, the perturbations will slightly change E and A during the time Δt. In order to simulate this effect in Monte Carlo fashion, we should first select at random the position of each star along its orbit. For the first time step, however, this is not necessary since the initial coordinates have been already chosen at random according to a given steady state distribution function. Thus we take each superstar where it is at $t = 0$. We shall compute the effect of the perturbations during a small time δt, satisfying:

$$\delta t \ll t_c. \tag{11}$$

Later on we shall multiply the effect by the correcting factor $\Delta t/\delta t$, in order to compensate for the rest of the time interval.

Next, we must consider each star in turn as a test-star, and then select at random another star as a representative of the field stars. We call $(\mathbf{r}, \mathbf{v}, m)$ and $(\mathbf{r}', \mathbf{v}', m')$ the coordinates of the test star and the field star respectively. As is well known, the total perturbation from the field stars up to a distance l is approximately proportional to $\ln(Nl/R)$, where R is a typical radius of the system; thus, although all stars contribute to the perturbation, the largest part of it comes from relatively near stars. For example, if $N = 10^5$, 80% of the perturbation comes from the field stars nearer than $R/10$. Therefore we shall make the simplifying assumption that the distribution of field stars with respect to mass and velocity is everywhere the same as where the test star is. In other words, we take the distribution function of field stars to be not $f(\mathbf{r}', \mathbf{v}', m')$, but $f(\mathbf{r}, \mathbf{v}', m')$. The advantage is that this last distribution is space-independent, so that we can now select separately the position of the field star on the one hand, its velocity and its mass on the other.

We select first the mass and the velocity, with a weight function proportional to the local distribution function. There is a very simple way to do this: we just select the mass and the velocity of the nearest star. A moment's reflection will show that the probability to obtain given values m' and $\mathbf{v}'$ is proportional to the local distribution function.

In fact, we do not know the positions of the stars, but only the radii of the superstars. Besides, it will be seen below that the choice of the nearest star leads to a divergence in the equations, which can be avoided by taking instead the p-th nearest star, with $p > 1$. Thus our practical rule will be: for a test star at r_k, take the mass and the velocity of the field star either at r_{k+p} or r_{k-p}. The probability to obtain given values of m' and $\mathbf{v}'$ is still proportional to the local distribution function, provided that p is not too large.

This procedure does not determine completely the velocity of the field star, because only v_r and v_t are specified for a superstar. More specifically, we do not know the angle φ between the orbital planes of the test star and of the field star, defined by $(\mathbf{r}, \mathbf{v})$ and $(\mathbf{r}, \mathbf{v}')$. However, because of the spherical symmetry, and the correlated assumption that the orbital planes of the stars belonging to a given superstar are randomly distributed, the angle φ will also be randomly distributed, and we simply select its value at random between 0 and 2π with a uniform probability.

We must now select the position of the field star. This position could be simply taken at random, with a uniform probability, inside the volume occupied by the system. However, a general rule in Monte Carlo computations says that whenever possible, a mean quantity should be computed exactly rather than estimated by sampling, in order to minimize the variance of the final results. It happens here that the mean effect of a field star with given mass and velocity can be computed rather simply. Therefore, instead of taking the position at random, we shall choose it in such a way that the effect of the field star will be equal to the computed mean effect.

We consider a moving frame of reference in which the centre of mass of the test star and the field star is at rest. In this frame, the modulus of the velocity of the test star does not change during the encounter; the direction of the velocity is deflected by an angle β, generally small, given by:

$$\tan\frac{\beta}{2} = \frac{G(m+m')}{w^2 l} = \frac{l_0}{l}, \tag{12}$$

where l is the impact parameter and w is the relative velocity:

$$w = |\mathbf{v}' - \mathbf{v}|\,; \tag{13}$$

thus l_0 defined by (12) is independent of the position of the field star. Let ν be the local number of field stars per unit volume. Then, since we sample one field star at random out of N, the probability to find it in a volume $\mathrm{d}V$ is: $\nu \mathrm{d}V/N$; and the probability for the test star to 'meet' the field star during the time δt, with an impact parameter between l and $\mathrm{d}l$, is:

$$2\pi l\, \mathrm{d}l w\, \delta t \nu N^{-1}. \tag{14}$$

Therefore the mean squared deflection is:

$$\langle \beta^2 \rangle = 8\pi l_0^2 w\, \delta t \nu N^{-1} \int (\mathrm{d}l/l)\,; \tag{15}$$

we have made here the approximation: $\beta/2 \approx \tan\beta/2$. The appropriate cut-offs for the integral are respectively of the order of R/N and R (Hénon, 1958), so that the integral is approximately equal to $\ln N$.

We multiply this now by the correcting factors $\Delta t/\delta t$ for the rest of the motion, and N for the other field stars, obtaining:

$$\langle\beta^2\rangle = 8\pi l_0^2 w \Delta t \nu \ln N . \tag{16}$$

This value will also result if we take a field star with a fixed impact parameter equal to:

$$l = (2\pi w \Delta t \nu \ln N)^{-1/2} . \tag{17}$$

There remains only the problem of estimating ν, the local number density of the field stars. It would be difficult to compute this density accurately, since there is only a limited number of superstars in the computation. Therefore we revert here to a sampling procedure: we shall estimate ν from the radial distance of the p-th nearest superstar, the same which has already been used to provide the mass and the velocity of the field star (this is not necessary, but convenient). Let Δr be this distance:

$$\Delta r = |r_{k\pm p} - r_k| . \tag{18}$$

Clearly the estimated density, ν^*, should be inversely proportional to Δr:

$$\nu^* = C/\Delta r . \tag{19}$$

Since ν enters linearly in (16), the constant C must be adjusted so that

$$\langle\nu^*\rangle = \nu . \tag{20}$$

Let σ be the number of superstars per unit interval in r; σ is related to ν by:

$$K\sigma = 4\pi r^2 \nu . \tag{21}$$

The probability for the p-th nearest superstar to be at a radial distance Δr is given by the Poisson formula:

$$e^{-x} \frac{x^{p-1}}{(p-1)!} \mathrm{d}x, \tag{22}$$

with: $x = \sigma \Delta r$. Therefore:

$$\langle\nu^*\rangle = C\sigma \int_0^\infty e^{-x} \frac{x^{p-2}}{(p-1)!} \mathrm{d}x = \frac{C\sigma}{p-1} . \tag{23}$$

From (20), (21) and (23) we obtain:

$$C = K \frac{p-1}{4\pi r^2} . \tag{24}$$

Using (19) and substituting ν^* for ν in (17), we finally have for the impact distance:

$$l = \left[\frac{(p-1)\, K w \Delta t \ln N}{2r^2 \Delta r}\right]^{-1/2} . \tag{25}$$

Equation (23) shows why we cannot use $p = 1$, i.e. the nearest superstar: $\langle\nu^*\rangle$ would be infinite. (This fact had been overlooked in Hénon (1966) and Equation (3) in that paper is wrong). We have used the next value, $p=2$, in the computations. Superstars are associated in pairs, according to their rank from the centre, as follows: 1–3, 2–4, 5–7, 6–8, etc. If n is not a multiple of 4, one or two superstars are left out at the outside; the corresponding error is negligible since n is large, and also because relaxation effects are very small in the outer parts of the system. For each pair, a single encounter is computed, and the velocities of both superstars are perturbed accordingly; in other words, each member of the pair acts as a field star for the other. This symmetrical scheme saves a factor 2 in computing time, and has also the advantage that the total energy of the system is exactly conserved during this stage of the computation.

In order to specify completely the encounter, one must know, in addition to the impact parameter l, the angle ψ of the plane of relative motion $(\mathbf{r}'-\mathbf{r}, \mathbf{v}'-\mathbf{v})$ with some reference plane. Because of the assumed homogeneous distribution of the field stars, ψ is uniformly distributed, and therefore is chosen at random between 0 and 2π.

The computation of an encounter for two superstars (r, v_r, v_t, m) and (r', v_r', v_t', m') proceeds in practice as follows. We consider a system of rectangular axes, x, y, z with the z axis parallel to $\mathbf{r}$ and the (x, z) plane parallel to $\mathbf{v}$. In this system the velocities of the two stars are:

$$\mathbf{v} = (v_t, 0, v_r), \quad \mathbf{v}' = (v_t' \cos\varphi, v_t' \sin\varphi, v_r'). \tag{26}$$

φ is computed by:

$$\varphi = 2\pi X, \tag{27}$$

where X is a random number between 0 and 1. We compute the relative velocity:

$$\mathbf{w} = (v_t' \cos\varphi - v_t, v_t' \sin\varphi, v_r' - v_r) = (w_x, w_y, w_z), \tag{28}$$

and its modulus w. We compute the impact parameter l from (25), with:

$$\Delta r = |r' - r|, \tag{29}$$

and r replaced by the mean value $(r+r')/2$ for the two superstars. We compute the deflection angle β from (12).

We compute the quantity:

$$w_p = (w_x^2 + w_y^2)^{1/2}. \tag{30}$$

Let us now consider the two vectors:

$$\begin{aligned} \mathbf{w}_1 &= (w_y w/w_p, -w_x w/w_p, 0), \\ \mathbf{w}_2 &= (-w_x w_z/w_p, -w_y w_z/w_p, w_p). \end{aligned} \tag{31}$$

It is readily verified that $\mathbf{w}_1$ and $\mathbf{w}_2$ have the same modulus as w and that the three vectors $\mathbf{w}$, $\mathbf{w}_1$, $\mathbf{w}_2$ are mutually perpendicular. We take the plane $(\mathbf{w}, \mathbf{w}_1)$ as origin for the angle ψ, which is computed by:

$$\psi = 2\pi X, \tag{32}$$

where X is a random number between 0 and 1. The new relative velocity of the two stars after the encounter is then:

$$\mathbf{w}^* = \mathbf{w}\cos\beta + \mathbf{w}_1 \sin\beta\cos\psi + \mathbf{w}_2 \sin\beta\sin\psi\,. \tag{33}$$

Therefore the new velocities of the two stars are:

$$\mathbf{v}^* = \mathbf{v} - \frac{m'}{m+m'}(\mathbf{w}^* - \mathbf{w}),$$

$$\mathbf{v}'^* = \mathbf{v}' + \frac{m}{m+m'}(\mathbf{w}^* - \mathbf{w}). \tag{34}$$

Using (33) and (34), we compute the three components of $\mathbf{v}^*$ and $\mathbf{v}'^*$. Finally, we compute the new radial and transverse velocities of the first star:

$$v_r^* = v_z^*\,, \quad v_t^* = (v_x^{*2} + v_y^{*2})^{1/2}\,, \tag{35}$$

the new energy and angular momentum of the first star:

$$E^* = U(r) + \tfrac{1}{2}(v_r^{*2} + v_t^{*2}), \quad A^* = rv_t^*\,, \tag{36}$$

and similar quantities for the second star.

6. Escape

We shall consider two cases:

A. ISOLATED SYSTEM

It may happen that the new energy of a star given by (36) is positive. In that case the star will escape from the system in a time of the order of t_c; on the relaxation time scale t_r, it can be considered as immediately lost. Therefore the superstar is removed from the computations. n is decreased by one unit.

B. NON-ISOLATED SYSTEM

If the system is subjected to an external gravitational field, escape is facilitated: in some directions the tidal force is away from the centre and exceeds the attraction of the system past a given distance. The tidal field is not spherically symmetrical in general; therefore it can only be represented in an approximate way in the present scheme which assumes spherical symmetry. We shall assume that all stars which go beyond a certain distance r_e from the centre escape from the system. r_e should be proportional to $M^{1/3}$ (von Hoerner, 1958), and we define it by:

$$r_e = r_{e0}(M/M_0)^{1/3}\,, \tag{37}$$

where M_0 is the initial mass of the system, and r_{e0} is a given constant, depending on the assumed strength of the tidal field. Then a star escapes iff:

$$2E - 2U(r_e) - A^2/r_e^2 > 0 \tag{38}$$

(see next paragraph). Since the whole system is contained inside r_e, this can also be written:

$$2E + 2GM/r_e - A^2/r_e^2 > 0. \tag{39}$$

7. New Positions

We consider now a new time step Δt, and this time we must begin by selecting a new position for each star.

Consider a particular star, which has values of E and A just computed by (36) (we drop the asterisks now). We assume that it moves in the smoothed-out potential $U(r)$ which has been computed above. The star describes a rosette orbit, with r oscillating between two extreme values r_{min} and r_{max}, which are the roots of:

$$Q(r) = 2E - 2U(r) - A^2/r^2 = 0. \tag{40}$$

We begin by computing r_{min} and r_{max}. If we introduce a new variable $z = 1/r$, then it is apparent from (40) and Figure 1 that d^2Q/dz^2 is always negative. On the other hand, Q is positive at the old position according to (36); for $z=0$, $Q=2E<0$; and for $z=\infty$, $Q=-\infty$. It follows that Equation (40) has always exactly two roots (cf. Contopoulos, 1954).

The interval in which r_{min} falls is found by looking up the tables of the ordered radii r_k and of the corresponding potentials U_k; that is, one determines k such that:

$$Q(r_k) < 0 < Q(r_{k+1}). \tag{41}$$

The potential is then given by (7), which can be written:

$$U = az + b, \tag{42}$$

with:

$$a = \frac{U_{k+1} - U_k}{z_{k+1} - z_k}, \quad b = \frac{U_k z_{k+1} - U_{k+1} z_k}{z_{k+1} - z_k}; \tag{43}$$

substituting into (40), we obtain a quadratic equation for z. Moreover, dQ/dr must be positive for $r = r_{min}$, and therefore the largest root for z must be taken, i.e.:

$$\frac{1}{r_{min}} = \frac{-a + [a^2 - 2A^2(b - E)]^{1/2}}{A^2}. \tag{44}$$

A similar computation gives r_{max}; the inequality signs in (41) are then reversed and the plus sign in (44) is replaced by a minus sign. For reasons of computational accuracy it is in fact preferable to write r_{max} in the equivalent form:

$$r_{max} = \frac{-a + [a^2 - 2A^2(b - E)]^{1/2}}{2(b - E)}. \tag{45}$$

Now we must select a position of the star between r_{min} and r_{max}. The probability to take it in an interval dr should be equal to the fraction of time spent by the star in dr, i.e.:

$$\frac{\mathrm{d}t}{T} = \frac{\mathrm{d}r/|v_r|}{\int\limits_{r_{\min}}^{r_{\max}} \mathrm{d}r/|v_r|}, \tag{46}$$

with the radial velocity v_r given by:

$$|v_r| = [2E - 2U(r) - A^2/r^2]^{1/2} = [Q(r)]^{1/2}. \tag{47}$$

The computation of the half-period T would be time-consuming; fortunately it can be avoided by the classical von Neumann rejection technique (Hammersley and Handscomb, 1964). We want a probability distribution proportional to a known function $f(r)$ (in the present case: $f(r)=1/|v_r|$), without knowing the constant of

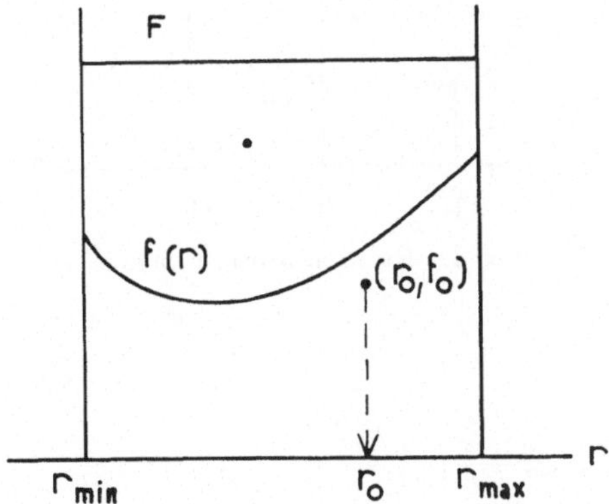

Fig. 2. The rejection technique.

proportionality. We take a number F which is everywhere larger than $f(r)$ (Figure 2). We select a point (r_0, f_0) at random in the rectangle $r_{\min} < r_0 < r_{\max}$, $0 < f_0 < F$, with a uniform distribution, i.e. we compute:

$$r_0 = r_{\min} + (r_{\max} - r_{\min})\, X, \quad f_0 = FX', \tag{48}$$

where X and X' are a pair of normalized random numbers. If the point is below the curve: $f_0 < f(r_0)$, we take $r = r_0$ as the selected value. In the opposite case, this value is rejected, a new point in the rectangle is tried with a fresh pair of random numbers, and so on until a point below the curve is obtained.

In the present case, however, this procedure cannot be applied directly because $f(r)=1/|v_r|$ becomes infinite at both ends of the interval. This difficulty is eliminated by the introduction of a new variable s, defined by an appropriate relation $r = r(s)$. The probability to sample a value in ds should be proportional to:

$$\frac{1}{|v_r|}\frac{\mathrm{d}r}{\mathrm{d}s} = g(s); \tag{49}$$

therefore $r(s)$ should be such that (49) remains finite in the whole interval. Since $\mathrm{d}Q/\mathrm{d}r$ has a finite slope, $|v_r|$ is proportional to $(r-r_{\min})^{1/2}$ for $r\to r_{\min}$; therefore there should be: $r-r_{\min}\propto(s-s_{\min})^2$ for $r\to r_{\min}$, and similarly for $r\to r_{\max}$. Also r should be an easily computable function of s. A function $r(s)$ satisfying these requirements is:

$$r = \alpha_0 + \alpha_1 s + \alpha_2 s^2 + \alpha_3 s^3, \tag{50}$$

with a minimum equal to $r_{\min}$ and a maximum equal to $r_{\max}$ (Figure 3). We normalize

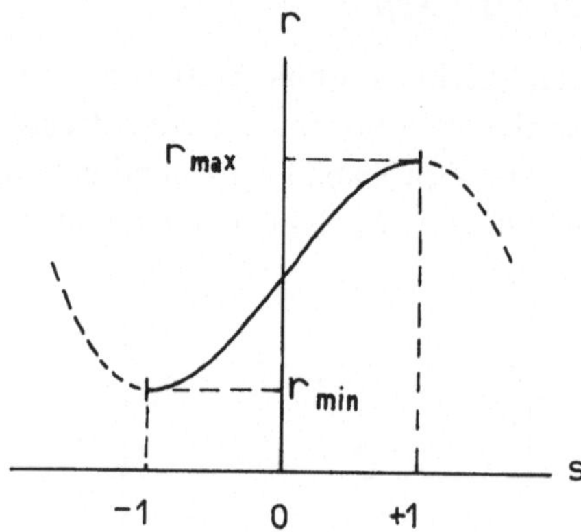

Fig. 3. Relation between r and s.

s by the conditions:

$$s_{\min} = -1, \quad s_{\max} = 1; \tag{51}$$

the coefficients in (50) are then determined and the relation becomes:

$$r = \tfrac{1}{2}(r_{\min} + r_{\max}) + \tfrac{1}{4}(r_{\max} - r_{\min})(3s - s^3). \tag{52}$$

For $s\to -1$ it is easily computed from (47), (49) and (52) that:

$$g(s) \to g(-1) = [3(r_{\max} - r_{\min})/(\mathrm{d}Q/\mathrm{d}r)_{r=r_{\min}}]^{1/2}, \tag{53}$$

and for $s\to +1$:

$$g(s) \to g(+1) = [-3(r_{\max} - r_{\min})/(\mathrm{d}Q/\mathrm{d}r)_{r=r_{\max}}]^{1/2}. \tag{54}$$

$\mathrm{d}Q/\mathrm{d}r$ is computed from (40) and (42):

$$\frac{\mathrm{d}Q}{\mathrm{d}r} = \frac{2A^2}{r^3} + \frac{2a}{r^2}. \tag{55}$$

An upper boundary for $g(s)$ is computed by the empirical formula:

$$F = 1.2 \max[g(-1), \quad g(+1)]. \tag{56}$$

A check is kept on all computed values of $g(s)$ to verify that they do not exceed F. Experience shows that formula (56) is entirely satisfactory in that respect.

In practice, the computation proceeds as follows. First r_{min} and r_{max} are computed. Next, F is computed from (53), (54), (55) and (56). Then the rejection procedure is applied: a pair of random values (s_0, g_0) is computed by:

$$s_0 = -1 + 2X, \quad g_0 = FX'; \tag{57}$$

corresponding values of r, dr/ds, $|v_r|$, and $g(s_0)$ are computed by (52), (47), and (49); and as soon as a pair is found for which there is: $g_0 < g(s_0)$, the corresponding value of r is taken as the new position of the star. The new radial velocity is: $v_r = \pm |v_r|$, since the motion can be outwards or inwards; the sign is chosen at random with a probability $\frac{1}{2}$ for each. Finally, the new transverse velocity is computed by:

$$v_t = A/r. \tag{58}$$

This procedure is repeated for all superstars. The same potential $U(r)$ is used throughout; it is not updated after the displacement of each superstar. The reason is that such an updating would mean changing part of the tables r_k and U_k after each displacement, and thus would involve a total number of operations proportional to n^2; this is to be avoided since in all other parts of the program the number of operations is proportional only to n or to $\ln n$. The effect is that the potential $U(r)$ used lags slightly behind the real potential, by a time equal to one half of the time step Δt in the mean. This effect probably has no serious consequences, since Δt is small compared to the relaxation time t_r, which is the time scale for the changes in the potential.

8. Continuation

Now that we have new positions and velocities for the superstars, the loop is closed by going back to the computation of the potential described above, and a new cycle of operations begins. The cycle is repeated again and again (typically a few hundred times), until the desired evolution of the system has been achieved.

9. Units

All symbols appearing in the equations so far are to be understood, as usual, as physical quantities. They will be related to the numbers appearing in the numerical computations by relations such as:

$$r = U_l [r], \tag{59}$$

where r is the physical quantity, $[r]$ is the corresponding number, and U_l is the adopted unit of length.

The units should be chosen in such a way that the numbers computed are the same whatever the actual mass of the system considered, its dimensions, and the number N of its stars. The first two requirements are easily met by an appropriate choice of the units of mass and length. The third one is a little more tricky, because

N appears explicitly in some equations, namely (6b), (6d) (remembering that $K=N/n$), (10) and (25); it can be met only by using a non-coherent system of units.

First, we define *basic units* of mass, length and time by the condition that the gravitational constant G, the total mass M_0 and the total energy $\mathscr{E}_0$ of the system at time $t=0$, expressed in these units, should be respectively:

$$[G] = 1, \quad [M_0] = 1, \quad [\mathscr{E}_0] = -\tfrac{1}{4}. \tag{60}$$

This gives the following expressions for the basic units:

$$U_m = M_0, \quad U_l = GM_0^2 (-4\mathscr{E}_0)^{-1}, \quad U_t = GM_0^{5/2} (-4\mathscr{E}_0)^{-3/2}. \tag{61}$$

U_l is of the order of the dimensions of the system, and U_t is of the order of the crossing time (see Equation (8)).

We shall use, for the various physical quantities appearing in the equations, the units given in Table I; N_0 is the number of stars at time $t=0$. The choice has been

TABLE I

Quantity	Unit
mass of the system, M; mass of a superstar; partial mass, M_k (Equation (6))	U_m
mass of a star, m	$U_m N_0^{-1}$
distance to centre, r; distance between shells, Δr	U_l
impact parameter, l	$U_l N_0^{-1}$
time step, Δt; evolution time, t	$U_t N_0/\ln N_0$
velocities v_r, v_t, w	$U_l U_t^{-1}$
energy per unit mass, E; potential, U	$U_l^2 U_t^{-2}$
angular momentum per unit mass, A	$U_l^2 U_t^{-1}$
gravitational constant, G	$U_m^{-1} U_l^3 U_t^{-2}$

guided by physical considerations, and it can be readily verified that it eliminates N (and also G) from all equations.

In fact, Equations (6), (10), (12), (25) become:

$$[U_{n+1}] = 0,$$

$$[M_n] = [M] = \frac{1}{n_0} \sum_{i=1}^{n} [m_i],$$

$$\left.\begin{aligned} [U_k] &= [U_{k+1}] - [M_k]\left(\frac{1}{[r_k]} - \frac{1}{[r_{k+1}]}\right), \\ [M_{k-1}] &= [M_k] - \frac{1}{n_0}[m_k]; \end{aligned}\right\} \quad (k = n, \ldots, 1) \tag{62}$$

$$[\Delta t] = b\frac{n}{n_0}[M]^{5/2}[|\mathscr{E}|]^{-3/2}; \tag{63}$$

$$\tan\frac{\beta}{2} = \frac{[m]+[m']}{[w]^2[l]}; \tag{64}$$

$$[l] = \left\{\frac{(p-1)[w][\Delta t]}{2[r]^2 n_0[\Delta r]}\right\}^{-1/2}. \tag{65}$$

In (63) and (65) we have made the approximation: $\ln N = \ln N_0$, which is acceptable since N is large. The other equations used in the computation are unaffected, i.e. one has simply to put each quantity into brackets.

The mean relaxation time is (Hénon, 1958):

$$t_r = \frac{1}{4\pi}\frac{(\overline{v^2})^{3/2}N}{\bar{\varrho}G^2 M \ln N}, \tag{66}$$

where $\bar{\varrho}$ is the mean density of the system. We estimate it by:

$$\bar{\varrho} = \frac{M/2}{4\pi R^3/3}, \tag{67}$$

where R is the mean radius of the system, and from the virial theorem we have:

$$R \approx \frac{GM^2}{-4\mathscr{E}}, \quad \overline{v^2} \approx \frac{-2\mathscr{E}}{M}. \tag{68}$$

Hence,

$$t_r = \frac{1}{24\sqrt{2}}\frac{N}{\ln N}GM^{5/2}(-\mathscr{E})^{-3/2}. \tag{69}$$

Thus, in our units and at time $t=0$, the mean relaxation time is:

$$[t_r(0)] = \frac{1}{3\sqrt{2}} = 0.236\ldots. \tag{70}$$

From (10) and (69) we derive also:

$$\frac{\Delta t}{t_r} = 24\sqrt{2}b. \tag{71}$$

Thus for $b=0.005$, the time step is equal to about $\frac{1}{6}$ of the mean relaxation time.

With the help of Equations (61) and Table I, the results of a Monte Carlo computation can be compared to any real system, or to any other theoretical model. If one considers for example a typical globular cluster with $N_0=10^5$, $\bar{m}=0.5m_\odot$ and $(\overline{v^2})^{1/2}=5\,\mathrm{km/s}$, there is: $U_m=M_0=5\times10^4\,m_\odot$, $-4\,\mathscr{E}_0=2.5\times10^6\,m_\odot\,\mathrm{km^2\,s^{-2}}$, $U_l=4.5$ pc, $U_t=6.4\times10^5$ yr, $U_tN_0/\ln N_0=5.5\times10^9$ yr.

10. Technical Problems

We describe now a few practical problems which have not been yet entirely solved.

A. DENSE CORE

The relaxation time at a given point varies as the inverse of the density. Thus, as a model evolves and begins to form a very dense core (see I), the relaxation time near the centre decreases sharply and eventually becomes shorter than the time step Δt. The computation then no longer represents the real evolution of the core. The main effect is to slow down artificially the evolution of the core.

The situation is aggravated, in the case of stars of equal masses, by the following subsidiary effect. For $t_r \ll \Delta t$, the computed impact distance l will be in general much smaller than l_0, and the deflection angle β computed from (12) will be close to 180°. Thus the two stars merely exchange their velocities, and from the point of view of the distribution function nothing happens. For stars of unequal masses, the effect is not so radical, but it still tends to slow down the computed relaxation.

This effect can be eliminated by limiting the deflection angle to 90°. Experience shows that this device effectively increases the rate of evolution of the core, and it has been incorporated into our procedure.

Still, the main problem remains. One could, of course, take a time step Δt equal to a fraction of the central relaxation time, rather than a fraction of the mean relaxation time. However, this would practically bring the computation to a standstill as the central relaxation time becomes very small.

A similar difficulty exists in the exact N-body computations, where it has been partially remedied by the introduction of individual time steps: stars are recomputed much more frequently in the core than in the halo. Perhaps a similar device could be introduced in the Monte Carlo scheme.

B. SPURIOUS RELAXATION

As already mentioned, the mean potential used to compute the orbital motion of the stars is not completely smooth: it contains random fluctuations in the radial direction. These fluctuations are time-dependent since at every new time step the potential is recomputed from the new positions of the stars. They produce a spurious relaxation, in addition to the normal relaxation which is already incorporated into the program. In order to estimate the importance of this effect, trial computations have been run with the normal relaxation suppressed: the computation of the encounters is simply by-passed. The initial steady state should then in principle maintain itself indefinitely. Figure 4 shows the results for two cases with 100 and 1000 superstars respectively. The initial state is a polytrope of index 1, with an isotropic distribution of velocities and with equal masses. Each curve represents the radius of the sphere containing a given fraction of the total mass. For $n=1000$, the steady state is very well conserved; comparing with I, Figure 2, which represents the same case with encounters, one can see that the spurious relaxation is entirely

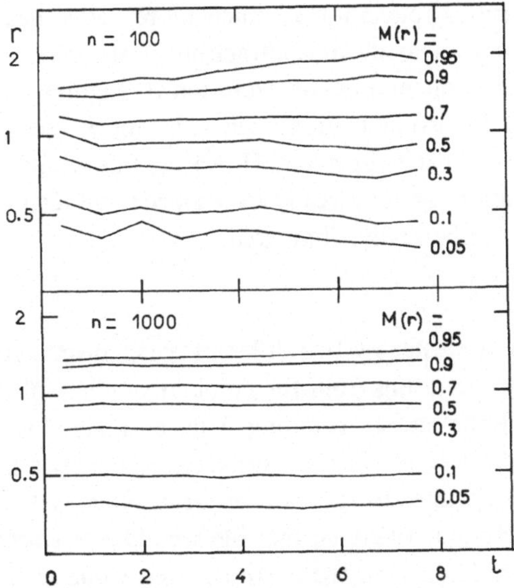

Fig. 4. Evolution of systems of 100 and 1000 superstars when the relaxation is suppressed.

negligible, at least during the first stages of the computation. For $n=100$, the spurious relaxation becomes visible (Figure 4). One can define a 'spurious relaxation time', which, from experimental results as well as from theoretical arguments, appears to be of the order of $n\Delta t$. Therefore the effect will be small compared to real relaxation if:

$$n\Delta t \gg t_r . \tag{72}$$

Since Δt is of the order of $t_r/6$, this criterion is easily satisfied. One can also require that the cumulative effect remains small to the end of the computation; this leads to the more stringent criterion:

$$n\Delta t \gg t_{\max} , \tag{73}$$

where $t_{\max}$ is the length of the computed evolution. Equation (73) means: the number of superstars should be large compared to the total number of time steps. A typical computation corresponds to 200 to 300 time steps; therefore the number of superstars should be at least 1000, and preferably more.

A particular consequence of the spurious relaxation is a slow but systematic drift of the total energy upwards, superimposed on random fluctuations. The explanation is simple: each star interacts with a fluctuating field and tends in the mean to be accelerated, just as with Fermi's mechanism for the acceleration of the cosmic rays. This effect becomes more serious as a dense core develops, because the fluctuations of the potential near the centre become quite wild. In order to neutralize it, we have sometimes adopted the procedure of renormalizing the system after each time step, so as to keep the total energy constant. This procedure, however, does not really eliminate the spurious relaxation, which is not uniform throughout the system; in fact, because the spurious relaxation is much more active near the centre, the renormalization produces an artificial contraction of the outer parts of the system, which becomes visible at the end of the evolution (I, Figures 2 and 6).

The total energy could also be made strictly constant by recomputing the potential after the displacement of each superstar. However, this would require a number of operations proportional to n^2, as already mentioned; and again the spurious relaxation would probably not be really eliminated.

C. SPURIOUS ESCAPE

Stars can escape from a cluster by two different mechanisms: either a slow, gradual variation of the energy resulting from the cumulative effect of many encounters, or a sudden variation of the energy resulting from a single close encounter (Hénon, 1960; Hayli, 1970). The second effect is normally smaller by a factor of the order of $\ln N$, and therefore negligible. In the case of an isolated system, however, the first effect is inoperative (Hénon, 1960), so that the second effect becomes important.

The first effect depends only on the mean squared value of the deflections, and it is correctly reproduced in our procedure where we compute the impact distance l so as to obtain the same mean squared deflection as in reality. The second effect, however, depends on the shape of the tail of the distribution of deflections, or roughly speaking, on the number of large deflections. The Monte Carlo procedure as described above does not reproduce exactly the tail of the distribution of deflections; therefore it does not represent correctly the mechanism of escape by a single close encounter.

Perhaps it would be possible to modify the procedure so as to reproduce the tail exactly. But, since the effect is proportional to $1/\ln N$, it would then be necessary to specify a value of N, thus restricting the generality of the results.

Note added in proof. A similar scheme has been recently developed by Spitzer and his collaborators (Spitzer and Hart, 1971a, b; Spitzer and Shapiro, 1971; Spitzer and Thuan, 1971).

References

Chandrasekhar, S.: 1942, *Principles of Stellar Dynamics*, Dover Publications.
Contopoulos, G.: 1954, *Z. Astrophys.* **35**, 67.
Coveyou, R. R. and Macpherson, R. D.: 1967, *J. Ass. Comput. Mach.* **14**, 100.
Hammersley, J. M. and Handscomb, D. C.: 1964, *Monte Carlo Methods*, Methuen, London.
Hayli, A.: 1970, *Astron. Astrophys.* **7**, 17.
Hénon, M.: 1958, *Ann. Astrophys.* **21**, 186.
Hénon, M.: 1960, *Ann. Astrophys.* **23**, 668.
Hénon, M.: 1966, *Compt. Rend. Acad. Sci. Paris* **262**, 666.
Hénon, M.: 1972, this volume, p. 44.
Spitzer, L., Jr. and Hart, M. H.: 1971a, *Astrophys. J.* **164**, 399.
Spitzer, L., Jr. and Hart, M. H.: 1971b, *Astrophys. J.* **166**, 483.
Spitzer, L., Jr. and Shapiro, S. L.: 1971, to be published.
Spitzer, L., Jr. and Thuan, T. X.: 1971, to be published.
van Albada, T. S.: 1968, *Bull. Astron. Inst. Neth.* **19**, 479.
von Hoerner, S.: 1958, *Z. Astrophys.* **44**, 221.

THE FLUID-DYNAMICAL METHOD

RICHARD B. LARSON
Yale University Observatory, New Haven, Conn., U.S.A.

The use of a fluid-dynamical technique for computing the evolution of stars clusters has been described briefly earlier in this volume, and in more detail by Larson (1970a, b). For convenience we repeat here the fundamental definitions and equations used in this method. We consider a spherical stellar system described by a distribution function of the form $f(r, u, v, w, t)$, where u is the velocity component in the r-direction and v and w are the two transverse velocity components. As the basic fluid-dynamical variables to be solved for, we define the following six moments of the velocity distribution at each point in space and time: the density of stars ϱ, the mean outward velocity $\langle u \rangle$, and the higher-order moments

$$\left.\begin{aligned} \alpha &\equiv \langle (u - \langle u \rangle)^2 \rangle \\ \beta &\equiv \langle v^2 \rangle = \langle w^2 \rangle \\ \varepsilon &\equiv \langle (u - \langle u \rangle)^3 \rangle \\ \xi &\equiv \langle (u - \langle u \rangle)^4 \rangle - 3\alpha^2 . \end{aligned}\right\} \tag{1}$$

Here α and β are the squares of the radial and transverse velocity dispersions, ε represents an outward energy flux or 'heat flow', and ξ represents an excess or deficiency of high velocity stars relative to a Maxwellian distribution.

For the variables defined above we obtain six fluid-dynamical equations by taking the corresponding moments of the Boltzmann equation. The system of moment equations has been closed by approximating the velocity distribution by a low-order expansion in Legendre polynomials; this allows the various unknown moments to be related to quantities already defined, as described by Larson (1970a). The resulting set of moment equations is as follows:

$$\frac{\partial \varrho}{\partial t} + \frac{1}{r^2} \frac{\partial}{\partial r} r^2 \varrho \langle u \rangle = 0 \tag{2}$$

$$\frac{\partial \langle u \rangle}{\partial t} + \langle u \rangle \frac{\partial \langle u \rangle}{\partial r} + \frac{1}{\varrho} \frac{\partial}{\partial r} \varrho \alpha + \frac{2}{r} (\alpha - \beta) + \frac{\partial \Phi}{\partial r} = 0 \tag{3}$$

$$\frac{\partial \alpha}{\partial t} + \langle u \rangle \frac{\partial \alpha}{\partial r} + 2\alpha \frac{\partial \langle u \rangle}{\partial r} + \frac{1}{\varrho} \frac{\partial}{\partial r} \varrho \varepsilon + \frac{2\varepsilon}{r} \left(1 - \frac{2}{3} \frac{\beta}{\alpha} \right) = - \frac{4}{5} \frac{(\alpha - \beta)}{T} \tag{4}$$

$$\frac{\partial \beta}{\partial t} + \langle u \rangle \frac{\partial \beta}{\partial r} + 2\beta \frac{\langle u \rangle}{r} + \frac{1}{3\varrho} \frac{\partial}{\partial r} \frac{\beta}{\alpha} \varrho \varepsilon + \frac{4}{3} \frac{\beta}{\alpha} \frac{\varepsilon}{r} = + \frac{2}{5} \frac{(\alpha - \beta)}{T} \tag{5}$$

$$\frac{\partial \varepsilon}{\partial t} + \langle u \rangle \frac{\partial \varepsilon}{\partial r} + 3\varepsilon \frac{\partial \langle u \rangle}{\partial r} + 3\alpha \frac{\partial \alpha}{\partial r} + \frac{1}{\varrho} \frac{\partial}{\partial r} \varrho \xi + \frac{2\xi}{r} \left(1 - \frac{\beta}{\alpha} \right) = - \frac{87}{160} \frac{\varepsilon}{T} \tag{6}$$

M. Lecar (ed.), Gravitational N-Body Problem, 423–427. All Rights Reserved

$$\frac{\partial \xi}{\partial t} + \langle u \rangle \frac{\partial \xi}{\partial r} + 4\xi \frac{\partial \langle u \rangle}{\partial r} + 6\varepsilon \frac{\partial \alpha}{\partial r} + 4\alpha \frac{\partial \varepsilon}{\partial r} = -\frac{3}{35} \frac{[7\xi - 15\alpha(\alpha - \beta)]}{T}. \tag{7}$$

The terms on the right-hand sides of the above equations represent the relaxational effects of encounters between the stars, and they have been evaluated from the Fokker-Planck equation using the assumption that deviations from a Maxwellian velocity distribution are small. The quantity

$$T = \frac{1}{16} \left(\frac{3}{\pi}\right)^{1/2} \frac{\langle V^2 \rangle^{3/2}}{G^2 m \varrho \ln (D_{\max} \langle V^2 \rangle / 2Gm)} \tag{8}$$

is the classical relation time defined by Chandrasekhar (1942). Here $\langle V^2 \rangle = \alpha + 2\beta$ is the mean squared random velocity, m is the stellar mass (all stars being assumed to have the same mass), and $D_{\max}$ is the dimension of the region over which relaxation effects are important.

The fluid-dynamical Equations (2)–(7) may be solved numerically by techniques similar to ones which have previously been used for gas-dynamical problems with spherical symmetry. For the present problem an Eulerian method is most suitable, since the boundary condition is specified in Eulerian form and since the use of an Eulerian grid ensures that we always have reasonable spatial resolution. The grid points r_i are most conveniently spaced at equal intervals in $\log r$; we have generally used a spacing of 0.1 in $\log r$, which is adequate to provide numerical accuracies of the order of 10%. Following a common procedure in numerical hydrodynamics, we suppose that the odd-order moments $\langle u \rangle$ and ε as well as the mass m inside radius r are assigned values at the grid points r_i, whereas the even-order moments ϱ, α, β, and ξ are assigned values at a second set of points $r_{i-1/2}$ half way between the regular grid points. To ensure stability for time steps which may be much greater than the dynamical time, the difference equations have been written in implicit form with backward time differences. Considerable arbitrariness is possible in the way the difference expressions are constructed, particularly in the method of averaging quantities in adjacent zones. After some experimentation, the equations given below appeared to represent a reasonable compromise between accuracy and stability requirements, although it is not claimed that this is the best possible set of difference equations.

To allow easy calculation of the mass variable m appearing in the gravitational acceleration term $\partial \Phi / \partial r = Gm/r^2$, the continuity Equation (2) has been replaced by the 2 equations

$$\frac{\partial m}{\partial t} = -4\pi r^2 \varrho u \tag{9}$$

$$\frac{\partial m}{\partial r} = 4\pi r^2 \varrho \tag{10}$$

(for convenience we henceforth write u in place of $\langle u \rangle$.) The differential Equations

(9), (10), and (3)–(7) are then approximated by the difference equations given below; here quantities with a superscript n refer to a time t^n, whereas quantities with no superscript all refer to the advanced time $t^{n+1}=t^n+\Delta t$.

$$m_i = m_i^n - 4\pi r_i^2 (\varrho_{i-1/2}\varrho_{i+1/2})^{1/2} u_i \Delta t \tag{11}$$

$$\frac{m_i - m_{i-1}}{r_i^3 - r_{i-1}^3} = \frac{4\pi}{3} \varrho_{i-1/2} \tag{12}$$

$$\frac{u_i - u_i^n}{\Delta t} + u_i \frac{u_{i+1} - u_i}{r_{i+1} - r_i} + \left(\frac{\alpha_{i-1/2} + \alpha_{i+1/2}}{2}\right) \frac{\ln(\varrho\alpha)_{i+1/2} - \ln(\varrho\alpha)_{i-1/2}}{r_{i+1/2} - r_{i-1/2}} + \frac{(\alpha-\beta)_{i-1/2} + (\alpha-\beta)_{i+1/2}}{r_i} + \frac{Gm_i}{r_i^2} = 0 \tag{13}$$

$$\frac{\alpha_{i-1/2} - \alpha_{i-1/2}^n}{\Delta t} + u_i \frac{\alpha_{i+1/2} - \alpha_{i-1/2}}{r_{i+1/2} - r_{i-1/2}} + 2\alpha_{i-1/2} \frac{u_i - u_{i-1}}{r_i - r_{i-1}} + \frac{1}{\varrho_{i-1/2}} \frac{(\varrho_{i-1/2}\varrho_{i+1/2})^{1/2} \varepsilon_i - (\varrho_{i-3/2}\varrho_{i-1/2})^{1/2} \varepsilon_{i-1}}{r_i - r_{i-1}} + 2\frac{\varepsilon_{i-1} + \varepsilon_i}{r_{i-1} + r_i}\left(1 - \frac{2}{3}\frac{\beta_{i-1/2}}{\alpha_{i-1/2}}\right) = -\frac{4}{5}\frac{(\alpha-\beta)_{i-1/2}}{T_{i-1/2}} \tag{14}$$

$$\frac{\beta_{i-1/2} - \beta_{i-1/2}^n}{\Delta t} + u_i \frac{\beta_{i+1/2} - \beta_{i-1/2}}{r_{i+1/2} - r_{i-1/2}} + 2\beta_{i-1/2} \frac{u_{i-1} + u_i}{r_{i-1} + r_i} + \frac{1}{3\varrho_{i-1/2}} \frac{[(\beta\varrho/\alpha)_{i-1/2}(\beta\varrho/\alpha)_{i+1/2}]^{1/2} \varepsilon_i - [(\beta\varrho/\alpha)_{i-3/2}(\beta\varrho/\alpha)_{i-1/2}]^{1/2} \varepsilon_{i-1}}{r_i - r_{i-1}} + \frac{4}{3}\frac{\beta_{i-1/2}}{\alpha_{i-1/2}}\frac{\varepsilon_{i-1} + \varepsilon_i}{r_{i-1} + r_i} = +\frac{2}{5}\frac{(\alpha-\beta)_{i-1/2}}{T_{i-1/2}} \tag{15}$$

$$\frac{\varepsilon_i - \varepsilon_i^n}{\Delta t} + u_i \frac{\varepsilon_{i+1} - \varepsilon_i}{r_{i+1} - r_i} + 3\varepsilon_i \frac{u_{i+1} - u_i}{r_{i+1} - r_i} + \frac{3}{2}\frac{\alpha_{i+1/2}^2 - \alpha_{i-1/2}^2}{r_{i+1/2} - r_{i-1/2}} + \frac{2}{\varrho_{i-1/2} + \varrho_{i+1/2}} \frac{\varrho_{i+1/2}\xi_{i+1/2} - \varrho_{i-1/2}\xi_{i-1/2}}{r_{i+1/2} - r_{i-1/2}} + \frac{\xi_{i-1/2} + \xi_{i+1/2}}{r_i}\left(1 - \frac{\beta_{i-1/2} + \beta_{i+1/2}}{\alpha_{i-1/2} + \alpha_{i+1/2}}\right) = -\frac{87}{160}\frac{\varepsilon_i}{(T_{i-1/2}T_{i+1/2})^{1/2}} \tag{16}$$

$$\frac{\xi_{i-1/2} - \xi_{i-1/2}^n}{\Delta t} + u_i \frac{\xi_{i+1/2} - \xi_{i-1/2}}{r_{i+1/2} - r_{i-1/2}} + 4\xi_{i-1/2} \frac{u_i - u_{i-1}}{r_i - r_{i-1}} + 3(\varepsilon_{i-1} + \varepsilon_i) \frac{(\alpha_{i-1/2}\alpha_{i+1/2})^{1/2} - (\alpha_{i-3/2}\alpha_{i-1/2})^{1/2}}{r_i - r_{i-1}} + 4\alpha_{i-1/2}\frac{\varepsilon_i - \varepsilon_{i-1}}{r_i - r_{i-1}} = -\frac{3}{35}\frac{[7\xi - 15\alpha(\alpha-\beta)]_{i-1/2}}{T_{i-1/2}}. \tag{17}$$

The difference approximations for terms of the form $u\partial u/\partial r$, $u\partial\alpha/\partial r$, etc. have been written in a form which is appropriate if $u<0$, i.e. if the flow is inward; for $u>0$, these expressions should be replaced by

$$u_i\frac{u_i - u_{i-1}}{r_i - r_{i-1}}, \quad u_{i-1}\frac{\alpha_{i-1/2} - \alpha_{i-3/2}}{r_{i-1/2} - r_{i-3/2}}, \quad \text{etc.}$$

to ensure stability. Unfortunately, the difference equations still exhibit unstable behavior in some circumstances, particularly in computing a solution which has large 'starting transients'. No cure was found for the instability; however, it was always possible to avoid it by choosing more realistic initial conditions which did not produce large transient effects.

In solving the difference equations, m_i may be treated as an auxiliary variable, leaving the six quantities $\varrho_{i-1/2}$, u_i, $\alpha_{i-1/2}$, $\beta_{i-1/2}$, ε_i, and $\xi_{i-1/2}$ to be solved for in each zone; thus if the numerical grid contains N zones we have a total of $6N$ unknowns to be solved for. Correspondingly we have the six difference Equations (12)–(17) for each of the N zones except the outermost one, where Equations (13) and (16) cannot be applied but must be replaced by boundary conditions for u_N and ε_N. The assumption of a perfectly absorbing boundary together with an assumed form of the velocity distribution leads to boundary conditions of the form

$$u_N \propto \alpha_N^{1/2}, \quad \varepsilon_N \propto \alpha_N^{3/2}$$

(Larson, 1970b). Incorporating these boundary conditions into the difference equations, we then have enough equations to allow a solution for all of the unknowns.

The standard method for solving a large number of simultaneous non-linear equations is the Newton-Raphson iterative technique, described for example by Larson and Demarque (1964). This method involves starting with an approximate solution for each of the unknowns; the exact solution is then expressed as the approximate value plus a correction, and the equations are linearized by expanding to first order in the correction terms, yielding a set of linear equations for the corrections. These linear equations are then solved by standard techniques of linear algebra, and the resulting corrections are used to obtain an improved solution to the difference equations. The correction procedure is repeated several times until adequate convergence is achieved. The details of application of the method are straightforward but lengthy, and will not be reproduced here. For the present problem, the first approximation to the value of any quantity at time t^{n+1} is conveniently taken as its value at time t^n. It was found that 3 iterations were generally sufficient to solve the difference equations with a fractional error less than 10^{-3}, which is considerably smaller than the truncation error of the difference equations for the fairly coarse grid used. With $N=65$, as used in calculating the evolution of globular clusters, the computing time on an IBM 7094 amounts to about 5 s per time step or about 5 min for a complete evolutionary calculation.

References

Chandrasekhar, S.: 1942, *Principles of Stellar Dynamics*, University of Chicago Press, Chicago.
Larson, R. B.: 1970a, *Monthly Notices Roy. Astron. Soc.* **147**, 323.
Larson, R. B.: 1970b, *Monthly Notices Roy. Astron. Soc.* **150**, 93.
Larson, R. B. and Demarque, P. R.: 1964, *Astrophys. J.* **140**, 524.

THE MODEL OF SPHERICAL CONCENTRIC SHELLS

G. JANIN

1. The Model of Spherical Concentric Shells

This model is applied to the study of the collisionless phase of a spherical star cluster evolution. The spherical symmetry of the system allows us to integrate the motion of spherical concentric shells instead of the motion of individual stars. Each shell of mass m, radius r, radial velocity u and tangential velocity v represents the set of stars having their coordinates between $(m, m+\mathrm{d}m)$, $(r, r+\mathrm{d}r)$, $(u, u+\mathrm{d}u)$ and $(v, v+\mathrm{d}v)$.

For a collisionless system, it is not restrictive to consider equal masses, this we shall do in the following.

The model, due to Campbell (1962), was numerically experimented by Hénon (1964, 1967a, 1967b) and Bouvier and Janin (1969) to study the initial evolution of spherical systems.

2. The Equations of Motion

Let us consider a system of N shells of equal mass m. A shell number i is characterised by three parameters r_i, u_i, v_i. The conservation of the angular momentum

$$J_i = m r_i v_i$$

reduces the number of equations of motion to two:

$$(\mathrm{d}r_i/\mathrm{d}t) = u_i \tag{1}$$

$$(\mathrm{d}u_i/\mathrm{d}t) = a_i \tag{2}$$

where a_i, the acceleration of the ith shell, is composed of a centrifugal component and a potential one:

$$a_i = (J_i^2/m^2 r_i^3) - G(M(r_i)/r_i^2), \tag{3}$$

G is the gravitational constant and $M(r_i)$ is the mass of the system inside the sphere of radius r_i (Hénon, 1964):

$$M(r_i) = m \sum_{j=1}^{N} H(r_i - r_j) \tag{4}$$

where

$$H(r_i - r_j) = \begin{cases} 0 & \text{if} \quad r_i < r_j \\ \frac{1}{2} & \text{if} \quad r_i = r_j \\ 1 & \text{if} \quad r_i > r_j . \end{cases}$$

M. Lecar (ed.), Gravitational N-Body Problem, 428–430.

3. Algorithm to Integrate the Equations of Motion

The dependence (3) between a_i and r_i is discontinuous, thus it would be unhandy to choose a high order numerical method. To take into account the fact that the unknown function u_i appears alone at the right side in the Equations (1), we choose a middle point method:

$$r_i(t + \mathrm{d}t) = r_i(t) + u_i(t + \mathrm{d}t/2)\,\mathrm{d}t$$
$$u_i(t + \mathrm{d}t/2) = u_i(t - \mathrm{d}t/2) + a_i(t)\,\mathrm{d}t$$

where the $a_i(t)$ are only functions of the $r_i(t)$. $\mathrm{d}t$ is the time step. The positions and corresponding velocities are known at times shifted by half a time step interval.

At the initial time t_0, when the initial coordinates are determined (by random numbers for instance, see Janin, 1970), half a time step is accomplished to define the initial velocities:

$$u_i(t_0 + \mathrm{d}t/2) = u_i(t_0) + a_i(t_0)\,\mathrm{d}t/2\,.$$

4. Calculation of the Accelerations

To calculate the acceleration through the expression (3), one must know $M(r_i)$. The computation of the expression (4) which defines $M(r_i)$ requires N operations, i.e. N^2 operations for the N shells per time step. This number may be reduced by sorting the shells at each time step with a high speed sorting algorithm (for instance the routine SORT, algorithm 347 of the ACM, Singleton, 1969). If N_i is the order of the ith shell,

$$M(r_i) = m(N_{i-1} + \tfrac{1}{2})$$

in the case of shells sorted with increasing order of their radius.

These methods are a little rough, because they suppose the order of the shells completely disturbed at each time step. We saw that the integration algorithm is a low-order one. Thus the time step must be small in order to preserve the computational precision, and at each step the number of shell crossings is small. An economic method must use this fact.

The simplest way to do this consists in having a subroutine which, at each time step, takes each shell and tests the positions of the neighbouring shells to calculate the crossings. The number of operations is thus proportional to N^2 again, but with a very low coefficient, so that for a thousand shells for instance, it is much faster than the use of a sorting routine.

5. Calculation of the Energy

The energy conservation is the usual test to check the integration of the motion of an isolated system. The kinetic energy of the system is

$$\tfrac{1}{2}\sum_{i=1}^{N}\left(mu_i^2+\frac{J_i^2}{mr_i^2}\right).$$

The potential energy of the system is the energy restored by the system by building it with elements coming from infinity. Its expression is

$$\frac{Gm^2}{2}\sum_{i=1}^{N}\frac{2i-1}{r_i}.$$

The total energy of the system is the sum of the kinetic and the potential energy.

6. Typical Run

A standard time step is 1/4000 of the crossing time τ_c, which is defined by

$$\tau_c=\sqrt{\frac{R^3}{GM}},$$

R is a typical radius of the cluster.

With a thousand shells system, 8 time steps per second can be computed without special programming tricks on a CDC 3800 computer. Thus the computing of ten crossing times of the system require 75 min of central processor time. The energy discrepancy from its initial value is always better then 1%.

References

Bouvier, P. and Janin, G.: 1969, *Astron. Astrophys.* **5**, 127.
Campbell, P. M.: 1962, *Proc. Nat. Acad. Sci. Am.* **48**, 1993.
Hénon, M.: 1964, *Ann. Astrophys.* **27**, 83.
Hénon, M.: 1967a, 14e Colloque de Liège, p. 243.
Hénon, M.: 1967b, NASA SP-153, p. 295.
Janin, G.: 1970, *Publ. Obs. Genève* **76**, s. A.
Singleton, R. C.: 1969, *Commun. ACM* **12**, 185.

INTEGRATION METHODS WHERE FORCE IS OBTAINED FROM THE SMOOTHED GRAVITATIONAL FIELD

FRANK HOHL

NASA, Langley Research Center, Hampton, Va.

1. Introduction

Many problems in stellar dynamics involve phenomena occurring in inhomogeneous systems in which the interaction between the particles is fully described by a self-consistent field operating in phase space. Because the particles interact by means of the long-range Coulomb force, each particle is under the simultaneous influence of a large number of other particles. Therefore, stellar systems will respond to any perturbation in a collective manner, and a study of such systems is concerned essentially with the N-body problem.

The collective phenomena do not depend on two-body collisions such as occur in ordinary gases, and therefore the collective effects will be present in collisionless systems. Since the number of particles in the system is large, a distribution function can be used to describe the density of particles in phase space. The distribution function must then satisfy the Vlasov equation (the self-consistent set of the Maxwell equations plus the collisionless Boltzmann equation). In using the Vlasov equation to describe a stellar system the number of masses which make up the system is assumed to become infinite while the total mass remains constant. Although such an approach allows description of the system by means of a distribution function which must satisfy the Vlasov equation, solutions to the time-dependent nonlinear Vlasov (or collisionless Boltzmann) equation are, in general, very difficult to obtain. An attempt is therefore made to condense the large number of stars which a galaxy or other stellar system may contain into a smaller number of superparticles. Numerical or computer models are then used to perform computer experiments simulating Vlasov phenomena by following the simultaneous motion of a large number of superparticles.

Computer models for collisionless systems have been extensively used to study plasmas and plasma flow problems. Much of this work is referred to in the proceedings of the two recent conferences (*Symposium on Computer Simulation of Plasma and Many-Body Problems*, Williamsburg, Virginia, April 1967, NASA SP-153, and *Proceedings of the APS Topical Conference on Numerical Simulation of Plasmas*, Los Alamos, New Mexico, Sept. 1968, LA-3990). The application of computer models to collisionless stellar systems has been more recent and we describe below the two-dimensional rod model and the model for disks of stars. For these two models the equations of motion used to advance the motion of the stars are the same. The

M. Lecar (ed.), Gravitational N-Body Problem, 431–441.

essential difference is in the determination of the gravitational field or force acting on a star.

At the present time there are no realistic three-dimensional computer simulations for collisionless systems in progress. The reason for this is simply a matter of economics, that is, limited computer storage and computer time. Nevertheless, the methods for performing three-dimensional calculations are available as a simple extension of the methods used in simulating disks of stars.

2. Computer Model for Disks of Stars

One of the more realistic models presently available for simulating stellar systems is the model which simulates the motion of large numbers of stars (point masses) that are confined to move in the plane of a highly flattened stellar system, such as a disk galaxy.

The model effectively simulates the evolution of an isolated disk of stars. Lindblad (1960) pioneered such calculations by following the motion of up to 192 mutually attracting mass points in the given central field of the Galaxy. By placing the mass points initially in a system of concentric rings with circular velocities, Lindblad investigated the mutual disturbances in such a system to simulate the spiral structure of galaxies. Because Lindblad was able to follow the motion of only a rather small number of stars, his model has limited applicability. Miller and Prendergast (1968) developed a model to study the motion of stars in a plane for systems which are doubly periodic and the forces, star positions, and velocities are allowed only discrete (integer) values which are less than some given maximum value.

The two computer models that are now in use for studying the evolution of self-consistent disk galaxies have recently been described by Miller and Prendergast (1968), and by Hohl and Hockney (1969). With the development of such models it becomes possible to simulate the dynamical evolution of galaxies (Miller, Prendergast and Quirk (1970) and Hohl (1970a, 1970b)) and to check some of the theoretical predictions for stellar systems (Hockney and Hohl (1969)). The dynamics of the gravitational two-stream instability and of the Jeans' instability in a plane stellar system have been investigated by Hohl (1970c). Miller (1970) has analyzed in detail some of the discretization properties of plane stellar systems.

A. COMPUTER MODEL

The computer model used for investigating the dynamics of disk galaxies is illustrated in Figure 1. The $N \times N$ array of cells shown in Figure 1 is superposed over the plane of the galactic disk. The array of cells is introduced only for the purpose of calculating the gravitational potential. The cells are identified by n, m with $n = 0, 1, 2, \ldots, N-1$ increasing in the x-direction and with $m = 0, 1, 2, \ldots, N-1$ increasing in the y-direction. The cell in the lower left-hand corner is 0, 0 and that in the upper right-hand corner of the array is $N-1$, $N-1$. The stars move over this imaginary array of cells. At the center of each cell a mass density is defined which is given by the number

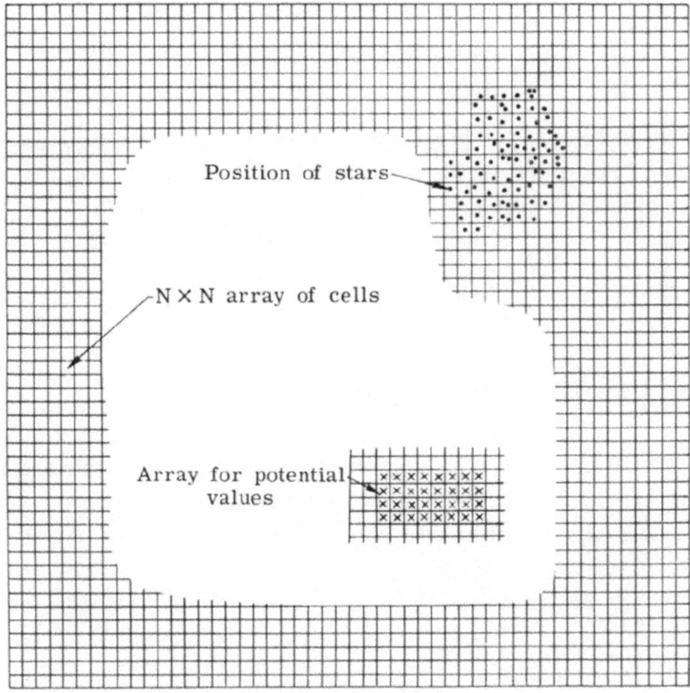

Fig. 1. Computer model illustrating the $N \times N$ array of cells used in calculating the gravitational potential.

of stars in that cell. The number of stars in a cell is usually of the order of 100 and can become much larger near condensations. The density distribution is used to obtain the gravitational potential at the center of each cell. From the gravitational potential, the force acting at the position of a star is computed. Newton's equations of motion are then used to advance the position and velocity of each star by a small time step. For the parameters of a typical galaxy, retardation or relativistic effects need not be considered.

One complete cycle for advancing the motion of the system by a time δt consists of the following procedure. First, the distribution of mass $\sigma_{n,m}$ is used to obtain the gravitational potential $\phi_{n,m}$ by effectively summing over the density. Second, the gravitational field at the position of the stars is computed from the potential $\phi_{n,m}$. Third, by applying Newton's laws of motion, the motion of all the stars is advanced for a small time step δt. This procedure represents one cycle and it is repeated until the desired evolution of the system is achieved.

B. EQUATIONS OF MOTION

The motion of the stars is described by the differential equations

$$\frac{\mathrm{d}V_x}{\mathrm{d}t} = \frac{\partial \phi}{\partial x} \qquad \frac{\mathrm{d}V_y}{\mathrm{d}t} = \frac{\partial \phi}{\partial y} \tag{1}$$

and

$$V_x = \frac{dx}{dt} \qquad V_y = \frac{dy}{dt}. \tag{2}$$

The variable ϕ represents the gravitational potential and the gravitational field is given by $K = \nabla\phi$. For a star in the (n, m)th cell, Equations (1) and (2) in the time-centered finite difference form are

$$\frac{1}{\delta t}\left[V_x\left(t + \frac{\delta t}{2}\right) - V_x\left(t - \frac{\delta t}{2}\right)\right] = (K_x(t))_{n,m} \tag{3}$$

and

$$\frac{1}{\delta t}[x(t + \delta t) - x(t)] = V_x\left(t + \frac{\delta t}{2}\right) \tag{4}$$

with similar equations for the y-components. The numerical calculations can be speeded up greatly by scaling the distance so that the cell dimensions are equal to unity, that is, $\Delta x^* = \Delta y^* = 1$. If, in addition, the velocity and mass of a star are scaled as

$$V_x^* = V_x \frac{\delta t}{\Delta x} \tag{5}$$

and

$$m^* = \frac{G(\delta t)^2}{2(\Delta x)^2} m \tag{6}$$

then the potential is scaled as

$$\phi^* = \frac{(\delta t)^2}{2(\Delta x)^2} \phi .$$

The equations of motion take on the simplified form

$$V_x^*\left(t + \frac{\delta t}{2}\right) = V_x^*\left(t - \frac{\delta t}{2}\right) + (K_x^*(t))_{n,m} \tag{7}$$

and

$$x^*(t + \delta t) = x^*(t) + V^*\left(t + \frac{\delta t}{2}\right). \tag{8}$$

Two methods can be used to obtain the gravitational field. The simple method is to let each star in a particular cell experience the same field components, namely,

$$(K_x^*)_{n,m} = \phi_{n+1,m}^* - \phi_{n-1,m}^* \tag{9}$$

for the x-component of the gravitational field in the n, m cell and

$$(K_y^*)_{n,m} = \phi_{n,m+1}^* - \phi_{n,m-1}^* \tag{10}$$

for the y-component of the field. Equations (9) and (10) show that all stars in the cell (n, m) experience the same gravitational field and the value of the field will jump in crossing the cell boundaries. A smoother variation of the field acting on a star is obtained by means of a bilinear interpolation of the fields (as given by Equations (9) and (10) at the four cell centers surrounding the position of a particular star. The two components of the field are then given by

$$K_x^* = (1 - \delta y)(1 - \delta x)(\phi_{n+1,m}^* - \phi_{n-1,m}^*) + \delta y(1 - \delta x) \times \times (\phi_{n+1,m+1}^* - \phi_{n-1,m+1}^*) + \delta x(1 - \delta y)(\phi_{n+2,m}^* - \phi_{n,m}^*) + + \delta x\, \delta y(\phi_{n+2,m+1}^* - \phi_{n,m+1}^*) \tag{11}$$

and

$$K_y^* = (1 - \delta y)(1 - \delta x)(\phi_{n,m+1}^* - \phi_{n,m-1}^*) + \delta x(1 - \delta y) \times \times (\phi_{n+1,m+1}^* - \phi_{n+1,m-1}^*) + \delta y(1 - \delta x)(\phi_{n,m+2}^* - \phi_{n,m}^*) + + \delta x\, \delta y(\phi_{n+1,m+2}^* - \phi_{n+1,m}^*) \tag{12}$$

where the pertinent parameters are defined in Figure 2. It is found that the bilinear interpolation gives a slightly more definite structure for the condensations which occurred during the initial evolution of a system. After about one rotation, the results obtained by the two methods display essentially the same structure.

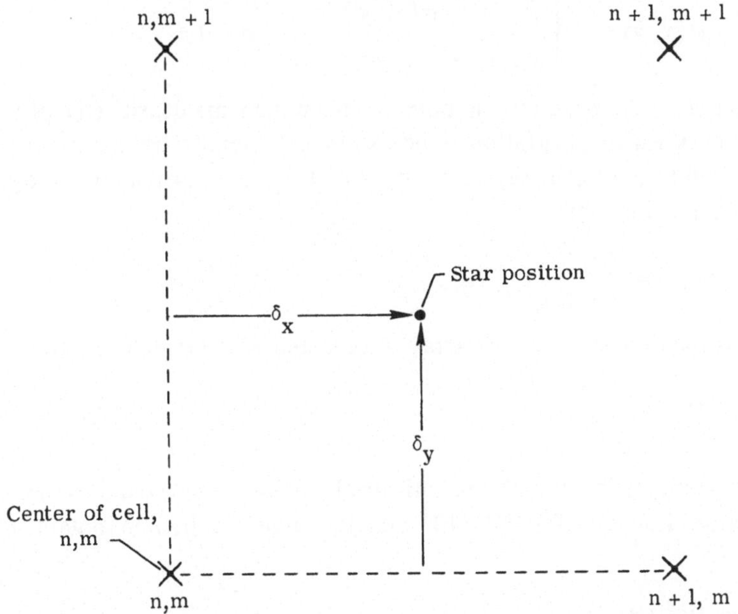

Fig. 2. Parameters used for the bilinear interpolation of the gravitational field.

If a star should leave the $N \times N$ array of cells, the field acting on it is calculated by placing all the mass remaining in the system at the center of the array. The stars outside the array will not interact among themselves, but they will be attracted by the central force due to the mass placed at the center of the array. Whenever an appreciable number of stars leave the array, the calculations are no longer accurate and the computer run should be repeated by either increasing the array or by changing the initial conditions.

C. THE POTENTIAL CALCULATION

In calculations with the two-dimensional rod model (Hohl, 1968: Hockney, 1967) the gravitational potential is easily obtained by solving the two-dimensional Poisson equation

$$\frac{\partial^2 \phi}{\partial x^2} + \frac{\partial^2 \phi}{\partial y^2} = 4\pi G \varrho (x, y). \tag{13}$$

An attempt therefore might be made to obtain the gravitational potential for the disk model by the same method. The Poisson equation then becomes

$$\frac{\partial^2 \phi}{\partial x^2} + \frac{\partial^2 \phi}{\partial y^2} + \frac{\partial^2 \phi}{\delta z^2} = 4\pi G \sigma (x, y)\, \delta (z) \tag{14}$$

where $\delta(z)$ is the Dirac delta function and $\sigma(x, y)$ is the surface density of stars in the plane of the disk. The difficulty is that no means are available to evaluate $\partial^2\phi/\partial z^2$. Therefore,

$$\phi(x, y) = G \iint \frac{\sigma(x', y')}{\sqrt{(x - x')^2 + (y - y')^2}}\, \mathrm{d}x'\, \mathrm{d}y' \tag{15}$$

is used to obtain the gravitational potential from the mass density (the primes denote variables over which integration is performed). Presently the density is given only at a finite number of cells separated by a unit distance so that the integral can be changed to a summation

$$\phi_{n,m} = \sum_{i=0}^{N-1} \sum_{j=0}^{N-1} \sigma_{i,j} H_{i-n,j-m} \tag{16}$$

where N is the dimension of the array of cells and H is Green's function defined by

$$H_{i,j} = \frac{1}{\sqrt{i^2 + j^2}}. \tag{17}$$

To perform directly the summation indicated by Equation (16) requires the summation of N^4 terms. For $N=100$, $N^4=10^8$ and the time required to obtain ϕ becomes excessive.

D. FOURIER METHOD

The method to obtain the gravitational potential makes use of the fast Fourier transform methods now available (Cooley and Tukey, 1965). The Fourier transform of the density is defined as

$$\tilde{\sigma}_{k,l} = \left(\frac{1}{N}\right)^2 \sum_{n=0}^{N-1} \sum_{m=0}^{N-1} \sigma_{n,m} \exp\left[i \frac{2\pi}{N}(kn + lm)\right]. \tag{18}$$

Similarly, the Fourier transform of Green's function is given by

$$\tilde{H}_{k,l} = \left(\frac{1}{N}\right)^2 \sum_{n=0}^{N-1} \sum_{m=0}^{N-1} H_{n,m} \exp\left[i \frac{2\pi}{N} (kn + lm)\right]. \tag{19}$$

Applying the finite convolution theorem to Equation (16) gives the result (Hohl and Hockney, 1969)

$$\begin{aligned} \phi_{n,m} &= \sum_{i=0}^{N-1} \sum_{j=0}^{N-1} \sigma_{i,j} H_{i-n,j-m} \\ &= \left(\frac{1}{N}\right)^2 \sum_{k=0}^{N-1} \sum_{l=0}^{N-1} \tilde{\sigma}_{k,l} \tilde{H}_{k,l} \exp\left[-i \frac{2}{N} (kn + lm)\right]. \end{aligned} \tag{20}$$

From Equation (22) and the definition of the Fourier transform, it is clear that

$$\tilde{\phi}_{k,l} = \tilde{\sigma}_{k,l} \tilde{H}_{k,l}.$$

Therefore, the potential $\phi_{n,m}$ is obtained directly from the inverse Fourier transform of $\tilde{\sigma} \cdot \tilde{H}$. Such a method gives a doubly periodic system. A similar method is used by Miller and Prendergast (1968) in their investigation of doubly periodic stellar systems.

The Fourier transform method just described can be modified to obtain the potential distribution for an isolated system. This modification is achieved by increasing the number of cells by a factor of 4 and by confining the system to one-quarter of the array of cells. The mass density in the remaining three-quarters of the array will then always be identically zero.

Consider now that in addition to the array under investigation, the summation in Equation (16) is extended over all the doubly infinite array of images. However, Green's function $H_{n,m}$ is now modified so that it corresponds to the correct single particle potential for particle separation r less than $N/2$ (one-half the dimension of the array) and to zero interaction for r greater than $N/2$. Even though the system is still doubly periodic, there is no longer any interaction between adjacent image systems because their masses are separated by at least $N/2$.

Thus, to get the correct potential for an isolated system at the expense of a fourfold increase in storage, Green's function to be used in Equation (20) is

$$H_{n,m} = \frac{1}{\sqrt{n^2 + m^2}} \qquad \begin{array}{c} \left(0 \leqq n, m \leqq \dfrac{N}{2}\right) \\ (n^2 + m^2 \neq 0) \end{array} \tag{21a}$$

$$H_{N-n,m} = H_{n,N-m} = H_{N-n,N-m} = H_{n,m} \tag{21b}$$

and

$$H_{0,0} = 1. \tag{21c}$$

As before, setting $H_{0,0}=1$ is equivalent to setting the self-potential of a star equal to unity. The Fourier transform of $H_{n,m}$ need be done only once. Also, because of the symmetry of $H_{n,m}$ only a finite cosine transform on a $(N/2+1)\times(N/2+1)$ mesh is required. The modified Fourier transform approach is described by Hohl and Hockney (1969). It should be pointed out that the Fourier transform method solves Equation (16) for the isolated system exactly (within computer rounding error). It should also be noted that since only one quarter of the $N\times N$ array contains the active potential, the potential calculations can be easily performed such that only storage for an $N\times N/2$ array is required. A listing of the computer program for obtaining the gravitational potential for isolated disk galaxies is given by Hohl (1970b) and a listing of the fast Fourier transform subroutine is given by Hockney (1970).

The two components of the gravitational field can also be directly computed by the Fourier method by performing the summations

$$(K_x)_{n,m} = -G\sum_{i,j}\frac{(x_n - x_i)\,\sigma_{i,j}}{[(x_n - x_i)^2 + (y_m - y_j)^2]^{3/2}} \tag{22}$$

and

$$(K_y)_{n,m} = -G\sum_{i,j}\frac{(y_m - y_j)\,\sigma_{i,j}}{[(x_n - x_i)^2 + (y_m - y_i)^2]^{3/2}}. \tag{23}$$

However, the computer storage and time required becomes much larger.

E. ARBITRARY FORCE LAW

The method presented for obtaining the gravitational potential can easily be extended to three-dimensional problems. Also, the force law between particles can easily be changed by simply changing Green's function H. For example, some of the effects of finite thickness of the galactic disk can be simulated by using finite-length mass rods instead of point masses to represent the stars. The force of attraction F between two mass rods of unit mass and of length l aligned perpendicular to the galactic plane with their centers in the galactic plane is

$$F = \frac{2G}{l^2}\left[\sqrt{1 + (l^2/r^2)} - 1\right]$$

where r is the separation of the two rods. Green's function then becomes

$$H_{n,m} = \frac{2}{l}\left\{\frac{r_{n,m}}{l}\left[1 - \sqrt{1 + (l^2/r_{n,m}^2)}\right] + \log_e\left[\frac{l}{r_{n,m}} + \sqrt{1 + (l^2/r_{n,m}^2)}\right]\right\}$$

where $r_{n,m}=n^2+m^2$. (24)

F. CALCULATION OF ENERGY AND MOMENTUM

The angular momentum of the disk at time t is given by

$$\Gamma(t) = \sum_i m_i[x_i\bar{V}_{x,i} - y_i\bar{V}_{y,i}] \tag{25}$$

where m_i is the mass of the ith star, $\bar{V}_{x,i}=[V_{x,i}(t+\delta t/2)+V_{x,i}(t-\delta t/2)]/2$ is the x-component of the velocity of that star, and the summation extends over all the stars.

The kinetic energy of the disk at time t is

$$T(t)=\tfrac{1}{2}\sum_i m_i[\bar{V}_{x,i}^2+\bar{V}_{y,i}^2]. \tag{26}$$

An approximate expression for the potential energy is

$$P=-\tfrac{1}{2}\sum_n\sum_m \sigma_{n,m}\phi_{n,m}. \tag{27}$$

A better definition of the potential energy should be devised by using the definition of potential energy as the work done on a test star and apply it to the present model. The difficulty in the use of Equation (27) is that in regions of large mass condensations this equation gives a value of ϕ which is too large for the potential energy.

3. Two-Dimensional Rod Model

Computer simulations of 'Cylindrical Galaxies' by means of two-dimensional mass rod models have been performed by Hockney (1967, 1968) and by Hohl (1968, 1969). The stability and dynamics of systems containing up to 100000 mass rods were investigated. However, the rod-star approximation is likely to be valid only for very few galaxies, such as possibly NGC 2685. The model is essentially identical to that for disk galaxies except that the gravitational potential is now strictly two dimensional and is obtained by solving the two-dimensional Poisson equation. The gravitational potential $\phi_{n,m}$ is then obtained from the density $\varrho_{n,m}$ by solving the two-dimensional Poisson equation

$$\frac{\partial^2\phi}{\partial x^2}+\frac{\partial^2\phi}{\partial y^2}=4\pi G\varrho(x,y) \tag{28}$$

by finite difference methods. The standard five-point difference equation

$$\phi_{n+1,m}+\phi_{n,m+1}+\phi_{n-1,m}+\phi_{n,m-1}-4\phi_{n,m}=4\pi G\varrho_{n,m} \tag{29}$$

is generally used to solve for the potential distribution. The cell dimensions Δx and Δy are taken to be equal to unity.

The potential at the boundary of the rectangular region is required in the solution of the Poisson equation. At an arbitrary boundary point a distance $z=x+iy\,(i=\sqrt{-1})$ from the center of the mesh, the potential is given by

$$\begin{aligned}\phi(z)&=2G\sum_{n,m}\varrho_{n,m}\log_e|z-z_{n,m}|=\\&=2GM\log_e|z|+2G\sum_{n,m}\varrho_{n,m}Re\left|\log_e\left(1-\frac{z_{n,m}}{z}\right)\right|\end{aligned} \tag{30}$$

where M is the total mass in the system and $z_{n,m}=x_{n,m}+iy_{n,m}$ is the coordinate of the cell n, m. Since $\varrho_{n,m}$ is nonzero only for $z_{n,m}/z<1$, Equation (30) can be written as

$$\phi(z) = 2GM \log_e|z| - 2GRe \sum_k \frac{a_k}{kz^k} \tag{31}$$

where

$$a_k = \sum_{n,m} \varrho_{n,m} z_{n,m}^k \tag{32}$$

and the series expansion for $\log_e(1-z_{n,m}/z)$ is truncated after 15 terms.

The set of simultaneous equations given by Equation (29) can be solved by an iteration of the form

$$\phi_{n,m}^{r+1} = \phi_{n,m}^r + \gamma(\phi_{n-1,m}^{r+1} + \phi_{n+1,m}^r + \phi_{n,m-1}^{r+1} + \\ + \phi_{n,m+1}^r - 4\phi_{n,m}^r - 4\pi G\varrho_{n,m}). \tag{33}$$

For the purpose of saving computer storage and increasing the convergence rate, the new values of ϕ (that is, ϕ^{r+1}) which have already been calculated during a particular iteration are used in the right-hand side of Equation (33). The superscript r refers to the rth iteration and the parameter γ is adjusted to give the maximum rate of convergence.

If the motion of all the stars in the system is advanced for a small time step δt, the mass distribution $\varrho_{n,m}$ will not change very much. The change in the gravitational potential will then also be very small. Thus, the solution of the finite difference form of the Poisson equation (Equation (33)) by an iteration method which uses the potential from the previous cycle as an initial guess will converge very rapidly. The accuracy of the iterative solution of the Poisson equation is easily checked during the calculations. This verification is made by obtaining the values of the potential and the field at several selected points by summing directly the contribution from each star. The values so obtained agree with those obtained from the solution of the Poisson equation to at least the first three digits. The number of iterations required for a 51×51 mesh was found to be 5 to 7 and for a 101×101 mesh, 12 iterations (Hohl, 1969).

The initial guess of the potential at $t=0$ is determined by using analytical expression for the potential of the initial cylinder (Hohl, 1969).

Since direct methods for solving the set of Equations (29) are now generally available it is preferable to use them for obtaining the potential.

The method described previously for disks of stars can of course also be used for obtaining the potential for the two-dimensional rod model. However, the method of solving the Poisson equation is faster and requires less computer storage.

4. Summary

The dynamics of collisionless stellar systems can be studied by representing the system by large numbers of representative stars. The numerical methods that are used to integrate the motion of the system in time are presented in some detail. Examples of actual computer experiments can be found in the literature cited.

References

Cooley, J. W. and Tukey, J. W.: 1965, *Math. Comput.* **19**, 297.
Hockney, R. W.: 1967, *Astrophys. J.* **150**, 797.
Hockney, R. W.: 1968, *Publ. Astron. Soc. Pacific* **80**, 662.
Hockney, R. W.: 1970, in *Methods in Computational Physics*, **9**, 135, Academic Press Inc., New York.
Hockney, R. W. and Hohl, F.: 1969, *Astron. J.* **74**, 1102.
Hohl, F.: 1968, *Bull. Astron.* **3**, 227.
Hohl, F.: 1969, 'Computer Simulation of a Cylindrical Galaxy', NASA TN D-5200.
Hohl, F.: 1970a, in W. Becker and G. Contopoulos (eds.), 'The Spiral Structure of our Galaxy', *IAU Symp.* **38**, 368.
Hohl, F.: 1970b, 'Dynamics of Disk Galaxies', NASA TR R-343.
Hohl, F.: 1970c, *Astron. J.* **76**, 202.
Hohl, F. and Hockney, R. W.: 1969, *J. Comput. Phys.* **4**, 306.
Lindblad, P. O.: 1960, *Stockholm Obser. Ann.* **21**, 3.
Miller, R. H.: 1970, *J. Comput. Phys.* **6**, 449.
Miller, R. H. and Prendergast, K. H.: 1968, *Astrophys. J.* **151**, 699.
Miller, R. H., Prendergast, K. H., and Quirk, W. J.: 1970, *Astrophys. J.* **161**, 903.

ASTROPHYSICS AND SPACE SCIENCE LIBRARY

Edited by

J. E. Blamont, R. L. F. Boyd, L. Goldberg, C. de Jager, Z. Kopal, G. H. Ludwig, R. Lüst, B. M. McCormac, H. E. Newell, L. I. Sedov, Z. Švestka, and W. de Graaff

1. C. de Jager (ed.), *The Solar Spectrum. Proceedings of the Symposium held at the University of Utrecht, 26–31 August, 1963.* 1965, XIV + 417 pp.
2. J. Ortner and H. Maseland (eds.), *Introduction to Solar Terrestrial Relations. Proceedings of the Summer School in Space Physics held in Alpbach, Austria, July 15–August 10, 1963 and Organized by the European Preparatory Commission for Space Research.* 1965, IX + 506 pp.
3. C. C. Chang and S. S. Huang (eds.), *Proceedings of the Plasma Space Science Symposium, held at the Catholic University of America, Washington, D.C., June 11–14, 1963.* IX + 377 pp.
4. Zdeněk Kopal, *An Introduction to the Study of the Moon.* 1966, XII + 464 pp.
5. Billy M. McCormac (ed.), *Radiation Trapped in the Earth's Magnetic Field. Proceedings of the Advanced Study Institute, held at the Chr. Michelsen Institute, Bergen, Norway, August 16–September 3, 1965.* 1966, XII + 901 pp.
6. A. B. Underhill, *The Early Type Stars.* 1966, XIII + 282 pp.
7. Jean Kovalevsky, *Introduction to Celestial Mechanics.* 1967, VIII + 427 pp.
8. Zdeněk Kopal and Constantine L. Goudas (eds.), *Measure of the Moon. Proceedings of the Second International Conference on Selenodesy and Lunar Topography held in the University of Manchester, England, May 30–June 4, 1966.* 1967, XVIII + 479 pp.
9. J. G. Emming (ed.), *Electromagnetic Radiation in Space. Proceedings of the Third ESRO Summer School in Space Physics, held in Alpbach, Austria, from 19 July to 13 August, 1965.* 1968, VII + 307 pp.
10. R. L. Carovillano, John F. McClay, and Henry R. Radoski (eds.), *Physics of the Magnetosphere. Based upon the Proceedings of the Conference held at Boston College, June 19–28, 1967.* 1968, X + 686 pp.
11. Syun-Ichi Akasofu, *Polar and Magnetospheric Substorms.* 1968, XVIII + 280 pp.
12. Peter M. Millman (ed.), *Meteorite Research. Proceedings of a Symposium on Meteorite Research held in Vienna, Austria, 7–13 August, 1968.* 1969, XV + 941 pp.
13. Margherita Hack (ed.), *Mass Loss from Stars. Proceedings of the Second Trieste Colloquium on Astrophysics, 12–17 September, 1968.* 1969, XV + 345 pp.
14. N. D'Angelo (ed.), *Low-Frequency Waves and Irregularities in the Ionosphere. Proceedings of the 2nd ESRIN-ESLAB Symposium, held in Frascati, Italy, 23–27 September, 1968.* 1969, VII + 218 pp.
15. G. A. Partel (ed.), *Space Engineering. Proceedings of the Second International Conference on Space Engineering, held at the Fondazione Giorgio Cini, Isola di San Giorgio, Venice, Italy, May 7–10, 1969.* 1970 XI + 728 pp.

p.t.o.

16. S. Fred Singer (ed.), *Manned Laboratories in Space. Second International Orbital Laboratory Symposium.* 1969, XIII + 133 pp.

17. B. M. McCormac (ed.), *Particles and Fields in the Magnetosphere. Symposium Organized by the Summer Advanced Study Institute, held at the University of California, Santa Barbara, Calif., August 4–15, 1969.* 1970, XI + 450 pp.

18. Jean-Claude Pecker, *Experimental Astronomy.* 1970, X + 105 pp.

19. V. Manno and D. E. Page (eds.), *Intercorrelated Satellite Observations related to Solar Events. Proceedings of the Third ESLAB/ESRIN Symposium held in Noordwijk, The Netherlands, September 16–19, 1969.* 1970, XVI + 627 pp.

20. L. Mansinha, D. E. Smylie and A. E. Beck, *Earthquake Displacement Fields and the Rotation of the Earth. A NATO Advanced Study Institute Conference Organized by the Department of Geophysics University of Western Ontario, London, Canada, 22 June–28 June, 1969.* 1970, XI + 308 pp.

21. Jean-Claude Pecker, *Space Observatories.* 1970, XI + 120 pp.

22. L. N. Mavridis (ed.), *Structure and Evolution of the Galaxy. Proceedings of the NATO Advanced Study Institute, held in Athens, September 8–19, 1969.* 1971, VII + 312 pp.

23. A. Muller (ed.), *The Magellanic Clouds. A European Southern Observatory Presentation: Principal Prospects, Current Observational and Theoretical Approaches, and Prospects for Future Research. Based on the Symposium on the Magellanic Clouds, held in Santiago de Chile, March 1969, on the Occasion of the Dedication of the European Southern Observatory.* 1971, XII + 189 pp.

24. B. M. McCormac (ed.), *The Radiating Atmosphere. Proceedings of a Symposium Organized by the Summer Advanced Study Institute, held at Queen's University, Kingston, Ontario, August 3–14, 1970.* 1971, XI + 455 pp.

25. G. Fiocco (ed.), *Mesospheric Models and Related Experiments. Proceedings of the 4th ESRIN-ESLAB Symposium, held at Frascati, Italy, July 6–10, 1970.* 1971, VIII + 298 pp.

26. I. Atanasijević, *Selected Exercises in Galactic Astronomy.* 1971, XII + 143 pp.

27. Constantin J. Macris (ed.), *Physics of the Solar Corona. Proceedings of the NATO Advanced Study Institute on Physics of the Solar Corona, held at Cavouri-Vouliagmeni, Athens, Greece, 6–17 September 1970.* 1971, XII + 345 pp.

28. F. Delobeau, *The Environment of the Earth.* 1971, IX + 113 pp.

29. E. R. Dyer (general ed.), *Solac-Terrestrial Physics/1970. Proceedings of the International Symposium on Solar-Terrestrial Physics, held in Leningrad, U.S.S.R., 12–19 May 1970.* 1972, VIII + 938 pp.

30. V. Manno and J. Ring (eds.), *Infrared Detection Techniques for Space Research. Proceedings of the Fifth ESLAB-ESRIN Symposium, held in Noordwijk, The Netherlands, June 8–11, 1971.* 1972, XII + 344 pp.

SOLE DISTRIBUTORS FOR U.S.A. AND CANADA:

Vols. 2–6, and 8: Gordon & Breach Inc., 150 Fith Ave., New York, N.Y. 10011
Vols. 7 and 9–28: Springer Verlag New York, Inc., 175 Fith Ave., New York, N.Y. 10011